Legal Notice

This book is copyright 2019 with all rights reserved. It is illegal to copy, distribute, or create derivative works from this book in whole or in part or to contribute to the copying, distribution, or creating of derivative works of this book.

For information on bulk purchases and licensing agreements, please email

support@SATPrepGet800.com

ISBN-13: 978-0-9998117-8-8

Also Available from Dr. Steve Warner

CONNECT WITH DR. STEVE WARNER

www.facebook.com/SATPrepGet800

www.youtube.com/TheSATMathPrep

www.twitter.com/SATPrepGet800

www.linkedin.com/in/DrSteveWarner

www.pinterest.com/SATPrepGet800

plus.google.com/+SteveWarnerPhD

Abstract Algebra for Beginners

A Rigorous Introduction to Groups, Rings, Fields, Vector Spaces, Modules, Substructures, Homomorphisms, Quotients, Permutations, Group Actions, Polynomials, and Galois Theory

Dr. Steve Warner

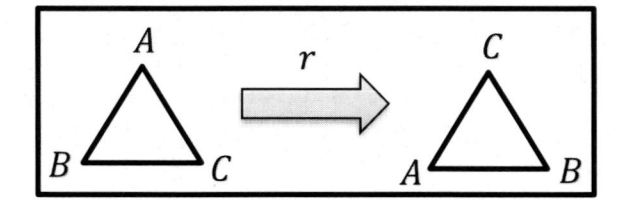

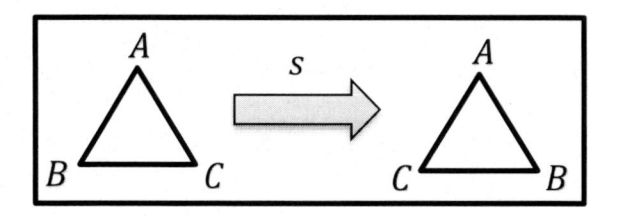

© 2019, All Rights Reserved

Table of Contents

Introduction — 7
    For students — 7
    For instructors — 8

Lesson 1 – Sets and Subsets — 9
    Describing Sets — 9
    Subsets and Proper subsets — 14
    Basic Theorems Involving Subsets — 16
    Power Sets — 18
    Transitivity of the Subset Relation — 19
    Equality of Sets — 19
    Cartesian Products — 22
    Basic Set Operations — 24
    Properties of Unions and Intersections — 27
    Arbitrary Unions and Intersections — 29
    Problem Set 1 — 31

Lesson 2 – Algebraic Structures — 36
    Binary Operations and Closure — 36
    Semigroups and Associativity — 38
    Monoids and Identity — 42
    Groups and Inverses — 43
    Rings and Distributivity — 50
    Fields — 57
    Vector Spaces Over Fields — 59
    Modules Over Rings — 65
    Problem Set 2 — 66

Lesson 3 – Relations and Partitions — 69
    Binary Relations — 69
    n-ary Relations — 71
    Orderings — 71
    Intervals — 74
    Equivalence Relations — 76
    Partitions — 78
    The Ring of Integers Mod n — 81
    Problem Set 3 — 83

Lesson 4 – Functions and Equinumerosity — 85
    Functions — 85
    Injections, Surjections, and Bijections — 88
    Inverse Functions — 89
    Composite Functions — 91
    Identity Functions — 93
    Images and Inverse Images — 93
    Groups and Monoids of Functions — 93

Equinumerosity	95
Countable and Uncountable Sets	96
Problem Set 4	102
Lesson 5 – Number Systems and Induction	**105**
The Natural Numbers	105
Well Ordering and the Principle of Mathematical Induction	107
The Integers	116
The Rational Numbers	117
The Real Numbers	118
The Complex Numbers	120
Exponential Form of a Complex Number	122
Problem Set 5	130
Lesson 6 – Substructures	**134**
Structures and Substructures	134
Subspaces of Vector Spaces	141
Substructures Generated by a Set	145
Problem Set 6	152
Lesson 7 – Homomorphisms and Isomorphisms	**155**
Homomorphisms	155
Isomorphisms	159
Linear Transformations	161
Matrices	166
The Matrix of a Linear Transformation	170
Images and Kernels	171
Normal Subgroups and Ring Ideals	174
Problem Set 7	177
Lesson 8 – Number Theory	**180**
Divisibility	180
Prime Numbers	181
The Division Algorithm	185
GCD and LCM	189
Problem Set 8	197
Lesson 9 – Number Theoretic Applications	**200**
Cyclic Groups	200
Modular Arithmetic	206
Solving Linear Congruences	210
Problem Set 9	218
Lesson 10 – Quotients	**220**
Cosets	220
Quotient Groups	227
Quotient Rings	231
Quotient Spaces	236
Problem Set 10	237
Lesson 11 – Structure Theorems	**239**
Isomorphism Theorems	239

Fundamental Theorem of Finite Commutative Groups	251
Problem Set 11	258
Lesson 12 – Permutations and Determinants	**260**
Permutations on Finite Sets	260
The Alternating Group	268
Permutation Matrices	269
Determinants	271
Matrix Groups	273
Problem Set 12	276
Lesson 13 – Sylow Theory and Group Actions	**278**
Conjugacy Classes	278
Groups of Prime Power Order	280
Sylow Subgroups	281
Group Actions	285
Problem Set 13	290
Lesson 14 – Polynomials	**292**
Polynomials Over a Ring	292
Polynomials Over a Field	298
The Division Algorithm for Polynomials	301
Principal Ideal Domains	303
Unique Factorization Domains	304
Polynomial Functions	308
Problem Set 14	310
Lesson 15 – Field Theory	**312**
Field Extensions	314
Algebraic Elements	309
Field Extensions as Vector Spaces	319
Splitting Fields	322
Problem Set 15	324
Lesson 16 – Galois Theory	**326**
Field Automorphisms	326
The Galois Correspondence	330
Solvability	333
Problem Set 16	337
Index	**339**
About the Author	**344**
Books by Dr. Steve Warner	**345**

INTRODUCTION
ABSTRACT ALGEBRA

This book was written to provide a basic but rigorous introduction to abstract algebra.

For students: There are no prerequisites for this book. The content is completely self-contained. Students with a bit of mathematical knowledge may have an easier time getting through some of the material, but no such knowledge is necessary to read this book.

More important than mathematical knowledge is "mathematical maturity." Although there is no single agreed upon definition of mathematical maturity, one reasonable way to define it is as "one's ability to analyze, understand, and communicate mathematics." A student with a higher level of mathematical maturity will be able to move through this book more quickly than a student with a lower level of mathematical maturity.

Whether your level of mathematical maturity is low or high, if you are just starting out in abstract algebra, then you're in the right place. If you read this book the "right way," then your level of mathematical maturity will continually be increasing. This increased level of mathematical maturity will not only help you to succeed in advanced math courses, but it will improve your general problem solving and reasoning skills. This will make it easier to improve your performance in college, in your professional life, and on standardized tests.

So, what is the "right way" to read this book? Simply reading each lesson from end to end without any further thought and analysis is not the best way to read the book. You will need to put in some effort to have the best chance of absorbing and retaining the material. When a new theorem is presented, don't just jump right to the proof and read it. Think about what the theorem is saying. Try to describe it in your own words. Do you believe that it is true? If you do believe it, can you give a convincing argument that it is true? If you do not believe that it is true, try to come up with an example that shows it is false, and then figure out why your example does not contradict the theorem. Pick up a pen or pencil. Draw some pictures, come up with your own examples, and try to write your own proof.

You may find that this book goes into more detail than other abstract algebra books when explaining examples, discussing concepts, and proving theorems. This was done so that any student can read this book, and not just students that are naturally gifted in mathematics. So, it is up to you as the student to try to answer questions before they are answered for you. When a new definition is given, try to think of your own examples before looking at those presented in the book. And when the book provides an example, do not just accept that it satisfies the given definition. Convince yourself. Prove it.

Each lesson is followed by a Problem Set. The problems in each Problem Set have been organized into five levels of difficulty, followed by one or more Challenge Problems. Level 1 problems are the easiest and Level 5 problems are the most difficult, except for the Challenge Problems. If you want to get just a small taste of abstract algebra, then you can work on the easier problems. If you want to achieve a deeper understanding of the material, take some time to struggle with the harder problems.

For instructors: This book can be used as an undergraduate text or an introductory graduate text in abstract algebra. The subject is developed slowly with an emphasis early on of developing skill with proof writing.

Lesson 1 provides the basic set theory that will be needed throughout the book. Sets, subsets, Cartesian products, and set operations are all covered.

Lesson 2 introduces the most important algebraic structures for the purposes of this book: semigroups, monoids, groups, rings, fields, vector spaces, and modules.

Lessons 3 and 4 provide a rigorous treatment of the intermediate set theory that anyone studying abstract algebra should know. All the basics of relations, partitions, functions, and equinumerosity are included. Lesson 4 also provides an introduction to symmetric groups.

Lesson 5 covers the Principle of Mathematical Induction and a provides a formal treatment of the natural numbers, integers, rational numbers, real numbers, and complex numbers.

Lesson 6 discusses substructures such as subsemigroups, submonoids, subgroups, subrings, subspaces, and submodules.

Lesson 7 covers homomorphisms and isomorphisms of algebraic structures. Other topics covered include matrix multiplication, the matrix of a linear transformation, kernels, images, normal subgroups, and ring ideals.

Lesson 8 reviews all the basic number theory that will be needed for the rest of the book. In particular, divisibility, the Division Algorithm, greatest common divisors, least common multiples, the Fundamental Theorem of Arithmetic, and the Euclidean Algorithm are all covered in detail here.

Lesson 9 applies the number theory from Lesson 8 to cyclic groups, finite fields, and modular arithmetic.

Lesson 10 covers cosets, quotient groups, quotient rings, and quotients of vector spaces.

Lesson 11 covers all the isomorphism theorems from group theory, ring theory, and linear algebra, followed by the Fundamental Theorem of Commutative Groups.

Lesson 12 discusses permutations of finite sets and their application to determinants and Lesson 13 discusses the Sylow theorems and group actions.

Lesson 14 introduces polynomials over rings as a prelude to field theory, which is covered in Lesson 15.

Finally, Lesson 16 provides an introduction to Galois theory.

The author welcomes all feedback from instructors. Any suggestions will be considered for future editions of the book. The author would also love to hear about the various courses that are created using these lessons. Feel free to email Dr. Steve Warner with any feedback at

steve@SATPrepGet800.com

LESSON 1
BASIC SET THEORY

Describing Sets

A **set** is simply a collection of "objects." These objects can be numbers, letters, shapes, birds, book titles, or just about anything else you can imagine. We will usually refer to the objects in a set as the **members** or **elements** of the set.

If a set consists of a small number of elements, we can describe the set simply by listing the elements in the set in curly braces, separating elements by commas.

Example 1.1:

1. {dragon, unicorn} is the set consisting of two elements: *dragon* and *unicorn*.

2. {bee, doctor, purple, sailboat, mailbox, briefcase} is the set consisting of six elements: *bee, doctor, purple, sailboat, mailbox,* and *briefcase*.

3. {0, 3, 6, 9, 12} is the set consisting of five elements: 0, 3, 6, 9, and 12. The elements in this set happen to be *numbers*.

A set is determined by its elements and not the order in which the elements are presented. For example, the set {3, 12, 6, 0, 9} is the same as the set {0, 3, 6, 9, 12}.

Also, the set {0, 0, 3, 6, 6, 6, 9, 9, 12} is the same as the set {0, 3, 6, 9, 12}. If we are describing a set by listing its elements, the most natural way to do so is to list each element just once.

We will usually name sets using capital letters such as $A, B, C,...$, and so on. For example, we might write $A = \{a, b, c, d\}$. So, A is the set consisting of the elements $a, b, c,$ and d.

Example 1.2: Consider the sets $X = \{x, y, z\}$, $Y = \{y, z, x\}$, $Z = \{x, y, x, z, x\}$. Then X, Y, and Z all represent the same set. We can write $X = Y = Z$.

We use the symbol $\in$ for the membership relation (we will define the term "relation" more carefully in Lesson 3). So, $x \in A$ means "x is an element of A," whereas $x \notin A$ means "x is **not** an element of A." We will often simply say "x is in A," and "x is not in A," respectively.

Example 1.3: Let $D = \{a, k, 11, \delta, \square\}$. Then $a \in D$, $k \in D$, $11 \in D$, $\delta \in D$, and $\square \in D$.

If a set consists of many elements, we can use **ellipses** (...) to help describe the set. For example, the set consisting of the natural numbers between 2 and 3000, inclusive, can be written {2, 3, 4, ..., 2999, 3000} ("inclusive" means that we include 2 and 3000). The ellipses between 4 and 2999 are there to indicate that there are elements in the set that we are not explicitly mentioning.

Ellipses can also be used to help describe **infinite sets**. The set of **natural numbers** can be written $\mathbb{N} = \{0, 1, 2, 3, ...\}$, and the set of **integers** can be written $\mathbb{Z} = \{..., -4, -3, -2, -1, 0, 1, 2, 3, 4, ...\}$.

Note: Some mathematicians exclude 0 from the set of natural numbers. In this book, 0 will always be included. Symbolically, $0 \in \mathbb{N}$.

Example 1.4:

1. The **even natural numbers** can be written $\mathbb{E} = \{0, 2, 4, 6, \ldots\}$. We may also use the notation $2\mathbb{N}$ to describe this set. So, $2\mathbb{N} = \mathbb{E} = \{0, 2, 4, 6, \ldots\}$.

2. The **odd natural numbers** can be written $\mathbb{O} = \{1, 3, 5, \ldots\}$. We may also use the notation $2\mathbb{N} + 1$ to describe this set. So, $2\mathbb{N} + 1 = \mathbb{O} = \{1, 3, 5, \ldots\}$.

3. The **positive integers** can be written $\mathbb{Z}^+ = \{1, 2, 3, 4, \ldots\}$ or $\mathbb{N}^+ = \{1, 2, 3, 4, \ldots\}$ (the positive integers and the positive natural numbers describe the same set).

4. The **even integers** can be written $2\mathbb{Z} = \{\ldots, -6, -4, -2, 0, 2, 4, 6, \ldots\}$.

5. The **integral multiples of 3** can be written $3\mathbb{Z} = \{\ldots, -9, -6, -3, 0, 3, 6, 9, \ldots\}$.

6. If $n \in \mathbb{Z}$, the **integral multiples of n** can be written $n\mathbb{Z} = \{\ldots, -3n, -2n, -n, 0, n, 2n, 3n, \ldots\}$. We saw the special cases where $n = 2$ and $n = 3$ in parts 4 and 5 above, respectively. For $n = 1$, we see that $1\mathbb{Z}$ is just the set of integers. So, $1\mathbb{Z} = \mathbb{Z}$. For $n = 0$, we see that $0\mathbb{Z} = \{0\}$, the set consisting of the single element 0. Also, note that for all integers n, $(-n)\mathbb{Z} = n\mathbb{Z}$. For example, we have

$$(-2)\mathbb{Z} = \{\ldots, 6, 4, 2, 0, -2, -4, -6, \ldots\} = \{\ldots, -6, -4, -2, 0, 2, 4, 6, \ldots\} = 2\mathbb{Z}.$$

7. The odd integers can be written $2\mathbb{Z} + 1 = \{\ldots, -5, -3, -1, 1, 3, 5, 7, \ldots\}$. Notice how we form the set $2\mathbb{Z} + 1$ by adding 1 to each element of the set $2\mathbb{Z}$.

8. If $n, m \in \mathbb{Z}$, we can form the set $n\mathbb{Z} + m$ by adding m to each element of the set $n\mathbb{Z}$. In other words, we have $n\mathbb{Z} + m = \{\ldots, -3n + m, -2n + m, -n + m, m, n + m, 2n + m, 3n + m, \ldots\}$. For example, if we let $n = 5$ and $m = 4$, then $5\mathbb{Z} + 4 = \{\ldots, -11, -6, -1, 4, 9, 14, \ldots\}$. As another example, we have $11\mathbb{Z} - 7 = 11\mathbb{Z} + (-7) = \{\ldots, -40, -29, -18, -7, 4, 15, \ldots\}$.

9. If $n, m \in \mathbb{Z}$, we can form the set $n\mathbb{N} + m$ in a way similar to how we formed $n\mathbb{Z} + m$. For example, $3\mathbb{N} + 0 = 3\mathbb{N} = \{0, 3, 6, 9, \ldots\}$, while $3\mathbb{N} + 2 = \{2, 5, 8, 11, \ldots\}$. As a more complex example, we have $(-6)\mathbb{N} = \{0, -6, -12, -18, \ldots\}$ and $(-6)\mathbb{N} + 4 = \{4, -2, -8, -14, \ldots\}$.

A set can also be described by a certain property P that all its elements have in common. In this case, we can use the **set-builder notation** $\{x|P(x)\}$ to describe the set. The expression $\{x|P(x)\}$ can be read "the set of all x such that the property $P(x)$ is true." Note that the symbol "|" is read as "such that."

As a simple example, let's consider the property of being a beverage. Formally, we can let $P(x)$ be the statement "x is a beverage." Symbolically, we can write $\{x|P(x)\} = \{x \mid x \text{ is a beverage}\}$. In words, this set can be described as "the set of all x such that x is a beverage." Water is an element of this set because water is a beverage. In other words, if we replace x by "water," then $P(x)$ is a true statement. Symbolically, we can write "water $\in \{x \mid x \text{ is a beverage}\}$." Tiger is not an element of this set because a tiger is not a beverage. Symbolically, we can write "tiger $\notin \{x \mid x \text{ is a beverage}\}$." However, tiger would be in the set $\{x \mid x \text{ is an animal}\}$. Tiger would also be in the set $\{x \mid x \text{ is a cat that roars}\}$.

Let's now turn to some examples involving numbers.

Example 1.5:

1. Let's look at a few different ways that we can describe the set $\{0, 2, 4, 6, 8, 10\}$. We have already seen that reordering and/or repeating elements does not change the set. For example, $\{2, 2, 0, 4, 10, 8, 8, 8, 6\}$ describes the same set. Here are a few more descriptions using set-builder notation:

   - $\{n \mid n \text{ is an even natural number less than or equal to } 10\}$
   - $\{n \in \mathbb{Z} \mid n \text{ is even}, 0 \leq n < 12\}$
   - $\{2k \mid k = 0, 1, 2, 3, 4, 5\}$

   The first expression in the bulleted list above can be read "the set of n such that n is an even natural number less than or equal to 10."

   The second expression can be read "the set of integers n such that n is even and n is between 0 and 12, including 0, but excluding 12." Note that the abbreviation "$n \in \mathbb{Z}$" can be read "n is in the set of integers," or more succinctly, "n is an integer."

   The third expression can be read "the set of $2k$ such that k is 0, 1, 2, 3, 4 or 5."

2. Recall from part 8 of Example 1.4 above that for $n, m \in \mathbb{Z}$, we have

   $$n\mathbb{Z} + m = \{\ldots, -3n + m, -2n + m, -n + m, m, n + m, 2n + m, 3n + m, \ldots\}$$

   Let's use the specific example $n = 3, m = 1$. So, we have

   $$3\mathbb{Z} + 1 = \{\ldots, -8, -5, -2, 1, 4, 7, 10, \ldots\}.$$

   Using set-builder notation, here are a few other ways to describe the set $3\mathbb{Z} + 1$:

   - $3\mathbb{Z} + 1 = \{t \mid t = 3k + 1 \text{ for some } k \in \mathbb{Z}\}$
   - $3\mathbb{Z} + 1 = \{3k + 1 \mid k \in \mathbb{Z}\}$
   - $3\mathbb{Z} + 1 = \{t \in \mathbb{Z} \mid \exists k \in \mathbb{Z} \, (t = 3k + 1)\}$

   The first expression in the bulleted list above can be read "the set of t such that $t = 3k + 1$ for some integer k."

   The second expression can be read "the set of $3k + 1$ such that k is an integer."

   The third expression contains the symbol $\exists$. This symbol is called an **existential quantifier**, and it is pronounced "there exists" or "there is." So, the third expression can be read "the set of integers t such that there is an integer k such that $t = 3k + 1$."

3. More generally, for $n, m \in \mathbb{Z}$, we can use set-builder notation to describe $n\mathbb{Z} + m$ in each of the following ways:

   - $n\mathbb{Z} + m = \{t \mid t = nk + m \text{ for some } k \in \mathbb{Z}\}$
   - $n\mathbb{Z} + m = \{nk + m \mid k \in \mathbb{Z}\}$
   - $n\mathbb{Z} + m = \{t \in \mathbb{Z} \mid \exists k \in \mathbb{Z} \, (t = nk + m)\}$

   The first expression in this bulleted list can be read "the set of t such that $t = nk + m$ for some integer k."

The second expression can be read "the set of $nk + m$ such that k is an integer."

The third expression can be read "the set of integers t such that there is an integer k such that $t = nk + m$."

4. For $n, m \in \mathbb{Z}$, we can use set-builder notation to describe $n\mathbb{N} + m$ in each of the following ways:

   - $n\mathbb{N} + m = \{t \mid t = nk + m \text{ for some } k \in \mathbb{N}\}$
   - $n\mathbb{N} + m = \{nk + m \mid k \in \mathbb{N}\}$
   - $n\mathbb{N} + m = \{t \in \mathbb{Z} \mid \exists k \in \mathbb{N} \, (t = nk + m)\}$

   Notice that although the variable k in each of the descriptions above must take on values that are natural numbers, the elements of the set $n\mathbb{N} + m$ do **not** need to be natural numbers (although they must all be integers). For example, $2\mathbb{N} - 1 = 2\mathbb{N} + (-1) = \{-1, 1, 3, 5, \ldots\}$ contains the integer -1, which is not a natural number. The set $(-1)\mathbb{N} - 3 = \{-3, -4, -5, \ldots\}$ doesn't contain any natural numbers.

In addition to the sets $\mathbb{N}$ (the natural numbers) and $\mathbb{Z}$ (the integers), let's look at a few more sets that will show up repeatedly throughout this book.

Example 1.6:

1. The set of **rational numbers** is $\mathbb{Q} = \left\{\frac{a}{b} \mid a, b \in \mathbb{Z} \text{ and } b \neq 0\right\}$. In words, $\mathbb{Q}$ is "the set of quotients a over b such that a and b are integers and b is not zero." Some examples of rational numbers are $\frac{0}{5}, \frac{1}{3}, \frac{2}{5},$ and $\frac{-6}{7}$. We identify rational numbers $\frac{a}{b}$ and $\frac{c}{d}$ whenever $ad = bc$. For example, $\frac{1}{2} = \frac{3}{6}$ because $1 \cdot 6 = 2 \cdot 3$. We also abbreviate the rational number $\frac{a}{1}$ as a. In this way, we can think of every integer as a rational number. For example, we have $\frac{0}{5} = \frac{0}{1}$ (because $0 \cdot 1 = 5 \cdot 0$), and therefore, we can abbreviate $\frac{0}{5}$ as 0. Similarly, we can abbreviate $\frac{15}{3}$ as 5 (because $\frac{15}{3} = \frac{5}{1}$).

2. Let's now give a naïve definition of the **real numbers**, $\mathbb{R}$. We first define a **digit** to be one of the symbols 0, 1, 2, 3, 4, 5, 6, 7, 8, or 9. We then define $\mathbb{R}$ to be the set of numbers of the form $x.y$, where $x \in \mathbb{Z}$ and y is an infinite "string" of digits without a **tail of 9's** (meaning there are infinitely many digits in the string that are **not** 9). Symbolically, we have

   $$\mathbb{R} = \{x.y \mid x \in \mathbb{Z} \text{ and } y \text{ is an infinite string of digits without a tail of 9's}\}.$$

   Some examples of real numbers are $0.000\ldots, 0.333\ldots, -16.000\ldots,$ and $1.010010001\ldots$ We will generally delete tails of 0's. So, we would write $0.000\ldots$ as 0 and $-16.000\ldots$ as -16. We will not consider $53.023999999\ldots$ to be a real number because of the tail of 9's (an alternative approach would be to identify $53.023999999\ldots$ with 53.024). We can visualize the set of real numbers with the **real line**.

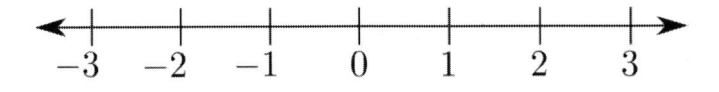

There is a fairly simple algorithm that allows us to identify every rational number as a real number. For example, the rational number $\frac{3}{2}$ can be represented as the real number 1.5 and the rational number $\frac{2}{3}$ can be represented as the real number 0.66666 ... I leave the details of this algorithm for the interested reader to explore (see Problem 35 below). Any real number that does **not** correspond to a rational number in this way is called an **irrational number**.

3. The **complex numbers** are defined as $\mathbb{C} = \{a + bi \mid a, b \in \mathbb{R}\}$. In words, $\mathbb{C}$ is "the set of $a + bi$ such that a and b are real numbers." Some examples of complex numbers are $0 + 0i$, $-2 + 0i$, $2.3 - 5i = 2.3 + (-5)i$, and $4.235235235... + 51.2020020002...i$. We will abbreviate $0 + 0i$ as 0, $a + 0i$ as a, and $0 + bi$ as bi. For example, $-2 + 0i = -2$. By identifying $a + 0i$ as a, we can think of every real number as a complex number. Complex numbers of the form bi are called **pure imaginary numbers**.

If we identify $1 = 1 + 0i$ with the ordered pair $(1, 0)$, and we identify $i = 0 + 1i$ with the ordered pair $(0, 1)$, then it is natural to write the complex number $a + bi$ as the point (a, b). Here is a reasonable justification for this: $a + bi = a(1,0) + b(0,1) = (a,0) + (0,b) = (a,b)$

In this way, we can visualize a complex number as a point in **The Complex Plane**. A portion of the Complex Plane is shown to the right with several complex numbers displayed as points of the form (x, y).

The Complex Plane is formed by taking two copies of the real line and placing one horizontally and the other vertically. The horizontal copy of the real line is called the x-axis or the **real axis** (labeled x in the figure) and the vertical copy of the real line is called the y-axis or **imaginary axis** (labeled y in the figure). The two axes intersect at the point $(0, 0)$. This point is called the **origin**.

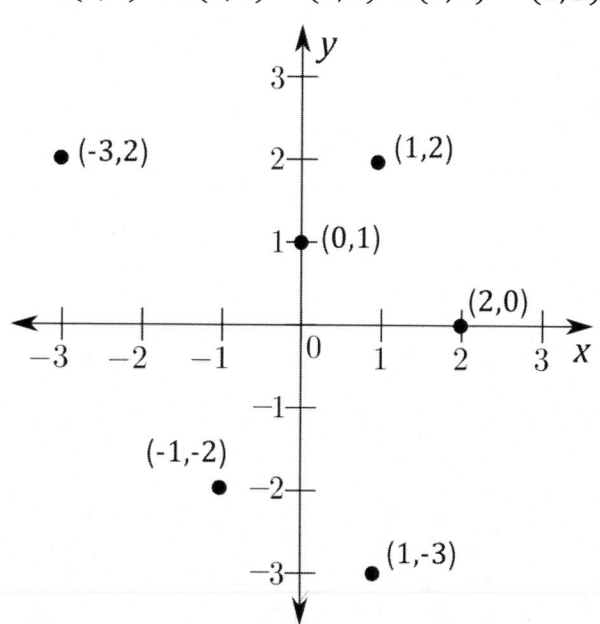

The **empty set** is the unique set with no elements. We use the symbol ∅ to denote the empty set (some authors use the symbol { } instead).

If A is a finite set, we define the **cardinality** of A, written $|A|$, to be the number of elements of A. For example, $|\{a, b, c\}| = 3$.

Example 1.7: Let $A = \{\text{giraffe, tulip, piano}\}$, $B = \{□, ∆, ∆\}$, $C = \{25, 26, 27, ..., 3167, 3168\}$, $D = \{\{x\}, \{x, x\}, \{x, x, x\}\}$, and $E = ∅$. Then $|A| = 3$, $|B| = 2$, $|C| = 3144$, $|D| = 1$, and $|E| = 0$.

Notes: (1) The set A consists of the three elements "giraffe," "tulip," and "piano."

(2) The set B consists of just two elements: □ and ∆. Remember that $\{□, ∆, ∆\} = \{□, ∆\}$.

(3) The number of consecutive integers from m to n, inclusive, is $\boldsymbol{n - m + 1}$. For set C, we have $m = 25$ and $n = 3168$. Therefore, $|C| = 3168 - 25 + 1 = 3144$.

(4) I call the formula "$n - m + 1$" the **fence-post formula**. If you construct a 3-foot fence by placing a fence-post every foot, then the fence will consist of 4 fence-posts ($3 - 0 + 1 = 4$).

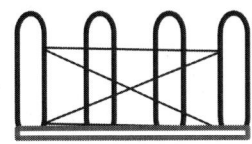

(5) Since $\{x, x\} = \{x\}$ and $\{x, x, x\} = \{x\}$, it follows that $D = \{\{x\}, \{x\}, \{x\}\} = \{\{x\}\}$. So, D consists of the single element $\{x\}$.

(6) Remember that $\emptyset$ (pronounced "the empty set") is the unique set with no elements.

Subsets and Proper Subsets

We say that a set A is a **subset** of a set B, written $A \subseteq B$, if every element of A is an element of B.

Example 1.8:

1. Let $A = \{x, y\}$ and $B = \{x, y, z\}$. The only elements of A are x and y. Since x and y are also elements of B, we see that $A \subseteq B$.

    Notice that $B \nsubseteq A$ (B is **not** a subset of A) because $z \in B$, but $z \notin A$.

2. Let $\mathbb{N} = \{0, 1, 2, 3, \ldots\}$ be the set of natural numbers and let $\mathbb{Z} = \{\ldots, -3, -2, -1, 0, 1, 2, 3, 4, \ldots\}$ be the set of integers. Since every natural number is an integer, $\mathbb{N} \subseteq \mathbb{Z}$.

3. As we saw in Example 1.6, by making appropriate identifications, we have the following sequence of inclusions:

    $$\mathbb{N} \subseteq \mathbb{Z} \subseteq \mathbb{Q} \subseteq \mathbb{R} \subseteq \mathbb{C}.$$

    We will see in Theorem 1.14 below that if $A \subseteq B$ and $B \subseteq C$, then $A \subseteq C$ (we say that $\subseteq$ is a **transitive** relation). In this way we see that we have many other inclusions such as $\mathbb{N} \subseteq \mathbb{Q}$, $\mathbb{N} \subseteq \mathbb{R}, \ldots$, and so on.

4. Recall that $2\mathbb{Z} = \{\ldots, -6, -4, -2, 0, 2, 4, 6, \ldots\}$ and $4\mathbb{Z} = \{\ldots, -12, -8, -4, 0, 4, 8, 12, \ldots\}$. It is easy to see that $4\mathbb{Z} \subseteq 2\mathbb{Z}$. You will be asked to prove this rigorously in Problem 14 below. Note that the opposite inclusion is false. That is, $2\mathbb{Z} \nsubseteq 4\mathbb{Z}$. To see this, we just need a single **counterexample** (a counterexample is an example that is used to show that a statement is false). Well, we have $2 \in 2\mathbb{Z}$, but $2 \notin 4\mathbb{Z}$.

5. Consider the sets $2\mathbb{Z} = \{\ldots, -6, -4, -2, 0, 2, 4, 6, \ldots\}$ and $3\mathbb{Z} = \{\ldots, -9, -6, -3, 0, 3, 6, 9, \ldots\}$. Neither of these sets is a subset of the other. To see that $2\mathbb{Z} \nsubseteq 3\mathbb{Z}$, observe that $2 \in 2\mathbb{Z}$, whereas $2 \notin 3\mathbb{Z}$. To see that $3\mathbb{Z} \nsubseteq 2\mathbb{Z}$, observe that $3 \in 3\mathbb{Z}$, whereas $3 \notin 2\mathbb{Z}$.

To the right we see a physical representation of $A \subseteq B$. This figure is called a **Venn diagram**. These types of diagrams are very useful to help visualize relationships among sets. Notice that set A lies completely inside set B. We assume that all the elements of A and B lie in some **universal set** U.

As an example, let's let U be the set of all species of animals. If we let A be the set of species of cats and we let B be the set of species of mammals, then we have $A \subseteq B \subseteq U$, and we see that the Venn diagram to the right gives a visual representation of this situation. (Note that every cat is a mammal and every mammal is an animal.)

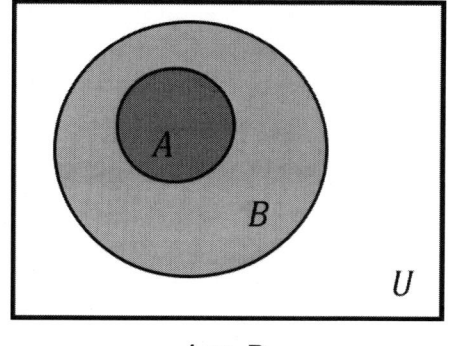

$A \subseteq B$

We say that A is a **proper subset** of B, written $A \subset B$ (or sometimes $A \subsetneq B$), if $A \subseteq B$, but $A \neq B$. For example, $\mathbb{N} \subset \mathbb{Z}$, whereas $\mathbb{N} \not\subset \mathbb{N}$ (although $\mathbb{N} \subseteq \mathbb{N}$).

Note: The definition of proper subset is not very important. It just gives us a convenient way to discuss all the subsets of a specific set except for the set itself. For example, it is quite cumbersome to say "Find all subsets of A, but exclude the set A." It's nice to be able to rephrase this as "Find all proper subsets of A."

Let's look at the definition of $\subseteq$ (subset) in a bit more detail.

Once again, we write $A \subseteq B$ if every element of A is an element of B. That is, $A \subseteq B$ if, for every x, $x \in A$ implies $x \in B$. Symbolically, we can write the following:

$$\forall x(x \in A \to x \in B)$$

Notes: (1) The symbol $\forall$ is called a **universal quantifier**, and it is pronounced "For all."

(2) The logical expression $\forall x(x \in A \to x \in B)$ can be translated into English as "For all x, if x is an element of A, then x is an element of B."

(3) To show that a set A is a subset of a set B, we need to show that the expression $\forall x(x \in A \to x \in B)$ is true. If the set A is finite and the elements are listed, we can just check that each element of A is also an element of B. However, if the set A is described by a property, say $A = \{x|P(x)\}$, we may need to craft an argument more carefully. We can begin by taking an **arbitrary but specific element** a from A and then arguing that this element a is in B.

What could we possibly mean by an arbitrary but specific element? Aren't the words "arbitrary" and "specific" antonyms? Well, by arbitrary, we mean that we don't know which element we are choosing – it's just some element a that satisfies the property P. So, we are just assuming that $P(a)$ is true. However, once we choose this element a, we use this same a for the rest of the argument, and that is what we mean by it being specific.

(4) "$p \to q$" is an example of a **statement** in **propositional logic**. It is usually read as "if p, then q" or "p implies q." The letters p and q are called **propositional variables**, and we generally assign a truth value of T (for true) or F (for false) to each propositional variable. Formally, we define a **truth assignment** of a list of propositional variables to be a choice of T or F for each propositional variable in the list.

The symbol → is called a **conditional** or **implication**. It is one example of a **logical connective** (it *connects* two propositional variables). The rules for determining the truth value for $p \to q$ are given by the following truth table:

p	q	p → q
T	T	T
T	F	F
F	T	T
F	F	T

For example, if p and q are both assigned the truth value T, then the truth value of $p \to q$ is also T, as can be seen by the first row of the above truth table. We can write T → T ≡ T. The symbol "≡" can be read "**is logically equivalent to**." Similarly, we have

$$T \to F \equiv F \qquad F \to T \equiv T \qquad F \to F \equiv T$$

Observe that the only time $p \to q$ can be false is if p is true and q is false. So, one way to prove that $p \to q$ is true is to assume that p is true and then provide a logically correct argument that q must also be true.

If we let p represent the statement "$x \in A$" and we let q represent the statement "$x \in B$," then $p \to q$ represents the statement "if $x \in A$, then $x \in B$." As stated in the last paragraph, one way to prove that this statement is true is to assume that $x \in A$ is true and then provide a logically correct argument that $x \in B$ must also be true.

Basic Theorems Involving Subsets

Let's try to prove our first theorem using the definition of a subset together with Note 3 above about arbitrary but specific elements.

Theorem 1.9: Every set A is a subset of itself.

Analysis: Before writing the proof, let's think about our strategy. We want to prove $A \subseteq A$. In other words, we want to show $\forall x (x \in A \to x \in A)$. So, we will take an arbitrary but specific $a \in A$ and then argue that $a \in A$. But that's pretty obvious, isn't it? In this case, the property we're describing is precisely the conclusion we are looking for. Here are the details.

Proof of Theorem 1.9: Let A be a set and let $a \in A$. Then $a \in A$. So, $a \in A \to a \in A$ is true. Since a was an arbitrary element of A, $\forall x(x \in A \to x \in A)$ is true. Therefore, $A \subseteq A$. □

Notes: (1) The proof begins with the **opening statement** "Let A be a set and let $a \in A$." In general, the opening statement states what is given in the problem and/or fixes any arbitrary but specific objects that we will need.

(2) The proof ends with the **closing statement** "Therefore, $A \subseteq A$." In general, the closing statement states the result.

(3) Everything between the opening statement and the closing statement is known as the **argument**.

(4) We place the symbol □ at the end of the proof to indicate that the proof is complete.

(5) Consider the logical statement $p \to p$. This statement is always true (T → T ≡ T and F → F ≡ T). $p \to p$ is an example of a tautology. A **tautology** is a statement that is true for every possible truth assignment of the propositional variables.

(6) If we let p represent the statement $a \in A$, by Note 5, we see that $a \in A \to a \in A$ is always true.

Alternate proof of Theorem 1.9: Let A be a set and let $a \in A$. Since $p \to p$ is a tautology, we have that $a \in A \to a \in A$ is true. Since a was arbitrary, $\forall x (x \in A \to x \in A)$ is true. Therefore, $A \subseteq A$. □

Let's prove another basic but important theorem.

Theorem 1.10: The empty set is a subset of every set.

Analysis: This time we want to prove $\emptyset \subseteq A$. In other words, we want to show $\forall x (x \in \emptyset \to x \in A)$. Since $x \in \emptyset$ is always false (the empty set has no elements), $x \in \emptyset \to x \in A$ is always true.

In general, if p is a false statement, then we say that $p \to q$ is **vacuously true**.

Proof of Theorem 1.10: Let A be a set. The statement $x \in \emptyset \to x \in A$ is vacuously true for any x, and so, $\forall x (x \in \emptyset \to x \in A)$ is true. Therefore, $\emptyset \subseteq A$. □

Note: The opening statement is "Let A be a set," the closing statement is "Therefore, $\emptyset \subseteq A$," and the argument is everything in between.

Example 1.11: Let $C = \{a, b, c\}$, $D = \{a, c\}$, $E = \{b, c\}$, $F = \{b, d\}$, and $G = \emptyset$. Then $D \subseteq C$ and $E \subseteq C$. Also, since **the empty set is a subset of every set**, we have $G \subseteq C$, $G \subseteq D$, $G \subseteq E$, $G \subseteq F$, and $G \subseteq G$. **Every set is a subset of itself**, and so, $C \subseteq C$, $D \subseteq D$, $E \subseteq E$, and $F \subseteq F$.

Note: Below are possible Venn diagrams for this problem. The diagram on the left shows the relationship between the sets C, D, E, and F. Notice how D and E are both subsets of C, whereas F is not a subset of C. Also, notice how D and E overlap, E and F overlap, but there is no overlap between D and F (they have no elements in common). The diagram on the right shows the proper placement of the elements. Here, I chose the universal set to be $U = \{a, b, c, d, e, f, g\}$. This choice for the universal set is somewhat arbitrary. Any set containing $\{a, b, c, d\}$ would do.

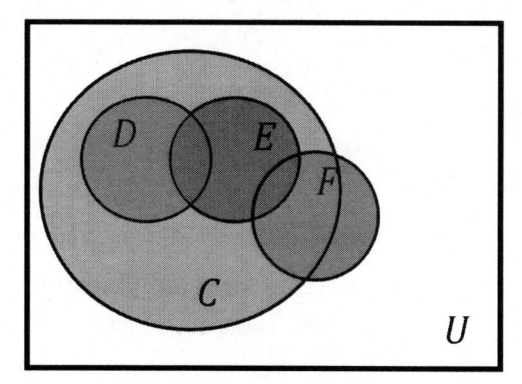

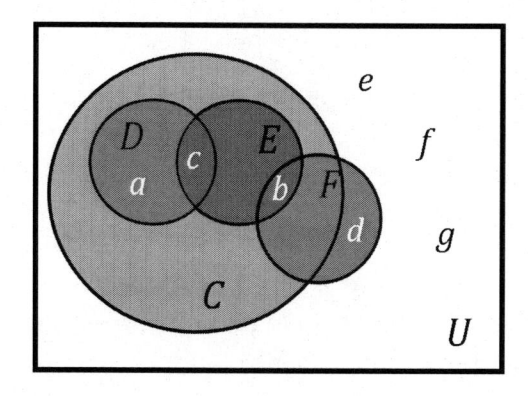

Power Sets

If A is a set, then the **power set** of A, written $\mathcal{P}(A)$, is the set of all subsets of A. In set-builder notation, we write $\mathcal{P}(A) = \{B \mid B \subseteq A\}$.

Example 1.12: The set $A = \{a, b\}$ has 2 elements and 4 subsets. The subsets of A are $\emptyset$, $\{a\}$, $\{b\}$, and $\{a, b\}$. It follows that $\mathcal{P}(A) = \{\emptyset, \{a\}, \{b\}, \{a, b\}\}$.

The set $B = \{a, b, c\}$ has 3 elements and 8 subsets. The subsets of B are $\emptyset$, $\{a\}$, $\{b\}$, $\{c\}$, $\{a, b\}$, $\{a, c\}$, $\{b, c\}$, and $\{a, b, c\}$. It follows that $\mathcal{P}(B) = \{\emptyset, \{a\}, \{b\}, \{c\}, \{a, b\}, \{a, c\}, \{b, c\}, \{a, b, c\}\}$.

Let's draw a **tree diagram** for the subsets of each of the sets A and B.

$$
\begin{array}{c}
\{a,b\} \\
/\ \ \backslash \\
\{a\}\ \ \ \{b\} \\
\backslash\ \ / \\
\emptyset
\end{array}
\qquad\qquad
\begin{array}{c}
\{a,b,c\} \\
/\ \ |\ \ \backslash \\
\{a,b\}\ \ \{a,c\}\ \ \{b,c\} \\
\backslash\ \times\ \times\ / \\
\{a\}\ \ \{b\}\ \ \{c\} \\
\backslash\ \ |\ \ / \\
\emptyset
\end{array}
$$

The tree diagram on the left is for the subsets of the set $A = \{a, b\}$. We start by writing the set $A = \{a, b\}$ at the top. On the next line we write the subsets of cardinality 1 ($\{a\}$ and $\{b\}$). On the line below that we write the subsets of cardinality 0 (just $\emptyset$). We draw a line segment between any two sets when the smaller (lower) set is a subset of the larger (higher) set. So, we see that $\emptyset \subseteq \{a\}$, $\emptyset \subseteq \{b\}$, $\{a\} \subseteq \{a, b\}$, and $\{b\} \subseteq \{a, b\}$. There is actually one more subset relationship, namely $\emptyset \subseteq \{a, b\}$ (and of course each set displayed is a subset of itself). We didn't draw a line segment from $\emptyset$ to $\{a, b\}$ to avoid unnecessary clutter. Instead, we can simply trace the path from $\emptyset$ to $\{a\}$ to $\{a, b\}$ (or from $\emptyset$ to $\{b\}$ to $\{a, b\}$). We are using a property called **transitivity** here (see Theorem 1.14 below).

The tree diagram on the right is for the subsets of $B = \{a, b, c\}$. Observe that from top to bottom we write the subsets of B of cardinality 3, then 2, then 1, and then 0. We then draw the appropriate line segments, just as we did for $A = \{a, b\}$.

How many subsets does a set of cardinality n have? Let's start by looking at some examples.

Example 1.13: A set with 0 elements must be $\emptyset$, and this set has exactly 1 subset (the only subset of the empty set is the empty set itself).

A set with 1 element has 2 subsets, namely $\emptyset$ and the set itself.

In Example 1.12, we saw that a set with 2 elements has 4 subsets, and we also saw that a set with 3 elements has 8 subsets.

Do you see the pattern yet? $1 = 2^0, 2 = 2^1, 4 = 2^2, 8 = 2^3$. So, we see that a set with 0 elements has 2^0 subsets, a set with 1 element has 2^1 subsets, a set with 2 elements has 2^2 subsets, and a set with 3 elements has 2^3 subsets.

A reasonable guess would be that a set with n elements has 2^n subsets. You will be asked to prove this result later (Problem 25 in Problem Set 5). We can also say that if $|A| = n$, then $|\mathcal{P}(A)| = 2^n$.

Transitivity of the Subset Relation

Let's get back to the transitivity mentioned above in part 3 of Example 1.8 and in our discussion of tree diagrams.

Theorem 1.14: Let A, B, and C be sets such that $A \subseteq B$ and $B \subseteq C$. Then $A \subseteq C$.

Proof: Suppose that A, B, and C are sets with $A \subseteq B$ and $B \subseteq C$, and let $a \in A$. Since $A \subseteq B$ and $a \in A$, it follows that $a \in B$. Since $B \subseteq C$ and $a \in B$, it follows that $a \in C$. Since a was an arbitrary element of A, we have shown that every element of A is an element of C. That is, $\forall x(x \in A \rightarrow x \in C)$ is true. Therefore, $A \subseteq C$. □

Note: To the right we have a Venn diagram illustrating Theorem 1.14.

Theorem 1.14 tells us that the relation $\subseteq$ is **transitive**. Since $\subseteq$ is transitive, we can write things like $A \subseteq B \subseteq C \subseteq D$, and without explicitly saying it, we know that $A \subseteq C$, $A \subseteq D$, and $B \subseteq D$.

Example 1.15: The membership relation $\in$ is an example of a relation that is **not** transitive. For example, let $A = \{0\}$, $B = \{0, 1, \{0\}\}$, and $C = \{x, y, \{0, 1, \{0\}\}\}$. Observe that $A \in B$ and $B \in C$, but $A \notin C$.

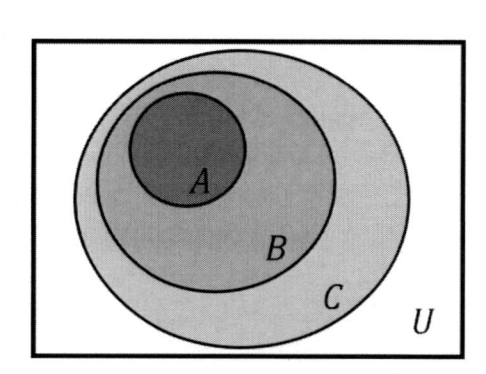

$A \subseteq B \subseteq C$

$\{0\} \in \{0, 1, \{0\}\} \in \{x, y, \{0, 1, \{0\}\}\}$

Notes: (1) The set A has just 1 element, namely 0.

(2) The set B has 3 elements, namely 0, 1, and $\{0\}$. But wait! $A = \{0\}$. So, $A \in B$. The set A is circled twice in the above image.

(3) The set C also has 3 elements, namely, x, y, and $\{0,1,\{0\}\}$. But wait! $B = \{0, 1, \{0\}\}$. So, $B \in C$. The set B has a rectangle around it twice in the above image.

(4) Since $A \neq x$, $A \neq y$, and $A \neq \{0, 1, \{0\}\}$, we see that $A \notin C$.

(5) Is it clear that $\{0\} \notin C$? $\{0\}$ is in a set that's in C (namely, B), but $\{0\}$ is not itself in C.

(6) Here is a more basic example showing that $\in$ is not transitive: $\emptyset \in \{\emptyset\} \in \{\{\emptyset\}\}$, but $\emptyset \notin \{\{\emptyset\}\}$
The only element of $\{\{\emptyset\}\}$ is $\{\emptyset\}$.

Equality of Sets

Two sets A and B are **equal**, written $A = B$, if they have the same elements. Symbolically, we can write the following:

$$\forall x(x \in A \leftrightarrow x \in B)$$

Notes: (1) "$p \leftrightarrow q$" is another example of a **statement** in **propositional logic**. It is usually read as "p if and only if q."

The logical connective $\leftrightarrow$ is called a **biconditional**. The rules for determining the truth value for $p \leftrightarrow q$ are given by the following truth table:

p	q	$p \leftrightarrow q$
T	T	T
T	F	F
F	T	F
F	F	T

In other words, $p \leftrightarrow q$ is true when p and q have the same truth value (both T or both F) and false when p and q have opposite truth values (one T and the other F).

If we let p represent the statement "$x \in A$" and we let q represent the statement "$x \in B$," then $p \leftrightarrow q$ represents the statement "$x \in A$ if and only if $x \in B$."

(2) In addition to the conditional ($\rightarrow$) and biconditional ($\leftrightarrow$), there are three other commonly used logical connectives: the **conjunction** $\wedge$ (pronounced "and"), the **disjunction** $\vee$ (pronounced "or"), and the **negation** $\neg$ (pronounced "not"). They have the following truth tables:

p	q	$p \wedge q$
T	T	T
T	F	F
F	T	F
F	F	F

p	q	$p \vee q$
T	T	T
T	F	T
F	T	T
F	F	F

p	$\neg p$
T	F
F	T

(3) We say that two statements are **logically equivalent** if every truth assignment of the propositional variables appearing in either statement (or both statements) leads to the same truth value for both statements.

It is easy to verify that $p \leftrightarrow q$ is logically equivalent to $(p \rightarrow q) \wedge (q \rightarrow p)$. To see this, we check that all possible truth assignments for p and q lead to the same truth value for the two statements. For example, if p and q are both true, then

$$p \leftrightarrow q \equiv T \leftrightarrow T \equiv T \quad \text{and} \quad (p \rightarrow q) \wedge (q \rightarrow p) \equiv (T \rightarrow T) \wedge (T \rightarrow T) \equiv T \wedge T \equiv T.$$

As another example, if p is true and q is false, then

$$p \leftrightarrow q \equiv T \leftrightarrow F \equiv F \quad \text{and} \quad (p \rightarrow q) \wedge (q \rightarrow p) \equiv (T \rightarrow F) \wedge (F \rightarrow T) \equiv F \wedge T \equiv F.$$

The reader should check the other two truth assignments for p and q.

(4) Letting p be the statement $x \in A$, letting q be the statement $x \in B$, and replacing $p \leftrightarrow q$ by the logically equivalent statement $(p \rightarrow q) \wedge (q \rightarrow p)$ gives us

$$\forall x (x \in A \leftrightarrow x \in B) \text{ if and only if } \forall x \big((x \in A \rightarrow x \in B) \wedge (x \in B \rightarrow x \in A)\big).$$

(5) It is also true that $\forall x(p(x) \wedge q(x))$ is logically equivalent to $\forall x(p(x)) \wedge \forall x(q(x))$. Therefore, we have the following: $\forall x(x \in A \leftrightarrow x \in B)$ if and only if $\forall x(x \in A \to x \in B)$ and $\forall x(x \in B \to x \in A)$.

In other words, to show that $A = B$, we can instead show that $A \subseteq B$ and $B \subseteq A$.

The statement "$A = B$ if and only if $A \subseteq B$ and $B \subseteq A$" is usually called the **Axiom of Extensionality**. It is often easiest to prove that two sets are equal by showing that each one is a subset of the other.

Example 1.16: Let's use the Axiom of Extensionality to prove that $5\mathbb{Z} + 3 = 5\mathbb{Z} + 18$.

Recall that $5\mathbb{Z} + 3 = \{t \in \mathbb{Z} \mid \exists k \in \mathbb{Z} \, (t = 5k + 3)\}$ and $5\mathbb{Z} + 18 = \{t \in \mathbb{Z} \mid \exists k \in \mathbb{Z} \, (t = 5k + 18)\}$ (see part 3 of Example 1.5).

In the proof that follows, we will use the fact that whenever we add two integers, we always get another integer (this property is known as **closure**, and we can say that $\mathbb{Z}$ is closed under addition).

Proof: We first prove that $5\mathbb{Z} + 3 \subseteq 5\mathbb{Z} + 18$. To see this, let $t \in 5\mathbb{Z} + 3$. Then there is $k \in \mathbb{Z}$ such that $t = 5k + 3$. Now, $5k + 3 = (5k - 15) + 18 = 5(k - 3) + 18$. It follows that $t = 5(k - 3) + 18$. Since k and 3 are integers, so is $k - 3 = k + (-3)$. Therefore, $t \in 5\mathbb{Z} + 18$. Since $t \in 5\mathbb{Z} + 3$ was arbitrary, $\forall t(t \in 5\mathbb{Z} + 3 \to t \in 5\mathbb{Z} + 18)$. Therefore, we have shown $5\mathbb{Z} + 3 \subseteq 5\mathbb{Z} + 18$.

We now prove that $5\mathbb{Z} + 18 \subseteq 5\mathbb{Z} + 3$. To see this, let $t \in 5\mathbb{Z} + 18$. Then there is $k \in \mathbb{Z}$ such that $t = 5k + 18$. Now, $5k + 18 = (5k + 15) + 3 = 5(k + 3) + 3$. It follows that $t = 5(k + 3) + 3$. Since k and 3 are integers, so is $k + 3$. Therefore, $t \in 5\mathbb{Z} + 3$. Since $t \in 5\mathbb{Z} + 18$ was arbitrary, $\forall t(t \in 5\mathbb{Z} + 18 \to t \in 5\mathbb{Z} + 3)$. Therefore, we have shown $5\mathbb{Z} + 18 \subseteq 5\mathbb{Z} + 3$.

Since $5\mathbb{Z} + 3 \subseteq 5\mathbb{Z} + 18$ and $5\mathbb{Z} + 18 \subseteq 5\mathbb{Z} + 3$, we see that $5\mathbb{Z} + 3 = 5\mathbb{Z} + 18$. □

Notes: (1) In the proof above, we used what I call the "**Standard Advanced Calculus Trick**." I sometimes abbreviate this as **SACT**. The trick is simple. If you need something to appear, just put it in. Then correct it by performing the opposite of what you just did.

In the first paragraph above, we wanted to rewrite $5k + 3$ as $5z + 18$ for some integer z. Using SACT, we can do the following:

- We simply put in what we need (and exactly where we need it): $5k + 3 + \mathbf{18}$
- Now, we undo the damage by performing the reverse operation: $5k + 3 - \mathbf{18} + 18$
- Finally, we leave the part we need as is and we simplify the rest: $(5k - 15) + 18$

We used the trick again in the second paragraph to rewrite $5k + 18$ as $5z + 3$. We simply added and subtracted 3 to get $5k + 18 = 5k + 18 - 3 + 3 = (5k + 15) + 3$.

(2) Also, in the proof above, we used the fact that multiplication is **distributive** over addition in $\mathbb{Z}$. This property says that if $x, y, z \in \mathbb{Z}$, then $x \cdot (y + z) = x \cdot y + x \cdot z$. We used this in the first paragraph of the proof to write $5k - 15 = 5k + (-15) = 5(k + (-3)) = 5(k - 3)$. We also used this in the second paragraph to write $5k + 15 = 5(k + 3)$. In Lesson 2, we will discuss distributivity in detail.

Cartesian Products

An **unordered pair** is a set with 2 elements. Recall, that a set doesn't change if we write the elements in a different order or if we write the same element multiple times. For example, $\{x, y\} = \{y, x\}$ and $\{x, x\} = \{x\}$.

We now define the **ordered pair** (x, y) in such a way that (y, x) will **not** be the same as (x, y). The simplest way to define a set with this property is as follows:

$$(x, y) = \{\{x\}, \{x, y\}\}$$

Let's show that with this definition, the ordered pair behaves as we would expect.

Theorem 1.17: $(x, y) = (z, w)$ if and only if $x = z$ and $y = w$.

Part of the proof of this theorem is a little trickier than expected. Assuming that $(x, y) = (z, w)$, there are actually two cases to consider: $x = y$ and $x \neq y$. If $x = y$, then (x, y) is a set with just one element. Indeed, $(x, x) = \{\{x\}, \{x, x\}\} = \{\{x\}, \{x\}\} = \{\{x\}\}$. So, the only element of (x, x) is $\{x\}$. Watch carefully how this plays out in the proof.

Proof of Theorem 1.17: First suppose that $x = z$ and $y = w$. Then by direct substitution, $\{x\} = \{z\}$ and $\{x, y\} = \{z, w\}$. So, $(x, y) = \{\{x\}, \{x, y\}\} = \{\{z\}, \{z, w\}\} = (z, w)$.

Conversely, suppose that $(x, y) = (z, w)$. Then $\{\{x\}, \{x, y\}\} = \{\{z\}, \{z, w\}\}$. There are two cases to consider.

Case 1: If $x = y$, then $\{\{x\}, \{x, y\}\} = \{\{x\}\}$. So, $\{\{x\}\} = \{\{z\}, \{z, w\}\}$. It follows that $\{z\} = \{x\}$ and $\{z, w\} = \{x\}$. Since $\{z, w\} = \{x\}$, we must have $z = x$ and $w = x$. Therefore, x, y, z, and w are all equal. In particular, $x = z$ and $y = w$.

Case 2: If $x \neq y$, then $\{x, y\}$ is a set with two elements. So, $\{x, y\}$ cannot be equal to $\{z\}$ (because $\{z\}$ has just one element). Therefore, we must have $\{x, y\} = \{z, w\}$. It then follows that $\{x\} = \{z\}$. So, we have $x = z$. Since $x = z$ and $\{x, y\} = \{z, w\}$, we must have $y = w$. □

Note: (x, y) is an abbreviation for the set $\{\{x\}, \{x, y\}\}$. In the study of set theory, every object can be written as a set like this. It's often convenient to use abbreviations, but we should always be aware that if necessary, we can write any object in its unabbreviated form.

We can extend the idea of an ordered pair to an **ordered k-tuple**. An ordered 3-tuple (also called an **ordered triple**) is defined by $(x, y, z) = ((x, y), z)$, an ordered 4-tuple is $(x, y, z, w) = ((x, y, z), w)$, and so on. For a general k-tuple, we will use a single letter with subscripts for the variable names. For example, using the letter x, we can write a k-tuple as $(x_1, x_2, \ldots, x_k)$.

Example 1.18: Let's write the ordered triple (x, y, z) in its unabbreviated form (take a deep breath!).

$$(x, y, z) = ((x, y), z) = \{\{(x, y)\}, \{(x, y), z\}\} = \{\{\{\{x\}, \{x, y\}\}\}, \{\{\{x\}, \{x, y\}\}, z\}\}$$

The **Cartesian product** of the sets A and B, written $A \times B$ is the set of ordered pairs (a, b) with $a \in A$ and $b \in B$. Symbolically, we have

$$A \times B = \{(a, b) \mid a \in A \land b \in B\}.$$

Observe that if A and B are finite sets with $|A| = m$ and $|B| = n$, then $|A \times B| = mn$.

Example 1.19:

1. Let $A = \{x, y\}$ and $B = \{0, 1, 2\}$. Then $A \times B = \{(x, 0), (x, 1), (x, 2), (y, 0), (y, 1), (y, 2)\}$. Note that $|A| = 2$, $|B| = 3$, and $|A \times B| = 2 \cdot 3 = 6$.

2. Let $C = \emptyset$ and $D = \{2, 4, 6, 8, 10\}$. Then $C \times D = \emptyset$ (since there are no elements in C, there can be no elements in $C \times D$). Note that $|C| = 0$, $|D| = 5$, and $|C \times D| = 0 \cdot 5 = 0$.

3. $\mathbb{N} \times \mathbb{Z} = \{(m, n) \mid m \in \mathbb{N} \land n \in \mathbb{Z}\}$. For example, $(5, -3) \in \mathbb{N} \times \mathbb{Z}$, whereas $(-3, 5) \notin \mathbb{N} \times \mathbb{Z}$ (although it is in $\mathbb{Z} \times \mathbb{N}$). We can visualize $\mathbb{N} \times \mathbb{Z}$ as follows:

   $\ldots, (0, -3), (0, -2), (0, -1), (0, 0), (0, 1), (0, 2), (0, 3), \ldots$

   $\ldots, (1, -3), (1, -2), (1, -1), (1, 0), (1, 1), (1, 2), (1, 3), \ldots$

   $\ldots, (2, -3), (2, -2), (2, -1), (2, 0), (2, 1), (2, 2), (2, 3), \ldots$

   $\vdots \qquad \vdots \qquad \vdots$

4. $\mathbb{R} \times \mathbb{R} = \{(x, y) \mid x, y \in \mathbb{R}\}$. We can visualize elements of $\mathbb{R} \times \mathbb{R}$ as points in the **Cartesian plane**. A portion of the Cartesian plane is shown to the right. The elements $(3, 2)$ and $(-1, -2)$ of $\mathbb{R} \times \mathbb{R}$ are displayed as points.

   We form the Cartesian plane by taking two copies of the real line and placing one horizontally and the other vertically, exactly as we did for the Complex Plane in part 3 of Example 1.6. The horizontal copy of the real line is called the x-axis (labeled x in the figure) and the vertical copy of the real line is called the y-axis (labeled y in the figure). The two axes intersect at the point $(0, 0)$. This point is called the **origin**.

   Notice that visually the Cartesian plane $\mathbb{R} \times \mathbb{R}$ is indistinguishable from the Complex Plane.

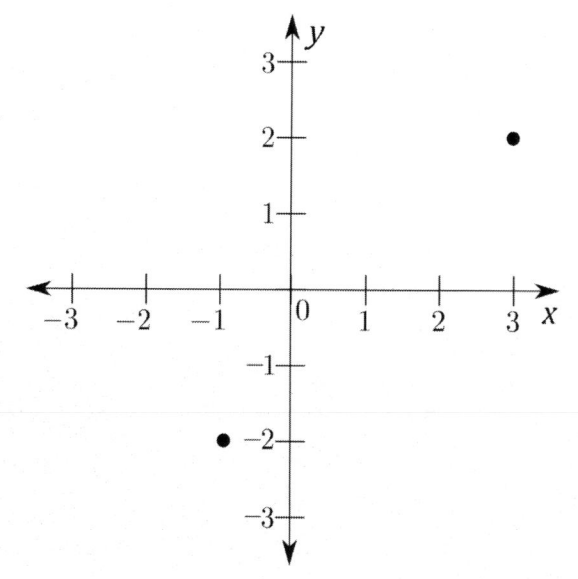

We can extend the definition of the Cartesian product to more than two sets in the obvious way:

$$A \times B \times C = \{(a, b, c) \mid a \in A \land b \in B \land c \in C\}$$

$$A \times B \times C \times D = \{(a, b, c, d) \mid a \in A \land b \in B \land c \in C \land d \in D\}$$

Observe that if A, B, and C are finite sets with $|A| = m$, $|B| = n$, and $|C| = k$, then we have $|A \times B \times C| = mnk$. Similarly, if A, B, C, and D are finite sets with $|A| = m$, $|B| = n$, $|C| = k$, and $|D| = j$, then we have $|A \times B \times C \times D| = mnkj$.

In general, the **Cartesian product** of the sets $A_1, A_2, \ldots, A_n$ is
$$A_1 \times A_2 \times \cdots \times A_n = \{(a_1, a_2, \ldots, a_n) \mid a_1 \in A_1 \wedge a_2 \in A_2 \wedge \cdots \wedge a_n \in A_n\}.$$

Note: Recall that $\wedge$ is just a symbol representing the word "and."

If $A_1, A_2, \ldots, A_n$ are finite, then the cardinality of $A_1 \times A_2 \times \cdots \times A_n$ is the product of the cardinalities of $A_1, A_2, \ldots, A_n$. Symbolically, we have $|A_1 \times A_2 \times \cdots \times A_n| = |A_1| \cdot |A_2| \cdots |A_n|$.

Example 1.20:

1. $\{a\} \times \{b\} \times \{c\} \times \{d\} = \{(a, b, c, d)\}$.

    Note that $|\{a\} \times \{b\} \times \{c\} \times \{d\}| = |\{a\}| \cdot |\{b\}| \cdot |\{c\}| \cdot |\{d\}| = 1 \cdot 1 \cdot 1 \cdot 1 = 1$.

2. $\{0\} \times \{0,1\} \times \{1\} \times \{0,1\} \times \{0\} = \{(0,0,1,0,0), (0,0,1,1,0), (0,1,1,0,0), (0,1,1,1,0)\}$.

    Note that $|\{0\} \times \{0,1\} \times \{1\} \times \{0,1\} \times \{0\}| = 1 \cdot 2 \cdot 1 \cdot 2 \cdot 1 = 4$.

3. $\mathbb{Z} \times \mathbb{N} \times \mathbb{R} = \{(m, n, x) \mid m \in \mathbb{Z} \wedge n \in \mathbb{N} \wedge x \in \mathbb{R}\}$.

We abbreviate Cartesian products of sets with themselves using exponents.

$$A^2 = A \times A \qquad A^3 = A \times A \times A \qquad A^4 = A \times A \times A \times A \qquad A^n = \underbrace{A \times A \times \cdots \times A}_{n \text{ times}}$$

Example 1.21:

1. $\mathbb{Z}^2 = \mathbb{Z} \times \mathbb{Z} = \{(x, y) \mid x, y \in \mathbb{Z}\}$ is the set of ordered pairs of integers. A few sample elements in $\mathbb{Z}^2$ are $(0, 0), (-1, 2), (15, -106)$ and $(-53, -53)$.

2. $\mathbb{N}^5 = \mathbb{N} \times \mathbb{N} \times \mathbb{N} \times \mathbb{N} \times \mathbb{N} = \{(a, b, c, d, e) \mid a, b, c, d, e \in \mathbb{N}\}$ is the set of ordered 5-tuples of natural numbers. A few sample elements in $\mathbb{N}^5$ are $(0, 0, 0, 0, 0), (1, 1, 2, 3, 3), (0, 1, 17, 86, 0)$ and $(1000, 2529, 8, 900, 106)$.

3. $\{0, 1\}^2 = \{0, 1\} \times \{0, 1\} = \{(0, 0), (0, 1), (1, 0), (1, 1)\}$.

4. $\{0, 1\}^3 = \{0, 1\} \times \{0, 1\} \times \{0, 1\}$
    $= \{(0,0,0), (0,0,1), (0,1,0), (0,1,1), (1,0,0), (1,0,1), (1,1,0), (1,1,1)\}$.

5. $\mathbb{R}^2 = \mathbb{R} \times \mathbb{R} = \{(x, y) \mid x, y \in \mathbb{R}\}$ was discussed in part 4 of Example 1.19 above. We will also be interested in larger Cartesian products of $\mathbb{R}$ such as $\mathbb{R}^3 = \{(x, y, z) \mid x, y, z \in \mathbb{R}\}$, and more generally, $\mathbb{R}^n = \{(x_1, x_2, \ldots, x_n) \mid x_1, x_2, \ldots, x_n \in \mathbb{R}\}$.

6. $\mathbb{C}^n = \{(z_1, z_2, \ldots, z_n) \mid z_1, z_2, \ldots, z_n \in \mathbb{C}\}$ is the set of ordered n-tuples of complex numbers. For example, $\left(3 + 2i, -1 + \frac{5}{7}i\right) \in \mathbb{C}^2$ and $\left(1 - i, 2 + \frac{1}{2}i, -5i, 0.253\right) \in \mathbb{C}^4$.

Basic Set Operations

The **union** of the sets A and B, written $A \cup B$, is the set of elements that are in A or B (or both).

$$A \cup B = \{x \mid x \in A \text{ or } x \in B\}$$

The **intersection** of A and B, written $A \cap B$, is the set of elements that are simultaneously in A and B.

$$A \cap B = \{x \mid x \in A \text{ and } x \in B\}$$

The following Venn diagrams for the union and intersection of two sets can be useful for visualizing these operations.

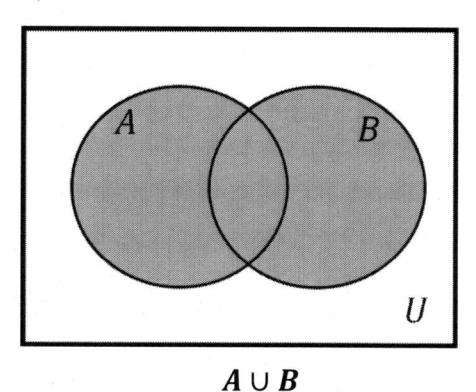

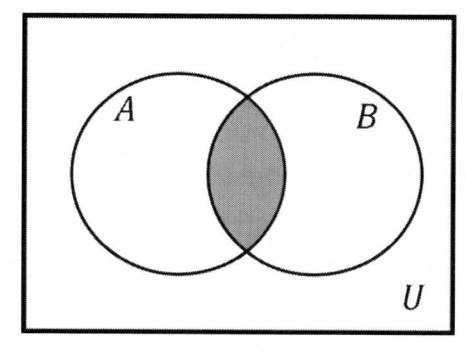

$A \cup B$

$A \cap B$

The **difference** $A \setminus B$ is the set of elements that are in A and not in B.

$$A \setminus B = \{x \mid x \in A \text{ and } x \notin B\}$$

The **symmetric difference** between A and B, written $A \Delta B$, is the set of elements that are in A or B, but not both.

$$A \Delta B = (A \setminus B) \cup (B \setminus A)$$

Let's also look at Venn diagrams for the difference and symmetric difference of two sets.

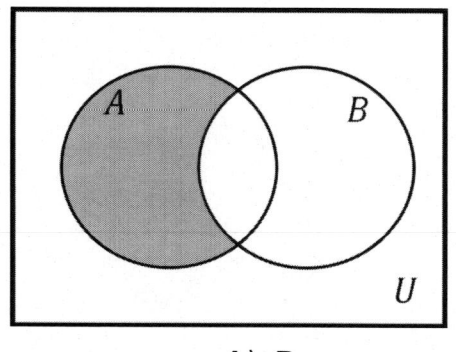

$A \setminus B$

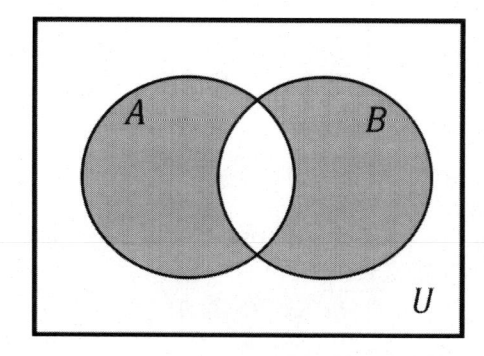

$A \Delta B$

Example 1.22: Let $A = \{0, 1, 2, 3, 4\}$ and $B = \{3, 4, 5, 6\}$. We have

1. $A \cup B = \{0, 1, 2, 3, 4, 5, 6\}$.
2. $A \cap B = \{3, 4\}$.
3. $A \setminus B = \{0, 1, 2\}$.
4. $B \setminus A = \{5, 6\}$.
5. $A \Delta B = \{0, 1, 2\} \cup \{5, 6\} = \{0, 1, 2, 5, 6\}$.

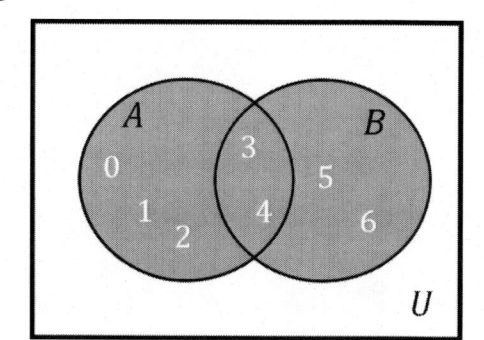

Example 1.23: Recall that the set of natural numbers is $\mathbb{N} = \{0, 1, 2, 3, \dots\}$ and the set of integers is $\mathbb{Z} = \{\dots, -4, -3, -2, -1, 0, 1, 2, 3, 4, \dots\}$. Observe that in this case, $\mathbb{N} \subseteq \mathbb{Z}$. We have

1. $\mathbb{N} \cup \mathbb{Z} = \mathbb{Z}$.
2. $\mathbb{N} \cap \mathbb{Z} = \mathbb{N}$.
3. $\mathbb{N} \setminus \mathbb{Z} = \emptyset$.
4. $\mathbb{Z} \setminus \mathbb{N} = \{\dots, -4, -3, -2, -1\} = \mathbb{Z}^-$. ($\mathbb{Z}^-$ is "the set of negative integers.")
5. $\mathbb{N} \triangle \mathbb{Z} = \emptyset \cup \mathbb{Z}^- = \mathbb{Z}^-$.

Note: Whenever A and B are sets and $B \subseteq A$, then $A \cup B = A$, $A \cap B = B$, and $B \setminus A = \emptyset$. We will prove the first and third of these three facts in Theorems 1.27 and 1.28 below, respectively. You will be asked to prove the second in Problem 24 below.

Example 1.24: Let $\mathbb{E} = 2\mathbb{N} = \{0, 2, 4, 6, \dots\}$ be the set of even natural numbers and let $\mathbb{O} = 2\mathbb{N} + 1 = \{1, 3, 5, 7, \dots\}$ be the set of odd natural numbers. We have

1. $\mathbb{E} \cup \mathbb{O} = \{0, 1, 2, 3, 4, 5, 6, 7, \dots\} = \mathbb{N}$.
2. $\mathbb{E} \cap \mathbb{O} = \emptyset$.
3. $\mathbb{E} \setminus \mathbb{O} = \mathbb{E}$.
4. $\mathbb{O} \setminus \mathbb{E} = \mathbb{O}$.
5. $\mathbb{E} \triangle \mathbb{O} = \mathbb{E} \cup \mathbb{O} = \mathbb{N}$.

In general, we say that sets A and B are **disjoint** or **mutually exclusive** if $A \cap B = \emptyset$. To the right is a Venn diagram for disjoint sets.

In Example 1.24 above, we saw that the sets $\mathbb{E} = 2\mathbb{N}$ and $\mathbb{O} = 2\mathbb{N} + 1$ are disjoint.

Example 1.25: Consider the sets $A = \{a + bi \in \mathbb{C} \mid a, b \in \mathbb{Z}\}$ and $B = \{a + bi \in \mathbb{C} \mid a \notin \mathbb{Q}\}$. Then A and B are disjoint. To see this, suppose that $a + bi \in A \cap B$. Then $a + bi \in A$, and so, $a \in \mathbb{Z}$. Since $\mathbb{Z} \subseteq \mathbb{Q}$, $a \in \mathbb{Q}$. Also, $a + bi \in B$. So, $a \notin \mathbb{Q}$. Since we cannot have both $a \in \mathbb{Q}$ and $a \notin \mathbb{Q}$, we must have $A \cap B = \emptyset$.

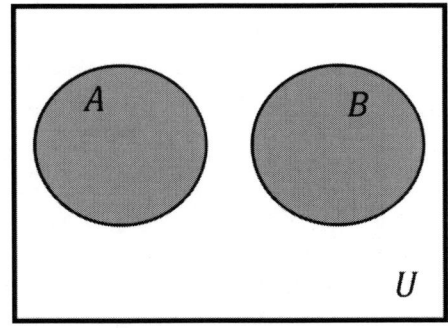

Let's prove some theorems involving unions of sets. You will be asked to prove the analogous results for intersections of sets in Problems 19 and 24 below.

Theorem 1.26: If A and B are sets, then $A \subseteq A \cup B$.

Before going through the proof, look once more at the Venn diagram above for $A \cup B$ and convince yourself that this theorem should be true.

Proof of Theorem 1.26: Suppose that A and B are sets and let $x \in A$. Then $x \in A$ or $x \in B$. Therefore, $x \in A \cup B$. Since x was an arbitrary element of A, we have shown that every element of A is an element of $A \cup B$. That is, $\forall x(x \in A \to x \in A \cup B)$ is true. Therefore, $A \subseteq A \cup B$. □

Note: If p is a true statement, then $p \vee q$ (p or q) is true no matter what the truth value of q is. You can see this by looking at the truth table for $\vee$ above (see Note 2 before Example 1.16). In the second sentence of the proof above, we are using this fact with p being the statement $x \in A$ and q being the statement $x \in B$.

We will use this same reasoning in the second paragraph of the next proof as well.

Theorem 1.27: $B \subseteq A$ if and only if $A \cup B = A$.

Before going through the proof, it's a good idea to draw a Venn diagram for $B \subseteq A$ and convince yourself that this theorem should be true.

Proof of Theorem 1.27: Suppose that $B \subseteq A$ and let $x \in A \cup B$. Then $x \in A$ or $x \in B$. If $x \in A$, then $x \in A$ (trivially). If $x \in B$, then since $B \subseteq A$, it follows that $x \in A$. Since x was an arbitrary element of $A \cup B$, we have shown that every element of $A \cup B$ is an element of A. That is, $\forall x(x \in A \cup B \to x \in A)$ is true. Therefore, $A \cup B \subseteq A$. By Theorem 1.26, $A \subseteq A \cup B$. Since $A \cup B \subseteq A$ and $A \subseteq A \cup B$, it follows that $A \cup B = A$.

Now, suppose that $A \cup B = A$ and let $x \in B$. Since $x \in B$, it follows that $x \in A$ or $x \in B$. Therefore, $x \in A \cup B$. Since $A \cup B = A$, we have $x \in A$. Since x was an arbitrary element of B, we have shown that every element of B is an element of A. That is, $\forall x(x \in B \to x \in A)$. Therefore, $B \subseteq A$. □

Theorem 1.28: Let A and B be sets. If $B \subseteq A$, then $B \setminus A = \emptyset$.

We will use an **indirect proof** to prove Theorem 1.28. Specifically, we will use a **proof by contrapositive**. The contrapositive of the conditional statement $p \to q$ is the statement $\neg q \to \neg p$. These two statements are logically equivalent. To see this, we check that all possible truth assignments for p and q lead to the same truth value for the two statements. For example, if p and q are both true, then $p \to q \equiv T \to T \equiv T$ and $\neg q \to \neg p \equiv F \to F \equiv T$. The reader should check the other three truth assignments for p and q.

The contrapositive of the statement "If $B \subseteq A$, then $B \setminus A = \emptyset$" is "If $B \setminus A \neq \emptyset$, then $B \not\subseteq A$." So, we will prove Theorem 1.28 by assuming that $B \setminus A \neq \emptyset$ and using this to show that $B \not\subseteq A$.

Proof of Theorem 1.28: Let A and B be sets such that $B \setminus A \neq \emptyset$. Since $B \setminus A \neq \emptyset$, there is $a \in B \setminus A$. Then $a \in B$ and $a \notin A$. So, $a \in B$ is true and $a \in A$ is false. So, $a \in B \to a \in A$ is false (because $T \to F \equiv F$). It follows that $\forall x(x \in B \to x \in A)$ is false, and therefore, $B \not\subseteq A$. □

Properties of Unions and Intersections

Unions, intersections, and set differences have many nice algebraic properties such as the following:

1. **Commutativity:** $A \cup B = B \cup A$ and $A \cap B = B \cap A$.
2. **Associativity:** $(A \cup B) \cup C = A \cup (B \cup C)$ and $(A \cap B) \cap C = A \cap (B \cap C)$.
3. **Distributivity:** $A \cap (B \cup C) = (A \cap B) \cup (A \cap C)$ and $A \cup (B \cap C) = (A \cup B) \cap (A \cup C)$.
4. **De Morgan's Laws:** $C \setminus (A \cup B) = (C \setminus A) \cap (C \setminus B)$ and $C \setminus (A \cap B) = (C \setminus A) \cup (C \setminus B)$.
5. **Idempotent Laws:** $A \cup A = A$ and $A \cap A = A$.

As an example, let's prove that the operation of forming unions is associative. You will be asked to prove that the other properties hold in the problems below.

Theorem 1.29: The operation of forming unions is associative.

Note: Before beginning the proof, let's draw Venn diagrams of the situation to convince ourselves that the theorem is true.

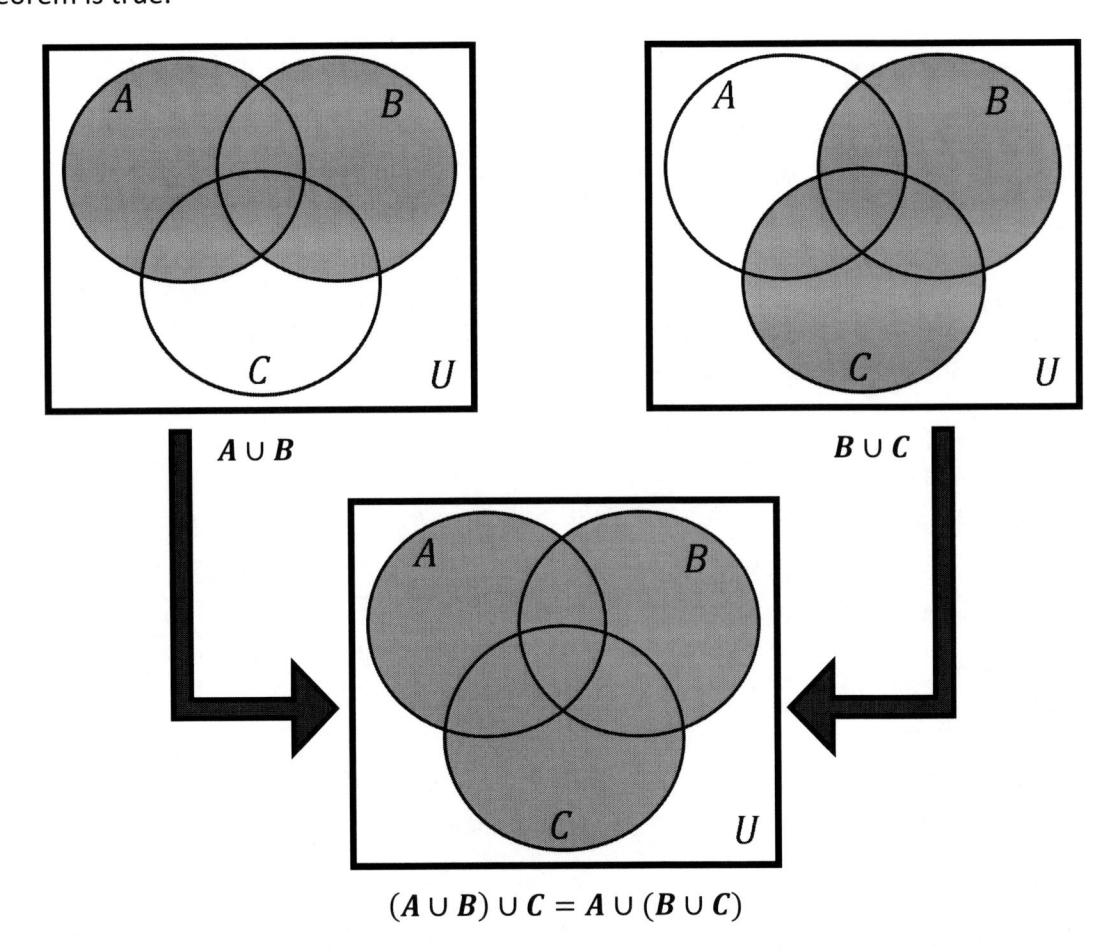

$(A \cup B) \cup C = A \cup (B \cup C)$

Proof of Theorem 1.29: Let A, B, and C be sets, and let $x \in (A \cup B) \cup C$. Then $x \in A \cup B$ or $x \in C$. If $x \in C$, then $x \in B$ or $x \in C$. So, $x \in B \cup C$. Then $x \in A$ or $x \in B \cup C$. So, $x \in A \cup (B \cup C)$. If, on the other hand, $x \in A \cup B$, then $x \in A$ or $x \in B$. If $x \in A$, then $x \in A$ or $x \in B \cup C$. So, $x \in A \cup (B \cup C)$. If $x \in B$, then $x \in B$ or $x \in C$. So, $x \in B \cup C$. Then $x \in A$ or $x \in B \cup C$. So, $x \in A \cup (B \cup C)$. Since x was arbitrary, we have shown $\forall x (x \in (A \cup B) \cup C \to x \in A \cup (B \cup C))$. Therefore, we have shown that $(A \cup B) \cup C \subseteq A \cup (B \cup C)$.

A similar argument can be used to show $A \cup (B \cup C) \subseteq (A \cup B) \cup C$ (the reader should write out the details).

Since $(A \cup B) \cup C \subseteq A \cup (B \cup C)$ and $A \cup (B \cup C) \subseteq (A \cup B) \cup C$, $(A \cup B) \cup C = A \cup (B \cup C)$, and therefore, the operation of forming unions is associative. □

Remember that associativity allows us to drop parentheses. So, we can now simply write $A \cup B \cup C$ when taking the union of the three sets A, B, and C.

Arbitrary Unions and Intersections

Many students find the definitions given in this section difficult to understand at first. If you have trouble grasping the material here, I wouldn't worry too much at first. I would suggest coming back and rereading this section as examples of infinite unions and intersections come up throughout the book. Try to understand it a little better each time and eventually it will all become clear.

We will often be interested in taking unions and intersections of more than two sets. Therefore, we make the following more general definitions.

Let X be a nonempty set of sets.

$$\cup X = \{y \mid \text{there is } Y \in X \text{ with } y \in Y\} \quad \text{and} \quad \cap X = \{y \mid \text{for all } Y \in X, y \in Y\}.$$

If you're having trouble understanding what these definitions are saying, you're not alone. The notation probably looks confusing, but the ideas behind these definitions are very simple. You have a whole bunch of sets (possibly infinitely many). To take the union of all these sets, you simply throw all the elements together into one big set. To take the intersection of all these sets, you take only the elements that are in every single one of those sets.

Example 1.30:

1. Let A and B be sets and let $X = \{A, B\}$. Then

   $$\cup X = \{y \mid \text{there is } Y \in X \text{ with } y \in Y\} = \{y \mid y \in A \text{ or } y \in B\} = A \cup B.$$
   $$\cap X = \{y \mid \text{for all } Y \in X, y \in Y\} = \{y \mid y \in A \text{ and } y \in B\} = A \cap B.$$

2. Let A, B, and C be sets, and let $X = \{A, B, C\}$. Then

   $$\cup X = \{y \mid \text{there is } Y \in X \text{ with } y \in Y\} = \{y \mid y \in A, y \in B, \text{ or } y \in C\} = A \cup B \cup C.$$
   $$\cap X = \{y \mid \text{for all } Y \in X, y \in Y\} = \{y \mid y \in A, y \in B, \text{ and } y \in C\} = A \cap B \cap C.$$

3. Let $X = \{\{-n, \ldots, -3, -2, -1, 0, 1, 2, 3, 4, \ldots, n\} \mid n \in \mathbb{N}\}$. The sets in X look as follows:

   $\{0\}, \quad \{-1, 0, 1\}, \quad \{-2, -1, 0, 1, 2\}, \quad \{-3, -2, -1, 0, 1, 2, 3\}, \quad \{-4, -3, -2, -1, 0, 1, 2, 3, 4\}, \ldots$

   We have
   $$\cup X = \{y \mid \text{there is } Y \in X \text{ with } y \in Y\}$$
   $$= \{y \mid \text{there is } n \in \mathbb{N} \text{ with } y \in \{-n, \ldots, -3, -2, -1, 0, 1, 2, 3, 4, \ldots, n\}\} = \mathbb{Z}.$$
   $$\cap X = \{y \mid \text{for all } Y \in X, y \in Y\}$$
   $$= \{y \mid \text{for all } n \in \mathbb{N}, y \in \{-n, \ldots, -3, -2, -1, 0, 1, 2, 3, 4, \ldots, n\}\} = \{0\}.$$

Notes: (1) Examples 1 and 2 give a good idea of what $\cup X$ and $\cap X$ look like when X is finite. More generally, if $X = \{A_1, A_2, \ldots, A_n\}$, then $\cup X = A_1 \cup A_2 \cup \cdots \cup A_n$ and $\cap X = A_1 \cap A_2 \cap \cdots \cap A_n$.

(2) As a specific example of Note 1, let $A_1 = \{\ldots, -3, -2, -1, 0, 1, 2\}$, $A_2 = \{0, 1, 2, 3, 4, 5\}$, $A_3 = \{1, 2\}$, and $A_4 = \{2, 3, 4, \ldots, 98, 99\}$. Let $X = \{A_1, A_2, A_3, A_4\}$. Then

$$\cup X = A_1 \cup A_2 \cup A_3 \cup A_4 = \{\ldots, -3, -2, -1, 0, 1, 2, 3, \ldots, 98, 99\}.$$

$$\cap X = A_1 \cap A_2 \cap A_3 \cap A_4 = \{2\}.$$

If you have trouble seeing how to compute the intersection, it may help to take the intersections two at a time:

$A_1 \cap A_2 = \{\ldots, -3, -2, -1, 0, 1, 2\} \cap \{0, 1, 2, 3, 4, 5\} = \{0, 1, 2\}$.

$\{0, 1, 2\} \cap A_3 = A_3 = \{1, 2\}$ because $A_3 \subseteq \{0, 1, 2\}$.

$\{1, 2\} \cap A_4 = \{1, 2\} \cap \{2, 3, 4, \ldots, 98, 99\} = \{2\}$.

(3) Let's prove carefully that $\{y \mid \text{there is } n \in \mathbb{N} \text{ with } y \in \{-n, \ldots, -3, -2, -1, 0, 1, 2, 3, 4, \ldots, n\}\} = \mathbb{Z}$.

For convenience, let's let $A = \{y \mid \text{there is } n \in \mathbb{N} \text{ with } y \in \{-n, \ldots, -3, -2, -1, 0, 1, 2, 3, 4, \ldots, n\}\}$.

If $y \in A$, then there is $n \in \mathbb{N}$ with $y \in \{-n, \ldots, -3, -2, -1, 0, 1, 2, 3, 4, \ldots, n\}$. In particular, $y \in \mathbb{Z}$. Since $y \in A$ was arbitrary, we have shown that $A \subseteq \mathbb{Z}$.

Let $y \in \mathbb{Z}$. Then $y \in \{-n, \ldots, -3, -2, -1, 0, 1, 2, 3, 4, \ldots, n\}$, where $n = y$ if $y \geq 0$ and $n = -y$ if $y < 0$ (in other words, n is the **absolute value** of y, written $n = |y|$). So, $y \in A$. Since $y \in \mathbb{Z}$ was arbitrary, we have shown that $\mathbb{Z} \subseteq A$.

Since $A \subseteq \mathbb{Z}$ and $\mathbb{Z} \subseteq A$, it follows that $A = \mathbb{Z}$.

(4) Let's also prove carefully that $\{y \mid \text{for all } n \in \mathbb{N}, y \in \{-n, \ldots, -3, -2, -1, 0, 1, 2, 3, 4, \ldots, n\}\} = \{0\}$.

For convenience, let's let $B = \{y \mid \text{for all } n \in \mathbb{N}, y \in \{-n, \ldots, -3, -2, -1, 0, 1, 2, 3, 4, \ldots, n\}\}$.

If $y \in B$, then for all $n \in \mathbb{N}$, $y \in \{-n, \ldots, -3, -2, -1, 0, 1, 2, 3, 4, \ldots, n\}$. In particular, $y \in \{0\}$. Since $y \in B$ was arbitrary, we have shown that $B \subseteq \{0\}$.

Now, let $y \in \{0\}$. Then $y = 0$. For all $n \in \mathbb{N}$, $0 \in \{-n, \ldots, -3, -2, -1, 0, 1, 2, 3, 4, \ldots, n\}$. So, $y \in B$. It follows that $\{0\} \subseteq B$.

Since $B \subseteq \{0\}$ and $\{0\} \subseteq B$, it follows that $B = \{0\}$.

(5) Note that the empty union is empty. Indeed, we have $\cup \emptyset = \{y \mid \text{there is } Y \in \emptyset \text{ with } y \in Y\} = \emptyset$.

If X is a nonempty set of sets, we say that X is **disjoint** if $\cap X = \emptyset$. We say that X is **pairwise disjoint** if for all $A, B \in X$ with $A \neq B$, A and B are disjoint. For example, if we let $X = \{(n, n+1) \mid n \in \mathbb{Z}\}$, then X is both disjoint and pairwise disjoint.

Are the definitions of disjoint and pairwise disjoint equivalent? You will be asked to answer this question in Problem 18 below.

Problem Set 1

Full solutions to these problems are available for free download here:
www.SATPrepGet800.com/AAFBTDW

LEVEL 1

1. Determine whether each of the following statements is true or false:

   (i) $x \in \{x\}$

   (ii) $\alpha \in \{\alpha, \beta, \gamma\}$

   (iii) $-5 \in \{5\}$

   (iv) $0 \in \mathbb{Z}$

   (v) $-27 \in \mathbb{N}$

   (vi) $\frac{11}{29} \in \mathbb{Q}$

   (vii) $\emptyset \subseteq \{a, b, c\}$

   (viii) $\{\Delta\} \subseteq \{\delta, \Delta\}$

   (ix) $\{x, y, z\} \subseteq \{x, y, z\}$

   (x) $\{1, 4, \{7, 9\}\} \subseteq \{1, 4, 7, 9\}$

2. Determine the cardinality of each of the following sets:

   (i) $\{\text{cat}, \text{dog}, \text{zebra}\}$

   (ii) $\{0, 2, 5, 11, 17\}$

   (iii) $\{1, 2, \ldots, 52\}$

   (iv) $\left\{\frac{1}{2}, \frac{1}{3}, \ldots, \frac{1}{11}\right\}$

3. List the elements of $\{k, x, t\} \times \{5, 6\}$.

4. Let $A = \{0\}$. Evaluate (i) A^2; (ii) A^3; (iii) $\mathcal{P}(A)$.

5. Let $A = \{a, b, \Delta, \delta\}$ and $B = \{b, c, \delta, \gamma\}$. Determine each of the following:

   (i) $A \cup B$

   (ii) $A \cap B$

   (iii) $A \setminus B$

   (iv) $B \setminus A$

   (v) $A \Delta B$

6. Draw Venn diagrams for $(A \setminus B) \setminus C$ and $A \setminus (B \setminus C)$. Are these two sets equal for all sets A, B, and C? If so, prove it. If not, provide a counterexample.

Level 2

7. Compute the power set of each of the following sets:
   (i) $\emptyset$
   (ii) $\{b\}$
   (iii) $\{\Delta, \Gamma\}$
   (iv) $\{\emptyset, \{\emptyset\}\}$
   (v) $\{\{\emptyset\}\}$

8. Determine whether each of the following statements is true or false:
   (i) $1 \in \emptyset$
   (ii) $\emptyset \in \emptyset$
   (iii) $\emptyset \in \{\emptyset, \{\emptyset\}\}$
   (iv) $\{\emptyset\} \in \emptyset$
   (v) $\{\emptyset\} \in \{\emptyset\}$
   (vi) $5 \in \{3k \mid k = 1, 2, 3, 4\}$
   (vii) $13 \in 3\mathbb{Z} + 1$
   (viii) $\emptyset \subseteq \emptyset$
   (ix) $\emptyset \subseteq \{\emptyset\}$
   (x) $\{\emptyset\} \subseteq \emptyset$
   (xi) $\{\emptyset\} \subseteq \{\emptyset\}$

9. Determine the cardinality of each of the following sets:
   (i) $\{x, x, y, z, z, z\}$
   (ii) $\{\{0, 1\}, \{2, 3, 4\}\}$
   (iii) $\{8, 9, 10, \ldots, 4226, 4227\}$

10. Compute $\{a, b\}^4$.

11. Let $A = \{\emptyset, \{\emptyset, \{\emptyset\}\}\}$ and $B = \{\emptyset, \{\emptyset\}\}$. Compute each of the following:
    (i) $A \cup B$
    (ii) $A \cap B$
    (iii) $A \setminus B$
    (iv) $B \setminus A$
    (v) $A \Delta B$

12. Prove the following:

    (i) The operation of forming unions is commutative.

    (ii) The operation of forming intersections is commutative.

    (iii) The operation of forming intersections is associative.

LEVEL 3

13. Determine the cardinality of each of the following sets:

    (i) $\{\{\{0,1\}\}\}$

    (ii) $\{\{x,y\},x,\{x\},\{x,\{x,y,z\}\}\}$

    (iii) $\{a,\{a\},\{a,a\},\{a,a,a,a\},\{a,a,\{a\}\},\{a,\{a\},\{a\}\}\}$

14. Prove that $4\mathbb{Z} \subseteq 2\mathbb{Z}$.

15. How many subsets does $\{a,b,c,d\}$ have? Draw a tree diagram for the subsets of $\{a,b,c,d\}$.

16. Let $A, B, C, D,$ and E be sets such that $A \subseteq B$, $B \subseteq C$, $C \subseteq D$, and $D \subseteq E$. Prove that $A \subseteq E$.

17. Let $A, B, C,$ and D be sets with $A \subseteq B$ and $C \subseteq D$. Prove that $A \times C \subseteq B \times D$.

18. Prove or provide a counterexample:

    (i) Every pairwise disjoint set of sets is disjoint.

    (ii) Every disjoint set of sets is pairwise disjoint.

19. Let A and B be sets. Prove that $A \cap B \subseteq A$.

LEVEL 4

20. A relation R is **reflexive** if $\forall x(xRx)$ and **symmetric** if $\forall x \forall y (xRy \to yRx)$. For example, the relation "=" is reflexive and symmetric because $\forall x(x=x)$ and $\forall x \forall y(x=y \to y=x)$. Show that $\subseteq$ is reflexive, but $\in$ is not. Then decide if each of $\subseteq$ and $\in$ is symmetric.

21. Determine whether each of the following statements is true or false:

    (i) $0 \in \{0,\{1\}\}$

    (ii) $\{b\} \in \{a,b\}$

    (iii) $\{1\} \in \{\{1\},x,2,y\}$

    (iv) $\emptyset \in \{\{\emptyset\}\}$

    (v) $\{\{\emptyset\}\} \in \emptyset$

22. We say that a set A is **transitive** if $\forall x(x \in A \to x \subseteq A)$. Determine if each of the following sets is transitive:

   (i) $\emptyset$
   (ii) $\{\emptyset\}$
   (iii) $\{\{\emptyset\}\}$
   (iv) $\{\emptyset, \{\emptyset\}\}$
   (v) $\{\emptyset, \{\emptyset\}, \{\{\emptyset\}\}\}$
   (vi) $\{\{\emptyset\}, \{\emptyset, \{\emptyset\}\}\}$

23. Let A, B, C, and D be sets. Determine if each of the following statements is true or false. If true, provide a proof. If false, provide a counterexample.

   (i) $(A \times B) \cap (C \times D) = (A \cap C) \times (B \cap D)$
   (ii) $(A \times B) \cup (C \times D) = (A \cup C) \times (B \cup D)$

24. Prove that $B \subseteq A$ if and only if $A \cap B = B$.

25. Let A, B, and C be sets. Prove each of the following:

   (i) $A \cap (B \cup C) = (A \cap B) \cup (A \cap C)$.
   (ii) $A \cup (B \cap C) = (A \cup B) \cap (A \cup C)$.
   (iii) $C \setminus (A \cup B) = (C \setminus A) \cap (C \setminus B)$.
   (iv) $C \setminus (A \cap B) = (C \setminus A) \cup (C \setminus B)$.

LEVEL 5

26. Let A and B be sets with $A \subseteq B$. Prove that $\mathcal{P}(A) \subseteq \mathcal{P}(B)$.

27. Prove that if A is a transitive set, then $\mathcal{P}(A)$ is also a transitive set (see Problem 22 above for the definition of a transitive set).

28. Let $A = \{a, b, c, d\}$, $B = \{X \mid X \subseteq A \wedge d \notin X\}$, and $C = \{X \mid X \subseteq A \wedge d \in X\}$. Show that there is a natural one-to-one correspondence between the elements of B and the elements of C. Then generalize this result to a set with $n + 1$ elements for $n > 0$.

29. Let X be a nonempty set of sets. Prove the following:

   (i) For all $A \in X$, $A \subseteq \bigcup X$.
   (ii) For all $A \in X$, $\bigcap X \subseteq A$.

30. Let A be a set and let $\mathbf{X}$ be a nonempty set of sets. Prove each of the following:
    - (i) $A \cap \cup \mathbf{X} = \cup \{A \cap B \mid B \in \mathbf{X}\}$
    - (ii) $A \cup \cap \mathbf{X} = \cap \{A \cup B \mid B \in \mathbf{X}\}$
    - (iii) $A \setminus \cup \mathbf{X} = \cap \{A \setminus B \mid B \in \mathbf{X}\}$
    - (iv) $A \setminus \cap \mathbf{X} = \cup \{A \setminus B \mid B \in \mathbf{X}\}$.

CHALLENGE PROBLEMS

31. Let $\mathbf{X}$ be a nonempty set of sets. Prove that $\mathcal{P}(\cap \mathbf{X}) = \cap \{\mathcal{P}(A) \mid A \in X\}$.

32. Let $\mathbf{X}$ be a nonempty set of sets. Prove that $\mathcal{P}(\cup \mathbf{X}) = \cup \{\mathcal{P}(A) \mid A \in X\}$ if and only if $\cup \mathbf{X} \in \mathbf{X}$.

33. Let A be a transitive set. Prove that $\cup A$ is a transitive set. (See Problem 22 above for the definition of a transitive set.)

34. Let A and B be sets with $A \subseteq B$ and B transitive. Prove that $\mathcal{P}(A) \subseteq \mathcal{P}(\mathcal{P}(B))$. (See Problem 22 above for the definition of a transitive set.)

35. Prove that there is a natural one-to-one correspondence between the elements of $\mathbb{Q}$ and the elements of a proper subset of $\mathbb{R}$.

LESSON 2
ALGEBRAIC STRUCTURES

Binary Operations and Closure

A **binary operation** on a set is a rule that combines two elements of the set to produce another element of the set.

Example 2.1: Let $S = \{0, 1\}$. Multiplication on S is a binary operation, whereas addition on S is **not** a binary operation (here we are thinking of multiplication and addition in the "usual" sense, meaning the way we would think of them in elementary school or middle school).

To see that multiplication is a binary operation on S, observe that $0 \cdot 0 = 0$, $0 \cdot 1 = 0$, $1 \cdot 0 = 0$, and $1 \cdot 1 = 1$. Each of the four computations produces 0 or 1, both of which are in the set S.

These computations can be summarized in the following **multiplication table**:

·	0	1
0	0	0
1	0	1

The multiplication table works as follows: For $a, b \in S$, we evaluate $a \cdot b$ by taking the entry in the row given by a and the column given by b. For example, the computation "$0 \cdot 1 = 0$" is illustrated below.

·	0	1
0	0	**0**
1	0	1

Multiplication ($\cdot$) is a binary operation on S because the only possible "outputs" are 0 and 1.

inputs $\longrightarrow \binom{0}{1}$
·	0	1
0	0	0
1	0	1
$\longleftarrow$ outputs

To see that addition is not a binary operation on S, just note that $1 + 1 = 2$, and $2 \notin S$.

Note that in the multiplication table for addition, the "output" 2 is not equal to either of the "inputs" 0 and 1 as we see below.

inputs $\longrightarrow \binom{0}{1}$
+	0	1
0	0	1
1	1	**2**
$\longleftarrow$ this output is not equal to either input

Note: It may seem odd that I used the expression "multiplication table" for the operation of addition. Nonetheless, "multiplication table" is a general expression that can be used for the table describing any binary operation. It would also be acceptable to call this last table an "addition table" instead.

Let's get a bit more technical and write down the formal definition of a binary operation. The terminology and notation used in this definition will be clarified in the notes below and formalized more rigorously later in Lesson 4.

Formally, a **binary operation** $\star$ on a set S is a **function** $\star : S \times S \to S$. So, if $a, b \in S$, then we have $\star(a, b) \in S$. For easier readability, we will usually write $\star(a, b)$ as $a \star b$.

Notes: (1) Recall from Lesson 1 that if A and B are sets, then $A \times B$ is called the **Cartesian product** of A and B. It consists of the **ordered pairs** (a, b), where $a \in A$ and $b \in B$. A **function** $f: A \times B \to C$ takes each such pair (a, b) to an element $f(a, b) \in C$.

As an example, let $A = \{\text{dog}, \text{fish}\}$, $B = \{\text{cat}, \text{snake}\}$, $C = \{0, 2, 4, 6, 8\}$, and define $f: A \times B \to C$ by $f(a, b) = $ the total number of legs that animals a and b have. Then we have $f(\text{dog}, \text{cat}) = 8$, $f(\text{dog}, \text{snake}) = 4$, $f(\text{fish}, \text{cat}) = 4$, $f(\text{fish}, \text{snake}) = 0$.

Ordered pairs and Cartesian products were defined in Lesson 1 and we will look at functions in more detail in Lesson 4.

(2) For a binary operation, all three sets A, B, and C in the expression $f: A \times B \to C$ are the same.

As we saw in Example 2.1 above, if we let $S = \{0, 1\}$, and we let $\star$ be multiplication, then $\star$ is a binary operation on S. Using function notation, we have $\star(0,0) = 0$, $\star(0,1) = 0$, $\star(1,0) = 0$, and $\star(1,1) = 1$.

As stated in the formal definition of a binary operation above, we will usually write the computations as $0 \star 0 = 0$, $0 \star 1 = 0$, $1 \star 0 = 0$, and $1 \star 1 = 1$.

We can use symbols other than $\star$ for binary operations. For example, if the operation is multiplication, we would usually use a dot ($\cdot$) for the operation as we did in Example 2.1 above. Similarly, for addition we would usually use $+$, for subtraction we would usually use $-$, and so on.

Example 2.2:

1. The operation of addition on the set of natural numbers is a binary operation because whenever we add two natural numbers we get another natural number. Here, the set S is $\mathbb{N}$ and the operation $\star$ is $+$. Observe that if $a \in \mathbb{N}$ and $b \in \mathbb{N}$, then $a + b \in \mathbb{N}$. For example, if $a = 1$ and $b = 2$ (both elements of $\mathbb{N}$), then $a + b = 1 + 2 = 3$, and $3 \in \mathbb{N}$. This will all be formalized in Lesson 5.

2. The operation of multiplication on the set of positive integers is a binary operation because whenever we multiply two positive integers we get another positive integer. Here, the set S is $\mathbb{Z}^+$ and the operation $\star$ is $\cdot$. Observe that if $a \in \mathbb{Z}^+$ and $b \in \mathbb{Z}^+$, then $a \cdot b \in \mathbb{Z}^+$. For example, if $a = 3$ and $b = 5$ (both elements of $\mathbb{Z}^+$), then $a \cdot b = 3 \cdot 5 = 15$, and $15 \in \mathbb{Z}^+$. Once again, this will all be formalized in Lesson 5.

3. Let $S = \mathbb{Z}$ and define $\star$ by $a \star b = \min\{a, b\}$, where $\min\{a, b\}$ is the smallest of a or b. Then $\star$ is a binary operation on $\mathbb{Z}$. For example, if $a = -5$ and $b = 3$ (both elements of $\mathbb{Z}$), then $a \star b = -5$, and $-5 \in \mathbb{Z}$.

4. Subtraction on the set of natural numbers is **not** a binary operation. To see this, we just need to provide a single **counterexample**. (Recall that a counterexample is an example that is used to prove that a statement is false.) If we let $a = 1$ and $b = 2$ (both elements of $\mathbb{N}$), then we see that $a - b = 1 - 2$ is not an element of $\mathbb{N}$.

5. Let $S = \{u, v, w\}$ and define $\star$ using the following multiplication table:

$\star$	u	v	w
u	v	w	w
v	w	u	u
w	u	v	v

For example, $v \star w = u$, as can be seen in the table below.

$\star$	u	v	w
u	v	w	w
v	w	u	u
w	u	v	v

$\star$ is a binary operation on S because the only possible "outputs" are $u, v,$ and w.

Note: In part 1 of Example 2.2 above, we provided evidence that addition on the set of natural numbers is a binary operation. However, we did not prove this result. The problem is that we do not currently have a rigorously defined description of the natural numbers. We will provide a formal definition of the natural numbers in Lesson 5. At that point we will be able to provide proofs of all the results that we mention in this lesson. We will provide a formal definition of the integers in Lesson 5 as well.

Some authors refer to a binary operation $\star$ on a set S even when the binary operation is not defined on all pairs of elements $a, b \in S$. We will always refer to these "false operations" as **partial binary operations**.

We say that the set S is **closed** under the partial binary operation $\star$ if whenever $a, b \in S$, we have $a \star b \in S$.

In Example 2.2, part 4 above, we saw that subtraction is a partial binary operation on $\mathbb{N}$ that is not a binary operation. In other words, $\mathbb{N}$ is not **closed** under subtraction.

Semigroups and Associativity

Let $\star$ be a binary operation on a set S. We say that $\star$ is **associative** in S if for all x, y, z in S, we have
$$(x \star y) \star z = x \star (y \star z)$$

A **semigroup** is a pair $(S, \star)$, where S is a set and $\star$ is an associative binary operation in S.

Example 2.3:

1. $(\mathbb{N}, +), (\mathbb{Z}, +), (\mathbb{N}, \cdot),$ and $(\mathbb{Z}, \cdot)$ are all semigroups. In other words, the operations of addition and multiplication are both associative in $\mathbb{N}$ and $\mathbb{Z}$.

2. Let $S = \mathbb{Z}$ and define $\star$ by $a \star b = \min\{a, b\}$, where $\min\{a, b\}$ is the smallest of a or b. Let's check that $\star$ is associative in $\mathbb{Z}$. Let a, b, and c be elements of $\mathbb{Z}$. There are actually 6 cases to consider (see Note 1 below). Let's go through one of these cases in detail. If we assume that $a \leq b \leq c$, then we have

$$(a \star b) \star c = \min\{a, b\} \star c = a \star c = \min\{a, c\} = a.$$
$$a \star (b \star c) = a \star \min\{b, c\} = a \star b = \min\{a, b\} = a.$$

Since both $(a \star b) \star c = a$ and $a \star (b \star c) = a$, we have $(a \star b) \star c = a \star (b \star c)$. After checking the other 5 cases, we can say the following: Since a, b, and c were arbitrary elements from $\mathbb{Z}$, we have shown that $\star$ is associative in $\mathbb{Z}$. It follows that $(\mathbb{Z}, \star)$ is a semigroup.

3. Subtraction is **not** associative in $\mathbb{Z}$. To see this, we just need to provide a single counterexample. If we let $a = 1$, $b = 2$, and $c = 3$, then $(a - b) - c = (1 - 2) - 3 = -1 - 3 = -4$ and $a - (b - c) = 1 - (2 - 3) = 1 - (-1) = 1 + 1 = 2$. Since $-4 \neq 2$, subtraction is not associative in $\mathbb{Z}$. It follows that $(\mathbb{Z}, -)$ is **not** a semigroup.

Note that $(\mathbb{N}, -)$ is also not a semigroup, but for a different reason. Subtraction is not even a binary operation on $\mathbb{N}$ (see part 4 in Example 2.2).

4. Let $S = \{u, v, w\}$ and define $\star$ using the following table (this is the same table from part 5 in Example 2.2):

$\star$	u	v	w
u	v	w	w
v	w	u	u
w	u	v	v

Notice that $(u \star v) \star w = w \star w = v$ and $u \star (v \star w) = u \star u = v$.

So, $(u \star v) \star w = u \star (v \star w)$. However, this single computation does **not** show that $\star$ is associative in S. In fact, we have the following counterexample: $(u \star w) \star v = w \star v = v$ and $u \star (w \star v) = u \star v = w$. Thus, $(u \star w) \star v \neq u \star (w \star v)$.

So, $\star$ is **not** associative in S, and therefore, $(S, \star)$ is **not** a semigroup.

5. Let $2\mathbb{Z} = \{\ldots, -6, -4, -2, 0, 2, 4, 6, \ldots\}$ be the set of even integers. When we multiply two even integers together, we get another even integer (we will prove this in Lesson 8—see Theorem 8.4). It follows that multiplication is a binary operation on $2\mathbb{Z}$. Since multiplication is associative in $\mathbb{Z}$ and $2\mathbb{Z} \subseteq \mathbb{Z}$, it follows that multiplication is associative in $2\mathbb{Z}$ (see Note 2 below). So, $(2\mathbb{Z}, \cdot)$ is a semigroup.

Similarly, $(2\mathbb{Z}, +)$ is a semigroup.

6. Let $2\mathbb{Z} + 1 = \{\ldots, -5, -3, -1, 1, 3, 5, 7, \ldots\}$ be the set of odd integers. When we multiply two odd integers together, we get another odd integer (you will be asked to prove this in Problem 15 in Problem Set 8). It follows that multiplication is a binary operation on $2\mathbb{Z} + 1$. Since multiplication is associative in $\mathbb{Z}$ and $2\mathbb{Z} + 1 \subseteq \mathbb{Z}$, it follows that multiplication is associative in $2\mathbb{Z} + 1$ (once again, see Note 2 below). So, $(2\mathbb{Z} + 1, \cdot)$ is a semigroup.

However, $(2\mathbb{Z} + 1, +)$ is **not** a semigroup. Associativity is not the issue here. The problem is that when we add two odd integers, we get an even integer. For example, $1 + 1 = 2$. It follows that $+$ is not a binary operation on $2\mathbb{Z} + 1$.

7. If $m \in \mathbb{N}$, then recall that $\mathbb{N} + m = \{m, 1 + m, 2 + m, 3 + m, ...\}$. $(\mathbb{N} + m, +)$ is a semigroup. To see that $\mathbb{N} + m$ is closed under addition, observe that if $n_1 + m, n_2 + m \in \mathbb{N} + m$, then we have $(n_1 + m) + (n_2 + m) = ((n_1 + n_2) + m) + m$. Since $\mathbb{N}$ is closed under addition, $(n_1 + n_2) + m \in \mathbb{N}$. Thus, we have $(n_1 + m) + (n_2 + m) = ((n_1 + n_2) + m) + m \in \mathbb{N} + m$. Since addition is associative in $\mathbb{N}$ and $\mathbb{N} + m \subseteq \mathbb{N}$, it follows that addition is associative in $\mathbb{N} + m$ (once again, see Note 2 below). So, $(\mathbb{N} + m, +)$ is a semigroup. As specific examples, we see that for $\mathbb{N} + 7 = \{7, 8, 9, 10, ...\}$ and $\mathbb{N} + 12 = \{12, 13, 14, 15, ...\}$, we have that $(\mathbb{N} + 7, +)$ and $(\mathbb{N} + 12, +)$ are semigroups.

   I leave it to the reader to show that for $m \in \mathbb{N}$, $(\mathbb{N} + m, \cdot)$ is also a semigroup. What about $n\mathbb{N} + m$ for arbitrary $n, m \in \mathbb{N}$? Are $(n\mathbb{N} + m, +)$ and $(n\mathbb{N} + m, \cdot)$ always semigroups? And what if we replace $\mathbb{N}$ by $\mathbb{Z}$? I leave these questions for the reader to think about.

Notes: (1) In part 2 above, we must prove the result for each of the following 6 cases:

$$a \leq b \leq c \quad a \leq c \leq b \quad b \leq a \leq c \quad b \leq c \leq a \quad c \leq a \leq b \quad c \leq b \leq a$$

The same basic argument can be used for all these cases. For example, we saw in the solution above that for the first case we get

$$(a \star b) \star c = \min\{a, b\} \star c = a \star c = \min\{a, c\} = a.$$
$$a \star (b \star c) = a \star \min\{b, c\} = a \star b = \min\{a, b\} = a.$$

Let's also do the last case $c \leq b \leq a$:

$$(a \star b) \star c = \min\{a, b\} \star c = b \star c = \min\{b, c\} = c.$$
$$a \star (b \star c) = a \star \min\{b, c\} = a \star c = \min\{a, c\} = c.$$

The reader should verify the other 4 cases to complete the proof.

(2) Associativity is **closed downwards**. By this, we mean that if $\star$ is associative in a set A, and $B \subseteq A$ (B is a **subset** of A), then $\star$ is associative in B.

The reason for this is that the definition of associativity involves only a **universal statement**—a statement that describes a property that is true for all elements without mentioning the existence of any new elements. A universal statement begins with the quantifier $\forall$ ("For all" or "Every") and never includes the quantifier $\exists$ ("There exists" or "There is").

As a simple example, if every object in set A is a fruit, and $B \subseteq A$, then every object in B is a fruit. The universal statement we are referring to might be $\forall x(P(x))$, where $P(x)$ is the property "x is a fruit."

In the case of associativity, the universal statement is $\forall x \forall y \forall z((x \star y) \star z = x \star (y \star z))$.

Associativity allows us to "add and drop parentheses" as we like. For example, since $+$ is associative in $\mathbb{N}$, the expression "$1 + 3 + 6$" makes sense. Indeed, it doesn't matter if we think about $1 + 3 + 6$ as $(1 + 3) + 6$ or $1 + (3 + 6)$. In each case, we get an answer of 10. More generally, if $\star$ is associative in a set S, then for $a, b, c \in S$, we can write $a \star b \star c$.

What if we are applying the operation more than twice? For example, if $a, b, c, d \in S$, does the expression $a \star b \star c \star d$ make sense? Well, there are five ways we can think about this product: $((a \star b) \star c) \star d$, $(a \star (b \star c)) \star d$, $(a \star b) \star (c \star d)$, $a \star ((b \star c) \star d)$, and $a \star (b \star (c \star d))$. Do all five of these yield the same result? We will now prove that they do!

Theorem 2.4: Let $(S, \star)$ be a semigroup and let $a, b, c, d \in S$. Then we have the following:
$$((a \star b) \star c) \star d = (a \star (b \star c)) \star d = (a \star b) \star (c \star d) = a \star ((b \star c) \star d) = a \star (b \star (c \star d)).$$

Analysis: The first and last equalities follow directly from associativity.

For the other equalities, it will help to make simple substitutions. For example, if we let $u = a \star b$, then $((a \star b) \star c) \star d$ can be written as $(u \star c) \star d$. By associativity, this expression is equal to $u \star (c \star d)$.

Let's write out the details of the proof.

Proof of Theorem 2.4: Let $(S, \star)$ be a semigroup, let $a, b, c, d \in S$, let $u = a \star b$, and let $v = c \star d$. Then associativity gives us $((a \star b) \star c) \star d = (a \star (b \star c)) \star d$ and $a \star ((b \star c) \star d) = a \star (b \star (c \star d))$. Also, we have the following:
$$((a \star b) \star c) \star d = (u \star c) \star d = u \star (c \star d) = (a \star b) \star (c \star d).$$
$$(a \star b) \star (c \star d) = (a \star b) \star v = a \star (b \star v) = a \star (b \star (c \star d)).$$

These equations yield the desired result. □

Associativity can be generalized even further to an arbitrary number of elements in a semigroup. In other words, associativity implies **generalized associativity**. We will prove this in Lesson 5 (Theorem 5.16) using the Principle of Mathematical Induction (see Theorem 5.6).

Let $\star$ be a binary operation on a set S. We say that $\star$ is **commutative** (or **Abelian**) in S if for all x, y in S, we have $x \star y = y \star x$.

Example 2.5:
1. $(\mathbb{N}, +)$, $(\mathbb{Z}, +)$, $(\mathbb{N}, \cdot)$, and $(\mathbb{Z}, \cdot)$ are all **commutative semigroups**. In other words, the operations of addition and multiplication are both commutative in $\mathbb{N}$ and $\mathbb{Z}$ (as well as being associative).

2. The semigroup $(\mathbb{Z}, \star)$, where $\star$ is defined by $a \star b = \min\{a, b\}$ is a commutative semigroup. Let's check that $\star$ is commutative in $\mathbb{Z}$. Let a and b be elements of $\mathbb{Z}$. This time there are just 2 cases to consider ($a \leq b$ and $b \leq a$). Let's do the first case in detail and assume that $a \leq b$. We then have $a \star b = \min\{a, b\} = a$ and $b \star a = \min\{b, a\} = a$. So, $a \star b = b \star a$. After verifying the other case (which you should do), we can say that $\star$ is commutative in $\mathbb{Z}$.

3. Define the binary operation $\star$ on $\mathbb{N}$ by $a \star b = a$. Then $(\mathbb{N}, \star)$ is a semigroup that is **not** commutative. For associativity, we have $(a \star b) \star c = a \star c = a$ and $a \star (b \star c) = a \star b = a$. Let's use a counterexample to show that $\star$ is not commutative. Well, $2 \star 5 = 2$ and $5 \star 2 = 5$.

Note: In part 3 above, the computation $a \star (b \star c)$ can actually be done in 1 step instead of 2. The way we did it above was to first compute $b \star c = b$, and then to replace $b \star c$ with b to get $a \star (b \star c) = a \star b = a$. However, the definition of $\star$ says that $a \star$ (anything) $= a$. In this case, the "anything" is $b \star c$. So, we have $a \star (b \star c) = a$ just by appealing to the definition of $\star$.

Monoids and Identity

Let $(S,\star)$ be a semigroup. An element e of S is called an **identity** with respect to the binary operation $\star$ if for all $a \in S$, we have $e \star a = a \star e = a$.

A **monoid** is a semigroup with an identity.

Example 2.6:

1. $(\mathbb{N}, +)$ and $(\mathbb{Z}, +)$ are commutative monoids with identity 0 (when we add 0 to any integer a, we get a). $(\mathbb{N}, \cdot)$ and $(\mathbb{Z}, \cdot)$ are commutative monoids with identity 1 (when we multiply any integer a by 1, we get a).

2. $(2\mathbb{Z}, +)$ is also a commutative monoid with identity 0. We already saw in part 5 of Example 2.3 that $(2\mathbb{Z}, +)$ is a semigroup. Commutativity of addition follows from the fact that addition is commutative in $\mathbb{Z}$, commutativity is closed downwards, and $2\mathbb{Z} \subseteq \mathbb{Z}$ (see Note 2 following Example 2.3).

    More generally, for any $n \in \mathbb{Z}$, $(n\mathbb{Z}, +)$ is a commutative monoid with identity 0.

    Similar reasoning can be used to show that $(n\mathbb{N}, +)$ is a commutative monoid for each $n \in \mathbb{Z}$.

3. The commutative semigroup $(\mathbb{Z},\star)$, where $\star$ is defined by $a \star b = \min\{a, b\}$ is **not** a monoid. To see this, let $a \in \mathbb{Z}$. Then $a + 1 \in \mathbb{Z}$ and $a \star (a + 1) = a \neq a + 1$. This shows that a is not an identity. Since a was an arbitrary element of $\mathbb{Z}$, we showed that there is no identity. It follows that $(\mathbb{Z},\star)$ is not a monoid.

4. The noncommutative semigroup $(\mathbb{N},\star)$, where $a \star b = a$ is also **not** a monoid. Use the same argument given in 3 above with $\mathbb{Z}$ replaced by $\mathbb{N}$.

5. $(2\mathbb{Z}, \cdot)$ is another example of a semigroup that is **not** a monoid. The identity element of $(\mathbb{Z}, \cdot)$ is 1, and this element is missing from $(2\mathbb{Z}, \cdot)$.

6. Let A be a nonempty set. Recall from Lesson 1 that $\mathcal{P}(A) = \{B \mid B \subseteq A\}$. Also, if $X, Y \in \mathcal{P}(A)$, then $X \cup Y = \{x \mid x \in X \text{ or } x \in Y\}$. Let's check that $(\mathcal{P}(A), \cup)$ is a monoid. If $X, Y \in \mathcal{P}(A)$, then $X \subseteq A$ and $Y \subseteq A$. If $x \in X \cup Y$, then $x \in X$ or $x \in Y$. If $x \in X$, then since $X \subseteq A$, $x \in A$. If $x \in Y$, then since $Y \subseteq A$, $x \in A$. So, $\forall x(x \in X \cup Y \to x \in A)$. It follows that $X \cup Y \subseteq A$, and therefore, $X \cup Y \in \mathcal{P}(A)$. This shows that $\cup$ is a binary operation on $\mathcal{P}(A)$. By Theorem 1.29, $\cup$ is associative in $\mathcal{P}(A)$. Finally, $\emptyset$ is an identity for $(\mathcal{P}(A), \cup)$ because if $X \in \mathcal{P}(A)$, then $X \cup \emptyset = X$ and $\emptyset \cup X = X$. It follows that $(\mathcal{P}(A), \cup)$ is a monoid. By part (i) of Problem 12 from Problem Set 1, $(\mathcal{P}(A), \cup)$ is a commutative monoid.

7. If A is a nonempty set, then $(\mathcal{P}(A), \cap)$ is a commutative monoid with identity A. The argument is similar to part 6 above, and so, it is left to the reader (see Problem 12 below).

By definition, every monoid has at least one identity. It is natural to ask if a monoid can have more than one identity. We will now prove that this cannot happen.

Theorem 2.7: In any monoid $(M, \star)$, the identity is unique.

Analysis: There is a standard way to prove that an object satisfying a certain property (or properties) is unique. You begin by assuming that you have two such objects (not necessarily distinct) and then provide a logically valid argument showing that they are equal. This is how we will prove this theorem.

So, we will start by assuming that e and f are both identities. The proof that e and f are equal is quite "obvious." Nonetheless, many students struggle to understand why it's so obvious. To help clarify this, let's write out all the equations that follow from e and f being identities.

Since e is an identity, for any $a \in M$, we have $e \star a = a$ and $a \star e = a$. In particular, since $f \in M$, we have $e \star f = f$ and $f \star e = f$.

Since f is an identity, for any $a \in M$, we have $f \star a = a$ and $a \star f = a$. In particular, since $e \in M$, we have $e \star f = e$ and $f \star e = e$.

Now, let's write down all these equations and choose two of them that will help us complete the proof of the theorem: $\boxed{e \star f = f}$ $\quad f \star e = f \quad$ $\boxed{e \star f = e}$ $\quad f \star e = e$

The two equations in rectangles show that e and f are equal to the same expression, namely $e \star f$. Therefore, they must be equal to each other. (Note that we could have also used the two equations not in rectangles.)

Let's write out the proof, which after this analysis, should no longer seem so mysterious.

Proof of Theorem 2.7: Let $(M, \star)$ be a monoid and suppose that e and f are both identities. Then we have $f = e \star f = e$. Therefore, there is only one identity. □

Note: $f = e \star f$ because e is an identity, whereas $e \star f = e$ because f is an identity.

Groups and Inverses

Let $(M, \star)$ be a monoid with identity e. An element $a \in M$ is called **invertible** if there is an element $b \in M$ such that $a \star b = b \star a = e$.

A **group** is a monoid in which every element is invertible.

Groups appear so often in mathematics that it's worth taking the time to explicitly spell out the full definition of a group.

A **group** is a pair $(G, \star)$ consisting of a set G together with a binary operation $\star$ satisfying:

(1) **(Associativity)** For all $x, y, z \in G$, $(x \star y) \star z = x \star (y \star z)$.

(2) **(Identity)** There exists an element $e \in G$ such that for all $x \in G$, $e \star x = x \star e = x$.

(3) **(Inverse)** For each $x \in G$, there is $y \in G$ such that $x \star y = y \star x = e$.

Notes: (1) If $y \in G$ is an inverse of $x \in G$, we will usually write $y = x^{-1}$.

(2) Recall that the definition of a binary operation already implies closure. However, many books on groups will mention this property explicitly:

(Closure) For all $x, y \in G$, $x \star y \in G$.

(3) A group is **commutative** or **Abelian** if for all $x, y \in G$, $x \star y = y \star x$.

(4) The properties that define a group are called the **group axioms**. In general, an **axiom** is a statement that is assumed to be true. So, the group axioms are the statements that are **given** to be true in all groups. There are many other statements that are true in groups. However, any additional statements need to be **proved** using the axioms.

Example 2.8:

1. $(\mathbb{Z}, +)$ is a commutative group with identity 0. The inverse of any integer k is the integer $-k$.

2. $(2\mathbb{Z}, +)$ is also a commutative group with identity 0. We already saw in part 2 of Example 2.6 that $(2\mathbb{Z}, +)$ is a commutative monoid. The inverse of $2k$ is $-2k$.

   More generally, for any $n \in \mathbb{Z}$, $(n\mathbb{Z}, +)$ is a commutative group. The inverse of nk is $-nk$.

3. $(\mathbb{N}, +)$ is a commutative monoid that is **not** a group. For example, the natural number 1 has no inverse in $\mathbb{N}$. In other words, the equation $x + 1 = 0$ has no solution in $\mathbb{N}$.

4. $(\mathbb{Z}, \cdot)$ is a commutative monoid that is **not** a group. For example, the integer 2 has no inverse in $\mathbb{Z}$. In other words, the equation $2x = 1$ has no solution in $\mathbb{Z}$.

5. Recall from part 1 of Example 1.6 that the set of rational numbers is
$$\mathbb{Q} = \left\{ \frac{a}{b} \mid a, b \in \mathbb{Z} \text{ and } b \neq 0 \right\}.$$
   Also, recall that we identify the rational number $\frac{a}{1}$ with the integer a. In this way, $\mathbb{Z} \subseteq \mathbb{Q}$.

   We add two rational numbers using the rule $\frac{a}{b} + \frac{c}{d} = \frac{a \cdot d + b \cdot c}{b \cdot d}$.

   Note that $0 = \frac{0}{1}$ is an identity for $(\mathbb{Q}, +)$ because $\frac{a}{b} + \frac{0}{1} = \frac{a \cdot 1 + b \cdot 0}{b \cdot 1} = \frac{a}{b}$ and $\frac{0}{1} + \frac{a}{b} = \frac{0 \cdot b + 1 \cdot a}{1 \cdot b} = \frac{a}{b}$.

   You will need to show that $(\mathbb{Q}, +)$ is a commutative group as part of Problem 19 below.

6. We multiply two rational numbers using the rule $\frac{a}{b} \cdot \frac{c}{d} = \frac{a \cdot c}{b \cdot d}$.

   Note that $1 = \frac{1}{1}$ is an identity for $(\mathbb{Q}, \cdot)$ because $\frac{a}{b} \cdot \frac{1}{1} = \frac{a \cdot 1}{b \cdot 1} = \frac{a}{b}$ and $\frac{1}{1} \cdot \frac{a}{b} = \frac{1 \cdot a}{1 \cdot b} = \frac{a}{b}$.

   Now, $0 \cdot \frac{a}{b} = \frac{0}{1} \cdot \frac{a}{b} = \frac{0 \cdot a}{1 \cdot b} = \frac{0}{b} = 0$. In particular, when we multiply 0 by any rational number, we can never get 1. So, 0 is a rational number with no multiplicative inverse. It follows that $(\mathbb{Q}, \cdot)$ is **not** a group.

   However, 0 is the **only** rational number without a multiplicative inverse. In fact, you will need to show that $(\mathbb{Q}^*, \cdot)$ is a commutative group as part of Problem 19 below, where $\mathbb{Q}^*$ is the set of rational numbers with 0 removed ($\mathbb{Q}^* = \mathbb{Q} \setminus \{0\}$).

7. Let A be a nonempty set. Recall from part 6 of Example 2.6 that $(\mathcal{P}(A), \cup)$ is a monoid. This monoid is **not** a group. For example, A has no inverse in $\mathcal{P}(A)$. Indeed, if $B \in \mathcal{P}(A)$, then since $B \subseteq A$, by Theorem 1.27, $A \cup B = A \neq \emptyset$.

8. If $(G, \star_G)$ and $(H, \star_H)$ are groups, then we can define an operation $\star$ on $G \times H$ by $(a, b) \star (c, d) = (a \star_G c, b \star_H d)$. The dedicated reader should check carefully that $(G \times H, \star)$ is a group (called the **direct product** of G and H). The identity is (e_G, e_H), where e_G is the identity of $(G, \star_G)$ and e_H is the identity of $(H, \star_H)$. The inverse of $(a, b) \in G \times H$ is (a^{-1}, b^{-1}).

   More generally, if $(G_1, \star_{G_1}), (G_2, \star_{G_2}), \ldots, (G_n, \star_{G_n})$ are groups, then $(G_1 \times G_2 \times \cdots \times G_n, \star)$ is a group, where the operation $\star$ is defined on $G_1 \times G_2 \times \cdots \times G_n$ by

   $$(a_1, a_2, \ldots, a_n) \star (b_1, b_2, \ldots, b_n) = (a_1 \star_{G_1} b_1, a_2 \star_{G_2} b_2, \ldots, a_n \star_{G_n} b_n).$$

   Once again, I leave the verification to the reader.

9. Let $C_{12} = \{1, 2, 3, 4, 5, 6, 7, 8, 9, 10, 11, 12\}$. We can define "**clock addition**" by performing addition the way we do it on a 12-hour clock. For example, using clock addition, we have $3 + 5 = 8$ and $10 + 3 = 1$ (If it is now 10:00, then in 3 hours, it will be 1:00). $(C_{12}, +)$ is a commutative group with identity 12. It's a bit more natural to name the identity 0 instead of 12, and so, we would normally write $C_{12} = \{0, 1, 2, 3, 4, 5, 6, 7, 8, 9, 10, 11\}$. Note that 1 and 11 are inverses of each other because $1 + 11 = 11 + 1 = 0$. Similarly, we have the following pairs of inverses: $\{2, 10\}, \{3, 9\}, \{4, 8\}, \{5, 7\}$. Also, 0 and 6 are each their own inverse.

   There is nothing special about the number 12 here (aside from the fact that many of us own 12-hour clocks). We can consider addition on a "5-hour clock." In this case, we would have $C_5 = \{0, 1, 2, 3, 4\}$ and we can perform computations such as $2 + 3 = 0$ and $3 + 4 = 2$. Once again, $(C_5, +)$ is a commutative group with identity 0. We have the following inverse pairs: $\{1, 4\}$ and $\{2, 3\}$. Once again, 0 is its own inverse.

   More generally, for each $n \in \mathbb{Z}^+$, we let $C_n = \{0, 1, 2, \ldots, n - 1\}$ and we define addition on C_n as if we are adding on an "n-hour clock." Just as in the two special cases above, $(C_n, +)$ is a commutative group with identity 0. If k is between 1 and $n - 1$, inclusive, then the inverse of k is $n - k$ (because $k + (n - k) = 0$). Once again, the inverse of 0 is 0. The formal name for "clock addition" in C_n is **addition modulo n**. This example will be treated more rigorously in Lesson 3.

10. We can also define "**clock multiplication**" on $C_n = \{0, 1, 2, \ldots, n - 1\}$ as repeated addition. For example, on C_{12} we have $2 \cdot 5 = 5 + 5 = 10$ and $3 \cdot 7 = (7 + 7) + 7 = 2 + 7 = 9$. We can also perform this last computation by multiplying 3 and 7 as we would "normally" to get $3 \cdot 7 = 21$ and then subtracting 12 to get $21 - 12 = 9$. We can subtract 12 as many times as we need to until we get a number between 0 and 11, inclusive. For example, $5 \cdot 8 = 40$, and subtracting 12 three times gives us a result of $40 - 36 = 4$ (we can also perform this computation more quickly by taking the remainder upon dividing 40 by 12). The formal name for "clock multiplication" in C_n is **multiplication modulo n**.

Now, $(C_n, \cdot)$ is a commutative monoid with identity 1. However, for each $n \in \mathbb{Z}^+$ with $n > 1$, $(C_n, \cdot)$ is **not** a group. After all, for each $k \in C_n$, we have $0 \cdot k = 0 \neq 1$, and therefore, 0 is **not** invertible.

For some values of n, if we remove 0 from C_n, we will get a group, and sometimes we will not. For example, $(C_3 \setminus \{0\}, \cdot) = (\{1, 2\}, \cdot)$ is a group (where the operation is multiplication modulo 3). 1 and 2 are each their own inverses ($1 \cdot 1 = 1$ and $2 \cdot 2 = 2 + 2 = 1$). As another example, $(C_5 \setminus \{0\}, \cdot) = (\{1, 2, 3, 4\}, \cdot)$ is a group (where the operation is multiplication modulo 5). 1 and 4 are each their own inverses ($1 \cdot 1 = 1$ and $4 \cdot 4 = 1$), and 2 and 3 are inverses of each other ($2 \cdot 3 = 1$ and $3 \cdot 2 = 1$). However, $(C_4 \setminus \{0\}, \cdot) = (\{1, 2, 3\}, \cdot)$ is **not** a group (here the operation is multiplication modulo 4). The problem is that $2 \cdot 2 = 0$, and so, the partial binary operation $\cdot$ is not a binary operation on $\{1, 2, 3\}$ (equivalently, $\{1, 2, 3\}$ is not closed under $\cdot$). If we delete the offending element 2 from this set, we see that $(\{1, 3\}, \cdot)$ is a group (where once again, the operation is multiplication modulo 4).

Note: When multiplying two numbers, we will sometimes drop the dot ($\cdot$) for easier readability. So, we may write $x \cdot y$ as xy. We may also use parentheses instead of the dot. For example, we might write $\frac{a}{b} \cdot \frac{c}{d}$ as $\left(\frac{a}{b}\right)\left(\frac{c}{d}\right)$, whereas we would probably write $\frac{a \cdot c}{b \cdot d}$ as $\frac{ac}{bd}$. We may even use this simplified notation for arbitrary group operations. So, we could write $a \star b$ as ab. However, we will avoid doing this if it would lead to confusion. For example, we will **not** write $a + b$ as ab.

By definition, in a group, each element has at least one inverse. It is natural to ask if a given element in a group can have more than one inverse. We will now prove that this cannot happen.

Theorem 2.9: Let $(G, \star)$ be a group and let $a \in G$. Then the inverse of a is unique.

The reader is encouraged to write out an analysis similar to what we did for Theorem 2.7 above before beginning the proof.

Proof of Theorem 2.9: Let $(G, \star)$ be a group with identity e, let $a \in G$, and suppose that b and c are both inverses of a. Then we have $c = c \star e = c \star (a \star b) = (c \star a) \star b = e \star b = b$. Therefore, a has just one inverse. □

Notes: (1) For the first and fifth equalities in the proof, we used the identity property.

(2) For the second and fourth equalities, we used our assumption that b and c are inverses of a.

(3) For the third equality, we used the associativity of $\star$ in G.

(4) The proof of Theorem 2.9 proves more than the theorem states. In the proof, we showed that if $a \star b = e$ (b is a "**right inverse**" of a) and $c \star a = e$ (c is a "**left inverse**" of a), then $c = b$. This provides us with the following additional results:

- If a is an element of a monoid with a right inverse b and a left inverse c, then $c = b$, a is invertible, and b is the unique inverse of a.
- If a is an element of a group with a right inverse b, then b is the inverse of a (to see this, just let c be the inverse of a and then repeat the proof of Theorem 2.9).

- If a is an element of a group with a left inverse c, then c is the inverse of a (to see this, just let b be the inverse of a and then repeat the proof of Theorem 2.9).

The second two bullet points in Note 4 above can save us quite a bit of time when looking for inverses of elements in a group $(G,\star)$. To determine if $y \in G$ is the inverse of $x \in G$, we need only show that $x \star y = e$ **or** $y \star x = e$. We do not need to verify both equations.

Example 2.10: Let $G = \{e, a, b\}$, where e, a, and b are distinct, and let $(G,\star)$ be a group with identity e. Let's construct the multiplication table for $(G,\star)$.

Since $e \star x = x \star e = x$ for all x in the group, we can easily fill out the first row and the first column of the table.

$\star$	e	a	b
e	e	a	b
a	a		
b	b		

Next, let's look at the entry corresponding to $a \star a$. We mark this entry with ⊡ below.

$\star$	e	a	b
e	e	a	b
a	a	⊡	
b	b		

The entry labeled with ⊡ must be either e or b because a is already in that row (see Note 1 below). If it were e, then the final entry in the row would be b, giving two b's in the last column. Therefore, the entry labeled with ⊡ must be b.

$\star$	e	a	b
e	e	a	b
a	a	b	
b	b		

Since the same element cannot be repeated in any row or column (once again, see Note 1 below), the rest of the table is now determined.

$\star$	e	a	b
e	e	a	b
a	a	b	e
b	b	e	a

Notes: (1) Why can't the same element appear twice in any row? Well if x appeared twice in the row corresponding to y, that would mean that there are elements z and w with $z \neq w$ such that $y \star z = x$ and $y \star w = x$. So, $y \star z = y \star w$. Then we have

$$z = e \star z = (y^{-1} \star y) \star z = y^{-1} \star (y \star z) = y^{-1} \star (y \star w) = (y^{-1} \star y) \star w = e \star w = w.$$

This **contradiction** establishes that no element x can appear twice in the same row of a group multiplication table.

We just proved that every group $(G,\star)$ satisfies the **left cancellation law**:

If $a, b, c \in G$ and $ab = ac$, then $b = c$.

A similar argument can be used to show that every group $(G,\star)$ satisfies the **right cancellation law**:

If $a, b, c \in G$ and $ba = ca$, then $b = c$.

By the right cancellation law, the same element in a group cannot appear twice in any column.

(2) The argument given in Note 1 used all the group properties (associativity, identity, and inverse). The dedicated reader should take the time to determine which group property was used for each equality (the first three Notes following the proof of Theorem 2.9 can be used as a guide). What if we remove one of the properties? For example, what about the multiplication table for a monoid? Can an element appear twice in a row or column? I leave this question for the reader to think about (The solution to Problem 2 below will provide an answer).

(3) In Note 1 above, we showed that in the multiplication table for a group, the same element cannot appear as the output more than once in any row or column. We can also show that every element must appear in every row and column. Let's show that the element y must appear in the row corresponding to x. We are looking for an element z such that $x \star z = y$. Well, $z = x^{-1} \star y$ works. Indeed, we have $x \star (x^{-1} \star y) = (x \star x^{-1}) \star y = e \star y = y$.

(4) Using Notes 1 and 3, we see that each element of a group appears exactly once in every row and column of the group's multiplication table.

(5) In the multiplication table we constructed, we can see that e is the identity because the row corresponding to e is the same as the "input row," and the column corresponding to e is the same as the "input column."

$\star$	e	a	b
e	e	a	b
a	a	b	e
b	b	e	a

$\star$	e	a	b
e	e	a	b
a	a	b	e
b	b	e	a

(6) Since $a \star b = e$ and $b \star a = e$, we see that a and b are inverses of each other. Also, since $e \star e = e$, we see that e is its own inverse.

(7) How do we know that $\star$ is associative in G? At this point, the only way we know how to verify this is by "brute force." This is quite tedious, as there are 27 equalities that need to be verified. For example, we have $(a \star b) \star b = e \star b = b$ and $a \star (b \star b) = a \star a = b$. Thus, $(a \star b) \star b = a \star (b \star b)$. The dedicated reader may want to check the remaining 26 equalities.

In practice, we rarely need to do these tedious computations to verify associativity. Most often, we will be able to recognize that a structure lies inside a larger structure for which associativity is already known to hold. We can then use Note 2 following example 2.3 above.

(8) We have shown that there is essentially just one group of size 3, namely the one given by the table that we produced. Any other group with 3 elements will look exactly like this one, except for possibly the names of the elements. In technical terms, we say that any two groups of order 3 are **isomorphic**. We will provide a formal definition of "isomorphic" in Lesson 7.

(9) This group is **commutative**. There is a very simple way to check commutativity using the multiplication table of the group. We simply check if the entries on opposite sides of the main diagonal are the same.

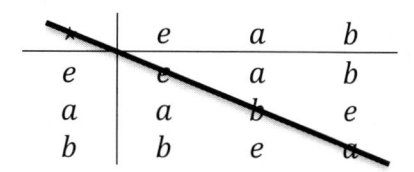

⋆	e	a	b
e	e	a	b
a	a	b	e
b	b	e	a

Notice how the entries in the table are symmetrical across the diagonal.

If $(G, \star)$ is a group and G is finite, then the **order** of G is simply $|G|$, the cardinality of G. If G is infinite, then G has **infinite order**.

Let $(G, \star)$ be a group and let $x \in G$. We define the **powers** of x as follows:

$$x^0 = e; \quad x^1 = x; \quad x^2 = x \star x; \quad x^3 = (x \star x) \star x = x \star (x \star x); \quad \ldots \text{ and so on.}$$

In general, for $n > 0$, x^n is the product of x with itself n times. We can write $x^{n+1} = x^n \star x$. Furthermore, for $n > 0$, we define x^{-n} to be $(x^n)^{-1}$. So, for example, x^{-1} is the inverse of x and $x^{-2} = (x^2)^{-1} = (x \star x)^{-1}$ is the inverse of $x^2 = x \star x$. We abbreviate the set of powers of x using the notation $\langle x \rangle$. So, we have $\langle x \rangle = \{x^n \mid n \in \mathbb{Z}\}$.

Note: When the group operation is addition, it is more natural to use **multiples** instead of powers. After all, we are all in the habit of writing $x + x = 2x$, and not $x + x = x^2$. In the case of addition, we write $0x = 0, 1x = 1, 2x = x + x, 3x = (x + x) + x = x + (x + x), \ldots$ and so on. In general, for $n > 0$, nx is the sum of x with itself n times and we can write $(n + 1)x = nx + x$. We define $-nx$ to be $-(nx)$. For example, $-x$ is the additive inverse of x and $-2x = -(2x) = -(x + x)$ is the additive inverse of $2x = x + x$. Once again, we use the notation $\langle x \rangle$ for the set of multiples of x. So, $\langle x \rangle = \{nx \mid n \in \mathbb{Z}\}$.

We say that a group $(G, \star)$ is **cyclic** if there is an element $x \in G$ such that $G = \langle x \rangle$. In this case, we can say that $(G, \star)$ is **generated by** x and we call x a **generator** of $(G, \star)$.

Example 2.11:

1. The group $(G, \star)$ that we produced in Example 2.10 is a cyclic group of order 3. To see this, let's look at the multiplication table once more:

⋆	e	a	b
e	e	a	b
a	a	b	e
b	b	e	a

Observe that in this multiplication table, we have $b = a \star a = a^2$. So, another way to draw the table is as follows:

$\star$	e	a	a^2
e	e	a	a^2
a	a	a^2	e
a^2	a^2	e	a

Notice how this group consists of all powers of the single element a (the elements are a, a^2, and $a^3 = a^2 \star a = e$). So, $(G, \star) = \langle a \rangle$. The element a^2 is also a generator of this group. To see this, note that $(a^2)^1 = a^2$, $(a^2)^2 = a^2 \star a^2 = a$, and $(a^2)^3 = (a^2)^2 \star a^2 = a \star a^2 = e$. So, we can also write $(G, \star) = \langle a^2 \rangle$. In this group, every power of a is equal to either e, a, or a^2. For example, $a^4 = a^3 \star a = e \star a = a$ and $a^{-1} = a^2$ (because $a^2 \star a = a \star a^2 = e$).

2. $(\mathbb{Z}, +)$ is a cyclic group of infinite order. As mentioned in the note above, to avoid confusion, it is best to write the "powers" x^n as multiples nx. It is easy to see that $(\mathbb{Z}, +)$ is generated by 1. So, we can write $(\mathbb{Z}, +) = \langle 1 \rangle$. $(\mathbb{Z}, +)$ actually has 2 possible generators. Another generator of $(\mathbb{Z}, +)$ is -1. So, we can also write $(\mathbb{Z}, +) = \langle -1 \rangle$, although 1 is the more "natural" choice of generator, and so, we will usually use 1 as the generator to "represent" this cyclic group.

3. $(2\mathbb{Z}, +)$ is also a cyclic group of infinite order. It also has two generators: 2 and -2. More generally, for any $n \in \mathbb{Z}$ with $n \neq 0$, $(n\mathbb{Z}, +)$ is a cyclic group with two generators: n and $-n$. For $n = 0$, $0\mathbb{Z} = \{0\}$. $(\{0\}, +)$ is called the **trivial group** and it is generated by 0.

4. For each $n \in \mathbb{Z}^+$, $(C_n, +)$ (defined in part 9 of Example 2.8) is a cyclic group of order n. It is generated by 1. For example, $(C_4, +) = (\{0, 1, 2, 3\}, +)$. Note that $(C_4, +)$ is also generated by 3. Indeed, we have $1 \cdot 3 = 3$, $2 \cdot 3 = 2$, $3 \cdot 3 = 1$, and $4 \cdot 3 = 0$.

5. $(\{1, 2, 3, 4\}, \cdot)$, where $\cdot$ is multiplication modulo 5 is a cyclic group of order 4. This group has 2 as a generator. To see this, observe that $2^1 = 2$, $2^2 = 4$, $2^3 = 2^2 \cdot 2 = 4 \cdot 2 = 3$, and $2^4 = 2^3 \cdot 2 = 3 \cdot 2 = 1$. I leave it to the reader to show that 3 is also a generator.

We will often refer to the group operation $\star$ as "multiplication," and as we mentioned in the Note following Example 2.8, we may adopt the usual abbreviations for multiplication of elements. For example, we may abbreviate $(a \star b) \star (c \star d)$ as $(ab)(cd)$. However, we will avoid such abbreviations when the group operation is addition. For example, in the group $(\mathbb{Z}, +)$, we will **not** abbreviate $a + b$ as ab. Also, it is conventional to use additive notation whenever the group we are considering is commutative.

Rings and Distributivity

Before giving the general definition of a ring, let's look at an important example.

Example 2.12: Recall that $\mathbb{Z} = \{\ldots, -4, -3, -2, -1, 0, 1, 2, 3, 4, \ldots\}$ is the set of integers. Let's review some of the properties of addition and multiplication on this set.

1. $\mathbb{Z}$ is **closed** under addition. In other words, whenever we add two integers, we get another integer. For example, 2 and 3 are integers, and we have $2 + 3 = 5$, which is also an integer. As another example, -8 and 6 are integers, and so is $-8 + 6 = -2$.

2. Addition is **commutative** in $\mathbb{Z}$. In other words, when we add two integers, it does not matter which one comes first. For example, $2 + 3 = 5$ and $3 + 2 = 5$. So, we see that $2 + 3 = 3 + 2$. As another example, $-8 + 6 = -2$ and $6 + (-8) = -2$. So, we see that $-8 + 6 = 6 + (-8)$.

3. Addition is **associative** in $\mathbb{Z}$. In other words, when we add three integers, it doesn't matter if we begin by adding the first two or the last two integers. For example, $(2 + 3) + 4 = 5 + 4 = 9$ and $2 + (3 + 4) = 2 + 7 = 9$. So, $(2 + 3) + 4 = 2 + (3 + 4)$. As another example, we have $(-8 + 6) + (-5) = -2 + (-5) = -7$ and $-8 + (6 + (-5)) = -8 + 1 = -7$. So, we see that $(-8 + 6) + (-5) = -8 + (6 + (-5))$.

4. $\mathbb{Z}$ has an **identity** for addition, namely 0. Whenever we add 0 to another integer, the result is that same integer. For example, we have $0 + 3 = 3$ and $3 + 0 = 3$. As another example, $0 + (-5) = -5$ and $(-5) + 0 = -5$.

5. Every integer has an additive **inverse**. This is an integer that we add to the original integer to get 0 (the additive identity). For example, the additive inverse of 5 is -5 because we have $5 + (-5) = 0$ and $-5 + 5 = 0$. Notice that the same two equations also show that the inverse of -5 is 5. We can say that 5 and -5 are additive inverses of each other.

We can summarize the five properties above by saying that $(\mathbb{Z}, +)$ is a **commutative group**.

6. $\mathbb{Z}$ is **closed** under multiplication. In other words, whenever we multiply two integers, we get another integer. For example, 2 and 3 are integers, and we have $2 \cdot 3 = 6$, which is also an integer. As another example, -3 and -4 are integers, and so is $(-3)(-4) = 12$.

7. Multiplication is **commutative** in $\mathbb{Z}$. In other words, when we multiply two integers, it does not matter which one comes first. For example, $2 \cdot 3 = 6$ and $3 \cdot 2 = 6$. So, $2 \cdot 3 = 3 \cdot 2$. As another example, $-8 \cdot 6 = -48$ and $6(-8) = -48$. So, we see that $-8 \cdot 6 = 6(-8)$.

8. Multiplication is **associative** in $\mathbb{Z}$. In other words, when we multiply three integers, it doesn't matter if we begin by multiplying the first two or the last two integers. For example, $(2 \cdot 3) \cdot 4 = 6 \cdot 4 = 24$ and $2 \cdot (3 \cdot 4) = 2 \cdot 12 = 24$. So, $(2 \cdot 3) \cdot 4 = 2 \cdot (3 \cdot 4)$. As another example, $(-5 \cdot 2) \cdot (-6) = -10 \cdot (-6) = 60$ and $-5 \cdot (2 \cdot (-6)) = -5 \cdot (-12) = 60$. So, we see that $(-5 \cdot 2) \cdot (-6) = -5 \cdot (2 \cdot (-6))$.

9. $\mathbb{Z}$ has an **identity** for multiplication, namely 1. Whenever we multiply 1 by another integer, the result is that same integer. For example, we have $1 \cdot 3 = 3$ and $3 \cdot 1 = 3$. As another example, $1 \cdot (-5) = -5$ and $(-5) \cdot 1 = -5$.

We can summarize the four properties above by saying that $(\mathbb{Z}, \cdot)$ is a **commutative monoid**.

10. Multiplication is **distributive** over addition in $\mathbb{Z}$. This means that whenever k, m, and n are integers, we have $k \cdot (m + n) = k \cdot m + k \cdot n$. For example, $4 \cdot (2 + 1) = 4 \cdot 3 = 12$ and $4 \cdot 2 + 4 \cdot 1 = 8 + 4 = 12$. So, $4 \cdot (2 + 1) = 4 \cdot 2 + 4 \cdot 1$. As another example, we have $-2 \cdot ((-1) + 3) = -2(2) = -4$ and $-2 \cdot (-1) + (-2) \cdot 3 = 2 - 6 = -4$. Therefore, we see that $-2 \cdot ((-1) + 3) = -2 \cdot (-1) + (-2) \cdot 3$.

Notes: (1) Since the properties listed in 1 through 10 above are satisfied, we say that $(\mathbb{Z}, +, \cdot)$ is a **commutative ring**. We will give the formal definitions of "ring" and "commutative ring" below.

(2) Observe that a ring consists of (i) a set (in this case $\mathbb{Z}$), and (ii) **two** binary operations on the set called **addition** and **multiplication**.

(3) $(\mathbb{Z}, +)$ is a commutative group and $(\mathbb{Z}, \cdot)$ is a commutative monoid. The distributive property is the only property mentioned that requires both addition and multiplication.

(4) We see that $\mathbb{Z}$ is missing one nice property—the inverse property for multiplication. For example, 2 has no multiplicative inverse in $\mathbb{Z}$. There is no integer n such that $2 \cdot n = 1$. So, the linear equation $2n - 1 = 0$ has no solution in $\mathbb{Z}$.

(5) If we replace $\mathbb{Z}$ by the set of natural numbers $\mathbb{N} = \{0, 1, 2, \ldots\}$, then all the properties mentioned above are satisfied **except** property 5—the inverse property for addition. For example, 1 has no additive inverse in $\mathbb{N}$. There is no natural number n such that $n + 1 = 0$.

(6) $\mathbb{Z}$ actually satisfies two distributive properties. **Left distributivity** says that whenever $k, m,$ and n are integers, we have $k \cdot (m + n) = k \cdot m + k \cdot n$. **Right distributivity** says that whenever $k, m,$ and n are integers, we have $(m + n) \cdot k = m \cdot k + n \cdot k$. Since multiplication is commutative in $\mathbb{Z}$, left distributivity and right distributivity are equivalent.

(7) Let's show that left distributivity together with commutativity of multiplication in $\mathbb{Z}$ implies right distributivity in $\mathbb{Z}$. If we assume that we have left distributivity and commutativity of multiplication, then for integers $k, m,$ and n, we have $(m + n) \cdot k = k(m + n) = k \cdot m + k \cdot n = m \cdot k + n \cdot k$.

(8) Once again, note that we will see a formal definition of the integers in Lesson 5. At that point, we will be able to provide proofs of the claims we made about the integers in Example 2.12 above.

We are now ready to give the more general definition of a ring.

A **ring** is a triple $(R, +, \cdot)$, where R is a set and $+$ and $\cdot$ are binary operations on R satisfying

(1) $(R, +)$ is a commutative group.

(2) $(R, \cdot)$ is a monoid.

(3) Multiplication is **distributive** over addition in R. That is, for all $x, y, z \in R$, we have

$$x \cdot (y + z) = x \cdot y + x \cdot z \quad \text{and} \quad (y + z) \cdot x = y \cdot x + z \cdot x.$$

We will always refer to the operation $+$ as addition and the operation $\cdot$ as multiplication. We will also adjust our notation accordingly. For example, we will refer to the identity for $+$ as 0, and the **additive inverse** of an element $x \in R$ as $-x$. Also, we will refer to the identity for $\cdot$ as 1, and the **multiplicative inverse** of an element $x \in R$ (if it exists) as x^{-1} or $\frac{1}{x}$.

Notes: (1) Recall that $(R, +)$ a commutative group means the following:

- **(Closure)** For all $x, y \in R$, $x + y \in R$.
- **(Associativity)** For all $x, y, z \in R$, $(x + y) + z = x + (y + z)$.
- **(Commutativity)** For all $x, y \in R$, $x + y = y + x$.
- **(Identity)** There exists an element $0 \in R$ such that for all $x \in R$, $0 + x = x + 0 = x$.
- **(Inverse)** For each $x \in R$, there is $-x \in R$ such that $x + (-x) = (-x) + x = 0$.

(2) Recall that $(R, \cdot)$ a monoid means the following:

- **(Closure)** For all $x, y \in R$, $x \cdot y \in R$.
- **(Associativity)** For all $x, y, z \in R$, $(x \cdot y) \cdot z = x \cdot (y \cdot z)$.
- **(Identity)** There exists an element $1 \in R$ such that for all $x \in R$, $1 \cdot x = x \cdot 1 = x$.

(3) Although commutativity of multiplication is not required for the definition of a ring, the most well-known example (the ring of integers) satisfies this condition. When multiplication is commutative in R, we call the ring a **commutative ring**. In this case we have the following additional property:

- **(Commutativity)** For all $x, y \in R$, $x \cdot y = y \cdot x$.

(4) Observe that we have two distributive properties in the definition for a ring. The first property is called **left distributivity** and the second is called **right distributivity**.

(5) In a commutative ring, left distributivity implies right distributivity and vice versa. In this case, the distributive property simplifies to

- **(Distributivity)** For all $x, y, z \in R$, $x \cdot (y + z) = x \cdot y + x \cdot z$

(6) Some authors leave out the multiplicative identity property in the definition of a ring and call such a ring a **unital ring** or a **ring with identity**. In this book, we will adopt the convention that a ring has a multiplicative identity. If we do not wish to assume that R has a multiplicative identity, then we will call the structure R "**almost a ring**" or a **rng** (note the missing "i"). If $(R, +, \cdot)$ is "almost a ring," then $(R, \cdot)$ is a semigroup.

(7) The properties that define a ring are called the **ring axioms**. In general, an **axiom** is a statement that is assumed to be true. So, the ring axioms are the statements that are **given** to be true in all rings. There are many other statements that are true in rings. However, any additional statements need to be **proved** using the axioms.

Example 2.13:

1. $(\mathbb{Z}, +, \cdot)$ is a commutative ring with additive identity 0 and multiplicative identity 1. The additive inverse of an integer a is the integer $-a$. See Example 2.12 above for more details.

2. $(2\mathbb{Z}, +, \cdot)$ is **not** a ring because $1 \notin 2\mathbb{Z}$. However, it is "almost a ring." In other words, the only property that fails is the multiplicative identity property. We already saw in part 2 of Example 2.8 that $(2\mathbb{Z}, +)$ is a commutative group and we saw in part 5 of Example 2.3 that $(2\mathbb{Z}, \cdot)$ is a semigroup.

    More generally, if $n \in \mathbb{Z}$ with $n \neq 0$, $n \neq \pm 1$ then $(n\mathbb{Z}, +, \cdot)$ is "almost a ring," but not quite a ring.

3. $(\mathbb{N}, +, \cdot)$ is **not** a ring because $(\mathbb{N}, +)$ is not a group. The only group property that fails is the additive inverse property. For example, the natural number 1 has no additive inverse. That is, $n + 1 = 0$ has no solution in $\mathbb{N}$. Note that $(\mathbb{N}, \cdot)$ is a commutative monoid and the distributive property holds in $\mathbb{N}$. Therefore, $(\mathbb{N}, +, \cdot)$ misses being a commutative ring by just that one property. $(\mathbb{N}, +, \cdot)$ is an example of a structure called a **semiring**.

4. Recall once again that the set of rational numbers is $\mathbb{Q} = \left\{\frac{a}{b} \mid a, b \in \mathbb{Z}, b \neq 0\right\}$ and we define addition and multiplication on $\mathbb{Q}$ by $\frac{a}{b} + \frac{c}{d} = \frac{ad+bc}{bd}$ and $\frac{a}{b} \cdot \frac{c}{d} = \frac{ac}{bd}$.

   $(\mathbb{Q}, +, \cdot)$ is a commutative ring with additive identity $0 = \frac{0}{1}$ and multiplicative identity $1 = \frac{1}{1}$. The additive inverse of a rational number $\frac{a}{b}$ is the rational number $\frac{-a}{b}$. You will need to verify all this as part of Problem 19 below.

5. Let $R = \{0\}$ and define addition and multiplication on R by $0 + 0 = 0$ and $0 \cdot 0 = 0$. Then $(R, +, \cdot)$ is a ring with additive and multiplicative identity both equal to 0. It is very easy to verify that each of the ring axioms hold, and so, I leave this verification to the reader. This ring is called the **zero ring**.

   It turns out that the zero ring is the **only** ring for which $0 = 1$. To prove this requires the use of Theorem 2.14 below. I leave the details to the reader.

6. For each $n \in \mathbb{Z}^+$, $(C_n, +, \cdot)$ is a commutative ring with additive identity 0 and multiplicative identity 1 (here $+$ and $\cdot$ are the operations of clock addition and multiplication on $C_n = \{0, 1, 2, \ldots, n-1\}$, as defined in parts 9 and 10 of Example 2.8). This example will be handled more rigorously in Lesson 3 (we will use the more formal names for clock addition and multiplication, namely addition and multiplication modulo n).

Note: A semiring $(R, +, \cdot)$ has one additional property called the **zero property**. It says that for all $x \in R$, $0 \cdot x = x \cdot 0 = 0$. In Theorem 2.14 below, we will see that this property is true in every ring. If we take away the additive inverse property, then this property does not follow from the others. So, it must be listed explicitly.

In the ring $(C_6, +, \cdot)$, there are two nonzero elements whose product is zero. Indeed, $2 \cdot 3 = 0$. In this case, we say that 2 and 3 are **zero divisors**.

Formally, if $(R, +, \cdot)$ is a ring, then $a \in R$ is a zero divisor if $a \neq 0$ and there is an element $b \in R$ with $b \neq 0$ such that $ab = 0$.

A ring that does **not** contain any zero divisors is called a **domain**, and a domain that is commutative is called an **integral domain**. For example, $(\mathbb{Z}, +, \cdot)$ and $(\mathbb{Q}, +, \cdot)$ are integral domains.

$(C_6, +, \cdot)$ is **not** an integral domain because 2 and 3 are zero divisors. Furthermore, if n is any positive integer such that $n = mk$ with $m, k \in \mathbb{Z}^+$ and $1 < m, k < n$ (such an n is called a **composite** number), then in C_n, we have $mk = 0$, and so, $(C_n, +, \cdot)$ is not an integral domain. Are there any positive integers $n > 1$ such that $(C_n, +, \cdot)$ is an integral domain? It turns out that for $n \in \mathbb{Z}^+$ with $n > 1$, $(C_n, +, \cdot)$ is an integral domain if and only if n is **prime** (a prime number is an integer greater than 1 that is not composite).

If $(R, +, \cdot)$ is a ring, then since $(R, +)$ is a group, it satisfies the cancellation laws mentioned in Note 1 following Example 2.10. However, we must use additive notation when expressing these laws. So, we have the following for $a, b, c \in R$: $a + b = a + c$ implies $b = c$ and $b + a = c + a$ implies $b = c$.

The corresponding cancellation laws for multiplication may fail in a ring, even if we exclude 0. For example, in $(C_6, +, \cdot)$, we have $1 \cdot 3 = 3$ and $5 \cdot 3 = 3$, but $1 \neq 5$.

The cancellation laws for multiplication (in a ring) can be stated as follows: if $a, b, c \in R$ with $a \neq 0$, then $ab = ac$ implies $b = c$ and $ba = ca$ implies $b = c$.

A ring satisfies the cancellation laws for multiplication if and only if it is a domain. The dedicated reader should prove this.

Subtraction and Division: Let $(R, +, \cdot)$ be a ring. If $a, b \in R$, we define $a - b = a + (-b)$ (this is called a **difference**) and if b has a multiplicative inverse in R, then we define $\frac{a}{b} = ab^{-1}$ (this is called a **quotient**).

From experience, one might be led to suspect that multiplying any element of a ring by 0 would result in 0. This turns out to be true. However, it should be noted that this is not a ring axiom, and so, we must prove it. The proof is not obvious, and it requires us to use several ring properties including the additive inverse property and distributivity.

Theorem 2.14: Let $(R, +, \cdot)$ be a ring with additive identity 0 and let a be any element of R. Then $a \cdot 0 = 0$ and $0 \cdot a = 0$.

Analysis: The proof of Theorem 2.14 requires just a bit of cleverness. Since 0 is the additive identity of R, we have $0 = 0 + 0$. This allows to write $a \cdot 0$ as $a(0 + 0)$. We can then use distributivity to rewrite $a(0 + 0)$ as $a \cdot 0 + a \cdot 0$. So, we have $a \cdot 0 = a \cdot 0 + a \cdot 0$. We would then like to "cancel" $a \cdot 0$ from each side of this last equation to get $0 = a \cdot 0$. By "canceling" $a \cdot 0$, we mean adding the additive inverse of $a \cdot 0$ to each side of the equation. There are at least two different ways to do this. One way is to use the additive identity property together with one of the cancellation laws. I leave this method as a straightforward exercise for the reader. I'm going to give a proof that uses only the ring axioms. In what follows, we use the standard abbreviation of $x - x$ for $x + (-x)$ (see the definition of subtraction above).

Proof of Theorem 2.14: Let $(R, +, \cdot)$ be a ring with additive identity 0 and let a be any element of R. Then we have
$$a \cdot 0 = a \cdot 0 + 0 = a \cdot 0 + (a \cdot 0 - a \cdot 0) = (a \cdot 0 + a \cdot 0) - a \cdot 0$$
$$= a(0 + 0) - a \cdot 0 = a \cdot 0 - a \cdot 0 = 0.$$

The proof that $0 \cdot a = 0$ is similar, and so, I leave it as an exercise for the reader. □

Notes: (1) For the first and fifth equalities in the proof, we used the additive identity property.

(2) For the second and sixth equalities, we used the additive inverse property. Note that we abbreviated the expression $a \cdot 0 + \bigl(-(a \cdot 0)\bigr)$ by $a \cdot 0 - a \cdot 0$.

(3) For the third equality, we used associativity of addition in R.

(4) For the fourth equality, we used left distributivity.

Example 2.15: Let $R = \{0, 1, x\}$, where 0, 1, and x are distinct, and let $(R, +, \cdot)$ be a ring with additive identity 0 and multiplicative identity 1. Let's construct the addition and multiplication tables for $(R, +, \cdot)$.

By Example 2.10, we already know that the addition table must look as follows (note that we are simply renaming e as 0, a as 1, and b as x):

+	0	1	x
0	0	1	x
1	1	x	0
x	x	0	1

Next, let's use Theorem 2.14 to fill in the first row and column of the multiplication table:

$\cdot$	0	1	x
0	0	0	0
1	0		
x	0		

We can now use the fact that 1 is the multiplicative identity to fill in the second row and column of the multiplication table:

$\cdot$	0	1	x
0	0	0	0
1	0	1	x
x	0	x	

Now, since $x = 1 + 1$ (from the addition table), we have

$$x \cdot x = (1 + 1) \cdot x = 1 \cdot x + 1 \cdot x = x + x = 1.$$

It follows that the only possible ring with the given conditions has the following addition and multiplication tables:

+	0	1	x
0	0	1	x
1	1	x	0
x	x	0	1

$\cdot$	0	1	x
0	0	0	0
1	0	1	x
x	0	x	1

It's standard to use the symbol "2" in place of "x," and so, we have the following tables:

+	0	1	2
0	0	1	2
1	1	2	0
2	2	0	1

$\cdot$	0	1	2
0	0	0	0
1	0	1	2
2	0	2	1

Notes: (1) To be certain that with the two tables constructed above, $(R, +, \cdot)$ is ring, we need to verify that multiplication is associative in G and that multiplication is distributive over addition in G. This can be done by brute force (doing every computation) or more simply by recognizing that these are the addition and multiplication tables of $(C_3, +, \cdot)$. In other words, these tables represent "clock addition and multiplication" on a 3-hour clock.

(2) It is easy to see that multiplication is commutative in R, as the multiplication table is symmetric about the main diagonal. It follows that $(R, +, \cdot)$ is a commutative ring. (Alternatively, we can again just use the fact that $(C_3, +, \cdot)$ is a commutative ring.)

Fields

A **field** is a triple $(F, +, \cdot)$, where F is a set and $+$ and $\cdot$ are binary operations on F satisfying

(1) $(F, +)$ is a commutative group.

(2) $(F^*, \cdot)$ is a commutative group.

(3) $\cdot$ is **distributive** over $+$ in F. That is, for all $x, y, z \in F$, we have
$$x \cdot (y + z) = x \cdot y + x \cdot z \quad \text{and} \quad (y + z) \cdot x = y \cdot x + z \cdot x.$$

(4) $0 \neq 1$.

We will refer to the operation $+$ as addition, the operation $\cdot$ as multiplication, the additive identity as 0, the multiplicative identity as 1, the additive inverse of an element $x \in F$ as $-x$, and the multiplicative inverse of an element $x \in F^*$ as x^{-1} or $\frac{1}{x}$. As usual, we will often abbreviate $x \cdot y$ as xy.

Notes: (1) Recall that $(F, +)$ a commutative group means the following:

- **(Closure)** For all $x, y \in F$, $x + y \in F$.
- **(Associativity)** For all $x, y, z \in F$, $(x + y) + z = x + (y + z)$.
- **(Commutativity)** For all $x, y \in F$, $x + y = y + x$.
- **(Identity)** There exists an element $0 \in F$ such that for all $x \in F$, $0 + x = x + 0 = x$.
- **(Inverse)** For each $x \in F$, there is $-x \in F$ such that $x + (-x) = (-x) + x = 0$.

(2) Similarly, $(F^*, \cdot)$ a commutative group means the following:

- **(Closure)** For all $x, y \in F^*$, $xy \in F^*$.
- **(Associativity)** For all $x, y, z \in F^*$, $(xy)z = x(yz)$.
- **(Commutativity)** For all $x, y \in F^*$, $xy = yx$.
- **(Identity)** There exists an element $1 \in F^*$ such that for all $x \in F^*$, $1x = x \cdot 1 = x$.
- **(Inverse)** For each $x \in F^*$, there is $x^{-1} \in F^*$ such that $xx^{-1} = x^{-1}x = 1$.

(3) Recall that F^* is the set of nonzero elements of F. We can write $F^* = \{x \in F \mid x \neq 0\}$ (pronounced "the set of x in F such that x is not equal to 0") or $F^* = F \setminus \{0\}$ (pronounced "F with 0 removed").

(4) The properties that define a field are called the **field axioms**. These are the statements that are **given** to be true in all fields. There are many other statements that are true in fields. However, any additional statements need to be **proved** using the axioms.

(5) If we replace the condition that "$(F^*, \cdot)$ is a commutative group" by "$(F, \cdot)$ is a monoid," then as we have seen above, the resulting structure is a ring.

The main difference between a ring and a field is that in a ring, there can be nonzero elements that do not have multiplicative inverses. For example, as we have seen, in the ring $\mathbb{Z}$, 2 has no multiplicative inverse. So, the equation $2x = 1$ has no solution.

(6) Every field is a commutative ring. Although this is not too hard to show (you will be asked to show this in Problem 14 below), it is worth observing that this is not completely obvious. For example, if $(F, +, \cdot)$ is a ring, then since $(F, \cdot)$ is a monoid with identity 1, it follows that $1 \cdot 0 = 0 \cdot 1 = 0$. However, in the definition of a field given above, this property of 0 is not given as an axiom. We **are** given that $(F^*, \cdot)$ is a commutative group, and so, it follows that 1 is an identity for F^*. But $0 \notin F^*$, and so, $1 \cdot 0 = 0 \cdot 1 = 0$ needs to be proved.

Similarly, in the definition of a field given above, 0 is excluded from associativity and commutativity. These need to be checked.

(7) Every field is an integral domain. This follows from Note 6 above and part (iv) of Problem 7 below.

Example 2.16:

1. $(\mathbb{Q}, +, \cdot)$ is a field. In particular, every nonzero element of $\mathbb{Q}$ has a multiplicative inverse. The inverse of the nonzero rational number $\frac{a}{b}$ is the rational number $\frac{b}{a}$. This is easy to verify: $\frac{a}{b} \cdot \frac{b}{a} = \frac{ab}{ba} = \frac{ab}{ab} = \frac{1}{1} = 1$ and $\frac{b}{a} \cdot \frac{a}{b} = \frac{ba}{ab} = \frac{ab}{ab} = \frac{1}{1} = 1$. You will be asked to provide the details that $(\mathbb{Q}, +, \cdot)$ is a field in Problem 19 below.

2. $(\mathbb{R}, +, \cdot)$ is a field. We will provide a formal definition of $\mathbb{R}$ in Lesson 5 and you will be asked to prove that $\mathbb{R}$ is a field in Problem 28 (part (iii)) in Problem Set 5.

3. $(\mathbb{C}, +, \cdot)$ is field, where addition and multiplication of the complex numbers $a + bi$ and $c + di$ are defined as follows:
$$(a + bi) + (c + di) = (a + c) + (b + d)i$$
$$(a + bi)(c + di) = (ac - bd) + (ad + bc)i$$
The verification of this is not too hard and mostly uses the fact that $(\mathbb{R}, +, \cdot)$ is a field. For example, to verify that addition is commutative in $\mathbb{C}$, we have
$$(a + bi) + (c + di) = (a + c) + (b + d)i = (c + a) + (d + b)i = (c + di) + (a + bi).$$
We have $a + c = c + a$ because $a, c \in \mathbb{R}$ and addition is commutative in $\mathbb{R}$. For the same reason, we have $b + d = d + b$.

4. $(\mathbb{Z}, +, \cdot)$ is a commutative ring that is **not** a field. The only integers with multiplicative inverses are 1 and −1 (they are each their own inverse). In particular, 2 has no multiplicative inverse in $\mathbb{Z}$, or equivalently, the equation $2x = 1$ has no solution in $\mathbb{Z}$.

5. Every finite integral domain is a field. To see this, let $(F, +, \cdot)$ be an integral domain such that $F = \{x_1, x_2, \ldots, x_n\}$ (note that 0 and 1 appear among these elements). Let $a \in F$ with $a \neq 0$. Then $ax_1, ax_2, \ldots, ax_n$ are all distinct because $ax_i = ax_j$ would imply that $x_i = x_j$ (the left cancellation law for multiplication holds in an integral domain). So, every element of F appears among $ax_1, ax_2, \ldots, ax_n$. In particular, for some i with $1 \leq i \leq n$, we have $ax_i = 1$. Therefore, a is invertible with inverse x_i.

6. In part 6 of Example 2.13, we saw that for each $n \in \mathbb{Z}^+$, $(C_n, +, \cdot)$ is a commutative ring with additive identity 0 and multiplicative identity 1 (here $+$ and $\cdot$ are the operations of clock addition and multiplication on $C_n = \{0, 1, 2, \ldots, n-1\}$). In part 10 of Example 2.8, we saw that for some values of n, $(C_n \setminus \{0\}, \cdot)$ is a group and for others it is not. For example, if $n = 2$ or $n = 5$, $(C_n \setminus \{0\}, \cdot)$ is a group, and therefore, $(C_2, +, \cdot)$ and $(C_5, +, \cdot)$ are fields. However, for $n = 4$, we saw that $(C_4 \setminus \{0\}, \cdot)$ is **not** a group, and therefore, $(C_4, +, \cdot)$ is not a field. In Lesson 9 (Theorem 9.14), we will see that $(C_n, +, \cdot)$ is a field if and only if n is a prime number.

7. In Example 2.15, we saw that $(\{0, 1, 2\}, +, \cdot)$ is a commutative ring, where addition and multiplication are defined by the following tables:

+	0	1	2
0	0	1	2
1	1	2	0
2	2	0	1

$\cdot$	0	1	2
0	0	0	0
1	0	1	2
2	0	2	1

In fact, with these tables, $(\{0, 1, 2\}, +, \cdot)$ is a field. The elements 1 and 2 are each their own multiplicative inverses. Once again, these are the addition and multiplication tables for the field $(C_3, +, \cdot)$.

Vector Spaces Over Fields

To give some motivation for the definition of a vector space, let's begin with an example.

Example 2.17: Consider the set $\mathbb{C}$ of complex numbers together with the usual definition of addition. Let's also consider another operation, which we will call **scalar multiplication**. For each $k \in \mathbb{R}$ and $z = a + bi \in \mathbb{C}$, we define kz to be $ka + kbi$.

The operation of scalar multiplication is a little different from other types of operations we have looked at previously because instead of multiplying two elements from $\mathbb{C}$ together, we are multiplying an element of $\mathbb{R}$ with an element of $\mathbb{C}$. In this case, we will call the elements of $\mathbb{R}$ **scalars**.

Let's observe that we have the following properties:

1. **$(\mathbb{C}, +)$ is a commutative group.** In other words, for addition in $\mathbb{C}$, we have closure, associativity, commutativity, an identity element (called 0), and the inverse property (the inverse of $a + bi$ is $-a - bi$). This follows immediately from the fact that $(\mathbb{C}, +, \cdot)$ is a field. When we choose to think of $\mathbb{C}$ as a vector space, we will "forget about" the multiplication in $\mathbb{C}$, and just consider $\mathbb{C}$ together with addition. In doing so, we lose much of the field structure of the complex numbers, but we retain the group structure of $(\mathbb{C}, +)$.

2. **$\mathbb{C}$ is closed under scalar multiplication.** That is, **for all $k \in \mathbb{R}$ and $z \in \mathbb{C}$, we have $kz \in \mathbb{C}$.** To see this, let $z = a + bi \in \mathbb{C}$ and let $k \in \mathbb{R}$. Then, by definition, $kz = ka + kbi$. Since $a, b \in \mathbb{R}$, and $\mathbb{R}$ is closed under multiplication, $ka \in \mathbb{R}$ and $kb \in \mathbb{R}$. It follows that $ka + kbi \in \mathbb{C}$.

3. **$1z = z$.** To see this, consider $1 \in \mathbb{R}$ and let $z = a + bi \in \mathbb{C}$. Then, since 1 is the multiplicative identity for $\mathbb{R}$, we have $1z = 1a + 1bi = a + bi = z$.

4. **For all $j, k \in \mathbb{R}$ and $z \in \mathbb{C}$, $(jk)z = j(kz)$ (Associativity of scalar multiplication).** To see this, let $j, k \in \mathbb{R}$ and $z = a + bi \in \mathbb{C}$. Then since multiplication is associative in $\mathbb{R}$, we have

$$(jk)z = (jk)(a + bi) = (jk)a + (jk)bi = j(ka) + j(kb)i = j(ka + kbi) = j(kz).$$

5. **For all $k \in \mathbb{R}$ and $z, w \in \mathbb{C}$, $k(z + w) = kz + kw$ (Distributivity of one scalar over two vectors).** To see this, let $k \in \mathbb{R}$ and $z = a + bi, w = c + di \in \mathbb{C}$. Then since multiplication distributes over addition in $\mathbb{R}$, we have

$$k(z + w) = k\big((a + bi) + (c + di)\big) = k\big((a + c) + (b + d)i\big) = k(a + c) + k(b + d)i$$
$$= (ka + kc) + (kb + kd)i = (ka + kbi) + (kc + kdi) = k(a + bi) + k(c + di) = kz + kw.$$

6. **For all $j, k \in \mathbb{R}$ and $z \in \mathbb{C}$, $(j + k)z = jz + kz$ (Distributivity of two scalars over one vector).** To see this, let $j, k \in \mathbb{R}$ and $z = a + bi \in \mathbb{C}$. Then since multiplication distributes over addition in $\mathbb{R}$, we have

$$(j + k)z = (j + k)(a + bi) = (j + k)a + (j + k)bi = (ja + ka) + (jb + kb)i$$
$$= (ja + jbi) + (ka + kbi) = j(a + bi) + k(a + bi) = jz + kz.$$

Notes: (1) Since the properties listed in 1 through 6 above are satisfied, we say that $\mathbb{C}$ is a **vector space** over $\mathbb{R}$. We will give the formal definition of a vector space below.

(2) Note that a vector space consists of (i) a set of **vectors** (in this case $\mathbb{C}$), (ii) a field (in this case $\mathbb{R}$), and (iii) two operations called **addition** and **scalar multiplication**.

(3) The operation of addition is a binary operation on the set of vectors, and the set of vectors together with this binary operation forms a commutative group. In the example above (Example 2.17), we have that $(\mathbb{C}, +)$ is a commutative group.

(4) Scalar multiplication is **not** a binary operation on the set of vectors. It takes pairs of the form (k, v), where k is in the field and v is a vector to a vector kv. Formally speaking, scalar multiplication is a function $f: F \times V \to V$, where F is the field of scalars and V is the set of vectors (see Lesson 4 for a more complete explanation of this notation).

(5) We started with the example of $\mathbb{C}$ as a vector space over $\mathbb{R}$ because it has a geometric interpretation where we can draw simple pictures to visualize what the vector space looks like. We can think of the complex number $a + bi$ as a directed line segment (which from now on we will call a **vector**) in the Complex Plane that begins at the origin and terminates at the point (a, b).

For example, pictured to the right, we can see the vectors $i = 0 + 1i$, $1 + 2i$, and $2 = 2 + 0i$ in the Complex Plane.

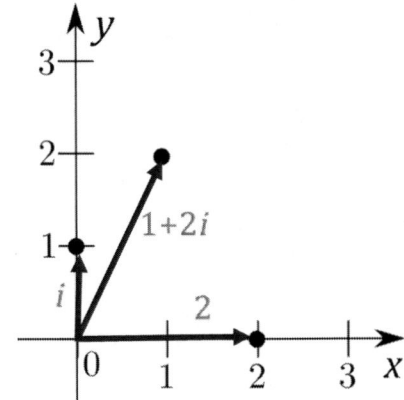

We can visualize the sum of two vectors as the vector starting at the origin that is the diagonal of the parallelogram formed from the original vectors. We see this in the first figure on the left below. In this figure, we have removed the Complex Plane and focused on the vectors $1 + 2i$ and 2, together with their sum $(1 + 2i) + (2 + 0i) = (1 + 2) + (2 + 0)i = 3 + 2i$.

A second way to visualize the sum of two vectors is to translate one of the vectors so that its initial point coincides with the terminal point of the other vector. The sum of the two vectors is then the vector whose initial point coincides with the initial point of the "unmoved" vector and whose terminal point coincides with the terminal point of the "moved" vector. We see two ways to do this in the center and rightmost figures below.

Technically speaking, the center figure shows the sum $(1 + 2i) + 2$ and the rightmost figure shows the sum $2 + (1 + 2i)$. If we superimpose one figure on top of the other, we can see strong evidence that commutativity holds for addition.

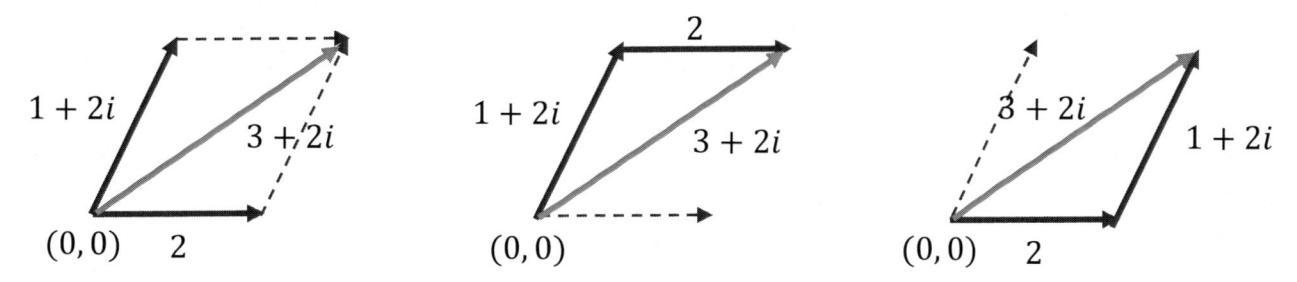

We can visualize a scalar multiple of a vector as follows: (i) if k is a positive real number and $z \in \mathbb{C}$, then the vector kz points in the same direction as z and has a length that is k times the length of z; (ii) if k is a negative real number and $z \in \mathbb{C}$, then the vector kz points in the direction opposite of z and has a length that is $|k|$ times the length of z; (iii) if $k = 0$ and $z \in \mathbb{C}$, then kz is a point.

In the figures below, we have a vector $z \in \mathbb{C}$, together with several scalar multiples of z.

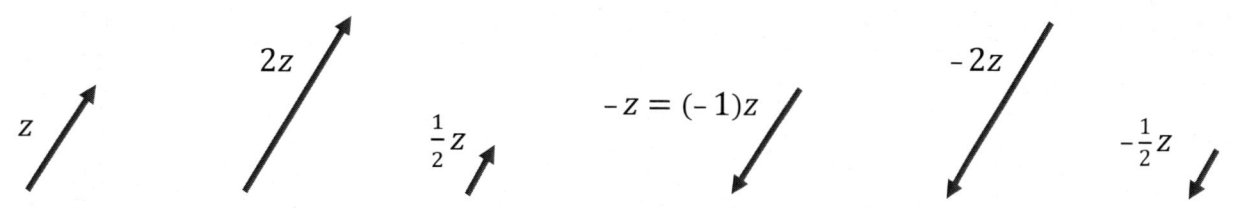

We are now ready for the general definition of a vector space. To avoid clutter, in what follows, we will often abbreviate a field $(F, +, \cdot)$ by simply referring to it as the field F.

A **vector space** over a field F (or an **F-vector space**) is a set V together with a binary operation $+$ on V (called **addition**) and an operation called **scalar multiplication** satisfying:

(1) $(V, +)$ is a commutative group.

(2) **(Closure under scalar multiplication)** For all $k \in F$ and $v \in V$, $kv \in V$.

(3) **(Scalar multiplication identity)** If 1 is the multiplicative identity of F and $v \in V$, then $1v = v$.

(4) **(Associativity of scalar multiplication)** For all $j, k \in F$ and $v \in V$, $(jk)v = j(kv)$.

(5) **(Distributivity of 1 scalar over 2 vectors)** For all $k \in F$ and $v, w \in V$, $k(v + w) = kv + kw$.

(6) **(Distributivity of 2 scalars over 1 vector)** For all $j, k \in F$ and $v \in V$, $(j + k)v = jv + kv$.

Notes: (1) Although we use the same symbol for scalar and vector addition, they are different operations. For example, in (6), $j + k$ is a sum of field elements, while $jv + kv$ is a sum of vectors.

(2) Recall that $(V, +)$ a commutative group means the following:

- **(Closure)** For all $v, w \in V$, $v + w \in V$.
- **(Associativity)** For all $v, w, u \in V$, $(v + w) + u = v + (w + u)$.
- **(Commutativity)** For all $v, w \in V$, $v + w = w + v$.
- **(Identity)** There exists an element $0 \in V$ such that for all $v \in V$, $0 + v = v + 0 = v$.
- **(Inverse)** For each $v \in V$, there is $-v \in V$ such that $v + (-v) = (-v) + v = 0$.

(3) The fields that we are most familiar with so far are $\mathbb{Q}$ (the field of rational numbers), $\mathbb{R}$ (the field of real numbers), and $\mathbb{C}$ (the field of complex numbers). We have also begun to study the fields C_p (C_p is the finite field with p elements that uses "clock addition and multiplication," where p is prime). In the next lesson we will formalize these finite fields, renaming them $\mathbb{Z}_p$ in the process.

Let's look at some basic examples of vector spaces.

Example 2.18:

1. Let $\mathbb{R}^2$ be the set of all ordered pairs of real numbers. That is, $\mathbb{R}^2 = \{(a, b) \mid a, b \in \mathbb{R}\}$. We define **addition** by $(a, b) + (c, d) = (a + c, b + d)$. We define **scalar multiplication** by $k(a, b) = (ka, kb)$ for each $k \in \mathbb{R}$. With these definitions, $\mathbb{R}^2$ is a vector space over $\mathbb{R}$.

    Notice that $\mathbb{R}^2$ looks just like $\mathbb{C}$. In fact, (a, b) is sometimes used as another notation for $a + bi$. Therefore, the verification that $\mathbb{R}^2$ is a vector space over $\mathbb{R}$ is nearly identical to what we did in Example 2.17 above.

    We can visualize elements of $\mathbb{R}^2$ as points or vectors in the Cartesian plane (see part 4 of Example 1.19) in exactly the same way that we visualize complex numbers in the Complex Plane.

2. $\mathbb{R}^3 = \{(a, b, c) \mid a, b, c \in \mathbb{R}\}$ is a vector space over $\mathbb{R}$, where we define addition and scalar multiplication by $(a, b, c) + (d, e, f) = (a + d, b + e, c + f)$ and $k(a, b, c) = (ka, kb, kc)$, respectively.

    We can visualize elements of $\mathbb{R}^3$ as points in space in a way similar to how we visualize elements of $\mathbb{R}^2$ and $\mathbb{C}$ as points in a plane.

3. More generally, we can let $\mathbb{R}^n = \{(a_1, a_2, \ldots, a_n) \mid a_i \in \mathbb{R} \text{ for each } i = 1, 2, \ldots, n\}$. Then $\mathbb{R}^n$ is a vector space over $\mathbb{R}$, where we define addition and scalar multiplication by

    $$(a_1, a_2, \ldots, a_n) + (b_1, b_2, \ldots, b_n) = (a_1 + b_1, a_2 + b_2, \ldots, a_n + b_n).$$
    $$k(a_1, a_2, \ldots, a_n) = (ka_1, ka_2, \ldots, ka_n).$$

4. More generally still, if F is any field (such as $\mathbb{Q}$, $\mathbb{R}$, $\mathbb{C}$, C_2, C_3, C_5, C_7, C_{11}, and so on), we let $F^n = \{(a_1, a_2, \ldots, a_n) \mid a_i \in F \text{ for each } i = 1, 2, \ldots, n\}$. Then F^n is a vector space over F, where we define addition and scalar multiplication by

    $$(a_1, a_2, \ldots, a_n) + (b_1, b_2, \ldots, b_n) = (a_1 + b_1, a_2 + b_2, \ldots, a_n + b_n).$$
    $$k(a_1, a_2, \ldots, a_n) = (ka_1, ka_2, \ldots, ka_n).$$

Notes: (1) Ordered pairs have the property that $(a, b) = (c, d)$ if and only if $a = c$ and $b = d$ (see Lesson 1). for example, $(1,2) \ne (2,1)$. Compare this to the unordered pair (or set) $\{1, 2\}$. Recall that a set is determined by its elements and not the order in which the elements are listed. So, $\{1, 2\} = \{2, 1\}$.

(2) Recall from Lesson 1 that $(a_1, a_2, \ldots, a_n)$ is called an **n-tuple**. So, $\mathbb{R}^n$ consists of all n-tuples of elements from $\mathbb{R}$, and more generally, F^n consists of all n-tuples of elements from the field F.

For example, $(3, 2 - i, \sqrt{2} + \sqrt{3}i, -3i) \in \mathbb{C}^4$, $\left(1, \frac{1}{2}, \frac{1}{3}, \frac{1}{4}, \frac{1}{5}, \frac{1}{6}, \frac{1}{7}, \frac{1}{8}\right) \in \mathbb{Q}^8$ (and since $\mathbb{Q}^8 \subseteq \mathbb{R}^8 \subseteq \mathbb{C}^8$, we can also say that this 8-tuple is in $\mathbb{R}^8$ or $\mathbb{C}^8$), and $(1, 1, 0, 0, 1, 0, 1, 1, 1) \in C_2^9$.

(3) Similar to what we said in Note 1, we have $(a_1, a_2, \ldots, a_n) = (b_1, b_2, \ldots, b_n)$ if and only if $a_i = b_i$ for all $i = 1, 2, \ldots, n$. So, for example, $(2, 5, \sqrt{2}, \sqrt{2})$ and $(2, \sqrt{2}, 5, \sqrt{2})$ are distinct elements from $\mathbb{R}^4$.

(4) You will be asked to verify that F^n is a vector space over the field F in Problem 9 below. Unless stated otherwise, from now on we will always consider the vector space F^n to be over the field F. For example, "Consider the vector space $\mathbb{R}^3$," means "Consider the vector space $\mathbb{R}^3$ over $\mathbb{R}$."

Let's look at a few more examples of vector spaces.

Example 2.19:

1. Let $M = \left\{ \begin{bmatrix} a & b \\ c & d \end{bmatrix} \,\middle|\, a, b, c, d \in \mathbb{R} \right\}$ be the set of all 2×2 matrices of real numbers. We add two matrices using the rule $\begin{bmatrix} a & b \\ c & d \end{bmatrix} + \begin{bmatrix} e & f \\ g & h \end{bmatrix} = \begin{bmatrix} a+e & b+f \\ c+g & d+h \end{bmatrix}$, and we multiply a matrix by a real number using the rule $k \begin{bmatrix} a & b \\ c & d \end{bmatrix} = \begin{bmatrix} ka & kb \\ kc & kd \end{bmatrix}$. It is straightforward to check that M is a vector space over $\mathbb{R}$.

2. For $m, n \in \mathbb{Z}^+$, an $m \times n$ **matrix** over a field F is a rectangular array with m rows and n columns, and entries in F. For example, the matrix $A = \begin{bmatrix} 5 & 2 & \frac{1}{5} \\ -3 & \sqrt{3} & 7 \end{bmatrix}$ is a 2×3 matrix over $\mathbb{R}$. We will generally use a capital letter to represent a matrix and the corresponding lowercase letter with double subscripts to represent the entries of the matrix. We use the first subscript for the row and the second subscript for the column. Using the matrix A above as an example, we see that $a_{21} = -3$ because the entry in row 2 and column 1 is -3. Similarly, we have $a_{11} = 5$, $a_{12} = 2$, $a_{13} = \frac{1}{5}$, $a_{22} = \sqrt{3}$, and $a_{23} = 7$.

Let $M_{m,n}^F$ be the set of all $m \times n$ matrices over the field F. For example, with A above, $A \in M_{2,3}^\mathbb{R}$. We add two matrices $A, B \in M_{m,n}^F$ to get $A + B \in M_{m,n}^F$ using the rule $(a + b)_{ij} = a_{ij} + b_{ij}$. We multiply a matrix $A \in M_{m,n}^F$ by a scalar $k \in \mathbb{F}$ using the rule $(ka)_{ij} = ka_{ij}$.

For example, if we let A be the matrix above and $B = \begin{bmatrix} 2 & -5 & \frac{4}{5} \\ -1 & -\sqrt{3} & 1 \end{bmatrix}$, then we have

$$A + B = \begin{bmatrix} 7 & -3 & 1 \\ -4 & 0 & 8 \end{bmatrix} \text{ and } 2A = \begin{bmatrix} 10 & 4 & \frac{2}{5} \\ -6 & 2\sqrt{3} & 14 \end{bmatrix}.$$

Notice that we get the entry in the first row and first column of $A + B$ as follows:

$$(a + b)_{11} = a_{11} + b_{11} = 5 + 2 = 7$$

Similarly, we get the other two entries in the first row like this:

$$(a+b)_{12} = a_{12} + b_{12} = 2 + (-5) = -3 \qquad (a+b)_{13} = a_{13} + b_{13} = \frac{1}{5} + \frac{4}{5} = \frac{5}{5} = 1$$

I leave it to the reader to write out the details for computing the entries in the second row of $A + B$.

We get the entries in the first row of $2A$ as follows:

$$(2a)_{11} = 2a_{11} = 2 \cdot 5 = 10 \qquad (2a)_{12} = 2a_{12} = 2 \cdot 2 = 4 \qquad (2a)_{13} = 2a_{13} = 2 \cdot \frac{1}{5} = \frac{2}{5}$$

I leave it to the reader to compute the entries in the second row of $2A$.

With the operations of addition and scalar multiplication defined as we have above, it is not too hard to show that M_{mn}^F is a vector space over F.

3. Let $P = \{ax^2 + bx + c \mid a, b, c \in \mathbb{R}\}$ be the set of **polynomials of degree at most 2 with real coefficients.** We define addition and scalar multiplication (with scalars in $\mathbb{R}$) on this set of polynomials as follows:

$$(ax^2 + bx + c) + (dx^2 + ex + f) = (a+d)x^2 + (b+e)x + (c+f).$$
$$k(ax^2 + bx + c) = (ka)x^2 + (kb)x + (kc).$$

For example, if $p(x) = 2x^2 + 3x - 5$ and $q(x) = -5x + 4$, then $p(x), q(x) \in P$ and we have

$$p(x) + q(x) = (2x^2 + 3x - 5) + (-5x + 4) = 2x^2 - 2x - 1.$$
$$3p(x) = 3(2x^2 + 3x - 5) = 6x^2 + 9x - 15.$$

Notice that we abbreviate the polynomial $2x^2 + 3x + (-5)$ by $2x^2 + 3x - 5$, and similarly for $2x^2 + (-2)x + (-1)$ and $6x^2 + 9x + (-15)$.

It is straightforward to check that P is a vector space over $\mathbb{R}$. As a hint, observe that a vector $(ax^2 + bx + c) \in P$ looks similar to a vector $(a, b, c) \in \mathbb{R}^3$. With this identification, addition and scalar multiplication behave exactly the same in both spaces.

4. Let $Q = \{ax^2 + bx + c \mid a, b, c \in \mathbb{R} \text{ and } a \neq 0\}$ be the set of **polynomials of degree exactly 2 with real coefficients**, with addition and scalar multiplication defined as in part 3 above. Then Q is **not** a vector space over $\mathbb{R}$. Indeed, we have $x^2 + x + 1, -x^2 + x + 1 \in Q$, but the sum $0x^2 + 2x + 2 = 2x + 2 \notin Q$, and so, Q is not closed under addition.

5. For each $n \in \mathbb{N}$, let $\mathbb{R}[x]_n = \{a_n x^n + a_{n-1} x^{n-1} + \cdots + a_1 x + a_0 \mid a_0, a_1, \ldots, a_n \in \mathbb{R}\}$ be the set of **polynomials of degree at most n with real coefficients.** We define addition and scalar multiplication (with scalars in $\mathbb{R}$) on this set of polynomials component by component, just as we did for $P = \mathbb{R}[x]_2$ in part 3 above. Notice that $\mathbb{R}[x]_n$ looks similar to $\mathbb{R}^{n+1}$ and the verification that $\mathbb{R}[x]_n$ is a vector space over $\mathbb{R}$ is nearly identical to the verification that $\mathbb{R}^{n+1}$ is a vector space over $\mathbb{R}$.

We can replace $\mathbb{R}$ by any field F in this example to get the vector space $F[x]_n$ of polynomials of degree at most n with coefficients in the field F.

6. Let $\mathbb{R}[x] = \{a_k x^k + a_{k-1} x^{k-1} + \cdots + a_1 x + a_0 \mid k \in \mathbb{N} \wedge a_0, a_1, \ldots, a_k \in \mathbb{R}\}$ be the set of **all polynomials with real coefficients.** With addition and scalar multiplication defined component by component, $\mathbb{R}[x]$ is a vector space over $\mathbb{R}$.

Modules Over Rings

If R is a ring, then the definition of a left module over R (or a left R-module) is **identical** to the definition of a vector space over a field F. The **only** difference here is that R is allowed to be a ring. Specifically, we have the following definition:

A **left module** over a ring R (or a **left R-module**) is a set M together with a binary operation $+$ on M (called **addition**) and an operation called **scalar multiplication** satisfying:

(1) $(M, +)$ is a commutative group.

(2) **(Closure under scalar multiplication)** For all $k \in R$ and $v \in M$, $kv \in M$.

(3) **(Scalar multiplication identity)** If 1 is the multiplicative identity of R and $v \in M$, then $1v = v$.

(4) **(Associativity of scalar multiplication)** For all $j, k \in R$ and $v \in M$, $(jk)v = j(kv)$.

(5) **(Distributivity of 1 scalar over 2 vectors)** For all $k \in R$ and $v, w \in M$, $k(v + w) = kv + kw$.

(6) **(Distributivity of 2 scalars over 1 vector)** For all $j, k \in R$ and $v \in M$, $(j + k)v = jv + kv$.

Example 2.20:

1. If V is a vector space over a field F, then V is also a left module over F. Indeed, the definition of a module over a field is the same as the definition of a vector space over a field.

2. Let $\mathbb{Z}^2$ be the set of all ordered pairs of integers. That is, $\mathbb{Z}^2 = \{(m, n) \mid m, n \in \mathbb{Z}\}$ We define **addition** by $(m, n) + (c, d) = (m + c, n + d)$. We define **scalar multiplication** by $k(m, n) = (km, kn)$ for each $k \in \mathbb{Z}$. With these definitions, $\mathbb{Z}^2$ is a left module over $\mathbb{Z}$.

    The verification that $\mathbb{Z}^2$ is a module over $\mathbb{Z}$ is similar to what we did in Example 2.17 above.

3. More generally, we can let $\mathbb{Z}^n = \{(m_1, m_2, \ldots, m_n) \mid m_i \in \mathbb{Z} \text{ for each } i = 1, 2, \ldots, n\}$. Then $\mathbb{Z}^n$ is a left module over $\mathbb{Z}$, where we define addition and scalar multiplication by

    $$(m_1, m_2, \ldots, m_n) + (b_1, b_2, \ldots, b_n) = (m_1 + b_1, m_2 + b_2, \ldots, m_n + b_n).$$
    $$k(m_1, m_2, \ldots, m_n) = (km_1, km_2, \ldots, km_n).$$

4. More generally still, $R^n = \{(a_1, a_2, \ldots, a_n) \mid a_i \in R \text{ for each } i = 1, 2, \ldots, n\}$ is a left module over R for any ring R, where we define addition and scalar multiplication by

    $$(a_1, a_2, \ldots, a_n) + (b_1, b_2, \ldots, b_n) = (a_1 + b_1, a_2 + b_2, \ldots, a_n + b_n).$$
    $$k(a_1, a_2, \ldots, a_n) = (ka_1, ka_2, \ldots, ka_n).$$

5. Let $(G, +)$ be a commutative group. Then G is a left module over $\mathbb{Z}$, where scalar multiplication is defined as follows: $0x = 0$, $1x = x$, $2x = x + x$, $3x = x + x + x$, and in general, if $n \in \mathbb{Z}^+$, $nx = x + x + \cdots + x$ (where x is being added to itself n times) and $(-n)x = -(nx)$. I leave it to the reader to verify that properties 1 through 6 above hold.

Note: If we change all the multiplications on the left by elements of the ring into multiplications on the right, then we will say that M is a **right module over R** (or a **right R-module**). For example, in the case of a right R-module, closure under scalar multiplication would be written "For all $k \in R$ and $v \in M$, $vk \in M$." The reader should rewrite axioms 3 through 6 for right R-modules as well.

Problem Set 2

Full solutions to these problems are available for free download here:

www.SATPrepGet800.com/AAFBTDW

LEVEL 1

1. For each of the following multiplication tables defined on the set $S = \{a, b\}$, determine if each of the following is true or false:

   (i) $\star$ defines a binary operation on S.

   (ii) $\star$ is commutative in S.

   (iii) a is an identity with respect to $\star$.

   (iv) b is an identity with respect to $\star$.

I

$\star$	a	b
a	a	a
b	a	a

II

$\star$	a	b
a	a	b
b	c	a

III

$\star$	a	b
a	a	b
b	b	a

IV

$\star$	a	b
a	a	a
b	b	b

2. Show that there are exactly two monoids on the set $S = \{e, a\}$, where e is the identity. Which of these monoids are groups? Which of these monoids are commutative?

3. The addition and multiplication tables below are defined on the set $S = \{0, 1\}$. Show that $(S, +, \cdot)$ does **not** define a ring.

+	0	1
0	0	1
1	1	0

$\cdot$	0	1
0	1	0
1	0	1

4. Let $S = \{0, 1\}$ and define addition (+) and multiplication ($\cdot$) so that $(S, +, \cdot)$ is a ring. Assume that 0 is the additive identity in S and 1 is the multiplicative identity in S. Draw the tables for addition and multiplication and verify that with these tables, $(S, +, \cdot)$ is a ring. Is $(S, +, \cdot)$ a field?

5. The addition and multiplication tables below are defined on the set $S = \{0, 1, 2\}$. Show that $(S, +, \cdot)$ does **not** define a field.

+	0	1	2
0	0	1	2
1	1	2	0
2	2	0	1

$\cdot$	0	1	2
0	0	0	0
1	0	1	2
2	0	2	2

6. Give an example of an integral domain that is **not** a field.

Level 2

7. Let $(R, +, \cdot)$ be a ring. Prove each of the following:
   (i) If $a, b \in R$ with $a + b = b$, then $a = 0$.
   (ii) If $a, b \in R$, b^{-1} exists, and $ab = b$, then $a = 1$.
   (iii) If $a, b \in R$, a^{-1} exists, and $ab = 1$, then $b = \frac{1}{a}$.
   (iv) If $(R, +, \cdot)$ is a field, $a, b \in R$ and $ab = 0$, then $a = 0$ or $b = 0$ (in other words, every field is an integral domain). Is this still true if we replace "field" by "ring?"
   (v) If $a \in R$, then $-a = -1a$
   (vi) $(-1)(-1) = 1$.

8. Let $(F, +, \cdot)$ be a field with $\mathbb{N} \subseteq F$. Prove that $\mathbb{Q} \subseteq F$.

9. Let F be a field. Prove that F^n is a vector space over F.

10. Let V be a vector space over F. Prove each of the following:
    (i) For every $v \in V$, $-(-v) = v$.
    (ii) For every $v \in V$, $0v = 0$.
    (iii) For every $k \in F$, $k \cdot 0 = 0$.
    (iv) For every $v \in V$, $-1v = -v$.

Level 3

11. Assume that a group $(G, \star)$ of order 4 exists with $G = \{e, a, b, c\}$, where e is the identity, $a^2 = b$ and $b^2 = e$. Construct the table for the operation of such a group.

12. Let A be a nonempty set. Prove that $(\mathcal{P}(A), \cap)$ is a commutative monoid that is not a group.

13. Let $(R, +, \cdot)$ be a ring and define addition and multiplication on $R \times R$ componentwise. That is, for $a, b, c, d \in R$, we define addition and multiplication by $(a, b) + (c, d) = (a + c, b + d)$ and $(a, b)(c, d) = (ac, bd)$. Prove that $(R \times R, +, \cdot)$ is a ring. If $(R, +, \cdot)$ is a domain, does it follow that $(R \times R, +, \cdot)$ must be a domain?

14. Let $(F, +, \cdot)$ be a field. Prove that $(F, \cdot)$ is a commutative monoid.

Level 4

15. Let $(G, \star)$ be a group with $a, b \in G$, and let a^{-1} and b^{-1} be the inverses of a and b, respectively. Prove
    (i) $(a \star b)^{-1} = b^{-1} \star a^{-1}$.
    (ii) the inverse of a^{-1} is a.

16. Let $(G,\star)$ be a group such that $a^2 = e$ for all $a \in G$. Prove that $(G,\star)$ is commutative.

17. Let $\mathbb{Z}[x]_n = \{a_n x^n + a_{n-1} x^{n-1} + \cdots + a_1 x + a_0 \mid a_0, a_1, \ldots, a_n \in \mathbb{Z}\}$. In other words, $\mathbb{Z}[x]_n$ consists of all polynomials of degree at most n with integer coefficients. Prove that $(\mathbb{Z}[x]_n, +)$ is a commutative group for $n = 0$, 1, and 2, where addition is defined in the "usual way." What if we replace "polynomials of degree at most n" by polynomials of degree exactly n?" Do we still get a commutative group?

LEVEL 5

18. Prove that there are exactly two groups of order 4, up to renaming the elements.

19. Prove that $(\mathbb{Q}, +, \cdot)$ is a field.

20. Prove that $(\mathbb{C}, +, \cdot)$ is a field. You may use the fact that $(\mathbb{R}, +, \cdot)$ is a field.

21. Let $S = \{a, b\}$, where $a \neq b$. How many binary operations are there on S? How many semigroups are there of the form $(S,\star)$, up to renaming the elements?

22. Let $R[x] = \{a_k x^k + a_{k-1} x^{k-1} + \cdots + a_1 x + a_0 \mid k \in \mathbb{N} \wedge a_0, a_1, \ldots, a_k \in R\}$, where R is a ring. Prove that $R[x]$ is both a left R-module and a right R-module.

CHALLENGE PROBLEMS

23. Let A be a nonempty set. Prove that $(\mathcal{P}(A), \Delta, \cap)$ is a commutative ring that is not a field (recall that $\mathcal{P}(A)$ is the power set of A and for sets X and Y, $X \Delta Y$ is the symmetric difference between X and Y). Is it an integral domain?

24. Let $S = \{a, b, c\}$, where a, b, and c are distinct. Find all semigroups of the form $(S,\star)$, up to renaming the elements. Which of these semigroups are monoids? Which of them are groups? Which of them are commutative?

25. Let $R = \{0, 1, a, b\}$. Find all rings of the form $(R, +, \cdot)$, where 0 is the additive identity in R and 1 is the multiplicative identity in R, up to renaming the elements. Which of these rings are commutative rings? Which of them are fields?

26. Let's call a structure $(R, +, \cdot)$ an **almost semiring** if it satisfies all the semiring axioms **except** for the zero property (for all $x \in R$, $0 \cdot x = x \cdot 0 = 0$). Describe an example of an almost semiring that is not a semiring. Prove that the zero property fails in this structure and that every other semiring property holds in this structure.

LESSON 3
RELATIONS AND PARTITIONS

Binary Relations

A **binary relation** on a set A is a subset of $A^2 = A \times A$. Symbolically, we have

R is a binary relation on A if and only if $R \subseteq A \times A$.

We will usually abbreviate $(a, b) \in R$ as aRb.

Remark: The statement $R \subseteq A \times A$ is equivalent to the statement $R \in \mathcal{P}(A \times A)$. It follows that for a finite set A, the number of binary relations on A is $|\mathcal{P}(A \times A)|$.

Example 3.1:

1. Let $R = \{(a, b) \in \mathbb{N} \times \mathbb{N} \mid a < b\}$. For example, we have $(0, 1) \in R$ because $0 < 1$. However, $(1, 1) \notin R$ because $1 \not< 1$. We abbreviate $(0, 1) \in R$ by $0R1$.

   Observe that $R \subseteq \mathbb{N} \times \mathbb{N}$, and so, R is a binary relation on $\mathbb{N}$.

   We would normally use the name $<$ for this relation R. So, we have $(0, 1) \in <$, which we abbreviate as $0 < 1$, and we have $(1, 1) \notin <$, which we abbreviate as $1 \not< 1$.

2. There are binary relations $<, \leq, >, \geq$ defined on $\mathbb{N}, \mathbb{Z}, \mathbb{Q}$, and $\mathbb{R}$. For example, if we consider $> \subseteq \mathbb{Z}^2$, we have $(13, -7) \in >$, or equivalently, $13 > -7$.

3. Let $A = \{a\}$. Since $|A| = 1$, we have $|A \times A| = 1 \cdot 1 = 1$. So, $|\mathcal{P}(A \times A)| = 2^1 = 2$. So, there are 2 binary relations on A. They are $R_1 = \emptyset$ and $R_2 = \{(a, a)\}$.

4. Let $B = \{0, 1\}$. Since $|B| = 2$, we have $|B \times B| = 2 \cdot 2 = 4$. So, $|\mathcal{P}(B \times B)| = 2^4 = 16$. So, there are 16 binary relations on B. A few examples are $R_1 = \emptyset$, $R_2 = \{(0,0)\}$, $R_3 = \{(0,1)\}$, and $R_4 = \{(0,0), (0,1)\}$. Can you list the rest of them?

5. Let A be a set and let R be the binary relation on A defined by $R = \{(a, b) \in A \times A \mid a \in b\}$. R is known as the **membership relation**, and it is usually denoted by $\in$. So, if $a, b \in A$ and a is a member of b, we can write $(a, b) \in \in$, which we will usually abbreviate as $a \in b$. As a specific example, let $A = \{\emptyset, \{\emptyset\}, \{\{\emptyset\}\}\}$. Then $(\emptyset, \{\emptyset\}) \in \in$, or equivalently, $\emptyset \in \{\emptyset\}$. Similarly, we have $(\{\emptyset\}, \{\{\emptyset\}\}) \in \in$, or equivalently, $\{\emptyset\} \in \{\{\emptyset\}\}$.

6. Let $R = \{((a, b), (c, d)) \in (\mathbb{N} \times \mathbb{N})^2 \mid a + d = b + c\}$. Then R is a binary relation on $\mathbb{N} \times \mathbb{N}$. For example, we have $(5, 0)R(6, 1)$ because $5 + 1 = 0 + 6$. However, we see that $(5, 0)\not R(6, 2)$ because $5 + 2 \neq 0 + 6$.

7. Let $R = \{((a, b), (c, d)) \in (\mathbb{Z} \times \mathbb{Z}^*)^2 \mid ad = bc\}$. (Recall that $\mathbb{Z}^*$ is the set of *nonzero* integers.) Then R is a binary relation on $\mathbb{Z} \times \mathbb{Z}^*$. For example, $(1, 2)R(2, 4)$ because $1 \cdot 4 = 2 \cdot 2$. However, $(1, 2)\not R(2, 5)$ because $1 \cdot 5 \neq 2 \cdot 2$. Compare this to the rational number system (see part 1 of Example 1.6), where we have $\frac{1}{2} = \frac{2}{4}$ because $1 \cdot 4 = 2 \cdot 2$, but $\frac{1}{2} \neq \frac{2}{5}$ because $1 \cdot 5 \neq 2 \cdot 2$.

69

The **domain** of a binary relation R, written dom R, is $\{x \mid \exists y(xRy)\}$. The **range** of a binary relation, written ran R, is $\{y \mid \exists x(xRy)\}$. The **field** of a binary relation R is dom $R \cup$ ran R.

Notes: (1) The symbol $\exists$ is called an **existential quantifier**, and it is pronounced "There exists" or "There is."

(2) The expression $\exists y(xRy)$ can be translated into English as "There exists a y such that xRy." Similarly, the expression $\exists x(xRy)$ can be translated into English as "There exists an x such that xRy." In general, if $P(x)$ is some property, then the expression $\exists x(P(x))$ can be translated into English as "There exists an x such that $P(x)$."

(3) The definition of field given above is not related to the algebraic structure that we learned about in Lesson 2. Unfortunately, we use the same word for these two unrelated definitions.

Example 3.2:

1. Let $R = \{(a,b) \in \mathbb{N} \times \mathbb{N} \mid a < b\}$. Then dom $R = \mathbb{N}$, ran $R = \mathbb{N}$, and field $R = \mathbb{N} \cup \mathbb{N} = \mathbb{N}$.

2. Let $B = \{0, 1, 2, 3\}$ and $R = \{(0, 2), (0, 3), (1, 3)\}$. Then dom $R = \{0, 1\}$, ran $R = \{2, 3\}$, and field $R = \{0, 1\} \cup \{2, 3\} = \{0, 1, 2, 3\} = B$.

3. Let $R = \{((a,b), (c,d)) \in (\mathbb{N} \times \mathbb{N})^2 \mid a + d = b + c\}$. Then dom $R = \mathbb{N} \times \mathbb{N}$, ran $R = \mathbb{N} \times \mathbb{N}$, and field $R = (\mathbb{N} \times \mathbb{N}) \cup (\mathbb{N} \times \mathbb{N}) = \mathbb{N} \times \mathbb{N}$.

We say that a binary relation R on a set A is

- **reflexive** if for all $a \in A$, aRa.
- **symmetric** if for all $a, b \in A$, aRb implies bRa.
- **transitive** if for all $a, b, c \in A$, aRb and bRc imply aRc.
- **antireflexive** if for all $a \in A$, $a\not{R}a$.
- **antisymmetric** if for all $a, b \in A$, aRb and bRa imply $a = b$.

Example 3.3:

1. Let A be any set and let $R = \{(a,b) \in A^2 \mid a = b\}$. Then R is reflexive ($a = a$), symmetric (if $a = b$, then $b = a$), transitive (if $a = b$ and $b = c$, then $a = c$), and antisymmetric (trivially). If $A \neq \emptyset$, then this relation is not antireflexive because $a \neq a$ is false for any $a \in A$.

2. The binary relations $\leq$ and $\geq$ defined in the usual way on $\mathbb{Z}$ are transitive (if $a \leq b$ and $b \leq c$, then $a \leq c$, and similarly for $\geq$), reflexive ($a \leq a$ and $a \geq a$), and antisymmetric (if $a \leq b$ and $b \leq a$, then $a = b$, and similarly for $\geq$). These relations are not symmetric. For example, $1 \leq 2$, but $2 \not\leq 1$). These relations are not antireflexive. For example, $1 \leq 1$ is true.

    Any relation that is transitive, reflexive, and antisymmetric is called a **partial ordering**.

3. The binary relations $<$ and $>$ defined on $\mathbb{Z}$ are transitive (if $a < b$ and $b < c$, then $a < c$, and similarly for $>$), antireflexive ($a \not< a$ and $a \not> a$), and antisymmetric (this is vacuously true because $a < b$ and $b < a$ can never occur). These relations are not symmetric (for example, $1 < 2$, but $2 \not< 1$). These relations are not reflexive (for example, $1 < 1$ is false).

Any relation that is transitive, antireflexive, and antisymmetric is called a **strict partial ordering**.

4. Let $R = \{(0,0), (0,2), (2,0), (2,2), (2,3), (3,2), (3,3)\}$ be a binary relation on $\mathbb{N}$. Then it is easy to see that R is symmetric. R is not reflexive because $1 \in \mathbb{N}$, but $(1,1) \notin R$ (however, if we were to consider R as a relation on $\{0, 2, 3\}$ instead of on $\mathbb{N}$, then R **would** be reflexive). R is not transitive because we have $(0,2), (2,3) \in R$, but $(0,3) \notin R$. R is not antisymmetric because we have $(2,3), (3,2) \in R$ and $2 \neq 3$. R is not antireflexive because $(0,0) \in R$ (and also, $(2,2) \in R$ and $(3,3) \in R$).

n-ary Relations

We can extend the idea of a binary relation on a set A to an **n-ary relation** on A. For example, a 3-ary relation (or **ternary relation**) on A is a subset of $A^3 = A \times A \times A$. More generally, we have that R is an n-ary relation on A if and only if $R \subseteq A^n$. A **1-ary relation** (or **unary relation**) on A is just a subset of A.

Example 3.4:

1. $\mathbb{R}$ is a unary relation on $\mathbb{C}$ because $\mathbb{R} \subseteq \mathbb{C}$.

2. Let $R = \{(x, y, z) \in \mathbb{Z}^3 \mid x + y = z\}$. Then R is a ternary (or 3-ary) relation on $\mathbb{Z}$. We have, for example, $(1, 2, 3) \in R$ (because $1 + 2 = 3$) and $(1, 2, 4) \notin R$ (because $1 + 2 \neq 4$).

3. Let C be the set of all colors. For example, blue $\in C$, pink $\in C$, and violet $\in C$. Let $S = \{(a, b, c) \in C^3 \mid$ when a and b are combined in equal quantities, the result is $c\}$. Then S is a ternary relation on C. We have, for example, (red, yellow, orange) $\in S$.

4. Let $T = \{(a, b, c, d, e) \in \mathbb{N}^5 \mid ab + c = de\}$. Then T is a 5-ary relation on $\mathbb{N}$. We have, for example, $(1, 2, 8, 5, 2) \in T$ $(1 \cdot 2 + 8 = 5 \cdot 2)$ and $(1, 1, 1, 1, 1) \notin T$ $(1 \cdot 1 + 1 \neq 1 \cdot 1)$.

Orderings

A binary relation $\leq$ on a set A is a **partial ordering** on A if $\leq$ is reflexive, antisymmetric, and transitive on A. If we replace "reflexive" by "antireflexive," then we call the relation a **strict partial ordering** on A (we would normally use the symbol $<$ instead of $\leq$ for a strict partial ordering).

A **partially ordered set** (or **poset**) is a pair $(A, \leq)$, where A is a set and $\leq$ is a partial ordering on A. Similarly, a **strict poset** is a pair $(A, <)$, where A is a set and $<$ is a strict partial ordering on A.

Note: If $<$ is a strict partial ordering on a set A, then $\forall a, b \in A (a < b \rightarrow b \not< a)$. To see this, let $a < b$. Since $<$ is antireflexive, $a \neq b$. Since $<$ is antisymmetric, $b < a$ would lead to $a = b$. Since we cannot have both $a \neq b$ and $a = b$, we must have $b \not< a$.

Example 3.5:

1. The usual ordering $\leq$ on $\mathbb{Z} = \{\ldots, -3, -2, -1, 0, 1, 2, 3, \ldots\}$ is a partial ordering, and the ordering $<$ on $\mathbb{Z}$ is a strict partial ordering. See Example 3.3 (parts 2 and 3).

2. If A is a set, then $(\mathcal{P}(A), \subseteq)$ is a poset. Since every set is a subset of itself, $\subseteq$ is reflexive (see Theorem 1.9). If $X, Y \in \mathcal{P}(A)$ with $X \subseteq Y$ and $Y \subseteq X$, then $X = Y$ (see the Axiom of Extensionality right before Example 1.16). So, $\subseteq$ is antisymmetric. By Theorem 1.14, $\subseteq$ is transitive.

See the tree diagrams at the end of Example 1.12 for visual representations of this poset when $A = \{a, b\}$ and $A = \{a, b, c\}$.

Similarly, $(\mathcal{P}(A), \subset)$ is a strict poset (the relation here is the **proper subset** relation). The relation $\subset$ is antireflexive because no set is a proper subset of itself. If $X \subset Y$ and $Y \subset X$, then $X \subseteq Y$ and $Y \subseteq X$, and so, $X = Y$ (again, see the Axiom of Extensionality right before Example 1.16). Finally, suppose that $X \subset Y$ and $Y \subset Z$. Then $X \subseteq Y$ and $Y \subseteq Z$, and so, by Theorem 1.14, $X \subseteq Z$. Suppose toward contradiction that $X = Z$. Then $Z \subseteq X$ and again, by Theorem 1.14, $Z \subseteq Y$. Since $Y \subseteq Z$ and $Z \subseteq Y$, it follows that $Y = Z$, contradicting our assumption that Y is a *proper* subset of Z. So, $X \neq Z$, and therefore, $X \subset Z$.

Note that in the argument above, to show that $X \neq Z$, we assumed that $X = Z$, and then used a logically valid argument to derive the statement $Y = Z$, which we already knew was false. This is known as a proof by contradiction. We will use this kind of argument in Theorems 3.7 and 3.8 below as well. See the notes after the proof of Theorem 4.16 for a more detailed explanation as to how this type of argument works. Theorem 5.6 provides another good example.

3. Let $(A, \leq_A)$ be a partially ordered set, $B \subseteq A$, and $\leq_B = \{(x, y) \mid x, y \in B \land x \leq_A y\}$. Then $(B, \leq_B)$ is also a partially ordered set. Let's check this carefully.

   To see that $\leq_B$ is reflexive on B, let $x \in B$. Since $B \subseteq A$, $x \in A$. Since $\leq_A$ is reflexive on A, $x \leq_A x$. Since $x \in B$, $x \leq_B x$.

   To see that $\leq_B$ is antisymmetric on B, let $x, y \in B$ with $x \leq_B y$ and $y \leq_B x$. Then $x \leq_A y$ and $y \leq_A x$. Since $\leq_A$ is antisymmetric on A, $x = y$.

   Finally, to see that $\leq_B$ is transitive on B, let $x, y, z \in B$ with $x \leq_B y$ and $y \leq_B z$. Then $x \leq_A y$ and $y \leq_A z$. Since $\leq_A$ is transitive on A, $x \leq_A z$. Since $x, z \in B$, $x \leq_B z$.

4. Let $X = \{x \mid x \text{ is a word in the English language}\}$ and define the **dictionary order** on X as follows: $x <_D y$ if x appears before y alphabetically. Then $(X, <_D)$ is a strict poset. In this poset, we have aardvark $<_D$ antelope (because a = a and a appears before n alphabetically), we have stranger $<_D$ violin (because s appears before v alphabetically), and we have dragon $<_D$ drainage (because d = d, r = r, a = a and g appears before i alphabetically).

   We can use a similar idea to define the **dictionary order** on Cartesian products of posets (and strict posets).

   For example, the dictionary order $<_D$ can be defined on $\mathbb{Z} \times \mathbb{Q}$ by $(a, b) <_D (c, d)$ if and only if either $a <_\mathbb{Z} c$ or both $a = c$ and $b <_\mathbb{Q} d$, where $<_\mathbb{Z}$ and $<_\mathbb{Q}$ are the usual strict partial orderings on $\mathbb{Z}$ and $\mathbb{Q}$, respectively. It is easy (but a bit tedious) to verify that $(\mathbb{Z} \times \mathbb{Q}, <_D)$ is a strict poset. I leave the verification to the reader.

   In the strict poset $(\mathbb{Z} \times \mathbb{Q}, <_D)$, we have $(4, 7) <_D (9, -2)$ because $4 <_\mathbb{Z} 9$ (notice that the second coordinates are irrelevant because the first coordinates are not equal). We also have $\left(3, \frac{1}{2}\right) <_D \left(3, \frac{3}{4}\right)$ because $3 = 3$ and $\frac{1}{2} <_\mathbb{Q} \frac{3}{4}$.

   We can visualize this particular dictionary order in a Cartesian plane as solid vertical lines passing through each integer value along the x-axis. Each point is less than any point higher than it on the same vertical line and each point is also less than any point on a vertical line to the right of that point (regardless of the height).

In the figure below, we see that $(1,-2) <_D \left(1, \frac{3}{2}\right)$ because $\left(1, \frac{3}{2}\right)$ is above $(1,-2)$ on the same vertical line. We also have $(1,-2) <_D (3,-3)$ because $(3,-3)$ is to the right of $(1,-2)$ (note that it does **not** matter that $(3,-3)$ is below $(1,-2)$).

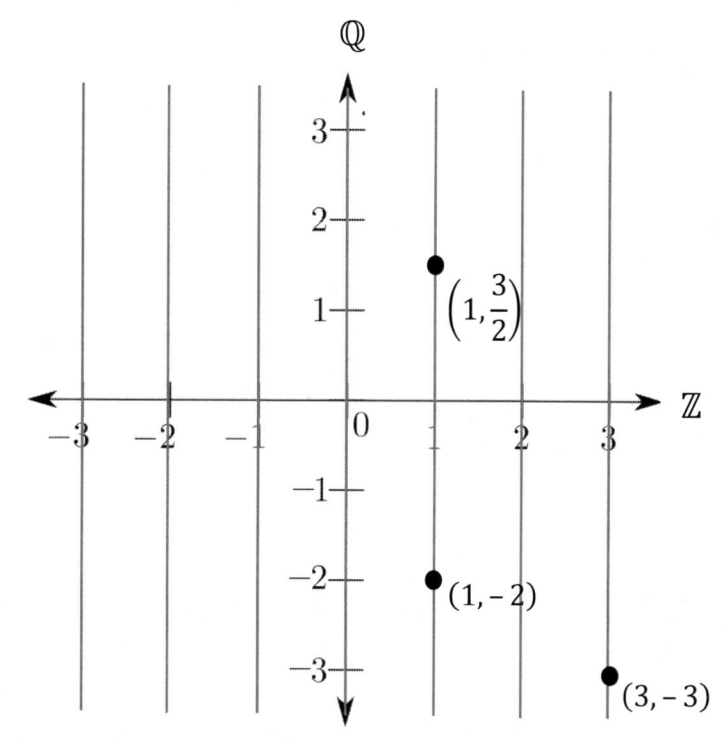

As another example, in the dictionary order of $\mathbb{R}^5$, we have $(1,3,5,2,11) <_D (1,3,5,4,10)$ because $1=1$, $3=3$, $5=5$, and $2 <_\mathbb{R} 4$.

As one more example, let $A = \{a,b\}$, consider the strict poset $(\mathcal{P}(A), \subset)$ (the relation here is the **proper subset** relation), and let $<_D$ be the corresponding dictionary order on $(\mathcal{P}(A))^2 = \mathcal{P}(A) \times \mathcal{P}(A)$. We have $(\emptyset, \{a\}) <_D (\{b\},\{b\})$ because $\emptyset \subset \{b\}$. Also, we have $(\{a\},\{b\}) <_D (\{a\},\{a,b\})$ because $\{a\} = \{a\}$ and $\{b\} \subset \{a,b\}$.

Let $(A, \leq)$ be a poset. We say that $a,b \in A$ are **comparable** if $a \leq b$ or $b \leq a$. The poset satisfies the **comparability condition** if every pair of elements in A are comparable. A poset that satisfies the comparability condition is called a **linearly ordered set** (or **totally ordered set**). Similarly, a **strict linearly ordered set** $(A, <)$ satisfies **trichotomy**: If $a,b \in A$, then $a < b$, $a = b$, or $b < a$.

Example 3.6:

1. $(\mathbb{N}, \leq)$, $(\mathbb{Z}, \leq)$, $(\mathbb{Q}, \leq)$, and $(\mathbb{R}, \leq)$ are linearly ordered sets. Similarly, $(\mathbb{N}, <)$, $(\mathbb{Z}, <)$, $(\mathbb{Q}, <)$, and $(\mathbb{R}, <)$ are strict linearly ordered sets.

2. If A has at least two elements, then $(\mathcal{P}(A), \subset)$ is **not** a strict linearly ordered set. Indeed, if $a, b \in A$ with $a \neq b$, then $\{a\} \not\subset \{b\}$ and $\{b\} \not\subset \{a\}$. See either of the tree diagrams at the end of Example 1.12.

   Similarly, $\left((\mathcal{P}(A))^2, <_D\right)$, where $<_D$ is the dictionary order is **not** a strict linearly ordered set. Indeed, $(\{a\}, \emptyset)$ and $(\{b\}, \emptyset)$ are not comparable with respect to $<_D$ because $\{a\} \not\subset \{b\}$ and $\{b\} \not\subset \{a\}$, just like we observed above. Similarly, $(\{a\},\{a\})$ and $(\{a\},\{b\})$ are not comparable.

3. If $(A, <_A)$ and $(B, <_B)$ are strict linearly ordered sets, then the dictionary order $<_D$ on $A \times B$ is also a strict linearly ordered set. To see this, let $(a,b), (c,d) \in A \times B$. If $a <_A c$, then we have $(a,b) <_D (c,d)$. If $c <_A a$, then $(c,d) <_D (a,b)$. If $a = c$, then we look at the second coordinates. In this case, if $b <_B d$, then $(a,b) <_D (c,d)$, and if $d <_B b$, then $(c,d) <_D (a,b)$. Otherwise, $a = c$ and $b = d$, and therefore, $(a,b) = (c,d)$.

We can modify a partial ordering slightly to get a strict partial ordering. We do this as follows:

Theorem 3.7: Let A be a set and let $\leq_A$ be a partial ordering on A. Define the binary relation $<_A$ on A by $a <_A b$ if and only if $a \leq_A b$ and $a \neq b$. Then $<_A$ is a strict partial ordering on A.

Proof: Suppose that $a <_A b$ and $b <_A c$. Then $a \leq_A b$ and $b \leq_A c$. Since $\leq_A$ is transitive, $a \leq_A c$. Also, $a \neq b$ and $b \neq c$. Suppose toward contradiction that $a = c$. Since $b \leq_A c$, we have $b \leq_A a$. Since $\leq_A$ is antisymmetric, $a = b$, contrary to our assumption. Therefore, $a \neq c$. Since $a \leq_A c$ and $a \neq c$, we have $a <_A c$. This shows that $<_A$ is transitive.

Now, suppose that $a <_A b$ and $b <_A a$. Then $a \leq_A b$ and $b \leq_A a$. Since $\leq_A$ is antisymmetric, $a = b$. Therefore, $<_A$ is antisymmetric.

Finally, by definition, $a <_A a$ is false. Therefore, $<_A$ is antireflexive.

Since $<_A$ is transitive, antisymmetric, and antireflexive, $<_A$ is a strict partial ordering on A. □

Similarly, we can modify a strict partial ordering to get a partial ordering.

Theorem 3.8: Let A be a set and let $<_A$ be a strict partial ordering on A. Define the binary relation $\leq_A$ on A by $a \leq_A b$ if and only if $a <_A b$ or $a = b$. Then $\leq_A$ is a partial ordering on A.

Proof: Suppose that $a \leq_A b$ and $b \leq_A c$. If $a = b$, then we have $a \leq_A c$ by direct substitution. Similarly, if $b = c$, we have $a \leq_A c$ by direct substitution. If $a <_A b$ and $b <_A c$, then $a <_A c$ because $<_A$ is transitive. It follows that $a \leq_A c$. This shows that $\leq_A$ is transitive.

Now, suppose that $a \leq_A b$ and $b \leq_A a$. Assume toward contradiction that $a \neq b$. Then $a <_A b$ and $b <_A a$. Since $<_A$ is antisymmetric, $a = b$, contrary to our assumption. Therefore, $\leq_A$ is antisymmetric.

Finally, by definition, $a \leq_A a$ is true. Therefore, $\leq_A$ is reflexive.

Since $\leq_A$ is transitive, antisymmetric, and reflexive, $\leq_A$ is a partial ordering on A. □

Intervals

A set I of real numbers is called an **interval** if any real number that lies between two numbers in I is also in I. Symbolically, we can write

$$\forall x, y \in I \, \forall z \in \mathbb{R} \, (x < z < y \rightarrow z \in I).$$

The expression above can be read "For all x, y in I and all $z \in \mathbb{R}$, if x is less than z and z is less than y, then z is in I."

A simple way to think of an interval is as a set of real numbers with no "gaps" or "holes." If x and y are in the interval, then everything between x and y is in the interval as well.

Example 3.9:

1. The set $A = \{0, 1\}$ is **not** an interval. A consists of just the two real numbers 0 and 1. There are infinitely many real numbers between 0 and 1. For example, the real number $\frac{1}{2}$ satisfies $0 < \frac{1}{2} < 1$, but $\frac{1}{2} \notin A$.

2. The set $(0, 1) = \{x \in \mathbb{R} \mid 0 < x < 1\}$ is an example of an **open interval**.

3. The set $[0, 1] = \{x \in \mathbb{R} \mid 0 \leq x \leq 1\}$ is an example of a **closed interval**.

4. $\mathbb{R}$ is an interval. This follows trivially from the definition. If we replace I by $\mathbb{R}$, we get $\forall x, y \in \mathbb{R} \, \forall z \in \mathbb{R} \, (x < z < y \rightarrow z \in \mathbb{R})$. In other words, if we start with two real numbers, and take a real number between them, then that number is a real number (which we already said).

When we are thinking of $\mathbb{R}$ as an interval, we sometimes use the notation $(-\infty, \infty)$ and refer to this as **the real line**. The following picture gives the standard geometric interpretation of the real line.

In addition to the real line, there are 8 other types of intervals.

Open Interval: $\quad (a, b) = \{x \in \mathbb{R} \mid a < x < b\}$

Closed Interval: $\quad [a, b] = \{x \in \mathbb{R} \mid a \leq x \leq b\}$

Half-open Intervals: $\quad (a, b] = \{x \in \mathbb{R} \mid a < x \leq b\} \qquad [a, b) = \{x \in \mathbb{R} \mid a \leq x < b\}$

Infinite Open Intervals: $\quad (a, \infty) = \{x \in \mathbb{R} \mid x > a\} \qquad (-\infty, b) = \{x \in \mathbb{R} \mid x < b\}$

Infinite Closed Intervals: $\quad [a, \infty) = \{x \in \mathbb{R} \mid x \geq a\} \qquad (-\infty, b] = \{x \in \mathbb{R} \mid x \leq b\}$

Warning: It is unfortunate that the notation "(a, b)" is used for both an open interval and an ordered pair of real numbers. Most of the time it will be clear which one of these we mean. However, occasionally we may be discussing open intervals and ordered pairs of real numbers at the same time. In these instances, it is important to pay attention to the details of the discussion.

Example 3.10:

1. The half-open interval $(-2, 1] = \{x \in \mathbb{R} \mid -2 < x \leq 1\}$ has the following graph:

2. The infinite open interval $(0, \infty) = \{x \in \mathbb{R} \mid x > 0\}$ has the following graph:

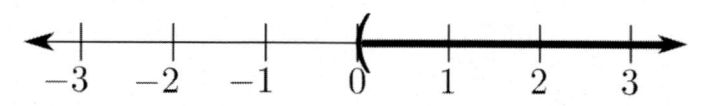

Note: If I is an interval of reals and $A \subseteq \mathbb{R}$, then we will use the notation $I \cap A$ for the corresponding interval as a subset of A. For example, $(-2, 1] \cap \mathbb{Q} = \{x \in \mathbb{Q} \mid -2 < x \leq 1\}$ is a half-open interval of rational numbers. It consists of only the rational numbers between -2 and 1, including 1 and excluding -2. We can visualize this interval of rational numbers the same way we visualize the corresponding interval of real numbers. If we wish, we may label the graph with a $\mathbb{Q}$ for extra clarification as follows:

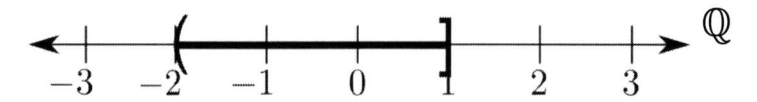

Example 3.11: Let $A = (-2, 1]$ and $B = (0, \infty)$. We have

1. $A \cup B = (-2, \infty)$
2. $A \cap B = (0, 1]$
3. $A \setminus B = (-2, 0]$
4. $B \setminus A = (1, \infty)$
5. $A \triangle B = (-2, 0] \cup (1, \infty)$

Note: If you have trouble seeing how to compute these, it may be helpful to draw the graphs of A and B lined up vertically, and then draw vertical lines through the endpoints of each interval.

The results follow easily by combining these graphs into a single graph using the vertical lines as guides. For example, let's look at $A \cap B$ in detail. We're looking for all numbers that are in both A and B. The two rightmost vertical lines drawn passing through the two graphs above isolate all those numbers nicely. We see that all numbers between 0 and 1 are in the intersection. We should then think about the two endpoints 0 and 1 separately. $0 \notin B$ and therefore, 0 cannot be in the intersection of A and B. On the other hand, $1 \in A$ and $1 \in B$. Therefore, $1 \in A \cap B$. So, we see that $A \cap B = (0, 1]$.

Equivalence Relations

A binary relation R on a set A is an **equivalence relation** if R is reflexive, symmetric, and transitive.

Example 3.12:

1. The most basic equivalence relation on a set A is the relation $R = \{(a, b) \in A^2 \mid a = b\}$ (the **equality relation**). We already saw in part 1 of Example 3.3 that this relation is reflexive, symmetric and transitive.

2. Another obvious equivalence relation on a set A is the set A^2. Since every ordered pair (a, b) is in A^2, reflexivity, symmetry, and transitivity can never fail. We will refer to A^2 as the **trivial equivalence relation** on A.

3. We say that integers a and b have the same **parity** if they are both even or both odd. Define $\equiv_2$ on $\mathbb{Z}$ by $\equiv_2 = \{(a,b) \in \mathbb{Z}^2 \mid a \text{ and } b \text{ have the same parity}\}$. It is easy to see that $\equiv_2$ is reflexive ($a \equiv_2 a$ because every integer has the same parity as itself), $\equiv_2$ is symmetric (if $a \equiv_2 b$, then a has the same parity as b, so b has the same parity as a, and therefore, $b \equiv_2 a$), and $\equiv_2$ is transitive (if $a \equiv_2 b$ and $b \equiv_2 c$, then a, b, and c all have the same parity, and so, $a \equiv_2 c$). Therefore, $\equiv_2$ is an equivalence relation. The relation $\equiv_2$ is called **congruence modulo 2**. If $a \equiv_2 b$, then we say that a is congruent to b modulo 2. Some authors write $a \equiv b \pmod 2$ instead of $a \equiv_2 b$.

4. An integer n is **divisible** by an integer m, written $m|n$, if there is another integer k such that $n = mk$. For example, 18 is divisible by 6 because $18 = 6 \cdot 3$ and -35 is divisible by 5 because $-35 = 5 \cdot (-7)$. Another way to say that a and b have the same parity is to say that $b - a$ is divisible by 2, or equivalently, $2|b - a$. This observation allows us to generalize the notion of having the same parity. For example, $\equiv_3 = \{(a,b) \in \mathbb{Z}^2 \mid 3|b - a\}$ is an equivalence relation, and more generally, for each $n \in \mathbb{Z}^+$, $\equiv_n = \{(a,b) \in \mathbb{Z}^2 \mid n|b - a\}$ is an equivalence relation. I leave the proof that $\equiv_n$ is reflexive, symmetric, and transitive on $\mathbb{Z}$ as an exercise (see Problem 5 below). The relation $\equiv_n$ is called **congruence modulo n**. If $a \equiv_n b$, then we say that a is congruent to b modulo n. Some authors write $a \equiv b \pmod n$ instead of $a \equiv_n b$.

5. Consider the relation $R = \{((a,b),(c,d)) \in (\mathbb{N} \times \mathbb{N})^2 \mid a + d = b + c\}$ defined in part 6 of Example 3.1. Since $a + b = b + a$, we see that $(a,b)R(a,b)$, and therefore, R is reflexive. If $(a,b)R(c,d)$, then $a + d = b + c$. Therefore, $c + b = d + a$, and so, $(c,d)R(a,b)$. Thus, R is symmetric. Finally, suppose that $(a,b)R(c,d)$ and $(c,d)R(e,f)$. Then $a + d = b + c$ and $c + f = d + e$. So, $a + d + c + f = b + c + d + e$. Therefore, $a + f = b + e$, and so, we have $(a,b)R(e,f)$. So, R is transitive. Since R is reflexive, symmetric, and transitive, it follows that R is an equivalence relation.

6. Consider the relation $R = \{((a,b),(c,d)) \in (\mathbb{Z} \times \mathbb{Z}^*)^2 \mid ad = bc\}$ defined in part 7 of Example 3.1. Since $ab = ba$, we see that $(a,b)R(a,b)$, and therefore, R is reflexive. If $(a,b)R(c,d)$, then $ad = bc$. Therefore, $cb = da$, and so, $(c,d)R(a,b)$. Thus, R is symmetric. Finally, suppose that $(a,b)R(c,d)$ and $(c,d)R(e,f)$. Then $ad = bc$ and $cf = de$. So, $adcf = bcde$. Therefore, $cd(af - be) = adcf - bcde = 0$. If $a = 0$, then $bc = 0$, and so, $c = 0$ (because $b \neq 0$). So, $de = 0$, and therefore, $e = 0$ (because $d \neq 0$). So, $af = be$ (because they're both 0). If $a \neq 0$, then $c \neq 0$. Therefore, $af - be = 0$, and so, $af = be$. Since $a = 0$ and $a \neq 0$ both lead to $af = be$, we have $(a,b)R(e,f)$. So, R is transitive. Since R is reflexive, symmetric, and transitive, it follows that R is an equivalence relation.

Let $\sim$ be an equivalence relation on a set S. If $x \in S$, the **equivalence class** of x, written $[x]$, is the set
$$[x] = \{y \in S \mid x \sim y\}.$$

Example 3.13:

1. Let A be a set and let $R = \{(a,b) \in A^2 \mid a = b\}$ be the equality relation. Then for each $a \in A$, $[a] = \{a\}$.

2. Let A be a set and let $R = A^2$ be the trivial equivalence relation. Then for each $a \in A$, $[a] = A$.

3. Consider the equivalence relation $\equiv_2 = \{(a,b) \in \mathbb{Z}^2 \mid a \text{ and } b \text{ have the same parity}\}$ on $\mathbb{Z}$ (see part 3 of Example 3.12). We have $[0]_2 = \{y \in \mathbb{Z} \mid 0 \equiv_2 y\} = 2\mathbb{Z}$ (we add the subscript 2 into the notation $[0]_2$ to emphasize that this is an equivalence class for the equivalence relation $\equiv_2$). Observe that $[2]_2 = [0]_2$, and in fact, if n is any even integer, then $[n]_2 = [0]_2 = 2\mathbb{Z}$. Similarly, if n is any odd integer, then $[n]_2 = [1]_2 = 2\mathbb{Z} + 1$.

4. More generally, consider the equivalence relation $\equiv_n = \{(a,b) \in \mathbb{Z}^2 \mid n \mid b - a\}$ (see part 4 of Example 3.12). Then $[a]_n = n\mathbb{Z} + a$ (again, we add the subscript n into the notation $[a]_n$ to emphasize that this is an equivalence class for the equivalence relation $\equiv_n$). To see this, note that $b \in [a]_n$ if and only if $a \equiv_n b$ if and only if $n \mid b - a$ if and only if there is an integer k such that $b - a = nk$ if and only if there is an integer k such that $b = nk + a$ if and only if $b \in n\mathbb{Z} + a$. Since b was arbitrary, we see that $[a]_n = n\mathbb{Z} + a$.

As a specific example, let's consider $\equiv_3$. Then $[0]_3 = 3\mathbb{Z}$, $[1]_3 = 3\mathbb{Z} + 1$, and $[2]_3 = 3\mathbb{Z} + 2$. Note that these are the only three equivalence classes. For example, we see that $[3]_3 = 3\mathbb{Z} + 3 = \{-3, -2, -1, 0, 3, 6, 9, \ldots\} = 3\mathbb{Z} = [0]_3$.

Theorem 3.14: Let $\sim$ be an equivalence relation on a set S and let $x, y \in S$. The following are equivalent:

1. $x \sim y$.
2. $[x] = [y]$.
3. $x \in [y]$.

This is the first theorem where we want to prove more than two statements equivalent. We will do this with the following chain: $1 \to 2 \to 3 \to 1$. In other words, we will assume statement 1 and use it to prove statement 2. We will then assume statement 2 and use it to prove statement 3. Finally, we will assume statement 3 and use it to prove statement 1.

Proof: ($1 \to 2$) Assume that $x \sim y$ and let $z \in [x]$. Then $x \sim z$. Since $\sim$ is symmetric, $y \sim x$. Since $\sim$ is transitive, $y \sim z$. Therefore, $z \in [y]$. Since $z \in [x]$ was arbitrary, we have shown that $[x] \subseteq [y]$. A symmetric argument shows that $[y] \subseteq [x]$. So, $[x] = [y]$.

($2 \to 3$) Assume $[x] = [y]$. Since $\sim$ is reflexive, $x \sim x$. So, $x \in [x]$. Since $x \in [x]$ and $[x] = [y]$, $x \in [y]$.

($3 \to 2$) Assume that $x \in [y]$. Then $y \sim x$. Since $\sim$ is symmetric, $x \sim y$. □

Partitions

Recall: (1) If X is a nonempty set of sets, we say that X is **pairwise disjoint** if for all $A, B \in X$ with $A \neq B$, A and B are disjoint ($A \cap B = \emptyset$).

(2) If X is a nonempty set of sets, then **union** X is defined by $\bigcup X = \{y \mid \text{there is } Y \in X \text{ with } y \in Y\}$.

A **partition** of a set S is a set of pairwise disjoint nonempty subsets of S whose union is S. Symbolically, X is a partition of S if and only if

$$\forall A \in X (A \neq \emptyset \land A \subseteq S) \land \forall A, B \in X (A \neq B \to A \cap B = \emptyset) \land \bigcup X = S.$$

Example 3.15:

1. Recall that $2\mathbb{Z} = \{2k \mid k \in \mathbb{Z}\}$ is the set of even integers and $2\mathbb{Z} + 1 = \{2k + 1 \mid k \in \mathbb{Z}\}$ is the set of odd integers. $\boldsymbol{X} = \{2\mathbb{Z}, 2\mathbb{Z} + 1\}$ is a partition of $\mathbb{Z}$. We can visualize this partition as follows:
$$\mathbb{Z} = \{\ldots, -4, -2, 0, 2, 4, \ldots\} \cup \{\ldots, -3, -1, 1, 3, 5, \ldots\}$$

2. Recall that $3\mathbb{Z} = \{3k \mid k \in \mathbb{Z}\}, 3\mathbb{Z} + 1 = \{3k + 1 \mid k \in \mathbb{Z}\}$, and $3\mathbb{Z} + 2 = \{3k + 2 \mid k \in \mathbb{Z}\}$. $\boldsymbol{X} = \{3\mathbb{Z}, 3\mathbb{Z} + 1, 3\mathbb{Z} + 2\}$ is a partition of $\mathbb{Z}$. We can visualize this partition as follows:
$$\mathbb{Z} = \{\ldots, -6, -3, 0, 3, 6, \ldots\} \cup \{\ldots, -5, -2, 1, 4, 7, \ldots\} \cup \{\ldots, -4, -1, 2, 5, 8, \ldots\}$$

3. More generally, if $n \in \mathbb{Z}^+$, then $\boldsymbol{X} = \{n\mathbb{Z}, n\mathbb{Z} + 1, n\mathbb{Z} + 2, \ldots, n\mathbb{Z} + (n - 1)\}$ is a partition of $\mathbb{Z}$. We can visualize this partition as follows:
$$\mathbb{Z} = \{\ldots, -2n, -n, 0, n, 2n, \ldots\} \cup \{\ldots, -2n + 1, -n + 1, 1, n + 1, 2n + 1, \ldots\}$$
$$\cup \{\ldots, -2n + 2, -n + 2, 2, n + 2, 2n + 2, \ldots\} \cup \cdots \cup \{\ldots, -n - 1, -1, n - 1, 2n - 1, 3n - 1, \ldots\}$$

4. For each $n \in \mathbb{N}$, let $A_n = \{2n, 2n + 1\}$. Then $\boldsymbol{X} = \{A_n \mid n \in \mathbb{N}\}$ is a partition of $\mathbb{N}$. We can visualize this partition as follows:
$$\mathbb{N} = \{0, 1\} \cup \{2, 3\} \cup \{4, 5\} \cup \{6, 7\} \cup \{8, 9\} \cup \cdots$$

5. For each $n \in \mathbb{Z}$, let $A_n = \{(n, m) \mid m \in \mathbb{Z}\}$. Then $\boldsymbol{X} = \{A_n \mid n \in \mathbb{Z}\}$ is a partition of $\mathbb{Z} \times \mathbb{Z}$. We can visualize this partition as follows:

$$\vdots \qquad \vdots \qquad \vdots$$
$$A_{-2} = \{\ldots, (-2, -3), (-2, -2), (-2, -1), (-2, 0), (-2, 1), (-2, 2), (-2, 3), \ldots\}$$
$$A_{-1} = \{\ldots, (-1, -3), (-1, -2), (-1, -1), (-1, 0), (-1, 1), (-1, 2), (-1, 3), \ldots\}$$
$$A_0 = \{\ldots, (0, -3), (0, -2), (0, -1), (0, 0), (0, 1), (0, 2), (0, 3), \ldots\}$$
$$A_1 = \{\ldots, (1, -3), (1, -2), (1, -1), (1, 0), (1, 1), (1, 2), (1, 3), \ldots\}$$
$$A_2 = \{\ldots, (2, -3), (2, -2), (2, -1), (2, 0), (2, 1), (2, 2), (2, 3), \ldots\}$$
$$\vdots \qquad \vdots \qquad \vdots$$

6. For each $a \in \mathbb{R}$, let $X_a = \{a + bi \mid b \in \mathbb{R}\}$. Then $\boldsymbol{X} = \{X_a \mid a \in \mathbb{R}\}$ is a partition of $\mathbb{C}$. This partition cannot be visualized as easily as the partition given in part 5 above. We will see in Lesson 4 that it is impossible to form a list of the real numbers and therefore, we cannot form a list of the elements of $\boldsymbol{X}$ or lists of the elements of each X_a.

7. The only partition of the one element set $\{a\}$ is $\{\{a\}\}$. The partitions of the two element set $\{a, b\}$ with $a \neq b$ are $\{\{a\}, \{b\}\}$ and $\{\{a, b\}\}$.

We will now explore the relationship between equivalence relations and partitions. Let's begin with an example.

Example 3.16: Consider the equivalence relation $\equiv_2$ from part 3 of Example 3.12, defined by $a \equiv_2 b$ if and only if a and b have the same parity, and the partition $\{2\mathbb{Z}, 2\mathbb{Z} + 1\}$ of $\mathbb{Z}$ from part 1 of Example 3.15. For this partition, we are thinking of $\mathbb{Z}$ as the union of the even and odd integers:
$$\mathbb{Z} = \{\ldots, -4, -2, 0, 2, 4, \ldots\} \cup \{\ldots, -3, -1, 1, 3, 5, \ldots\}$$

Observe that a and b are in the same member of the partition if and only if $a \equiv_2 b$ if and only if $[a]_2 = [b]_2$. If n is any even integer, then we have $[n]_2 = [0]_2 = 2\mathbb{Z}$ and if n is any odd integer, then we have $[n]_2 = [1]_2 = 2\mathbb{Z} + 1$.

Similarly, the equivalence classes of $\equiv_3$ are $[0]_3 = 3\mathbb{Z}$, $[1]_3 = 3\mathbb{Z} + 1$, and $[2]_3 = 3\mathbb{Z} + 2$, and so, we see that $\{[0]_3, [1]_3, [2]_3\} = \{3\mathbb{Z}, 3\mathbb{Z} + 1, 3\mathbb{Z} + 2\}$ is a partition of $\mathbb{Z}$.

More generally, for $n \in \mathbb{Z}^+$, the equivalence classes of $\equiv_n$, namely $[0]_n, [1]_n, [2]_n, \ldots, [n-1]_n$, yield the following partition of $\mathbb{Z}$: $\{[0]_n, [1]_n, [2]_n, \ldots, [n-1]_n\} = \{n\mathbb{Z}, n\mathbb{Z} + 1, n\mathbb{Z} + 2, \ldots, n\mathbb{Z} + (n-1)\}$. A formal proof of this requires the Division Algorithm (see Theorems 8.13 and 9.11). We will name this partition $\mathbb{Z}_n$. So, we have

$$\mathbb{Z}_1 = \{[0]_1\} = \{\mathbb{Z}\}$$
$$\mathbb{Z}_2 = \{[0]_2, [1]_2\} = \{2\mathbb{Z}, 2\mathbb{Z} + 1\}$$
$$\mathbb{Z}_3 = \{[0]_3, [1]_3, [2]_3\} = \{3\mathbb{Z}, 3\mathbb{Z} + 1, 3\mathbb{Z} + 2\}$$
$$\vdots \qquad \vdots \qquad \vdots$$
$$\mathbb{Z}_n = \{[0]_n, [1]_n, [2]_n, \ldots, [n-1]_n\} = \{n\mathbb{Z}, n\mathbb{Z} + 1, n\mathbb{Z} + 2, \ldots, n\mathbb{Z} + (n-1)\}$$

Note: These sets provide more formal descriptions of the sets C_n from parts 9 and 10 of Example 2.8. We will formally put a ring structure on each $\mathbb{Z}_n$ below.

Example 3.17: Recall that the **power set** of A, written $\mathcal{P}(A)$, is the set consisting of all subsets of A.

$$\mathcal{P}(A) = \{X \mid X \subseteq A\}$$

For example, if $A = \{a, b, c\}$, then $\mathcal{P}(A) = \{\emptyset, \{a\}, \{b\}, \{c\}, \{a, b\}, \{a, c\}, \{b, c\}, \{a, b, c\}\}$. We can define a binary relation $\sim$ on $\mathcal{P}(A)$ by $X \sim Y$ if and only if $|X| = |Y|$ (X and Y have the same number of elements). It is easy to see that $\sim$ is an equivalence relation on $\mathcal{P}(A)$. There are four equivalence classes.

$[\emptyset] = \{\emptyset\}$ $\qquad\qquad$ $[\{a\}] = \{\{a\}, \{b\}, \{c\}\}$

$[\{a, b\}] = \{\{a, b\}, \{a, c\}, \{b, c\}\}$ $\qquad\qquad$ $[\{a, b, c\}] = \{\{a, b, c\}\}$

Notes: (1) $\{a\} \sim \{b\} \sim \{c\}$ because each of these sets has one element. It follows that $\{a\}, \{b\}$, and $\{c\}$ are all in the same equivalence class. Above, we chose to use $\{a\}$ as the **representative** for this equivalence class. This is an arbitrary choice. In fact, $[\{a\}] = [\{b\}] = [\{c\}]$.

Similarly, $[\{a, b\}] = [\{a, c\}] = [\{b, c\}]$.

(2) The empty set is the only subset of A with 0 elements. Therefore, the equivalence class of $\emptyset$ contains only itself. Similarly, the equivalence class of $A = \{a, b, c\}$ contains only itself.

(3) Notice that the four equivalence classes are pairwise disjoint, nonempty, and their union is $\mathcal{P}(A)$. In other words, the equivalence classes form a partition of $\mathcal{P}(A)$.

Theorem 3.18: Let P be a partition of a set S. Then there is an equivalence relation $\sim$ on S for which the elements of P are the equivalence classes of $\sim$. Conversely, if $\sim$ is an equivalence relation on a set S, then the equivalence classes of $\sim$ form a partition of S.

You will be asked to prove Theorem 3.18 in Problem 12 below.

Important note: We will sometimes want to define relations or operations on equivalence classes. When we do this, we must be careful that what we are defining is **well-defined**. For example, consider the equivalence relation $\equiv_2$ on $\mathbb{Z}$, and let $X = \{[0]_2, [1]_2\}$ be the set of equivalence classes.

Let's attempt to define a relation on X by $[x]_2 R [y]_2$ if and only if $x < y$. Is $[0]_2 R [1]_2$ true? It looks like it is because $0 < 1$. But this isn't the end of the story. Since $[0]_2 = [2]_2$, if $[0]_2 R [1_2]$, then we must also have $[2]_2 R [1]_2$ (by a direct substitution). But $2 \not< 1$! So, $[2]_2 R [1]_2$ is false. To summarize, $[0]_2 R [1]_2$ should be true and $[2]_2 R [1]_2$ should be false, but $[0]_2 R [1]_2$ and $[2]_2 R [1]_2$ represent the same statement. So, R is **not** a well-defined relation on X.

As another example, let's attempt to define an operation $+: X \times X \to X$ by $[x]_2 + [y]_2 = [x+y]_2$. This **is** a well-defined operation. To see this, suppose that $[x]_2 = [z]_2$ and $[y]_2 = [w]_2$. By Theorem 3.14, $x \equiv_2 z$ and $y \equiv_2 w$. So, there are integers a and b such that $z - x = 2a$ and $w - y = 2b$. Adding these equations gives us $(z - x) + (w - y) = 2a + 2b$, and so, $(z + w) - (x + y) = 2(a + b)$. Therefore, $x + y \equiv_2 z + w$. Again, by Theorem 3.14, $[x + y]_2 = [z + w]_2$.

The Ring of Integers Mod n

Recall that $\mathbb{Z}_n$ is the set of equivalence classes of the equivalence relation $\equiv_n = \{(a, b) \in \mathbb{Z}^2 \mid n \mid b - a\}$. This equivalence relation is called the **congruence relation mod n**. It turns out that $\mathbb{Z}_n$ contains exactly n elements. Specifically, $\mathbb{Z}_n = \{[0]_n, [1]_n, [2]_n, \ldots, [n-1]_n\}$ (see Example 3.16 above for an informal explanation and Theorem 9.11 for a rigorous proof).

We now define a sum and product on $\mathbb{Z}_n$ as follows: For $x, y \in \mathbb{Z}_n$,

$$[x]_n + [y]_n = [x + y]_n \quad \text{and} \quad [x]_n \cdot [y]_n = [xy]_n$$

We call these operations **addition modulo n** and **multiplication modulo n, respectively.**

We need to check that these operations are well-defined. To do this, we assume that $[x]_n = [z]_n$ and $[y]_n = [w]_n$. By Theorem 3.14, $x \equiv_n z$ and $y \equiv_n w$. So, there are integers a and b such that $z - x = na$ and $w - y = nb$. Adding these equations gives us $(z - x) + (w - y) = na + nb$, and so, $(z + w) - (x + y) = n(a + b)$. Since the set of integers is closed under addition, $a + b$ is an integer. So, $x + y \equiv_n z + w$. Again, by Theorem 3.14, $[x + y]_n = [z + w]_n$. Thus, addition is well-defined.

For multiplication, we use the equations $z - x = na$ and $w - y = nb$ (and SACT – see Note 1 after Example 1.16) to get

$$zw - xy = zw - xw + xw - xy = (z - x)w + x(w - y) = (na)w + x(nb) = n(aw + xb).$$

Since the set of integers is closed under addition and multiplication, $aw + xb$ is an integer. So, $xy \equiv_n zw$, and by Theorem 3.14, $[xy]_n = [zw]_n$. It follows that multiplication is well-defined.

Note: At first glance, many students find the application of SACT used above very mysterious. We subtracted and added the expression xw, and then almost magically, everything worked out exactly the way we wanted it to. How did we know to use the expression xw here? Well, we want to use the equation $z - x = na$. So, we need to make the expression $z - x$ appear. The "w" attached to the "z" in the expression zw suggests that we use xw, so that when we subtract, we get $zw - xw = (z - x)w$.

With the definitions of addition and multiplication just given, $(\mathbb{Z}_n, +, \cdot)$ is a commutative ring. The reader will be asked to verify this in Problem 10 below.

Observe that $(\mathbb{Z}_n, +, \cdot)$ behaves just like $(C_n, +, \cdot)$ from Lesson 2 (see parts 9 and 10 of Example 2.8). Here we have $\mathbb{Z}_n = \{[0]_n, [1]_n, [2]_n, \ldots, [n-1]_n\}$ and $C_n = \{0, 1, 2, \ldots, n-1\}$. We have just renamed the elements. Furthermore, the operation of addition modulo n on $\mathbb{Z}_n$ works in exactly the same way that "clock addition" works on C_n. Similarly, multiplication modulo n on $\mathbb{Z}_n$ works in exactly the same way that "clock multiplication" works on C_n. Therefore, any theorems we prove about $(\mathbb{Z}_n, +, \cdot)$ will be true for $(C_n, +, \cdot)$ as well.

It is standard practice to identify the elements of $\mathbb{Z}_n$ with the elements of C_n. So, we will often write $\mathbb{Z}_n = \{0, 1, 2, \ldots, n-1\}$ instead of the more correct $\mathbb{Z}_n = \{[0]_n, [1]_n, [2]_n, \ldots, [n-1]_n\}$. We can then make statements such as the following:

- In $\mathbb{Z}_2$, we have $1 + 1 = 0$.
- In $\mathbb{Z}_3$, we have $1 + 2 = 0$, $2 + 2 = 1$, and $2 \cdot 2 = 1$.
- In $\mathbb{Z}_6$, we have $5 + 2 = 1$, $4 + 5 = 3$, and $2 \cdot 3 = 0$.
- In $\mathbb{Z}_n$, we have $1 + (n - 1) = 0$ and more generally, for $0 < m < n$, $m + (n - m) = 0$ (so we see that m and $n - m$ are additive inverses of each other in $\mathbb{Z}_n$).

It is important to recognize that when we abbreviate an element $[m]_n \in \mathbb{Z}_n$ as m, that the definition has not changed. m is still the equivalence class $n\mathbb{Z} + m = \{nk + m \mid k \in \mathbb{Z}\}$.

For example, when working in $\mathbb{Z}_2$, if we write $1 + 1 = 0$, we mean $(2\mathbb{Z} + 1) + (2\mathbb{Z} + 1) = 2\mathbb{Z}$, or equivalently,

$$\{\ldots, -5, -3, -1, 1, 3, 5, 7, \ldots\} + \{\ldots, -5, -3, -1, 1, 3, 5, 7, \ldots\} = \{\ldots, -6, -4, -2, 0, 2, 4, 6, \ldots\}.$$

Observe that in the ring $(\mathbb{Z}_2, +, \cdot)$, we have $1 + 1 = 0$. We will say that this ring has characteristic 2.

In general, the **characteristic** of a ring is the least positive integer n such that $\underbrace{1 + 1 + \cdots + 1}_{n \text{ times}} = 0$. If $1 + 1 + \cdots + 1$ is never 0, then the characteristic of the ring is 0.

Example 3.19:

1. For each $n \in \mathbb{Z}^+$, $(\mathbb{Z}_n, +, \cdot)$ is a ring with characteristic n.

2. $(\mathbb{Q}, +, \cdot)$, $(\mathbb{R}, +, \cdot)$, and $(\mathbb{C}, +, \cdot)$ are fields with characteristic 0. $(\mathbb{Z}, +, \cdot)$ is a ring with characteristic 0 that is not a field.

3. The characteristic of a field is either 0 or a prime number (Prove this!). Are there rings with prime characteristic that are not fields? (Hint: $(\mathbb{Z}_p \times \mathbb{Z}_p, +, \cdot)$ has characteristic p.)

Problem Set 3

Full solutions to these problems are available for free download here:

www.SATPrepGet800.com/AAFBTDW

LEVEL 1

1. Let $C = (-\infty, 2]$ and $D = (-1, 3]$. Compute each of the following:

    (i) $C \cup D$

    (ii) $C \cap D$

    (iii) $C \setminus D$

    (iv) $D \setminus C$

    (v) $C \triangle D$

2. Find all partitions of the three-element set $\{a, b, c\}$ and the four-element set $\{a, b, c, d\}$.

3. Let $A = \{1, 2, 3, 4\}$ and let $R = \{(1,1), (1,3), (2,2), (2,4), (3,1), (3,3), (4,2), (4,4)\}$. Note that R is an equivalence relation on A. Find the equivalence classes of R.

LEVEL 2

4. Find the domain, range, and field of each of the following relations:

    (i) $R = \{(a, b), (c, d), (e, f), (f, a)\}$

    (ii) $S = \{(2k, 2t + 1) \mid k, t \in \mathbb{Z}\}$

5. Prove that for each $n \in \mathbb{Z}^+$, $\equiv_n$ (see part 4 of Example 3.12) is an equivalence relation on $\mathbb{Z}$.

LEVEL 3

6. Prove that there do not exist sets A and B such that the relation $<$ on $\mathbb{R}$ is equal to $A \times B$.

7. Let X be a set of equivalence relations on a nonempty set A. Prove that $\cap X$ is an equivalence relation on A.

LEVEL 4

8. Let $R = \{(x, y) \in \mathbb{R} \times \mathbb{R} \mid x - y \in \mathbb{Z}\}$. Prove that R is an equivalence relation on $\mathbb{R}$ and describe the equivalence classes of R.

9. Let R be a relation on a set A. Determine if each of the following statements is true or false. If true, provide a proof. If false, provide a counterexample.

    (i) If R is symmetric and transitive on A, then R is reflexive on A.

    (ii) If R is antisymmetric on A, then R is not symmetric on A.

10. Define a sum and product on $\mathbb{Z}_n$ as follows: For $x, y \in \mathbb{Z}_n$, let $[x]_n + [y]_n = [x+y]_n$ and let $[x]_n \cdot [y]_n = [xy]_n$. Prove that $(\mathbb{Z}_n, +, \cdot)$ is a commutative ring.

LEVEL 5

11. For $a, b \in \mathbb{N}$, we will say that a divides b, written $a|b$, if there is a natural number k such that $b = ak$. Notice that $|$ is a binary relation on $\mathbb{N}$. Prove that $(\mathbb{N}, |)$ is a partially ordered set, but it is not a linearly ordered set.

12. Let P be a partition of a set S. Prove that there is an equivalence relation $\sim$ on S for which the elements of P are the equivalence classes of $\sim$. Conversely, if $\sim$ is an equivalence relation on a set S, prove that the equivalence classes of $\sim$ form a partition of S.

CHALLENGE PROBLEMS

13. Let $a, b, c, d \in \mathbb{N}$, let A and B be finite sets with $|A| = a$ and $|B| = b$, let $C = \left(\mathcal{P}(A \times B)\right)^c$, let $D = \left(\mathcal{P}(C^2)\right)^d$, and let Z be the set of relations on $C \times D$. Evaluate $|Z|$.

14. Let R and S be binary relations on a set A. The composition of R and S, written $R \circ S$, is defined as $R \circ S = \{(a, b) \mid \exists c \in A\bigl((a, c) \in R \wedge (c, b) \in S\bigr)\}$. Suppose that R and S are equivalence relations on A. Prove that $R \circ S$ is an equivalence relation if and only if $R \circ S = S \circ R$.

LESSON 4
FUNCTIONS AND EQUINUMEROSITY

Functions

Let A and B be sets. f is a **function** from A to B, written $f: A \to B$, if the following two conditions hold.

1. $f \subseteq A \times B$.
2. For all $a \in A$, there is a unique $b \in B$ such that $(a, b) \in f$.

Notes: (1) A function $f: A \to B$ is a binary relation on $A \cup B$.

(2) Not every binary relation on $A \cup B$ is a function from A to B. See part 2 of Example 4.1 below.

(3) The uniqueness in the second clause in the definition of a function above is equivalent to the statement "if $(a, b), (a, c) \in f$, then $b = c$."

(4) When we know that f is a function, we will abbreviate $(a, b) \in f$ by $f(a) = b$.

If $f: A \to B$, the **domain** of f, written dom f, is the set A, and the **range** of f, written ran f, is the set $\{f(a) \mid a \in A\}$. Observe that ran $f \subseteq B$. The set B is sometimes called the **codomain** of f.

Example 4.1:

1. $f = \{(0, a), (1, a)\}$ is a function with dom $f = \{0, 1\}$ and ran $f = \{a\}$. Instead of $(0, a) \in f$, we will usually write $f(0) = a$. Similarly, instead of $(1, a) \in f$, we will write $f(1) = a$. Here is a visual representation of this function.

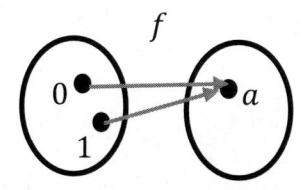

   This function $f: \{0, 1\} \to \{a\}$ is called a **constant function** because the range of f consists of a single element.

   Note also that f is a binary relation on the set $\{0, 1, a\}$.

2. If $a \neq b$, then $g = \{(0, a), (0, b)\}$ is **not** a function because it violates the second clause in the definition of being a function. It is, however, a binary relation on $\{0, a, b\}$.

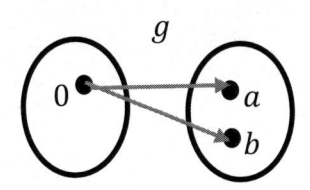

3. $h = \{(a,b) \mid a, b \in \mathbb{R} \land a > 0 \land a^2 + b^2 = 2\}$ is a relation on $\mathbb{R}$ that is **not** a function. $(1, 1)$ and $(1, -1)$ are both elements of h, violating the second clause in the definition of a function. See the figure below on the left. Notice how a vertical line hits the graph twice.

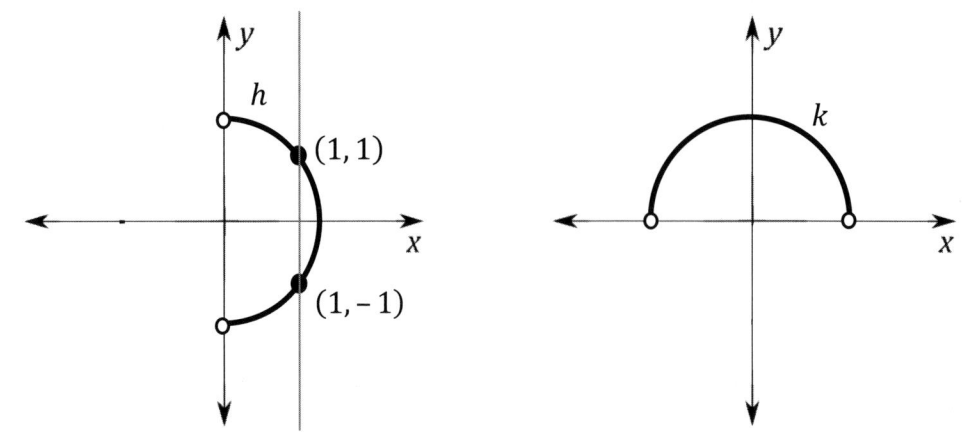

4. $k = \{(a,b) \mid a, b \in \mathbb{R} \land b > 0 \land a^2 + b^2 = 2\}$ **is** a function. See the figure above on the right. To see that the second clause in the definition of a function is satisfied, suppose that (a, b) and (a, c) are both in f. Then $a^2 + b^2 = 2$, $a^2 + c^2 = 2$, and b and c are both positive. It follows that $b^2 = c^2$, and since b and c are both positive, we have $b = c$.

   We have dom $k = (-\sqrt{2}, \sqrt{2})$ and ran $k = (0, \sqrt{2}\,]$. So, $k: (-\sqrt{2}, \sqrt{2}) \to (0, \sqrt{2}\,]$.

   Note that if $a \in \mathbb{R}$, then $a^2 = a \cdot a$. Also, if $a \geq 0$ and $a^2 = b$, then $\sqrt{b} = a$.

5. $b = \{(z, w) \mid z, w \in \mathbb{C} \land w = z - 1\}$ is a function. By Note 4 above, we can describe b using the notation $b(z) = z - 1$.

   If we write $z = x + yi$ and $b(z) = u + vi$, then $b(x + yi) = x + yi - 1 = (x - 1) + yi$ and we see that $u(x, y) = x - 1$ and $v(x, y) = y$.

   b is an example of a complex-valued function. One way to visualize this function is to simply stay in the same plane and to analyze how a typical point moves or how a certain set is transformed. The function b takes the point (x, y) to the point $(x - 1, y)$. That is, each point is shifted one unit to the left. Similarly, if $S \subseteq \mathbb{C}$, then each point of the set S is shifted one unit to the left by the function b. Both these situations are demonstrated in the figure to the right. Observe how the point $(2, 2)$ is shifted to the point $(1, 2)$ and how each point of the rightmost rectangle is shifted one point to the left to form the leftmost rectangle.

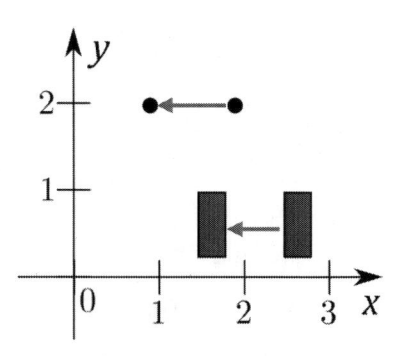

   A second way to visualize this function is to draw two separate planes: an xy-plane and a uv-plane. We can then draw a point or a set in the xy-plane and its image under b in the uv-plane. In the figure below, we do this for the same point and the same rectangle as we did in the previous figure.

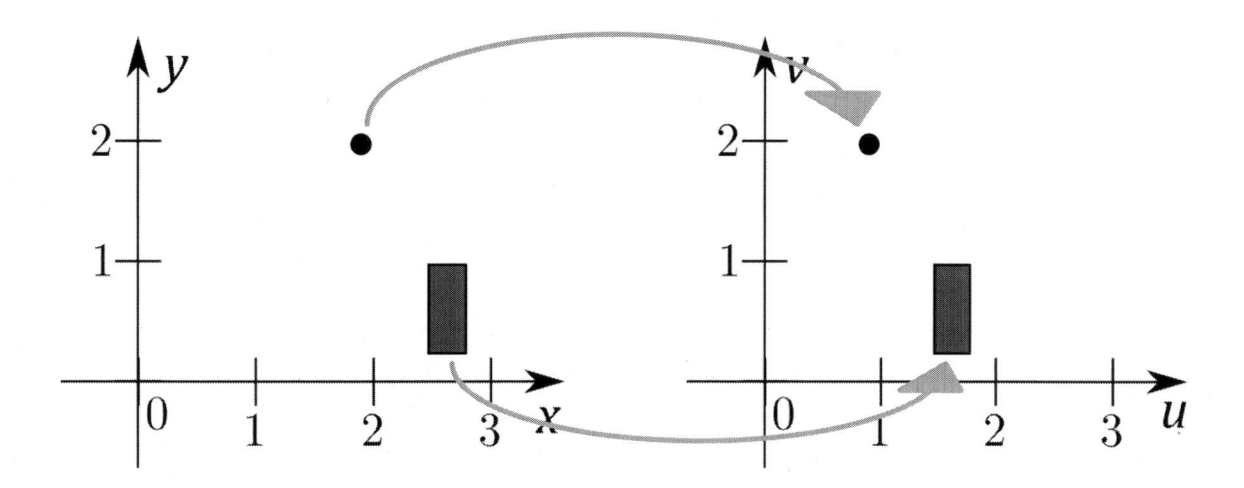

A function with domain $\mathbb{N}$ is called an **infinite sequence**. For example, let $f: \mathbb{N} \to \{0, 1\}$ be defined by $f(n) = \begin{cases} 0 \text{ if } n \text{ is even.} \\ 1 \text{ if } n \text{ is odd.} \end{cases}$ A nice way to visualize an infinite sequence is to list the "outputs" of the sequence in order in parentheses. So, we may write f as $(0, 1, 0, 1, 0, 1, \ldots)$. In general, if A is a nonempty set and $f: \mathbb{N} \to A$ is a sequence, then we can write f as $(f(0), f(1), f(2), \ldots)$.

Similarly, a **finite sequence** is a function with domain $\{0, 1, \ldots, n-1\}$ for some n. For example, the sequence $(0, 2, 4, 6, 8, 10)$ is the function $g: \{0, 1, 2, 3, 4, 5\} \to \mathbb{N}$ defined by $g(k) = 2k$. If the domain of a finite sequence is $\{0, 1, \ldots, n-1\}$, we say that n is the **length** of the sequence.

Observe how a finite sequence with domain $\{0, 1, \ldots, n-1\}$ and range A looks just like an n-tuple in A^n. In fact, it's completely natural to identify a finite sequence of length n with the corresponding n-tuple. So, $(0, 2, 4, 6, 8, 10)$ can be thought of as a 6-tuple from $\mathbb{N}^6$, or as the function $g: \{0, 1, 2, 3, 4, 5\} \to \mathbb{N}$ defined by $g(k) = 2k$.

Informally, we can think of an infinite sequence as an infinite length tuple. As one more example, $(1, 2, 4, 8, 16, 32, \ldots)$ represents the sequence $h: \mathbb{N} \to \mathbb{N}$ defined by $h(n) = 2^n$. If $f: \mathbb{N} \to X$ is defined by $f(n) = x_n$, then we may represent the sequence f using the notation $(x_n)_{n \in \mathbb{N}}$ or simply (x_n). For example, the sequence h can be represented as $(2^n)_{n \in \mathbb{N}}$ or (2^n). x_n is called the nth term of the sequence. The 0th term of the sequence h is 1, the first term of the sequence h is 2, and so on. In general, the nth term of the sequence h is 2^n.

Note: In the study of set theory, we define the natural numbers by letting $0 = \emptyset$, $1 = \{0\}$, $2 = \{0, 1\}$, $3 = \{0, 1, 2\}, \ldots$ and so on. In general, the natural number n is the set of all its predecessors. Specifically, $n = \{0, 1, 2, \ldots, n-1\}$. Using this notation, we can say that a finite sequence of length n is a function $f: n \to A$ for some set A. For example, the function g above has domain 6, so that $g: 6 \to \mathbb{N}$. This notation will be formalized in Lesson 5.

Example 4.2:

1. There is exactly one finite sequence of length 0, namely the empty sequence, $\emptyset$. The empty sequence can be described using function notation as $f: 0 \to X$, where X is any set.

2. The infinite sequence $(0, -1, 2, -3, 4, -5, ...)$ is a function from $\mathbb{N}$ to $\mathbb{Z}$. If we name this function g, then we have that $g: \mathbb{N} \to \mathbb{Z}$ and g is defined by $g(n) = (-1)^n n$. So, the nth term of the sequence g is $(-1)^n n$ and we can represent the sequence as $((-1)^n n)$. This function g is an example of an *integer-valued function* (or *$\mathbb{Z}$-valued function*) because the codomain of g consists of only integers.

3. The infinite sequence $\left(\frac{1}{n+1}\right)$ is a function from $\mathbb{N}$ to $\mathbb{Q}$. The nth term of this sequence is $\frac{1}{n+1}$. If we name this function h, then we have that $h: \mathbb{N} \to \mathbb{Q}$ and h is defined by $h(n) = \frac{1}{n+1}$. This function h is an example of a *rational-valued function* (or *$\mathbb{Q}$-valued function*) because the codomain of h consists of only rational numbers. Since the "outputs" of h take on only positive values, we can "shrink" the codomain of h to $\mathbb{Q}^+$, the set of positive rational numbers. So, we can write $h: \mathbb{N} \to \mathbb{Q}^+$. We can visualize this sequence as follows:

$$\left(1, \frac{1}{2}, \frac{1}{3}, \frac{1}{4}, \frac{1}{5}, ...\right)$$

4. The infinite sequence $\left(n^2 + \sqrt{n}i\right)$ is a function from $\mathbb{N}$ to $\mathbb{C}$. The nth term of this sequence is $n^2 + \sqrt{n}i$. If we name this function k, then we have that $k: \mathbb{N} \to \mathbb{C}$ and k is defined by $k(n) = n^2 + \sqrt{n}i$. This function k is an example of a *complex-valued function* (or *$\mathbb{C}$-valued function*) because the codomain of k consists of only complex numbers. We can visualize this sequence as follows: $\left(0, 1 + i, 4 + \sqrt{2}i, 9 + \sqrt{3}i, 16 + 2i, ...\right)$

Injections, Surjections, and Bijections

A function $f: A \to B$ is **injective** (or **one-to-one**), written $f: A \hookrightarrow B$, if for all $a, b \in A$, if $a \neq b$, then $f(a) \neq f(b)$. In this case, we call f an **injection**.

Notes: (1) The contrapositive of the conditional statement $p \to q$ is the statement $\neg q \to \neg p$. These two statements are logically equivalent (see the analysis after the statement of Theorem 1.28).

(2) The contrapositive of the statement "If $a \neq b$, then $f(a) \neq f(b)$" is "If $f(a) = f(b)$, then $a = b$." So, we can say that a function $f: A \to B$ is injective if for all $a, b \in A$, if $f(a) = f(b)$, then $a = b$.

A function $f: A \to B$ is **surjective** (or **onto B**), written $f: A \mapsto B$, if for all $b \in B$, there is an $a \in A$ such that $f(a) = b$. In this case, we call f a **surjection**.

A function $f: A \to B$ is **bijective**, written $f: A \cong B$ if f is both injective and surjective. In this case, we call f a **bijection**.

Example 4.3:

1. $f = \{(0, a), (1, a)\}$ from part 1 of Example 4.1 is **not** an injective function because $f(0) = a$, $f(1) = a$, and $0 \neq 1$. If we think of f as $f: \{0, 1\} \to \{a\}$, then f is surjective. However, if we think of f as $f: \{0, 1\} \to \{a, b\}$, then f is **not** surjective. So, surjectivity depends upon the codomain of the function.

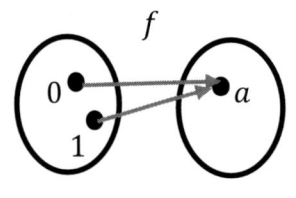

2. $k = \{(a,b) \mid a, b \in \mathbb{R} \wedge b > 0 \wedge a^2 + b^2 = 2\}$ from part 4 of Example 4.1 is **not** an injective function. For example, $(1,1) \in k$ because $1^2 + 1^2 = 1 + 1 = 2$ and $(-1,1) \in k$ because $(-1)^2 + 1^2 = 1 + 1 = 2$. Notice how a horizontal line hits the graph twice. If we think of k as a function from $(-\sqrt{2}, \sqrt{2})$ to $\mathbb{R}^+$, then k is **not** surjective. For example, $2 \notin \operatorname{ran} k$ because for any $a \in \mathbb{R}$, $a^2 + 2^2 = a^2 + 4 \geq 4$, and so, $a^2 + 2^2$ cannot be equal to 2. However, if instead we consider k as a function with codomain $(0, \sqrt{2}]$, that is $k: (-\sqrt{2}, \sqrt{2}) \to (0, \sqrt{2}]$, then k **is** surjective. Indeed, if $0 < b \leq \sqrt{2}$, then $0 < b^2 \leq 2$, and so, $a^2 = 2 - b^2 \geq 0$. Therefore, $a = \sqrt{2 - b^2}$ is a real number such that $k(a) = b$.

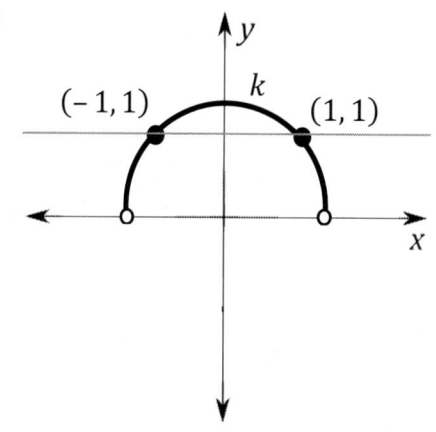

3. Define $g: \mathbb{R} \to \mathbb{R}$ by $g(x) = 7x - 3$. Then g is injective because if $g(a) = g(b)$, we then have $7a - 3 = 7b - 3$. Adding 3 to each side of this equation, we get $7a = 7b$, and then multiplying each side of this last equation by $\frac{1}{7}$, we get $a = b$ (see Lesson 5 for justification of these cancellation rules). Also, g is surjective because if $b \in \mathbb{R}$, then $\frac{b+3}{7} \in \mathbb{R}$ and

$$g\left(\frac{b+3}{7}\right) = 7\left(\frac{b+3}{7}\right) - 3 = (b+3) - 3 = b + (3-3) = b + 0 = b$$

Therefore, g is bijective. See the image to the right for a visual representation of $\mathbb{R}^2$ and the graph of the function g.

Notice that any vertical line will hit the graph of g exactly once because g is a function with domain $\mathbb{R}$. Also, any horizontal line will hit the graph exactly once because g is bijective. Injectivity ensures that each horizontal line hits the graph *at most* once and surjectivity ensures that each horizontal line hits the graph *at least* once.

This function g is an example of a *real-valued function* (or $\mathbb{R}$-*valued function*) because the codomain of g consists of only real numbers.

Inverse Functions

If $f: A \to B$ is bijective, we define $f^{-1}: B \to A$, the **inverse** of f, by $f^{-1} = \{(b, a) \mid (a, b) \in f\}$. In other words, for each $b \in B$, $f^{-1}(b) = $ "the unique $a \in A$ such that $f(a) = b$."

Notes: (1) Let $f: A \to B$ be bijective. Since f is surjective, for each $b \in B$, there is an $a \in A$ such that $f(a) = b$. Since f is injective, there is only one such value of a.

(2) The inverse of a bijective function is also bijective.

Example 4.4:

1. Define $f:\{0,1\} \to \{a,b\}$ by $f = \{(0,a),(1,b)\}$. Then f is a bijection and $f^{-1}:\{a,b\} \to \{0,1\}$ is defined by $f^{-1} = \{(a,0),(b,1)\}$. Observe that f^{-1} is also a bijection.

2. Let $2\mathbb{N} = \{0,2,4,6,8,\ldots\}$ be the set of even natural numbers and let $2\mathbb{N}+1 = \{1,3,5,7,9\ldots\}$ be the set of odd natural numbers. The function $f:2\mathbb{N} \to 2\mathbb{N}+1$ defined by $f(n) = n+1$ is a bijection with inverse $f^{-1}:2\mathbb{N}+1 \to 2\mathbb{N}$ defined by $f(n) = n-1$.

3. If X and Y are sets, we define ${}^X Y$ to be the set of functions from X to Y. Symbolically, we have
$$^X Y = \{f \mid f:X \to Y\}.$$

   For example, if $A = \{a,b\}$ and $B = \{0,1\}$, then ${}^A B$ has 4 elements (each element is a function from A to B). The elements are $f_1 = \{(a,0),(b,0)\}, f_2 = \{(a,0),(b,1)\}, f_3 = \{(a,1),(b,0)\}$, and $f_4 = \{(a,1),(b,1)\}$. Here is a visual representation of these four functions.

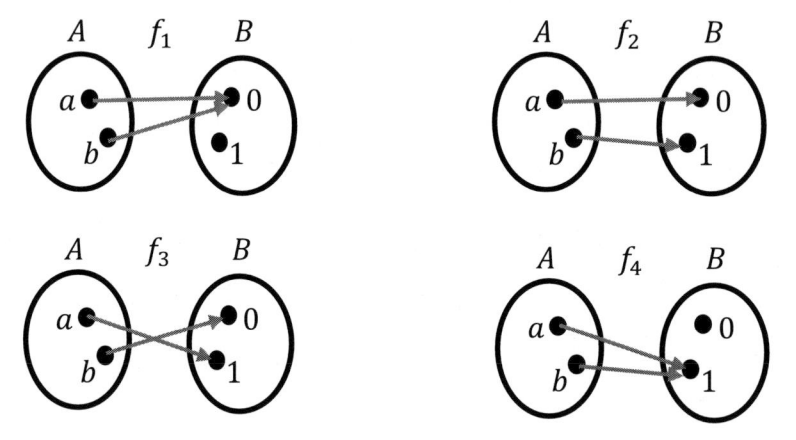

   Define $F: {}^A B \to \mathcal{P}(A)$ by $F(f) = \{x \in A \mid f(x) = 1\}$.

   So, $F(f_1) = \emptyset, F(f_2) = \{b\}, F(f_3) = \{a\}$, and $F(f_4) = \{a,b\}$.

   Since $\mathcal{P}(A) = \{\emptyset, \{a\}, \{b\}, \{a,b\}\}$, we see that F is a bijection from ${}^A B$ to $\mathcal{P}(A)$.

   The inverse of F is the function $F^{-1}:\mathcal{P}(A) \to {}^A B$ defined by $F^{-1}(C)(x) = \begin{cases} 0 & \text{if } x \notin C. \\ 1 & \text{if } x \in C. \end{cases}$

   So, we see that $F^{-1}(\emptyset) = f_1, F^{-1}(\{b\}) = f_2, F^{-1}(\{a\}) = f_3$, and $F^{-1}(\{a,b\}) = f_4$.

4. For $A \neq \emptyset$ and $B = \{0,1\}$, the function $F: {}^A B \to \mathcal{P}(A)$ defined by $F(f) = \{x \in A \mid f(x) = 1\}$ is always a bijection.

   To see that F is injective, let $f,g \in {}^A B$ with $f \neq g$. Since f and g are different, there is some $a \in A$ such that either $f(a) = 0, g(a) = 1$ or $f(a) = 1, g(a) = 0$. **Without loss of generality**, (see Note 1 below) assume that $f(a) = 0$ and $g(a) = 1$. Since $f(a) = 0$, $a \notin F(f)$. Since $g(a) = 1, a \in F(g)$. So, $F(f) \neq F(g)$. Since $f \neq g$ implies $F(f) \neq F(g)$, F is injective.

   To see that F is surjective, let $C \in \mathcal{P}(A)$, so that $C \subseteq A$. Define $f \in {}^A B$ by $f(x) = \begin{cases} 0 & \text{if } x \notin C. \\ 1 & \text{if } x \in C. \end{cases}$ Then $x \in F(f)$ if and only if $f(x) = 1$ if and only if $x \in C$. So, $F(f) = C$. Since $C \in \mathcal{P}(A)$ was arbitrary, F is surjective.

As in 3, the inverse of F is the function $F^{-1}: \mathcal{P}(A) \to {}^A B$ defined by $F^{-1}(C)(x) = \begin{cases} 0 & \text{if } x \notin C. \\ 1 & \text{if } x \in C. \end{cases}$

Notes: (1) In part 4 of Example 4.4, we used the expression "Without loss of generality." This expression can be used when an argument can be split up into 2 or more cases, and the proof of each of the cases is nearly identical.

In the example above, the two cases are (i) $f(a) = 0, g(a) = 1$ and (ii) $f(a) = 1, g(a) = 0$. The argument for case (ii) is the same as the argument for case (i), essentially word for word—only the roles of f and g are interchanged.

(2) Using the definition $n = \{0, 1, 2, \ldots, n-1\}$, we have just shown that for any nonempty set A, there is a bijection $f: {}^A 2 \to \mathcal{P}(A)$.

Composite Functions

Given functions $f: A \to B$ and $g: B \to C$, the **composite** (or **composition**) of f and g, written $g \circ f: A \to C$, is defined by $(g \circ f)(a) = g(f(a))$ for all $a \in A$. Symbolically, we have

$$g \circ f = \{(a,c) \in A \times C \mid \text{There is a } b \in B \text{ such that } (a,b) \in f \text{ and } (b,c) \in g\}.$$

We can visualize the composition of two functions f and g as follows.

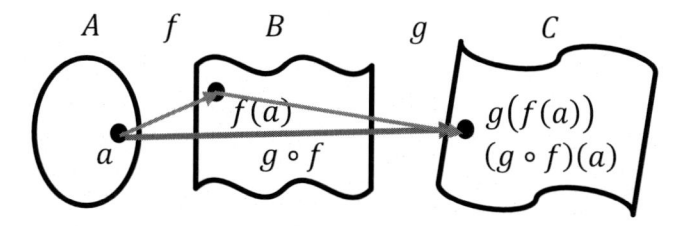

In the picture above, sets A, B, and C are drawn as different shapes simply to emphasize that they can all be different sets. Starting with an arbitrary element $a \in A$, we have an arrow showing a being mapped by f to $f(a) \in B$ and another arrow showing $f(a)$ being mapped by g to $g(f(a)) \in C$. There is also an arrow going directly from $a \in A$ to $(g \circ f)(a) = g(f(a))$ in C. However, note that the only way we know how to get from a to $(g \circ f)(a)$ is to first travel from a to $f(a)$, and then to travel from $f(a)$ to $g(f(a))$.

Example 4.5:

1. Define $f: \mathbb{R} \to \mathbb{R}$ by $f(x) = x + 3$ and $g: \mathbb{R} \to \mathbb{R}$ by $g(x) = (x-1)^2$ Then $g \circ f: \mathbb{R} \to \mathbb{R}$ is defined by $(g \circ f)(x) = g(f(x)) = g(x+3) = ((x+3) - 1)^2 = (x+2)^2$.

2. Define $f: \mathbb{Z} \to \mathbb{Q}$ by $f(n) = \frac{n}{2}$ and define $g: \mathbb{Q} \to \{0,1\}$ by $g(n) = \begin{cases} 0 & \text{if } n \in \mathbb{Z}. \\ 1 & \text{if } n \in \mathbb{Q} \setminus \mathbb{Z}. \end{cases}$ We will show that $g \circ f: \mathbb{Z} \to \{0,1\}$ is defined by $(g \circ f)(n) = \begin{cases} 0 & \text{if } n \text{ is even.} \\ 1 & \text{if } n \text{ is odd.} \end{cases}$

To see this, observe that if $n \in \mathbb{Z}$ is even, then there is an integer k such that $n = 2k$. It follows that $(g \circ f)(n) = g(f(n)) = g\left(\frac{n}{2}\right) = g\left(\frac{2k}{2}\right) = g(k) = 0$ because $k \in \mathbb{Z}$. If n is odd, then there is an integer k such that $n = 2k + 1$. So, $(g \circ f)(n) = g(f(n)) = g\left(\frac{n}{2}\right) = g\left(\frac{2k+1}{2}\right) = 1$ because $\frac{2k+1}{2} \in \mathbb{Q} \setminus \mathbb{Z}$. To see that $\frac{2k+1}{2} \in \mathbb{Q}$, simply observe that $2k + 1 \in \mathbb{Z}$, $2 \in \mathbb{Z}$, and $2 \neq 0$. To see that $\frac{2k+1}{2} \notin \mathbb{Z}$, first note that, $\frac{2k+1}{2} = \frac{2k}{2} + \frac{1}{2} = k + \frac{1}{2}$ (Check this!). Now, let $m = k + \frac{1}{2}$. If $m \in \mathbb{Z}$, then we would have $m - k \in \mathbb{Z}$ because when we subtract one integer from another, we always get an integer. It would then follow that $\frac{1}{2} \in \mathbb{Z}$, which we know it is not. Since assuming that $\frac{2k+1}{2} \in \mathbb{Z}$ would lead to the false statement $\frac{1}{2} \in \mathbb{Z}$, we know that the statement $\frac{2k+1}{2} \in \mathbb{Z}$ must be false.

Note: In part 2 of Example 4.5 above, to show that $m \notin \mathbb{Z}$, we began with the assumption that $m \in \mathbb{Z}$, and then used a logically valid argument to derive a false statement. This is known as a proof by contradiction. See the proof of Theorem 4.16 below for a formal proof by contradiction and the notes following the proof for a detailed explanation as to how it works. Theorem 5.6 provides another example.

It will be important to know that when we take the composition of bijective functions, we always get a bijective function. We will prove this in two steps. We will first show that the composition of injective functions is injective. We will then show that the composition of surjective functions is surjective.

Theorem 4.6: If $f: A \hookrightarrow B$ and $g: B \hookrightarrow C$, then $g \circ f: A \hookrightarrow C$.

Note: We are given that f and g are injections, and we want to show that $g \circ f$ is an injection. We can show this directly using the definition of injectivity, or we can use the contrapositive of the definition of injectivity. Let's do it both ways.

Direct proof of Theorem 4.6: Suppose that $f: A \hookrightarrow B$ and $g: B \hookrightarrow C$, and let $x, y \in A$ with $x \neq y$. Since f is injective, $f(x) \neq f(y)$. Since g is injective, $g(f(x)) \neq g(f(y))$. So, $(g \circ f)(x) \neq (g \circ f)(y)$. Since $x, y \in A$ were arbitrary, $g \circ f: A \hookrightarrow C$. □

Contrapositive proof of Theorem 4.6: Suppose that $f: A \hookrightarrow B$ and $g: B \hookrightarrow C$, let $x, y \in A$ and suppose that $(g \circ f)(x) = (g \circ f)(y)$. Then $g(f(x)) = g(f(y))$. Since g is injective, $f(x) = f(y)$. Since f is injective, $x = y$. Since $x, y \in A$ were arbitrary, $g \circ f: A \hookrightarrow C$. □

Theorem 4.7: If $f: A \mapsto B$ and $g: B \mapsto C$, then $g \circ f: A \mapsto C$.

Proof: Suppose that $f: A \mapsto B$ and $g: B \mapsto C$, and let $c \in C$. Since g surjective, there is $b \in B$ with $g(b) = c$. Since f is surjective, there is $a \in A$ with $f(a) = b$. So, $(g \circ f)(a) = g(f(a)) = g(b) = c$. Since $c \in C$ was arbitrary, $g \circ f$ is surjective. □

Corollary 4.8: If $f: A \cong B$ and $g: B \cong C$, then $g \circ f: A \cong C$.

Proof: Suppose that $f: A \cong B$ and $g: B \cong C$. Then f and g are injective. By Theorem 4.6, $g \circ f$ is injective. Also, f and g are surjective. By Theorem 4.7, $g \circ f$ is surjective. Since $g \circ f$ is both injective and surjective, $g \circ f$ is bijective. □

Note: A **corollary** is a theorem that follows easily from a theorem or theorems that have already been proved.

Identity Functions

If A is any set, then we define the **identity function** on A, written $i_A: A \to A$ by $i_A(a) = a$ for all $a \in A$. Note that the identity function on A is a bijection from A to itself.

Theorem 4.9: If $f: A \cong B$, then $f^{-1} \circ f = i_A$ and $f \circ f^{-1} = i_B$.

Proof: Let $a \in A$ with $f(a) = b$. Then $f^{-1}(b) = a$, and so, $(f^{-1} \circ f)(a) = f^{-1}(f(a)) = f^{-1}(b) = a$. Since $i_A(a) = a$, we see that $(f^{-1} \circ f)(a) = i_A(a)$. Since $a \in A$ was arbitrary, $f^{-1} \circ f = i_A$.

Now, let $b \in B$. Since $f: A \cong B$, there is a unique $a \in A$ with $f(a) = b$. Equivalently, $f^{-1}(b) = a$. We have $(f \circ f^{-1})(b) = f(f^{-1}(b)) = f(a) = b$. Since $i_B(b) = b$, we see that $(f \circ f^{-1})(b) = i_B(b)$. Since $b \in B$ was arbitrary, $f \circ f^{-1} = i_B$. □

Images and and Inverse Images

If $f: X \to Y$ and $A \subseteq X$, then the **image of A under f** is the set $f[A] = \{f(x) \mid x \in A\}$. Similarly, if $B \subseteq Y$, then the **inverse image of B under f** is the set $f^{-1}[B] = \{x \in X \mid f(x) \in B\}$.

Example 4.10:

1. Let $f: \{a, b, c, d\} \to \{0, 1, 2\}$ be defined by $f = \{(a, 0), (b, 0), (c, 1), (d, 2)\}$. Let $A = \{a\}$, $B = \{a, b\}$, $C = \{a, c\}$, and $D = \{b, c, d\}$. Then $f[A] = \{0\}$, $f[B] = \{0\}$, $f[C] = \{0, 1\}$, and $f[D] = \{0, 1, 2\}$. Now, let $X = \{0\}$, $Y = \{0, 1\}$, $Z = \{0, 2\}$, and $W = \{0, 1, 2\}$. Then we have $f^{-1}[X] = \{a, b\}$, $f^{-1}[Y] = \{a, b, c\}$, $f^{-1}[Z] = \{a, b, d\}$, and $f^{-1}[W] = \{a, b, c, d\}$.

2. Define $f: \mathbb{R} \to \mathbb{R}$ by $f(x) = x^4$. Then we have $f[\mathbb{R}] = [0, \infty)$, $f(\{-2, 0, 3\}) = \{0, 16, 81\}$, $f[(-3, 2]] = [0, 81)$, $f^{-1}[\mathbb{R}] = \mathbb{R}$, $f^{-1}[\{16\}] = \{-2, 2\}$, $f^{-1}[[0, \infty)] = \mathbb{R}$, $f^{-1}[(-\infty, 0)] = \emptyset$, and $f^{-1}[(0, \infty)] = (-\infty, 0) \cup (0, \infty) = \mathbb{R} \setminus \{0\}$.

Groups and Monoids of Functions

Let A be a nonempty set. A bijection from A to itself is called a **permutation** of A. We define $S(A)$ to be the set of permutations of A. Symbolically, $S(A) = \{f \mid f: A \cong A\}$.

Theorem 4.11: $(S(A), \circ)$ is a group, where $\circ$ is the operation of composition.

$(S(A), \circ)$ is called the **Symmetric Group on A**.

Proof of Theorem 4.11: By Corollary 4.8, $S(A)$ is closed under $\circ$.

To see that $\circ$ is associative, let $f, g, h \in S(A)$ and let $a \in A$. Then

$$((f \circ g) \circ h)(a) = (f \circ g)(h(a)) = f(g(h(a))) = f((g \circ h)(a)) = (f \circ (g \circ h))(a).$$

Since $a \in A$ was arbitrary, $(f \circ g) \circ h = f \circ (g \circ h)$. So, $\circ$ is associative in $S(A)$.

The identity function on A, $i_A: A \to A$ (defined by $i_A(a) = a$ for all $a \in A$) is a permutation. We have
$$(i_A \circ f)(a) = i_A(f(a)) = f(a) = f(i_A(a)) = (f \circ i_A)(a).$$

Since $a \in A$ was arbitrary, we see that $i_A \circ f = f$ and $f \circ i_A = f$.

For any permutation f on A, the inverse function f^{-1} satisfies $f^{-1} \circ f = f \circ f^{-1} = i_A$ for each $f \in S(A)$ (by Theorem 4.9).

So, we have verified that $(S(A), \circ)$ is a group. □

If $A = \{1, 2, \ldots, n\}$, then we define S_n to be $(S(A), \circ)$.

Example 4.12: Consider the group $S_3 = (S(\{1, 2, 3\}), \circ)$. We can visualize each element of S_3 with a **cycle diagram**. Here are the six elements of S_3 visualized this way.

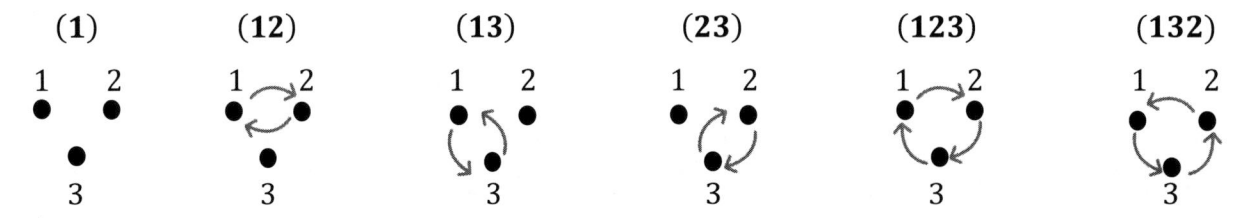

| (1) | (12) | (13) | (23) | (123) | (132) |

The first diagram represents the identity permutation $\{(1, 1), (2, 2), (3, 3)\}$, where each element is being mapped to itself. Technically, we should have an arrow from each point looping back to itself. However, to avoid unnecessary clutter, we leave out arrows for elements that are mapping to themselves. In **cycle notation**, we have $(1)(2)(3)$, which we abbreviate as (1).

The second diagram represents the permutation $\{(1, 2), (2, 1), (3, 3)\}$, where 1 is being mapped to 2, 2 is being mapped to 1, and 3 is being mapped to itself. Again, we leave out the arrow from 3 to itself to avoid clutter, and we just put in the arrows from 1 to 2 and from 2 to 1. In cycle notation, we have $(12)(3)$, which we abbreviate as (12). In this notation, (12) represents a **cycle**. The cycle moves from left to right and the last element in the cycle connects to the first. So, 1 maps to 2 and 2 maps to 1. Any element that does not appear in the cycle notation maps to itself.

As one more example, in the cycle (123), 1 maps to 2, 2 maps to 3, and 3 maps to 1.

To compose two permutations in cycle notation, we write the one we want to apply first on the right (just as we do in function notation). For example, let's simplify $(12)(13)$. Starting with 1, we see that the rightmost cycle sends 1 to 3. The leftmost cycle sends 3 to itself, and so, the composition sends 1 to 3. Let's do 2 next. The rightmost cycle sends 2 to itself, and then the leftmost cycle sends 2 to 1. So, the composition sends 2 to 1. And finally, let's look at 3. The rightmost cycle sends 3 to 1, and then the leftmost cycle sends 1 to 2. So, the composition sends 3 to 2. It follows that $(12)(13) = (132)$.

Observe that the group $(S_3, \circ)$ is not commutative. For example, $(12)(13) = (132)$, whereas $(13)(12) = (123)$.

Now, what if we consider the set of all functions from a set to itself. In other words, let's let A be a nonempty set and $^A A = \{f \mid f: A \to A\}$. Is $(^A A, \circ)$ a group?

Example 4.13: Let $A = \{a, b\}$. Then $^AA = \{f_1, f_2, f_3, f_4\}$, where f_1, f_2, f_3, and f_4 are defined as follows: $f_1 = \{(a,a), (b,b)\}, f_2 = \{(a,a), (b,a)\}, f_3 = \{(a,b), (b,b)\}$, and $f_4 = \{(a,b), (b,a)\}$. $(^AA, \circ)$ has the following multiplication table:

$\circ$	f_1	f_2	f_3	f_4
f_1	f_1	f_2	f_3	f_4
f_2	f_2	f_2	f_2	f_2
f_3	f_3	f_3	f_3	f_3
f_4	f_4	f_3	f_2	f_1

As a sample computation, observe that

$$(f_3 \circ f_2)(a) = f_3(f_2(a)) = f_3(a) = b \quad \text{and} \quad (f_3 \circ f_2)(b) = f_3(f_2(b)) = f_3(a) = b.$$

So, $f_3 \circ f_2 = f_3$. I leave it to the reader to verify that the rest of the multiplication table is correct. Observe from the table that $f_1 = i_A$ is an identity for $(^AA, \circ)$ and f_1 and f_4 are each their own inverse. However, each of f_2 and f_3 have no inverse. It follows that $(^AA, \circ)$ is **not** a group. Also, notice that $f_4 \circ f_3 = f_2$, whereas $f_3 \circ f_4 = f_3$, showing that $\circ$ is **not** commutative in ^AA.

Although $(^AA, \circ)$ is not usually a group, it is a monoid as we now prove.

Theorem 4.14: $(^AA, \circ)$ is a monoid, where $\circ$ is the operation of composition.

Proof: If $f: A \to A$ and $g: A \to A$, then clearly $g \circ f: A \to A$. So, ^AA is closed under $\circ$.

The arguments that $\circ$ is associative in ^AA and that $i_A: A \to A$ (defined by $i_A(a) = a$ for all $a \in A$) is an identity of ^AA are identical to the arguments given in the proof of Theorem 4.11.

So, we have verified that $(^AA, \circ)$ is a monoid. □

Equinumerosity

We say that two sets A and B are **equinumerous**, written $A \sim B$ if there is a bijection $f: A \cong B$. In this case, we may also say that A and B have the same **cardinality**, and we can write $|A| = |B|$.

It is easy to see that $\sim$ **is an equivalence relation**. For any set A, the identity function $i_A: A \to A$ is a bijection, showing that $\sim$ is reflexive. For sets A and B, if $f: A \cong B$, then $f^{-1}: B \cong A$, showing that $\sim$ is symmetric. For sets A, B, and C, if $f: A \cong B$ and $g: B \cong C$, then $g \circ f: A \cong C$ by Corollary 4.8, showing that $\sim$ is transitive.

Example 4.15:

1. Let $A = \{\text{anteater, elephant, giraffe}\}$ and $B = \{\text{apple, banana, orange}\}$. Then $A \sim B$. We can define a bijection $f: A \cong B$ by $f(\text{anteater}) = \text{apple}$, $f(\text{elephant}) = \text{banana}$, and $f(\text{giraffe}) = \text{orange}$. This is not the only bijection from A to B (there are 6 distinct bijections from A to B), but we need only find one (or prove one exists) to show that the sets are equinumerous.

2. At this point it should be easy to see that two finite sets are equinumerous if and only if they have the same number of elements. It should also be easy to see that a finite set can never be equinumerous with an infinite set.

3. Let $\mathbb{N} = \{0, 1, 2, 3, 4 \ldots\}$ be the set of natural numbers and $2\mathbb{N} = \{0, 2, 4, 6, 8 \ldots\}$ the set of even natural numbers. Then $\mathbb{N} \sim 2\mathbb{N}$. We can actually see a bijection between these two sets just by looking at the sets themselves.

$$0 \quad 1 \quad 2 \quad 3 \quad 4 \quad 5 \quad 6 \ldots$$
$$0 \quad 2 \quad 4 \quad 6 \quad 8 \quad 10 \quad 12 \ldots$$

The function $f: \mathbb{N} \to 2\mathbb{N}$ defined by $f(n) = 2n$ is an explicit bijection. To see that f maps $\mathbb{N}$ into $2\mathbb{N}$, just observe that if $n \in \mathbb{N}$, then $2n \in 2\mathbb{N}$ by the definition of an even integer ($a \in \mathbb{N}$ is even if there is $b \in \mathbb{N}$ with $a = 2b$). f is injective because if $f(n) = f(m)$, then $2n = 2m$, and so, $n = m$ (see part (viii) of Problem 4 from Problem Set 5 for formal details). Finally, f is surjective because if $n \in 2\mathbb{N}$, then there is $k \in \mathbb{N}$ such that $n = 2k$. So, $f(k) = 2k = n$.

4. $\mathbb{N} \sim \mathbb{Z}$ via the bijection $f: \mathbb{N} \cong \mathbb{Z}$ defined by $f(n) = \begin{cases} \frac{n}{2} & \text{if } n \text{ is even.} \\ -\frac{n+1}{2} & \text{if } n \text{ is odd.} \end{cases}$

You will be asked to show that f is a bijection in Problem 11 below. Let's look at this correspondence visually:

$$0 \quad 1 \quad 2 \quad 3 \quad 4 \quad 5 \quad 6 \ldots$$
$$0 \quad -1 \quad 1 \quad -2 \quad 2 \quad -3 \quad 3 \ldots$$

Many students get confused here because they are under the misconception that the integers should be written "in order." However, when checking to see if two sets are equinumerous, we **do not** include any other structure. In other words, we are just trying to "pair up" elements—it does not matter how we do so.

5. For A any nonempty set, $^A 2 \sim \mathcal{P}(A)$. We showed this in part 4 of Example 4.4.

6. $[0, 1] \sim [0, 5]$ via the bijection $f: [0, 1] \to [0, 5]$ defined by $f(x) = 5x$. To see that f maps into $[0, 5]$, note that if $0 \leq x \leq 1$, then $0 \leq 5x \leq 5$. To see that f is injective, note that if $x \neq y$, then $5x \neq 5y$. Finally, to see that f is surjective, let $y \in [0, 5]$. Then $0 \leq y \leq 5$ and so, it follows that $0 \leq \frac{y}{5} \leq 1$ and $f\left(\frac{y}{5}\right) = 5 \cdot \frac{y}{5} = y$.

Countable and Uncountable Sets

We say that a set is **countable** if it is equinumerous with a subset of $\mathbb{N}$. It's easy to visualize a countable set because a bijection from a subset of $\mathbb{N}$ to a set A generates a list. For example, the set $2\mathbb{N}$ can be listed as $0, 2, 4, 6, \ldots$ and the set $\mathbb{Z}$ can be listed as $0, -1, 1, -2, 2, \ldots$ (see Example 4.15 above).

There are two kinds of countable sets: finite sets and **denumerable** sets. We say that a set is denumerable if it is countably infinite.

At this point, you may be asking yourself if all infinite sets are denumerable. If this were the case, then we would simply have finite sets and infinite sets, and that would be the end of it. However, there are in fact infinite sets that are **not** denumerable. An infinite set that is not denumerable is **uncountable**.

Theorem 4.16 (Cantor's Theorem): If A is any set, then A is **not** equinumerous with $\mathcal{P}(A)$.

Analysis: How can we prove that A is not equinumerous with $\mathcal{P}(A)$? Well, we need to show that there **does not** exist a bijection from A to $\mathcal{P}(A)$. Recall that a bijection is a function that is both an injection and a surjection. So, we will attempt to show that there do not exist any surjections from A to $\mathcal{P}(A)$. To do this, we will take an arbitrary function $f: A \to \mathcal{P}(A)$, and then argue that f is not surjective. We will show that ran $f \neq \mathcal{P}(A)$ by finding a set $B \in \mathcal{P}(A) \setminus \text{ran } f$. In words, we will find a subset of A that is **not** in the range of f.

Let's begin by looking at $\mathbb{N}$, the set of natural numbers. Given a specific function $f: \mathbb{N} \to \mathcal{P}(\mathbb{N})$, it's not too hard to come up with a set $B \in \mathcal{P}(\mathbb{N}) \setminus \text{ran } f$. Let's choose a specific such f and use this example to try to come up with a procedure for describing the set B.

$$f(0) = \{\mathbf{0}, 1, 2, 3, 4, 5, 6, 7, 8, 9, 10, \ldots\}$$
$$f(1) = \{0, \mathbf{1}, 3, 4, 5, 6, 7, 8, 9, 10, \ldots\}$$
$$f(2) = \{0, 1, 4, 5, 6, 7, 8, 9, 10, \ldots\}$$
$$f(3) = \{0, 1, 4, 6, 7, 8, 9, 10, \ldots\}$$
$$f(4) = \{0, 1, \mathbf{4}, 6, 8, 9, 10, \ldots\}$$
$$\ldots$$

Technical note: Recall that a **prime number** is a natural number with **exactly** two factors, 1 and itself. The set of prime numbers looks like this: $\{2, 3, 5, 7, 11, 13, 17, \ldots\}$. The function $f: \mathbb{N} \to \mathcal{P}(\mathbb{N})$ that we chose to use here is defined by $f(n) = \{k \in \mathbb{N} \mid k \text{ is not equal to one of the first } n \text{ prime numbers}\}$. Notice how $f(0)$ is just the set $\mathbb{N}$ of all natural numbers, $f(1)$ is the set of all natural numbers except 2 (we left out the first prime), $f(2)$ is the set of all natural numbers except 2 and 3 (we left out the first two primes), and so on. We will look at prime numbers in detail in Lesson 8.

Observe that the "inputs" of our function are natural numbers, and the "outputs" are sets of natural numbers. So, it's perfectly natural to ask the question "Is n in $f(n)$?"

For example, we see that $0 \in f(0)$, $1 \in f(1)$, and $4 \in f(4)$ (indicated in bold in the definition of the function above). However, we also see that $2 \notin f(2)$ and $3 \notin f(3)$.

Let's let B be the set of natural numbers n that are **not** inside their images. Symbolically, we have
$$B = \{n \in \mathbb{N} \mid n \notin f(n)\}.$$

Which natural numbers are in the set B? Well, we already said that $0 \in f(0)$. It follows that $0 \notin B$. Similarly, $1 \notin B$ and $4 \notin B$, but $2 \in B$ and $3 \in B$.

Why did we choose to define B this way? The reason is because we are trying to make sure that B cannot be equal to $f(n)$ for every n. Since $0 \in f(0)$, but $0 \notin B$, it follows that $f(0)$ and B are different sets because they differ by at least one element, namely 0. Similarly, since $1 \in f(1)$, but $1 \notin B$, B cannot be equal to $f(1)$. What about 2? Well $2 \notin f(2)$, but $2 \in B$. Therefore, $B \neq f(2)$ as well... and so on down the line. We intentionally chose to make B disagree with $f(n)$ for every natural number n, ensuring that B will not be in the range of f.

I think we are now ready to prove the theorem.

Proof of Theorem 4.16: Let $f: A \to \mathcal{P}(A)$, and let $B = \{a \in A \mid a \notin f(a)\}$. Suppose toward contradiction that $B \in \operatorname{ran} f$. Then there is $a \in A$ with $f(a) = B$. But then we have $a \in B$ if and only if $a \notin f(A)$ if and only if $a \notin B$. This contradiction tells us that $B \notin \operatorname{ran} f$, and so, f is not surjective. Since $f: A \to \mathcal{P}(A)$ was arbitrary, there does not exist a surjection from A to $\mathcal{P}(A)$, and therefore, there is no bijection from A to $\mathcal{P}(A)$. So, A is not equinumerous with $\mathcal{P}(A)$. □

Notes: (1) The proof given here is a **proof by contradiction**. A proof by contradiction works as follows:

1. We assume the negation of what we are trying to prove.
2. We use a logically valid argument to derive a statement which is false.
3. Since the argument was logically valid, the only possible error is our original assumption. Therefore, the negation of our original assumption must be true.

(2) In this problem we are trying to prove that A is **not** equinumerous with $\mathcal{P}(A)$. The negation of this statement is that A **is** equinumerous with $\mathcal{P}(A)$, and so that is what we assume. Since $A \sim \mathcal{P}(A)$, there is a bijection $f: A \to \mathcal{P}(A)$. So, f is a surjection, which means that every subset of $\mathcal{P}(A)$ is in the range of f. In particular, the set B described in the proof is a subset of $\mathcal{P}(A)$, and therefore it is in the range of f. We then use a logically valid argument to derive the obviously false statement "$a \in B$ if and only if $a \notin B$."

By Theorem 4.16, $\mathbb{N}$ is not equinumerous with $\mathcal{P}(\mathbb{N})$. Which of these two sets is the "bigger" one? Let's consider the function $f: \mathbb{N} \to \mathcal{P}(\mathbb{N})$ defined by $f(n) = \{n\}$. This function looks like this:

$$0 \quad 1 \quad 2 \quad 3 \quad 4 \ldots$$
$$\{0\} \; \{1\} \; \{2\} \; \{3\} \; \{4\} \ldots$$

Observe that we are matching up each natural number with a subset of natural numbers (a very simple subset consisting of just one natural number) in a way so that different natural numbers get matched with different subsets. In other words, we defined an injective function from $\mathbb{N}$ to $\mathcal{P}(\mathbb{N})$. It seems like there are lots of subsets of $\mathbb{N}$ that didn't get mapped to (for example, all infinite subsets of $\mathbb{N}$). So, it seems that $\mathbb{N}$ is a "smaller" set than $\mathcal{P}(\mathbb{N})$.

We use the notation $A \preccurlyeq B$ if there is an injective function from A to B.

$$A \preccurlyeq B \text{ if and only if } \exists f (f: A \hookrightarrow B)$$

Recall: The symbol $\exists$ is called an **existential quantifier**, and it is pronounced "There exists" or "There is." The expression $\exists f (f: A \hookrightarrow B)$ can be translated into English as "There exists an f such that f is an injective function from A to B."

We write $A \prec B$ if $A \preccurlyeq B$ and $A \nsim B$.

So, for example, $\mathbb{N} \prec \mathcal{P}(\mathbb{N})$.

Theorem 4.17: If A is any set, then $A \prec \mathcal{P}(A)$.

Proof: The function $f: A \to \mathcal{P}(A)$ defined by $f(a) = \{a\}$ is injective. So, $A \preccurlyeq \mathcal{P}(A)$. By Theorem 4.16, $A \not\sim \mathcal{P}(A)$. It follows that $A \prec \mathcal{P}(A)$. □

Example 4.18: If we let $A = \mathcal{P}(\mathbb{N})$, we can apply Theorem 4.17 to this set A to see that $\mathcal{P}(\mathbb{N}) \prec \mathcal{P}(\mathcal{P}(\mathbb{N}))$. Continuing in this fashion, we get a sequence of increasingly larger sets.

$$\mathbb{N} \prec \mathcal{P}(\mathbb{N}) \prec \mathcal{P}(\mathcal{P}(\mathbb{N})) \prec \mathcal{P}(\mathcal{P}(\mathcal{P}(\mathbb{N}))) \prec \cdots$$

If A and B are arbitrary sets, in general it can be difficult to determine if A and B are equinumerous by producing a bijection. Luckily, the next theorem provides an easier way.

Theorem 4.19 (The Cantor-Schroeder-Bernstein Theorem): If A and B are sets such that $A \preccurlyeq B$ and $B \preccurlyeq A$, then $A \sim B$.

Note: At first glance, many students think that Theorem 4.19 is obvious and that the proof must be trivial. This is not true. The theorem says that if there is an injective function from A to B and another injective function from B to A, then there is a bijective function from A to B. This is a deep result, which is far from obvious. Constructing a bijection from two arbitrary injections is not an easy thing to do. I suggest that the reader takes a few minutes to try to do it, if for no other reason than to convince themselves that the proof is difficult. I leave the proof itself as an exercise (see Problem 20 below).

Example 4.20: Let's use Theorem 4.19 to prove that the open interval of real numbers $(0, 1)$ is equinumerous with the closed interval of real numbers $[0, 1]$.

Analysis: Since $(0, 1) \subseteq [0, 1]$, there is an obvious injective function $f: (0, 1) \to [0, 1]$ (just send each element to itself).

The harder direction is finding an injective function g from $[0, 1]$ into $(0, 1)$. We will do this by drawing a line segment with endpoints $\left(0, \frac{1}{4}\right)$ and $\left(1, \frac{3}{4}\right)$. This will give us a bijection from $[0, 1]$ to $\left[\frac{1}{4}, \frac{3}{4}\right]$. We can visualize this bijection using the graph to the right. We will write an equation for this line segment in the slope-intercept form $y = mx + b$. Here m is the slope of the line and b is the y-intercept of the line. We can use the graph to see that $b = \frac{1}{4}$ and $m = \frac{\text{rise}}{\text{run}} = \frac{\frac{3}{4} - \frac{1}{4}}{1 - 0} = \frac{\frac{2}{4}}{1} = \frac{1}{2}$. So, we define $g: [0, 1] \to (0, 1)$ by $g(x) = \frac{1}{2}x + \frac{1}{4}$.

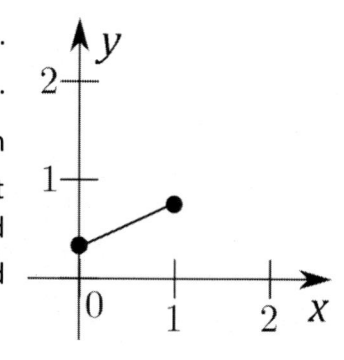

Let's write out the details of the proof.

Proof: Let $f: (0, 1) \to [0, 1]$ be defined by $f(x) = x$. Clearly, f is injective, so that $(0, 1) \preccurlyeq [0, 1]$.

Next, we define $g: [0, 1] \to \mathbb{R}$ by $g(x) = \frac{1}{2}x + \frac{1}{4}$. If $0 \le x \le 1$, then $0 \le \frac{1}{2}x \le \frac{1}{2}$, and therefore, $\frac{1}{4} \le \frac{1}{2}x + \frac{1}{4} \le \frac{3}{4}$. Since $0 < \frac{1}{4}$ and $\frac{3}{4} < 1$, we have $0 < g(x) < 1$. Therefore, $g: [0, 1] \to (0, 1)$. If $x \ne x'$, then $\frac{1}{2}x \ne \frac{1}{2}x'$, and so, $g(x) = \frac{1}{2}x + \frac{1}{4} \ne \frac{1}{2}x' + \frac{1}{4} = g(x')$. This shows that g is injective. It follows that $[0, 1] \preccurlyeq (0, 1)$.

Since $(0, 1) \preccurlyeq [0, 1]$ and $[0, 1] \preccurlyeq (0, 1)$, it follows from the Cantor-Schroeder-Bernstein Theorem that $(0, 1) \sim [0, 1]$. □

Notes: (1) If $A \subseteq B$, then the function $f : A \to B$ defined by $f(a) = a$ for all $a \in A$ is always injective. It is called the **inclusion map**.

(2) It is unfortunate that the same notation is used for points (ordered pairs of real numbers) and open intervals. Normally this isn't an issue, but in this particular example both usages of this notation appear. Take another look at the analysis above and make sure you can see when the notation (a, b) is being used for a point and when it is being used for an open interval.

(3) We could have used any closed interval $[a, b]$ with $0 < a < b < 1$ in place of $\left[\frac{1}{4}, \frac{3}{4}\right]$.

Example 4.21: Let's use Theorem 4.19 to prove that the half-open interval of real numbers $[0, 1)$ is equinumerous with $\mathcal{P}(\mathbb{N})$.

Proof: Let $f : \mathcal{P}(\mathbb{N}) \to [0, 1)$ be defined by $f(A) = 0.a_0 a_1 \ldots$, where $a_n = \begin{cases} 0 & \text{if } n \notin A. \\ 1 & \text{if } n \in A. \end{cases}$

If $A, B \in \mathcal{P}(\mathbb{N})$ with $A \neq B$, then there is $n \in \mathbb{N}$ with $n \in A \setminus B$ or $n \in B \setminus A$. **Without loss of generality**, assume that $n \in A \setminus B$. Assuming that $f(A) = 0.a_0 a_1 \ldots$ and $f(B) = 0.b_0 b_1 \ldots$, we have $a_n = 1$ and $b_n = 0$. Therefore, $f(A) \neq f(B)$. This shows that f is injective. It follows that $\mathcal{P}(\mathbb{N}) \preccurlyeq [0, 1)$.

Next, we define $g : [0, 1) \to \mathcal{P}(\mathbb{N})$. Let $a \in [0, 1)$. We define $A_0 = \left[0, \frac{1}{2}\right)$ and $A_1 = \left[\frac{1}{2}, 1\right)$. Then $\{A_0, A_1\}$ is a partition of $[0, 1)$, and so, it follows that $a \in A_0$ or $a \in A_1$, but not both. Similarly, we define $A_{00} = \left[0, \frac{1}{4}\right)$, $A_{01} = \left[\frac{1}{4}, \frac{1}{2}\right)$, $A_{10} = \left[\frac{1}{2}, \frac{3}{4}\right)$, and $A_{11} = \left[\frac{3}{4}, 1\right)$. Then $\{A_{00}, A_{01}, A_{10}, A_{11}\}$ is a partition of $[0, 1)$, and so, it follows that a is in exactly one of the four sets $A_{00}, A_{01}, A_{10},$ or A_{11}. Note that $\{A_{00}, A_{01}\}$ is a partition of A_0 and $\{A_{10}, A_{11}\}$ is a partition of A_1. It follows that if $a \in A_0$, then a must be in either A_{00} or A_{01}, and similarly if $a \in A_1$. Continuing in this fashion, we get a sequence of subscripts $a_0, a_0 a_1, a_0 a_1 a_2, \ldots$, where each subscript a_n is either 0 or 1. We let $g(a) = \{n \in \mathbb{N} \mid a_n = 1\}$. Now, suppose that $a \neq b$. Without loss of generality assume that $a < b$. If $a \in A_0$ and $b \in A_1$, then $0 \notin g(a)$, $0 \in g(b)$. Otherwise, there must be n such that $a, b \in A_{a_0 a_1 a_2 \cdots a_n}$, $a \in A_{a_0 a_1 a_2 \cdots a_n 0}$, and $b \in A_{a_0 a_1 a_2 \cdots a_n 1}$. It follows that $n + 1 \notin g(a)$ and $n + 1 \in g(b)$. Therefore, $g(a) \neq g(b)$. This shows that g is injective. It follows that $[0, 1) \preccurlyeq \mathcal{P}(\mathbb{N})$.

Since $\mathcal{P}(\mathbb{N}) \preccurlyeq [0, 1)$ and $[0, 1) \preccurlyeq \mathcal{P}(\mathbb{N})$, it follows from the Cantor-Schroeder-Bernstein Theorem that $\mathcal{P}(\mathbb{N}) \sim [0, 1)$. □

Notes: (1) In the second paragraph of the proof above, we used the expression "Without loss of generality." Recall from Note 1 following Example 4.4 that this expression can be used when an argument can be split up into 2 or more cases, and the proof of each of the cases is nearly identical.

For the proof above, the two cases are (i) $n \in A \setminus B$ and (ii) $n \in B \setminus A$. The argument for case (ii) is the same as the argument for case (i) except that the roles of A and B (and the roles of a_n and b_n for each n) are reversed.

(2) Given $a \in [0, 1)$, the number $.a_0 a_1 a_2 \ldots$ is called the **binary expansion** of a.

As an example, let's find the binary expansion of $\frac{1}{8}$.

Since $0 \leq \frac{1}{8} < \frac{1}{2}$, we have $\frac{1}{8} \in \left[0, \frac{1}{2}\right) = A_0$. So, $a_0 = 0$. (Here we are in the "leftmost" set in the partition of $[0, 1) = \left[0, \frac{1}{2}\right) \cup \left[\frac{1}{2}, 1\right)$.)

Next, since $0 \leq \frac{1}{8} < \frac{1}{4}$, we have $\frac{1}{8} \in \left[0, \frac{1}{4}\right) = A_{00}$. So, $a_1 = 0$. (Here we are once again in the "leftmost" set in the partition of $\left[0, \frac{1}{2}\right) = \left[0, \frac{1}{4}\right) \cup \left[\frac{1}{4}, \frac{1}{2}\right)$. It may help to rewrite this using only denominators of 4 as follows: $\left[0, \frac{2}{4}\right) = \left[0, \frac{1}{4}\right) \cup \left[\frac{1}{4}, \frac{2}{4}\right)$.)

Now, since $\frac{1}{8} \leq \frac{1}{8} < \frac{1}{4}$, we have $\frac{1}{8} \in \left[\frac{1}{8}, \frac{1}{4}\right) = A_{001}$. So, $a_2 = 1$. (This time we are in the "rightmost" set in the partition of $\left[0, \frac{1}{4}\right) = \left[0, \frac{1}{8}\right) \cup \left[\frac{1}{8}, \frac{1}{4}\right)$. Once again, it may help to rewrite this using only denominators of 8 as follows: $\left[0, \frac{2}{8}\right) = \left[0, \frac{1}{8}\right) \cup \left[\frac{1}{8}, \frac{2}{8}\right)$.)

Since $\frac{1}{8} \leq \frac{1}{8} < \frac{3}{16}$, we have $\frac{1}{8} \in \left[\frac{1}{8}, \frac{3}{16}\right) = A_{0010}$. So, $a_3 = 0$. (This time we are back in the "leftmost" set in the partition of $\left[\frac{1}{8}, \frac{1}{4}\right) = \left[\frac{1}{8}, \frac{3}{16}\right) \cup \left[\frac{3}{16}, \frac{1}{4}\right)$. Again, it may help to rewrite this using only denominators of 16 as follows: $\left[\frac{2}{16}, \frac{4}{16}\right) = \left[\frac{2}{16}, \frac{3}{16}\right) \cup \left[\frac{3}{16}, \frac{4}{16}\right)$.)

From this point on, it should be easy to see that we will always be choosing the "leftmost" set in the partition. So, the remaining digits will all be 0. It follows that the binary expansion of $\frac{1}{8}$ is 0.001.

(3) Using this method of writing a binary expansion, we will never get a tail of 1's. Therefore, using this method, each real number corresponds to a unique binary expansion without a tail of 1's. See part 2 of Example 1.6 for the definition of a "tail."

(4) We can use this same method to form the **decimal expansion** of a real number as well. In this case, we partition $[0, 1)$ into $A_0 = \left[0, \frac{1}{10}\right)$, $A_1 = \left[\frac{1}{10}, \frac{2}{10}\right)$, $A_2 = \left[\frac{2}{10}, \frac{3}{10}\right)$,..., and $A_9 = \left[\frac{9}{10}, 1\right)$. Using $\frac{1}{8}$ as an example again, we have $\frac{1}{8} \in \left[\frac{1}{10}, \frac{2}{10}\right) = A_1$. So, $a_0 = 1$. We then partition $\left[\frac{1}{10}, \frac{2}{10}\right) = \left[\frac{10}{100}, \frac{20}{100}\right)$ into $A_{10} = \left[\frac{10}{100}, \frac{11}{100}\right)$, $A_{11} = \left[\frac{11}{100}, \frac{12}{100}\right)$, $A_{12} = \left[\frac{12}{100}, \frac{13}{100}\right)$,..., and $A_{19} = \left[\frac{19}{100}, \frac{20}{100}\right)$. Now, we have $\frac{1}{8} \in \left[\frac{12}{100}, \frac{13}{100}\right) = A_{12}$, and so, $a_1 = 2$. You should now check that $\frac{1}{8} \in \left[\frac{125}{1000}, \frac{126}{1000}\right) = A_{125}$, and so, $a_2 = 5$. It is then not too hard to check that the remaining digits will be 0. It follows that the decimal expansion of $\frac{1}{8}$ is 0.125.

(5) We can use this same method to form n-ary expansions for any integer $n \geq 2$. If $n = 2$, we get the binary expansion, as defined in Note 2. If $n = 10$, we get the decimal expansion, as defined in Note 4. If $n = 3$, we get the **ternary expansion**. The reader may want to check that the ternary expansion of $\frac{1}{8}$ is $0.010101010101\ldots$

Problem Set 4

Full solutions to these problems are available for free download here:
www.SATPrepGet800.com/AAFBTDW

LEVEL 1

1. Determine if each of the following relations are functions. For each such function, determine if it is injective. State the domain and range of each function.

    (i) $R = \{(a,b), (b,b), (c,d), (e,a)\}$

    (ii) $S = \{(a,a), (a,b), (b,a)\}$

    (iii) $T = \{(a,b) \mid a, b \in \mathbb{R} \wedge b < 0 \wedge a^2 + b^2 = 9\}$

2. Define $f: \mathbb{Z} \to \mathbb{Z}$ by $f(n) = n^2$. Let $A = \{0, 1, 2, 3, 4\}$, $B = \mathbb{N}$, and $C = \{-2n \mid n \in \mathbb{N}\}$. Evaluate each of the following:

    (i) $f[A]$

    (ii) $f^{-1}[A]$

    (iii) $f^{-1}[B]$

    (iv) $f[B \cup C]$

3. Let A, B, and C be sets. Prove the following:

    (i) $\preccurlyeq$ is transitive.

    (ii) $\prec$ is transitive.

    (iii) If $A \preccurlyeq B$ and $B \prec C$, then $A \prec C$.

    (iv) If $A \prec B$ and $B \preccurlyeq C$, then $A \prec C$.

4. Write the elements of S_4 in cycle notation.

5. Draw a group multiplication table for S_3.

LEVEL 2

6. Find sets A and B and a function f such that $f[A \cap B] \neq f[A] \cap f[B]$.

7. Let $f: A \to B$ and let $V \subseteq B$. Prove that $f[f^{-1}[V]] \subseteq V$.

8. Define $\mathcal{P}_k(\mathbb{N})$ for each $k \in \mathbb{N}$ by $\mathcal{P}_0(\mathbb{N}) = \mathbb{N}$ and $\mathcal{P}_{k+1}(\mathbb{N}) = \mathcal{P}(\mathcal{P}_k(\mathbb{N}))$ for $k > 0$. Find a set B such that for all $k \in \mathbb{N}$, $\mathcal{P}_k(\mathbb{N}) \prec B$.

9. Prove that if $A \sim B$ and $C \sim D$, then $A \times C \sim B \times D$.

Level 3

10. For $f, g \in {}^{\mathbb{R}}\mathbb{R}$, define $f \preccurlyeq g$ if and only if for all $x \in \mathbb{R}$, $f(x) \leq g(x)$. Is $({}^{\mathbb{R}}\mathbb{R}, \preccurlyeq)$ a poset? Is it a linearly ordered set? What if we replace $\preccurlyeq$ by $\preccurlyeq^*$, where $f \preccurlyeq^* g$ if and only if there is an $x \in \mathbb{R}$ such that $f(x) \leq g(x)$?

11. Prove that the function $f: \mathbb{N} \to \mathbb{Z}$ defined by $f(n) = \begin{cases} \dfrac{n}{2} & \text{if } n \text{ is even} \\ -\dfrac{n+1}{2} & \text{if } n \text{ is odd} \end{cases}$ is a bijection.

12. Define a partition P of $\mathbb{N}$ such that $P \sim \mathbb{N}$ and for each $X \in P$, $X \sim \mathbb{N}$.

13. Prove that a countable union of countable sets is countable.

14. Let A and B be sets such that $A \sim B$. Prove that $\mathcal{P}(A) \sim \mathcal{P}(B)$.

Level 4

15. Prove the following:
    (i) $\mathbb{N} \times \mathbb{N} \sim \mathbb{N}$.
    (ii) $\mathbb{Q} \sim \mathbb{N}$.
    (iii) Any two intervals of real numbers are equinumerous (including $\mathbb{R}$ itself).
    (iv) ${}^{\mathbb{N}}\mathbb{N} \sim \mathcal{P}(\mathbb{N})$.

16. Prove that if $A \sim B$ and $C \sim D$, then ${}^{A}C \sim {}^{B}D$.

Level 5

17. Let X be a nonempty set of sets and let f be a function such that $\bigcup X \subseteq \text{dom } f$. Prove each of the following:
    (i) $f[\bigcup X] = \bigcup \{f[A] \mid A \in X\}$
    (ii) $f[\bigcap X] \subseteq \bigcap \{f[A] \mid A \in X\}$
    (iii) $f^{-1}[\bigcup X] = \bigcup \{f^{-1}[A] \mid A \in X\}$
    (iv) $f^{-1}[\bigcap X] = \bigcap \{f^{-1}[A] \mid A \in X\}$

18. Prove that for any sets A, B, and C, ${}^{B \times C}A \sim {}^{C}({}^{B}A)$.

19. Prove the following:
    (i) $\mathcal{P}(\mathbb{N}) \sim S(\mathbb{N})$ (recall that $S(\mathbb{N}) = \{f \in {}^{\mathbb{N}}\mathbb{N} \mid f \text{ is a bijection}\}$).
    (ii) ${}^{\mathbb{N}}\mathbb{R} \not\sim {}^{\mathbb{R}}\mathbb{N}$.

Challenge Problem

20. Prove the Cantor-Schroeder-Bernstein Theorem.

21. If S is a set and $(G,\star)$ is a group, define an operation $\circ$ on ^SG so that $(^SG,\circ)$ is a group.

22. If S is a set and $(R, +, \cdot)$ is a ring, define operations $+$ and $\cdot$ on ^SR so that $(^SR, +, \cdot)$ is a ring. With the same operations, if $(R, +, \cdot)$ is a domain, will $(^SR, +, \cdot)$ be a domain?

23. If S is a set and F is a field, define addition and scalar multiplication on ^SF so that ^SF is a vector space over F. In what ways can this definition be generalized further?

LESSON 5
NUMBER SYSTEMS AND INDUCTION

The Natural Numbers

At this point, let's provide more formal definitions of the number systems we use most frequently, beginning with the set of natural numbers.

We define the following:

$$0 = \emptyset$$
$$1 = \{\emptyset\} = \{0\}$$
$$2 = \{\emptyset, \{\emptyset\}\} = \{0, 1\}$$
$$3 = \{\emptyset, \{\emptyset\}, \{\emptyset, \{\emptyset\}\}\} = \{0, 1, 2\}$$

In general, we let $n = \{0, 1, 2, \ldots, n-1\}$ and we define the **natural numbers** to be the set

$$\mathbb{N} = \{0, 1, 2, 3, 4, \ldots\}$$

If n is a natural number, we define the **successor** of n, written n^+, to be the natural number $n^+ = n \cup \{n\}$. Note that $n^+ = \{0, 1, 2, \ldots, n-1, n\}$.

Example 5.1:

1. $0^+ = 0 \cup \{0\} = \{0\} = 1$.
2. $1^+ = 1 \cup \{1\} = \{0\} \cup \{1\} = \{0, 1\} = 2$.
3. $2^+ = 2 \cup \{2\} = \{0, 1\} \cup \{2\} = \{0, 1, 2\} = 3$.

If n is a natural number such that $n \neq 0$, we define the **predecessor** of n, written n^-, to be the natural number k such that $n = k^+$ (by Problem 4, part (vii) below, k is unique). Thus, $n = n^- \cup \{n^-\}$.

Note that for all $n \in \mathbb{N}$, $(n^+)^- = n$ and for all $n \in \mathbb{N}$ with $n \neq 0$, $(n^-)^+ = n$.

Example 5.2:

1. Since $1 = 0^+$, $1^- = 0$.
2. Since $2 = 1^+$, $2^- = 1$.
3. Since $3 = 2^+$, $3^- = 2$.

We define the ordering $<_\mathbb{N}$ on $\mathbb{N}$ by $n <_\mathbb{N} m$ if and only if $n \in m$. We will usually abbreviate $<_\mathbb{N}$ simply by $<$, especially if it is already clear that we are working with the natural numbers.

With this definition, $<_\mathbb{N}$ is a strict linear ordering on $\mathbb{N}$. You will be asked to prove this as part of Problem 31 below.

Example 5.3:
1. $0 \in \{0\}$. Therefore, $0 \in 0 \cup \{0\} = 1$. So, $0 < 1$.
2. $1 = \{0\} \notin \emptyset = 0$. So, $1 \not< 0$.
3. $1 = \{0\} \notin 0$ and $1 = \{0\} \notin \{0\}$. Therefore, $1 = \{0\} \notin 0 \cup \{0\} = 1$. So, $1 \not< 1$.
4. If $n \neq 0$, then $n^- \in \{n^-\}$. Therefore, $n^- \in n^- \cup \{n^-\} = n$. So, $n^- < n$.

We add the natural numbers m and n as follows:
   (i) If $n = 0$, then $m + n = m$.
   (ii) If $n = k^+$, then $m + n = (m + k)^+$.

We multiply the natural numbers m and n as follows:
   (i) If $n = 0$, then $mn = 0$.
   (ii) If $n = k^+$, then $mn = (mk) + m$.

Notes: (1) With the definition of addition given above, we see that for all natural numbers m, we have $m + 1 = m + 0^+ = (m + 0)^+ = m^+$. So, whenever we wish, we can write $m + 1$ instead of m^+. In particular, clause (ii) in the definition of the sum of two natural numbers can be written as follows:

   (ii) If $n = k + 1$, then $m + n = (m + k) + 1$ (or $m + (k + 1) = (m + k) + 1$).

(2) Whenever it will not cause confusion, we will use the usual order of operations that we learned in elementary school. So, for example, we can abbreviate $(mk) + m$ as $mk + m$.

(3) Addition and multiplication of natural numbers are examples of **recursive definitions**. A full exploration of recursive definitions lies outside the scope of this book. Here we will just accept the existence of recursively defined functions without further mention.

(4) We define the **sum** of natural numbers m and n to be $m + n$. We define the **product** of natural numbers m and n to be $m \cdot n$.

Example 5.4:
1. $0 + 0 = 0$ by the first part of the definition of addition of natural numbers.
2. $0 + 1 = 0 + 0^+ = (0 + 0)^+ = 0^+ = 0 \cup \{0\} = \emptyset \cup \{\emptyset\} = \{\emptyset\} = 1$.
3. $0 + 2 = 0 + 1^+ = (0 + 1)^+ = 1^+ = 1 \cup \{1\} = \{0\} \cup \{1\} = \{0, 1\} = 2$.
4. $2 + 0 = 2$ by the first part of the definition of addition of natural numbers.
5. $2 + 1 = 2 + 0^+ = (2 + 0)^+ = 2^+ = 2 \cup \{2\} = \{0, 1\} \cup \{2\} = \{0, 1, 2\} = 3$.
6. $2 + 2 = 2 + 1^+ = (2 + 1)^+ = 3^+ = 3 \cup \{3\} = \{0, 1, 2\} \cup \{3\} = \{0, 1, 2, 3\} = 4$.
7. $2 \cdot 0 = 0$ by the first part of the definition of multiplication of natural numbers.
8. $2 \cdot 1 = 2 \cdot 0^+ = 2 \cdot 0 + 2 = 0 + 2 = 2$ (by part 3 above).
9. $2 \cdot 2 = 2 \cdot 1^+ = 2 \cdot 1 + 2 = 2 + 2 = 4$ (by part 6 above).

Well Ordering and the Principle of Mathematical Induction

If A is a subset of natural numbers, then we say that a is the **least** element of A if a is less than every other element of A. Symbolically, $\forall b \in A(b \neq a \to a < b)$.

Example 5.5:

1. The least element of $\mathbb{N}$ itself is 0.
2. The least element of $\mathbb{N} \setminus \{0\}$ is 1.
3. The least element of $\mathbb{N} \setminus \{0, 1\}$ is 2.

We now describe two important principles that are equivalent to each other.

The **Well Ordering Principle** (abbreviated **WOP**) is the following statement: Every nonempty subset of natural numbers has a least element.

The **Principle of Mathematical Induction** (abbreviated **POMI**) is the following statement: Let S be a set of natural numbers such that (i) $0 \in S$ and (ii) for all $k \in \mathbb{N}$, $k \in S \to k^+ \in S$. Then $S = \mathbb{N}$.

Notes: (1) The Principle of Mathematical Induction works like a chain reaction. We know that $0 \in S$ (this is condition (i)). Substituting 0 in for k in the expression "$k \in S \to k^+ \in S$" (condition (ii)) gives us $0 \in S \to 1 \in S$. So, we have that 0 is in the set S, and "if 0 is in the set S, then 1 is in the set S." So, $1 \in S$ must also be true.

(2) In terms of logic, if we let p be the statement $0 \in S$ and q the statement $1 \in S$, then we are given that $p \wedge (p \to q)$ is true. Observe that the only way that this statement can be true is if q is also true. Indeed, we must have both $p \equiv \text{T}$ and $p \to q \equiv \text{T}$. If q were false, then we would have $p \to q \equiv \text{T} \to \text{F} \equiv \text{F}$. So, we must have $q \equiv \text{T}$.

(3) Now that we showed $1 \in S$ is true (from Note 1 above), we can substitute 1 for k in the expression "$k \in S \to k^+ \in S$" (condition (ii)) to get $1 \in S \to 2 \in S$. So, we have $1 \in S \wedge (1 \in S \to 2 \in S)$ is true. So, $2 \in S$ must also be true.

(4) In general, we get the following chain reaction:
$$0 \in S \to 1 \in S \to 2 \in S \to 3 \in S \to \cdots$$

I hope that the "argument" presented in Notes 1 through 4 above convinces you that the Principle of Mathematical Induction is a reasonable one.

Now let's show that the Principle of Mathematical Induction follows from the Well Ordering Principle (In Problem 11 below, you will be asked to show that the Well Ordering Principle follows from the Principle of Mathematical Induction, thus showing that the two statements are equivalent). Proofs involving the Well Ordering Principle are generally done by contradiction.

Theorem 5.6: WOP $\to$ POMI.

Proof: Assume WOP and let S be a set of natural numbers such that $0 \in S$ (condition (i)), and such that whenever $k \in S$, $k^+ \in S$ (condition (ii)). Assume toward contradiction that $S \neq \mathbb{N}$. Let $A = \{k \in \mathbb{N} \mid k \notin S\}$ (so, A is the set of natural numbers **not** in S). Since $S \neq \mathbb{N}$, A is nonempty. So, by the Well Ordering Principle, A has a least element, let's call it a. $a \neq 0$ because $0 \in S$ and $a \notin S$. So, $a^- \in \mathbb{N}$. Letting $k = a^-$, we have $a^- \in S \to k \in S \to k^+ \in S \to (a^-)^+ \in S \to a \in S$. But $a \in A$, which means that $a \notin S$. This is a contradiction, and so, $S = \mathbb{N}$. □

Recall: A proof by contradiction works as follows:

1. We assume the negation of what we are trying to prove.

2. We use a logically valid argument to derive a statement which is false.

3. Since the argument was logically valid, the only possible error is our original assumption. Therefore, the negation of our original assumption must be true.

In this problem we are trying to prove that $S = \mathbb{N}$. The negation of this statement is that $S \neq \mathbb{N}$, and so that is what we assume.

We then define a set A which contains elements of $\mathbb{N}$ that are not in S. In reality, this set is empty (because the conclusion of the theorem is $S = \mathbb{N}$). However, our (wrong!) assumption that $S \neq \mathbb{N}$ tells us that this set A actually has something in it. Saying that A has something in it is an example of a false statement that was derived from a logically valid argument. This false statement occurred not because of an error in our logic, but because we started with an incorrect assumption ($S \neq \mathbb{N}$).

The Well Ordering Principle then allows us to pick out the least element of this set A. Note that we can do this because A is a subset of $\mathbb{N}$. This wouldn't work if we knew only that A was a subset of $\mathbb{Z}$, as $\mathbb{Z}$ does **not** satisfy the Well Ordering Principle (for example, $\mathbb{Z}$ itself has no least element).

Again, although the argument that A has a least element is logically valid, A does not actually have any elements at all. We are working from the (wrong!) assumption that $S \neq \mathbb{N}$.

Once we have our hands on this least element a, we can get our contradiction. What can this least element a be? Well a was chosen to **not** be in S, so a cannot be 0 (because 0 **is** in S). Also, we know that $a^- \in S$ (because a is the **least** element not in S and $a^- < a$). But condition (ii) then forces a to be in S (because $a = (a^-)^+$).

So, we wind up with $a \in S$, contradicting the fact that a is the least element **not** in S.

The Principle of Mathematical Induction is often written in the following way:

($\star$) Let $P(n)$ be a statement and suppose that (i) $P(0)$ is true and (ii) for all $k \in \mathbb{N}$, $P(k) \to P(k^+)$. Then $P(n)$ is true for all $n \in \mathbb{N}$.

In Problem 14 below, you will be asked to show that statement ($\star$) is equivalent to POMI.

There are essentially two steps involved in a proof by mathematical induction:

The first step is to prove that $P(0)$ is true (this is called the **base case**).

The second step is to assume that $P(k)$ is true, and use this to show that $P(k^+)$ is true (this is called the **inductive step**). While doing the inductive step, the statement "$P(k)$ is true" is often referred to as the **inductive hypothesis**.

Theorem 5.7: The sum of two natural numbers is a natural number.

Note: Consider the sum $m + n$ of the natural numbers m and n. We will prove this theorem by induction on n (and **not** by induction on m). We will start by letting m be an arbitrary natural number.

The base case will be to show that $m + 0$ is a natural number.

The inductive step will be to assume that $m + k$ is a natural number and then to prove that $m + k^+$ (or equivalently, $m + (k + 1)$) is a natural number.

Proof of Theorem 5.7: Assume that m is a natural number.

Base Case ($k = 0$): $m + 0 = m$ (by clause (i) in the definition of the sum of two natural numbers), which we assumed was a natural number. Thus, we have shown that $m + 0$ is a natural number.

Inductive Step: Let k be a natural number and assume that $m + k$ is also a natural number. Then $m + k^+ = (m + k)^+$ (or equivalently, $m + (k + 1) = (m + k) + 1$), which is also a natural number.

Here we used the fact that the successor of a natural number is a natural number.

By the Principle of Mathematical Induction, $m + n$ is a natural number for all natural numbers n.

Since m was an arbitrary natural number, we have shown that the sum of any two natural numbers is a natural number. □

Note: Since the sum of two natural numbers is always a natural number (by Theorem 5.7), we have now formally proved that $\mathbb{N}$ is **closed** under addition, or equivalently, $+$ is a **binary operation** on $\mathbb{N}$.

Theorem 5.8: For all natural numbers n, $0 + n = n$.

Proof: Base Case ($k = 0$): $0 + 0 = 0$ by the definition of addition of natural numbers.

Inductive Step: Let $k \in \mathbb{N}$ and assume that $0 + k = k$. Then $0 + (k + 1) = (0 + k) + 1 = k + 1$, as desired.

For the first equality, we used the definition of addition of natural numbers. For the second equality, we used the inductive hypothesis.

By the Principle of Mathematical Induction, for all natural numbers n, $0 + n = n$. □

Notes: (1) It's also true that for all natural numbers n, $n + 0 = n$. This follows right from the definition of addition of natural numbers.

(2) Since for all natural numbers n, we have $0 + n = n + 0 = 0$, we have now formally proved that 0 is an **identity** with respect to addition, or that 0 is an **additive identity** in $\mathbb{N}$.

Theorem 5.9: For all natural numbers m, n, and t, $(m + n) + t = m + (n + t)$.

Proof: Let m and n be natural numbers.

Base Case ($k = 0$): $(m + n) + 0 = m + n = m + (n + 0)$ by the definition of addition in $\mathbb{N}$.

Inductive Step: Let $k \in \mathbb{N}$ and assume that $(m + n) + k = m + (n + k)$. Then we have
$$(m + n) + (k + 1) = ((m + n) + k) + 1 = (m + (n + k)) + 1$$
$$= m + ((n + k) + 1) = m + (n + (k + 1)).$$

For the first, third, and fourth equalities, we used the definition of addition of natural numbers. For the second equality, we used the inductive hypothesis.

By the Principle of Mathematical Induction, for all natural numbers t, $(m + n) + t = m + (n + t)$.

Since m and n were arbitrary natural numbers, we have shown that for all natural numbers m, n, and t, $(m + n) + t = m + (n + t)$. □

Notes: (i) Since for all natural numbers m, n, and t, we have $(m + n) + t = m + (n + t)$, we have now formally proved that addition is **associative** in $\mathbb{N}$.

(ii) Recall from Lesson 2 that a **monoid** is a pair $(S,\star)$, where S is a set, $\star$ is an associative binary operation on S, and there is an identity $e \in S$ with respect to the operation $*$. Theorems 5.7, 5.8, and 5.9 together provide us with a formal proof that $(\mathbb{N}, +)$ is a monoid.

(iii) Addition is also **commutative** in $\mathbb{N}$. That is, for all natural numbers m and n, $m + n = n + m$. You will be asked to prove this in part (i) of Problem 4 below. So, $(\mathbb{N}, +)$ is a **commutative monoid**.

Just like the sum operation, the product operation on the set of natural numbers has many nice algebraic properties such as

1. **Closure:** For all natural numbers m and n, mn is a natural number.
2. **Identity:** For all natural numbers n, $1 \cdot n = n \cdot 1 = n$.
3. **Associativity:** For all natural numbers m, n, and t, $(mn)t = m(nt)$.
4. **Commutativity:** For all natural numbers m and n, $mn = nm$.
5. **Distributivity:** For all natural numbers m, n, and t, $m(n + t) = mn + mt$.

The proofs that these properties hold are very similar to the proofs already given for addition. You will be asked to provide detailed proofs in Problem 4 below.

Theorem 5.10: For every natural number n, there is a natural number j such that $n = 2j$ or $n = 2j + 1$.

If there is a natural number j such that $n = 2j$, then we say that n is **even**. If there is a natural number j such that $n = 2j + 1$, then we say that n is odd. So, Theorem 5.10 says that every natural number is either even or odd.

Proof of Theorem 5.10: Base Case ($k = 0$): $0 = 2 \cdot 0$ by definition of mutiplication in $\mathbb{N}$.

Inductive Step: Suppose that $k \in \mathbb{N}$ and there is $j \in \mathbb{N}$ such that $k = 2j$ or $k = 2j + 1$. If $k = 2j$, then $k + 1 = 2j + 1$. If $k = 2j + 1$, then $k + 1 = (2j + 1) + 1 = 2j + (1 + 1) = 2j + 2 = 2(j + 1)$. Here we used associativity of addition in $\mathbb{N}$ and distributivity of multiplication over addition in $\mathbb{N}$. Since $\mathbb{N}$ is closed under addition, $j + 1 \in \mathbb{N}$.

By the Principle of Mathematical Induction, for every natural number n, there is a natural number j such that $n = 2j$ or $n = 2j + 1$. □

Sometimes a statement involving the natural numbers may be false for 0, but true from some natural number on. In this case, we can still use induction. We just need to adjust the base case (see Problem 15 below).

Theorem 5.11: $n^2 > 2n + 1$ for all natural numbers $n \geq 3$.

Proof: Base Case ($k = 3$): $3^2 = 9$ and $2 \cdot 3 + 1 = 6 + 1 = 7$. So, $3^2 > 2 \cdot 3 + 1$.

Inductive Step: Let $k \in \mathbb{N}$ with $k \geq 3$ and assume that $k^2 > 2k + 1$. Then we have

$$(k + 1)^2 = (k + 1)(k + 1) = (k + 1)k + (k + 1)(1) = k^2 + k + k + 1 > (2k + 1) + k + k + 1$$
$$= 2k + 2 + k + k = 2(k + 1) + k + k \geq 2(k + 1) + 1 \text{ (because } k + k \geq 3 + 3 = 6 \geq 1\text{)}.$$

By the Principle of Mathematical Induction, $n^2 > 2n + 1$ for all $n \in \mathbb{N}$ with $n \geq 3$. □

Notes: (1) If we have a sequence of equations and inequalities of the form =, ≥, and > (with at least one inequality symbol appearing), beginning with a and ending with b, then the final result is $a > b$ if > appears at least once and $a \geq b$ otherwise.

For example, if $a = j = h = m > n = p = q \geq b$, then $a > b$. The sequence that appears in the solution above has this form.

(2) By definition, $x^2 = x \cdot x$. We used this in the first equality in the inductive step to write $(k + 1)^2$ as $(k + 1)(k + 1)$.

(3) For the second equality in the inductive step, we used distributivity to write $(k + 1)(k + 1)$ as $(k + 1)k + (k + 1)(1)$.

(4) For the third equality in the inductive step, we used commutativity, distributivity, and the multiplicative identity property to write $(k + 1)k$ as $k(k + 1) = k \cdot k + k \cdot 1 = k^2 + k$. We also used the multiplicative identity property to write $(k + 1)(1) = k + 1$.

(5) Associativity of addition is being used when we write the expression $k^2 + k + k + 1$. Notice the lack of parentheses. Technically speaking, we should have written $(k^2 + k) + (k + 1)$ and then taken another step to rewrite this as $k^2 + \big(k + (k + 1)\big)$. However, since we have associativity, we can simply drop all those parentheses.

(6) The inequality "$k^2 + k + k + 1 > (2k + 1) + k + k + 1$" was attained by using the inductive hypothesis "$k^2 > 2k + 1$" together with part (x) from Problem 4 below.

(7) The dedicated reader should verify that the remaining equalities and inequalities in the proof are valid by determining which properties were used at each step.

Let $(G,\star)$ be a group and let $x \in G$. Recall that we define the **powers** of x as follows:

$x^0 = e$, $x^1 = x$, and for $n > 0$, $x^{n+1} = x^n \cdot x$ and $x^{-n} = (x^n)^{-1}$. So, for example, x^{-1} is the inverse of x and $x^{-2} = (x^2)^{-1}$ is the inverse of $x^2 = x \cdot x$. We will now use POMI to help us prove that the "usual laws of exponents" hold in groups (in the expression x^n, n is called an **exponent**).

Theorem 5.12: Let $(G, \cdot)$ be a group, let $x \in G$, and let $r, s \in \mathbb{Z}$. Then $x^r x^s = x^{r+s}$.

Analysis: We will need to prove this theorem in 5 parts:

1. We will first use induction on s to prove that the theorem is true whenever r and s are positive integers. This is the only part that we will use induction for.
2. We will then prove that the theorem is true if $r = 0$ or $s = 0$. This part is trivial.
3. We will then prove that the theorem is true when both exponents are negative.
4. Next we will assume that the first exponent is positive and the second is negative. This part is a bit tedious and will need to be split up into 3 cases.
5. Finally, we will assume that the first exponent is negative and the second is positive. This part is similar to part 4 (and equally tedious), and so, it will be left to the reader.

Proof of Theorem 5.12: We label each part of the proof as we did in the analysis above.

1. We first assume that $r, s \geq 1$ and proceed by induction on $s \geq 1$.

Base Case ($k = 1$): $x^r x^1 = x^r \cdot x = x^{r+1}$ by definition.

Inductive Step: Let $k \in \mathbb{N}$ with $k \geq 1$ and assume that $x^r x^k = x^{r+k}$. Then we have

$$x^r x^{k+1} = x^r(x^k \cdot x) = (x^r x^k) \cdot x = x^{r+k} \cdot x = x^{(r+k)+1} = x^{r+(k+1)}.$$

For the first equality, we used the definition of x^{k+1}. For the second equality, we used associativity of the group operation. For the third equality, we used the inductive hypothesis. For the fourth equality, we used the definition of $x^{(r+k)+1}$. For the fifth equality, we used associativity of addition in $\mathbb{N}$.

By the Principle of Mathematical Induction, $x^r x^s = x^{r+s}$ for all $s \in \mathbb{N}$ with $n \geq 1$.

2. If $r = 0$, then $x^r x^s = x^0 x^s = e x^s = x^s = x^{0+s}$. If $s = 0$, then $x^r x^s = x^r x^0 = x^r e = x^r = x^{r+0}$.

3. If $r, s > 0$, then we have $x^{-r} x^{-s} = (x^r)^{-1}(x^s)^{-1} = (x^s x^r)^{-1} = (x^{s+r})^{-1} = x^{-(s+r)} = x^{-r+(-s)}$.

For the first equality, we used the definitions of x^{-r} and x^{-s}. For the second equality, we used part (i) of Problem 15 from Problem Set 2. For the third equality, we used part 1 that we proved above. For the fourth equality, we used the definition of $x^{-(s+r)}$. For the fifth equality, we used distributivity of multiplication over addition in $\mathbb{N}$ and commutativity of addition in $\mathbb{N}$.

4. For $x^r x^{-s}$, we will need 3 cases. For each case, we assume that $r, s > 0$.

Case 1: $(r = s)$: $x^r x^{-s} = x^r x^{-r} = x^r(x^r)^{-1} = e = x^0 = x^{r-r} = x^{r+(-r)} = x^{r+(-s)}$.

Case 2: $(r > s)$: $x^{r-s}x^s = x^{(r-s)+s}$ (by part 1 above) $= x^r$. So,
$$x^r x^{-s} = (x^{r-s}x^s)(x^{-s}) = x^{r-s}(x^s \cdot (x^s)^{-1}) = x^{r-s}e = x^{r-s} = x^{r+(-s)}.$$

Case 3: $(r < s)$: $x^{s-r}x^r = x^{(s-r)+r}$ (by part 1 above) $= x^s = (x^{-s})^{-1}$. So,
$$(x^r x^{-s})^{-1} = (x^{-s})^{-1}(x^r)^{-1} = (x^{s-r}x^r)(x^r)^{-1} = x^{s-r}(x^r(x^r)^{-1}) = x^{s-r}e = x^{-(r-s)} = (x^{r-s})^{-1}.$$

Therefore, $x^r x^{-s} = x^{r-s} = x^{r+(-s)}$.

5. $x^{-r}x^s$ requires 3 cases that are similar to the previous 3 cases. So, I leave them to the reader. □

Recall that a group $(G,\star)$ is **cyclic** if there is an element $x \in G$ such that $G = \langle x \rangle = \{x^n \mid n \in \mathbb{Z}\}$. In this case, x is called a generator of $(G,\star)$.

Theorem 5.12 allows us to prove the following really nice result quite easily.

Corollary 5.13: All cyclic groups are commutative.

Proof: Let $(G,\star)$ be a cyclic group, let x be a generator of $(G,\star)$, and let $a, b \in G$. There are $r, s \in \mathbb{Z}$ such that $a = x^r$ and $b = x^s$. Then, by Theorem 5.12, $ab = x^r x^s = x^{r+s} = x^{s+r} = x^s x^r = ba$. Since a and b were arbitrary elements of G, we see that $(G,\star)$ is commutative. □

Theorem 5.14: Let $(G,\star)$ be a group, let $x \in G$, and let $r, s \in \mathbb{Z}$. Then $(x^r)^s = x^{rs}$.

Analysis: We will need to prove this theorem in 3 parts:

1. We will first use induction on s to prove that the theorem is true whenever r and s are positive integers. This is the only part that we will use induction for.
2. We will then prove that the theorem is true if $s = 0$ (r can be any integer). This part is trivial.
3. We will then prove that the theorem is true if s is negative (once again, r can be any integer).

Proof: We label each part of the proof as we did in the analysis above.

1. We first assume that $r, s \geq 1$ and proceed by induction on $s \geq 1$.

Base Case $(k = 1)$: $(x^r)^1 = x^r = x^{r \cdot 1}$.

Inductive Step: Let $k \in \mathbb{N}$ with $k \geq 1$ and assume that $(x^r)^k = x^{rk}$. Then we have
$$(x^r)^{k+1} = (x^r)^k x^r = x^{rk} x^r = x^{rk+r} = x^{r(k+1)}.$$

For the first equality, we used the definition of $(x^r)^{k+1}$. For the second equality, we used the inductive hypothesis. For the third equality, we used Theorem 5.12. For the fourth equality, we used distributivity of multiplication over addition in $\mathbb{N}$.

By the Principle of Mathematical Induction, $(x^r)^s = x^{rs}$ for all $s \in \mathbb{N}$ with $n \geq 1$.

2. Now, let $r \in \mathbb{Z}$ and $s = 0$. Then $(x^r)^0 = e = x^0 = x^{r \cdot 0}$.

3. Finally, let $r, s \in \mathbb{Z}$ with $s < 0$. Then $(x^r)^s = ((x^r)^{-s})^{-1} = \left(x^{r(-s)}\right)^{-1} = \left(x^{-(rs)}\right)^{-1} = x^{rs}$.

For the first equality, we used the definition of $((x^r)^{-s})^{-1}$. For the second equality, we used part 1 that we proved above. For the third equality, we used associativity and commutativity of multiplication in $\mathbb{N}$. For the fourth equality, we used the definition of $\left(x^{-(rs)}\right)^{-1}$. $\square$

As a nice Corollary to Theorem 5.14, in a group, interchanging powers with inverses does not change the result of a computation.

Corollary 5.15: Let $(G, \star)$ be a group, let $x \in G$, and let $n \in \mathbb{Z}$. Then $(x^{-1})^n = (x^n)^{-1}$.

By Corollary 5.15, we can write x^{-n} as $(x^n)^{-1}$ or $(x^{-1})^n$. In other words, we can raise x to the power of n first and then take the inverse or we can take the inverse of x first and then raise to the power of n. We will get the same result either way.

Proof of Theorem 5.15: Let $x \in G$ and let $n \in \mathbb{Z}$. Then $(x^{-1})^n = x^{-1n} = x^{-n} = (x^n)^{-1}$.

For the first equality, we used Theorem 5.14. For the second equality, we used the fact that $-1n = -n$ in $\mathbb{N}$. For the third equality, we used the definition of x^{-n}. $\square$

We will now use the Principle of Mathematical Induction to prove that associativity implies generalized associativity. Recall that a binary operation $\star$ is associative in a set S if for all $x, y, z \in S$, we have

$$(x \star y) \star z = x \star (y \star z)$$

Note: It is not essential that you understand the following definitions and the proof of Theorem 5.16. If you find any of it to be too difficult, then you can feel free to skip ahead to the next section on the integers without missing out on anything important. You will probably want to skip Problem 26 below as well.

Before giving a formal definition of generalized associativity, let's introduce some new notation. We will be using the capital greek letter $\prod$ (pronounced "pi") to represent products.

$$\prod_{i=1}^{1} x_i = x_1, \quad \prod_{i=1}^{2} x_i = x_1 \star x_2, \ldots \quad \prod_{i=1}^{n} x_i = x_1 \star x_2 \star \cdots \star x_n = (x_1 \star \cdots \star x_{n-1}) \star x_n \ (n \geq 3)$$

For example, for $n = 3$ and $n = 4$, we have the following:

$$\prod_{i=1}^{3} x_i = (x_1 \star x_2) \star x_3, \quad \prod_{i=1}^{4} x_i = ((x_1 \star x_2) \star x_3) \star x_4$$

So, when writing a product, of more than two elements without parentheses, we will assume we are associating toward the left.

To help ensure that the proof of Theorem 5.16 below goes smoothly, we also make the following definition:

$$\prod_{i=1}^{0} x_i = e$$

Notes: (1) If S has an identity with respect to $\star$ (for example, if $(S,\star)$ is a monoid), then e written above is that identity. Otherwise, just think of e as a new symbol **not** in S such that whenever $x \in S$, then $e \star x$ and $x \star e$ are both equal to x. In other words, e is behaving just like an identity would, regardless of whether e is actually an element of the set S.

(2) For all $n \in \mathbb{N}^+$, we have

$$\prod_{i=1}^{n} x_i = (x_1 \star \cdots \star x_{n-1}) \star x_n = \prod_{i=1}^{n-1} x_i \star \prod_{i=1}^{1} x_{(n-1)+i} = \left(\prod_{i=1}^{n-1} x_i\right) \star x_n.$$

Observe that in the special case where $n = 1$, we have

$$\prod_{i=1}^{n} x_i = \prod_{i=1}^{1} x_i = x_1 = e \star x_1 = \prod_{i=1}^{0} x_i \star \prod_{i=1}^{1} x_{0+i} = \prod_{i=1}^{n-1} x_i \star \prod_{i=1}^{1} x_{(n-1)+i} = \left(\prod_{i=1}^{n-1} x_i\right) \star x_n$$

We can now say that an operation $\star$ satisfies **generalized associativity** in a set S if for all $m, n \in \mathbb{N}^+$ and for all $x_1, x_2, \ldots, x_{m+n} \in S$, we have

$$\prod_{i=1}^{m} x_i \star \prod_{i=1}^{n} x_{m+i} = \prod_{i=1}^{m+n} x_i.$$

For example, if $m = 2$ and $n = 3$, then we get the following instance of generalized associativity:

$$(x_1 \star x_2) \star \big((x_3 \star x_4) \star x_5\big) = \prod_{i=1}^{2} x_i \star \prod_{i=1}^{3} x_{2+i} = \prod_{i=1}^{5} x_i = \big(((x_1 \star x_2) \star x_3) \star x_4\big) \star x_5$$

Theorem 5.16: Let $(S,\star)$ be a semigroup. Then $\star$ satisfies generalized associativity in S.

Proof: Let $(S,\star)$ be a semigroup. We will prove this theorem by induction on n, where n is the number of elements being multiplied. The case $k = 1$ is vacuously satisfied and the case $k = 2$ is trivial. So, we begin with $k = 3$ as the base case.

Base Case ($k = 3$): Since $\star$ is associative in S, using Note 2 above, we have

$$\prod_{i=1}^{3} x_i = \prod_{i=1}^{2} x_i \star \prod_{i=1}^{1} x_{2+i} = (x_1 \star x_2) \star x_3 = x_1 \star (x_2 \star x_3) = \prod_{i=1}^{1} x_i \star \prod_{i=1}^{2} x_{1+i}.$$

Inductive Step: Assume that $k \in \mathbb{N}$ with $k \geq 3$ and that for all $m, n \in \mathbb{N}^+$ with $m + n = k$, and for all $x_1, x_2, \ldots, x_{m+n} \in S$, we have

$$\prod_{i=1}^{m} x_i \star \prod_{i=1}^{n} x_{m+i} = \prod_{i=1}^{m+n} x_i.$$

Let $m, n \in \mathbb{N}^+$ with $m + n = k + 1$. Then we have

$$\prod_{i=1}^{m} x_i \star \prod_{i=1}^{n} x_{m+i} = \prod_{i=1}^{m} x_i \star \left(\left(\prod_{i=1}^{n-1} x_{m+i}\right) \star x_{k+1}\right) = \left(\prod_{i=1}^{m} x_i \star \prod_{i=1}^{n-1} x_{m+i}\right) \star x_{k+1}$$

$$= \left(\prod_{i=1}^{m+n-1} x_i\right) \star x_{k+1} = \prod_{i=1}^{m+n} x_i.$$

For the first equality, we used Note 2 above. For the second equality, we used the associativity of $\star$ in S. For the third equality, we used the inductive hypothesis (note that $m + (n - 1) = k$). For the fourth equality, we again used Note 2 above.

By the Principle of Mathematical Induction, for all $m, n \in \mathbb{N}^+$, and for all $x_1, x_2, \ldots, x_{m+n} \in S$, we have

$$\prod_{i=1}^{m} x_i \star \prod_{i=1}^{n} x_{m+i} = \prod_{i=1}^{m+n} x_i. \qquad \square$$

The Integers

To motivate the definition of the integers, note that we can think of every integer as a difference of two natural numbers. For example, the integer -3 can be thought of as $1 - 4$. However, -3 can also be thought of as $2 - 5$. So, we must insist that $1 - 4 = 2 - 5$, or equivalently, $1 + 5 = 2 + 4$.

We define a relation R on $\mathbb{N} \times \mathbb{N}$ by $R = \{((a,b),(c,d)) \in (\mathbb{N} \times \mathbb{N})^2 \mid a + d = b + c\}$. In part 5 of Example 3.12, we showed that this relation is an equivalence relation. We can now define the set of integers to be the set of equivalence classes for this equivalence relation. That is, we define the set of integers to be $\mathbb{Z} = \{[(a,b)] \mid (a,b) \in \mathbb{N} \times \mathbb{N}\}$.

We identify the integer $[(n, 0)]$ with the natural number n. In this way, we have $\mathbb{N} \subseteq \mathbb{Z}$.

We define the ordering $<_\mathbb{Z}$ on $\mathbb{Z}$ by $[(a,b)] <_\mathbb{Z} [(c,d)]$ if and only if $a + d <_\mathbb{N} b + c$, where $<_\mathbb{N}$ is the usual ordering on $\mathbb{N}$ ($n <_\mathbb{N} m$ if and only if $n \in m$).

In Problem 12 below, you will be asked to show that $<_\mathbb{Z}$ is a well-defined strict linear ordering on $\mathbb{Z}$.

We add and multiply two integers using the following rules:
$$[(a,b)] + [(c,d)] = [(a+c, b+d)]$$
$$[(a,b)] \cdot [(c,d)] = [(ac + bd, ad + bc)]$$

In Problems 16 and 29 below, you will be asked to show that these two operations are well-defined.

Notes: (1) We will usually abbreviate $<_\mathbb{Z}$ simply by $<$, especially if it is already clear that we are working with the integers.

(2) If $a, b \in \mathbb{N}$ with $a \geq b$, then $[(a, b)] = [(a - b, 0)]$. If $a < b$, then $[(a, b)] = [(0, b - a)]$. In this way, we see that every integer can be written in the form $[(n, 0)]$ or $[(0, n)]$ for some $n \in \mathbb{N}$. We abbreviate $[(n, 0)]$ by n and we abbreviate $[(0, n)]$ by $-n$. For example, $[(2, 7)] = [(0, 5)] = -5$.

Example 5.17:

1. $[(k, k)] = [(0, 0)]$ for all $k \in \mathbb{N}$ because $k + 0 = k + 0$.

2. $[(5, 0)] = [(6, 1)]$ because $5 + 1 = 0 + 6$. Similarly, we have $[(5, 0)] = [(7, 2)] = [(8, 3)]$, and in general $[(5, 0)] = [(5 + k, k)]$ for any natural number k. $[(5, 0)]$ is the most "natural" way to express the natural number 5 as an integer. More generally, the natural number n can be expressed as an integer as $[(n, 0)]$.

3. We usually abbreviate the integer $[(0, k)]$ as $-k$. For example, -3 is an abbreviation for $[(0, 3)]$. We can also write -3 as $[(1, 4)]$ because $1 + 3 = 4 + 0$.

4. $[(0, 0)] < [(4, 0)]$ because $0 + 0 < 0 + 4$. More generally, for any natural number $k \neq 0$, we have $[(0, 0)] < [(k, 0)]$ because $0 + 0 < 0 + k$. This shows that for any natural number $k \neq 0$, the natural number k satisfies $0 < k$.

5. $[(0, 4)] < [(0, 0)]$ because $0 + 0 < 4 + 0$. More generally, for any natural number $k \neq 0$, we have $[(0, k)] < [(0, 0)]$ because $0 + 0 < k + 0$. This shows that for any natural number $k \neq 0$, the integer $-k$ satisfies $-k < 0$.

6. $7 + (-2) = [(7, 0)] + [(0, 2)] = [(7, 2)] = [(5, 0)] = 5$.

7. $-3 \cdot 5 = [(0, 3)] \cdot [(5, 0)] = [(0 \cdot 5 + 3 \cdot 0, 0 \cdot 0 + 3 \cdot 5)] = [(0, 15)] = -15$.

The Rational Numbers

In part 6 of Example 3.12, we showed that $R = \{((a, b), (c, d)) \in (\mathbb{Z} \times \mathbb{Z}^*)^2 \mid ad = bc\}$ is an equivalence relation on $\mathbb{Z} \times \mathbb{Z}^*$. For each $a \in \mathbb{Z}$ and $b \in \mathbb{Z}^*$, we define the **rational number** $\frac{a}{b}$ to be the equivalence class of (a, b). So, $\frac{a}{b} = [(a, b)]$, and we have $\frac{a}{b} = \frac{c}{d}$ if and only if $(a, b)R(c, d)$ if and only if $ad = bc$. The set of rational numbers is $\mathbb{Q} = \{\frac{a}{b} \mid a \in \mathbb{Z} \land b \in \mathbb{Z}^*\}$. In words, $\mathbb{Q}$ is "the set of quotients a over b such that a and b are integers and b is not zero."

We identify the rational number $\frac{a}{1}$ with the integer a. In this way, we have $\mathbb{Z} \subseteq \mathbb{Q}$.

We define $<_\mathbb{Q}$ on $\mathbb{Q}$ by $\frac{a}{b} <_\mathbb{Q} \frac{c}{d}$ if and only if $ad <_\mathbb{Z} bc$, where $<_\mathbb{Z}$ is the usual ordering on $\mathbb{Z}$.

In Problem 31 below, you will be asked to show that $<_\mathbb{Q}$ is a well-defined strict linear ordering on $\mathbb{Q}$.

We add and multiply two rational numbers using the following rules:

$$\frac{a}{b} + \frac{c}{d} = \frac{a \cdot d + b \cdot c}{b \cdot d} \qquad \frac{a}{b} \cdot \frac{c}{d} = \frac{a \cdot c}{b \cdot d}$$

In Problem 17 below, you will be asked to show that these two operations are well-defined.

Example 5.18:

1. $\frac{2}{3} < \frac{5}{4}$ because $2 \cdot 4 < 3 \cdot 5$. Also, $\frac{2}{3} + \frac{5}{4} = \frac{2 \cdot 4 + 3 \cdot 5}{3 \cdot 4} = \frac{23}{12}$ and $\frac{2}{3} \cdot \frac{5}{4} = \frac{2 \cdot 5}{3 \cdot 4} = \frac{10}{12} = \frac{5}{6}$ (because $10 \cdot 6 = 12 \cdot 5$).

2. $\frac{-3}{4} < \frac{-5}{7}$ because $-3 \cdot 7 < 4(-5)$, $\frac{-3}{4} + \frac{-5}{7} = \frac{-3 \cdot 7 + 4(-5)}{4 \cdot 7} = \frac{-41}{28}$, and $\frac{-3}{4} \cdot \frac{-5}{7} = \frac{-3(-5)}{4 \cdot 7} = \frac{15}{28}$.

The Real Numbers

There are several equivalent ways to define the set of real numbers. We will define this set here as equivalence classes of **Cauchy sequences**. Informally, a Cauchy sequence is a rational-valued sequence whose values get "closer and closer to each other" as we go further out into the sequence.

We can recognize a Cauchy sequence as one that seems to be "converging" to a fixed value.

For example, it looks like the sequence $(x_n) = \left(\frac{1}{n+1}\right) = \left(1, \frac{1}{2}, \frac{1}{3}, \frac{1}{4}, \frac{1}{5}, \ldots\right)$ is a Cauchy sequence, as it seems to be converging to 0. The sequence $(x_n) = (n) = (0, 1, 2, 3, 4, 5, \ldots)$ is not a Cauchy sequence, as it does not seem to be converging to a fixed number. See Example 5.19 below for more details.

Note: Cauchy sequences will not be of much importance in this book. I am defining them for the sole purpose of ensuring that we have a concrete definition of the real numbers. If you have trouble understanding the formal definition presented below, you should not be concerned. If you continue to think of the real numbers informally, as we have been doing up until this point, you will not be at a disadvantage when reading any other part of this book.

Let's now define Cauchy sequence more formally. We first make a few preliminary definitions.

If $x = \frac{a}{b}$ is a rational number, then we define $-x$ to be $\frac{-a}{b}$.

The **absolute value** of the rational number x is then defined by $|x| = \begin{cases} x & \text{if } x \geq 0. \\ -x & \text{if } x < 0. \end{cases}$

For example, we have $\left|\frac{2}{3}\right| = \frac{2}{3}$, $|0| = 0$, and $\left|\frac{-5}{7}\right| = -\left(\frac{-5}{7}\right) = \frac{-(-5)}{7} = \frac{5}{7}$.

The **distance** between rational numbers x and y is $|x - y|$. For example, the distance between 3 and 5 is $|3 - 5| = |-2| = 2$ and the distance between $\frac{1}{5}$ and $\frac{1}{7}$ is $\left|\frac{1}{5} - \frac{1}{7}\right| = \left|\frac{7-5}{5 \cdot 7}\right| = \left|\frac{2}{35}\right| = \frac{2}{35}$.

Notice that the distance between 5 and 3 is the same as the distance between 3 and 5. Indeed, we also have $|5 - 3| = |2| = 2$. In general, for rational numbers x and y, we have $|x - y| = |y - x|$.

Let $f = (x_n)$ be a rational-valued sequence. We say that f is a **Cauchy sequence** if

for every $k \in \mathbb{N}^+$, there is $K \in \mathbb{N}$ such that $m \geq n > K$ implies $|x_m - x_n| < \frac{1}{k}$.

The idea is that we can make the distance between any two terms of the sequence as small as we choose by deleting a finite portion of the beginning of the sequence. If we wish to make the distance between any two terms less than $\frac{1}{k}$, we delete the first $K + 1$ terms of the sequence.

Example 5.19:

1. The sequence $(x_n) = \left(\frac{1}{n+1}\right)$ is a Cauchy sequence. To see this, let $k \in \mathbb{N}$, let $K = k$, and let $m \geq n > K$. Then
$$|x_m - x_n| = \left|\frac{1}{m+1} - \frac{1}{n+1}\right| = \left|\frac{n-m}{(m+1)(n+1)}\right| \leq \left|\frac{m}{(m+1)(n+1)}\right| = \left|\frac{m}{mn+m+n+1}\right| \leq \left|\frac{m}{mn}\right| = \frac{1}{n} < \frac{1}{K} = \frac{1}{k}.$$

2. The sequence $(x_n) = ((-1)^n n)$ is **not** a Cauchy sequence. To see this, let $k = 1$ and let $K \in \mathbb{N}$. Then $K + 2 \geq K + 1 > K$, and
$$|x_{K+2} - x_{K+1}| = |(-1)^{K+2}(K+2) - (-1)^{K+1}(K+1)| = |(K+2) + (K+1)| = 2K + 3 \geq 3$$
(See the Note below). However, $\frac{1}{k} = \frac{1}{1} = 1$ and $3 \not< 1$.

3. For each $q \in \mathbb{Q}$, the **constant sequence** $(x_n) = (q)$ is a Cauchy sequence. To see this, let $k \in \mathbb{N}$, let $K = 0$, and let $m \geq n > 0$. Then
$$|x_m - x_n| = |q - q| = |0| = 0 < \frac{1}{k}.$$

Note: If K is even, then $|(-1)^{K+2}(K+2) - (-1)^{K+1}(K+1)| = (K+2) + (K+1) = 2K + 3$, whereas, if K is odd, then $(-1)^{K+2}(K+2) - (-1)^{K+1}(K+1) = -(K+2) - (K+1) = -(2K+3)$.

Next, we would like to identify Cauchy sequences that seem to be converging to the same value. For example, we will identify the Cauchy sequences $(x_n) = \left(\frac{1}{n+1}\right)$ and $(y_n) = (0)$.

An equivalent way of saying that (x_n) and (y_n) converge to the same value is to say that $(x_n - y_n)$ converges to 0.

Let $A = \{(x_n) \mid (x_n) \text{ is a Cauchy sequence of rational numbers}\}$. We define a relation R on A by:

$(x_n) R (y_n)$ if and only if for every $k \in \mathbb{N}^+$, there is $K \in \mathbb{N}$ such that $n > K$ implies $|x_n - y_n| < \frac{1}{k}$.

In Problem 18 below, you will be asked to show that R is an equivalence relation on A.

Note: To show that R is transitive, you will need to use the **Triangle Inequality**. The Triangle Inequality says that if $a, b \in \mathbb{Q}$, then $|a + b| \leq |a| + |b|$.

One way to prove the Triangle Inequality is to analyze several cases separately. As an example of one such case, suppose that a and b are both nonnegative. Then $|a| = a$, $|b| = b$, and $|a + b| = a + b$. So, we get $|a + b| = a + b = |a| + |b|$ (in this case we get equality). I leave it to the reader to describe the other cases and to prove that the Triangle Inequality is true for each of these cases.

We can now define the set of real numbers as follows:
$$\mathbb{R} = \{[(x_n)] \mid (x_n) \text{ is a Cauchy sequence of rational numbers}\}.$$

We identify the real number $[(q)]$ with the rational number q. In this way, we have $\mathbb{Q} \subseteq \mathbb{R}$.

We define the ordering $\leq_\mathbb{R}$ on $\mathbb{R}$ by

$[(x_n)] \leq_\mathbb{R} [(y_n)]$ if and only if there is $K \in \mathbb{N}$ such that $n > K$ implies $x_n \leq y_n$.

We can then define $<_{\mathbb{R}}$ on $\mathbb{R}$ by $[(x_n)] <_{\mathbb{R}} [(y_n)]$ if and only if $[(x_n)] \leq_{\mathbb{R}} [(y_n)]$ and $[(x_n)] \neq [(y_n)]$.

In Problem 31 below, you will be asked to show that $<_{\mathbb{R}}$ is a well-defined strict linear ordering on $\mathbb{R}$.

We add and multiply two real numbers using the following rules:
$$[(x_n)] + [(y_n)] = [(x_n + y_n)]$$
$$[(x_n)] \cdot [(y_n)] = [(x_n \cdot y_n)]$$

In Problems 27 and 30 below, you will be asked to show that $+$ and $\cdot$ are well-defined operations on the real numbers and that the sum and product of two real numbers are real numbers.

Note: We will generally drop the subscript from $<_{\mathbb{R}}$ and simply write $<$.

The Complex Numbers

The set of complex numbers is $\mathbb{C} = \{a + bi \mid a, b \in \mathbb{R}\}$.

See part 3 of Example 1.6 to see how to visualize the complex number $a + bi$ as a point in the Complex Plane and see Note 5 following Example 2.17 to see how to visualize $a + bi$ as a directed line segment (or **vector**) starting at the origin and ending at the point (a, b).

If $z = a + bi$ is a complex number, we call a the **real part** of z and b the **imaginary part** of z, and we write $a = \text{Re } z$ and $b = \text{Im } z$.

Two complex numbers are **equal** if and only if they have the same real part and the same imaginary part. In other words,
$$a + bi = c + di \quad \text{if and only if} \quad a = c \text{ and } b = d.$$

We add two complex numbers by simply adding their real parts and adding their imaginary parts. So,
$$(a + bi) + (c + di) = (a + c) + (b + d)i.$$

As a point, this sum is $(a + c, b + d)$. We can visualize this sum as the vector starting at the origin that is the diagonal of the parallelogram formed from the vectors $a + bi$ and $c + di$. Here is an example showing that $(1 + 2i) + (-3 + i) = -2 + 3i$.

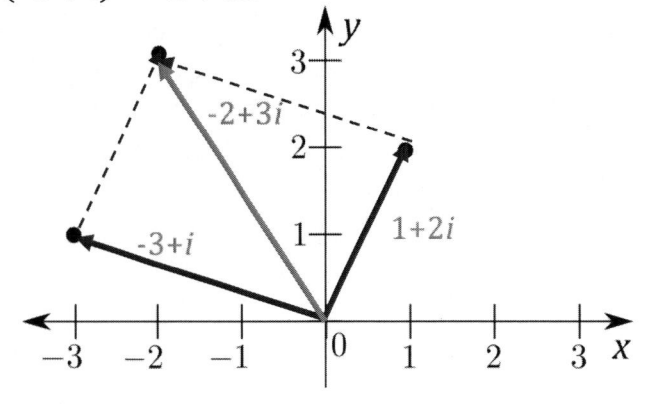

The definition for multiplying two complex numbers is a bit more complicated:
$$(a + bi)(c + di) = (ac - bd) + (ad + bc)i.$$

Notes: (1) If $b = 0$, then we call $a + bi = a + 0i = a$ a **real number**. Note that when we add or multiply two real numbers, we always get another real number.
$$(a + 0i) + (b + 0i) = (a + b) + (0 + 0)i = (a + b) + 0i = a + b.$$
$$(a + 0i)(b + 0i) = (ab - 0 \cdot 0) + (a \cdot 0 + 0b)i = (ab - 0) + (0 + 0)i = ab + 0i = ab.$$

In this way, we have $\mathbb{R} \subseteq \mathbb{C}$.

(2) If $a = 0$, then we call $a + bi = 0 + bi = bi$ a **pure imaginary number**.

(3) $i^2 = -1$. To see this, note that $i^2 = i \cdot i = (0 + 1i)(0 + 1i)$, and we have
$$(0 + 1i)(0 + 1i) = (0 \cdot 0 - 1 \cdot 1) + (0 \cdot 1 + 1 \cdot 0)i = (0 - 1) + (0 + 0)i = -1 + 0i = -1.$$

(4) The definition of the product of two complex numbers is motivated by how we expect multiplication should behave, together with replacing i^2 by -1. If we were to naïvely multiply the two complex numbers, we would have
$$(a + bi)(c + di) = (a + bi)c + (a + bi)(di) = ac + bci + adi + bdi^2$$
$$= ac + bci + adi + bd(-1) = ac + (bc + ad)i - bd = (ac - bd) + (ad + bc)i.$$

Those familiar with the mnemonic FOIL may notice that "FOILing" will always work to produce the product of two complex numbers, provided we replace i^2 by -1 and simplify.

Example 5.20: Let $z = 2 - 3i$ and $w = -1 + 5i$. Then
$$z + w = (2 - 3i) + (-1 + 5i) = (2 + (-1)) + (-3 + 5)i = \mathbf{1 + 2i}.$$
$$zw = (2 - 3i)(-1 + 5i) = \bigl(2(-1) - (-3)(5)\bigr) + \bigl(2 \cdot 5 + (-3)(-1)\bigr)i$$
$$= (-2 + 15) + (10 + 3)i = \mathbf{13 + 13i}.$$

The **absolute value** or **modulus** of the complex number $z = a + bi$ is the nonnegative real number
$$|z| = \sqrt{a^2 + b^2} = \sqrt{(\text{Re } z)^2 + (\text{Im } z)^2}$$

Continuing Example 5.20, we have

$|z| = |2 - 3i| = \sqrt{2^2 + (-3)^2} = \sqrt{13}$ $\qquad$ $|w| = |-1 + 5i| = \sqrt{(-1)^2 + 5^2} = \sqrt{26}$

Note: If $z = a + 0i = a$ is a real number, then $|a| = \sqrt{a^2}$. This is equal to a if $a \geq 0$ and $-a$ if $a < 0$.

For example, $|4| = \sqrt{4^2} = \sqrt{16} = 4$ and $|-4| = \sqrt{(-4)^2} = \sqrt{16} = 4 = -(-4)$.

The statement "$|a| = -a$ for $a < 0$" often confuses students. This confusion is understandable, as a minus sign is usually used to indicate that an expression is negative, whereas here we are negating a negative number to make it positive. Unfortunately, this is the simplest way to say, "delete the minus sign in front of the number" using basic notation.

Exponential Form of a Complex Number

A **circle** in the Complex Plane is the set of all points that are at a fixed distance (called the **radius** of the circle) from a fixed point (called the **center** of the circle).

The **circumference** of a circle is the distance around the circle.

If C and C' are the circumferences of two circles with radii r and r', respectively, then it turns out that $\frac{C}{2r} = \frac{C'}{2r'}$. In other words, the value of the ratio $\frac{\text{Circumference}}{2(\text{radius})}$ is independent of the circle that we use to form this ratio. We leave the proof of this fact for the interested reader to investigate themselves. We call the common value of this ratio π (pronounced "pi"). So, we have $\frac{C}{2r} = \pi$, or equivalently, $C = 2\pi r$.

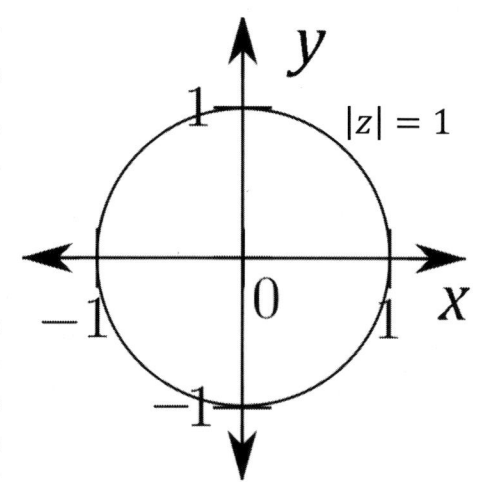

Example 5.21: The **unit circle** is the circle with radius 1 and center $(0,0)$. The equation of this circle is $|z| = 1$. If we write z in the standard form $z = x + yi$, we see that $|z| = \sqrt{x^2 + y^2}$, and so, the equation of the unit circle can also be written $x^2 + y^2 = 1$. To the right is a picture of the unit circle in the Complex Plane.

The circumference of the unit circle is $2\pi \cdot 1 = 2\pi$.

An **angle in standard position** consists of two **rays**, both of which have their initial point at the origin, and one of which is the positive x-axis. We call the positive x-axis the **initial ray** and we call the second ray the **terminal ray**. The **radian measure** of the angle is the part of the circumference of the unit circle beginning at the point $(1, 0)$ on the positive x-axis and *eventually* ending at the point on the unit circle intercepted by the second ray. If the motion is in the counterclockwise direction, the radian measure is positive and if the motion is in the clockwise direction, the radian measure is negative.

Example 5.22: Let's draw a few angles where the terminal ray lies along the line $y = x$.

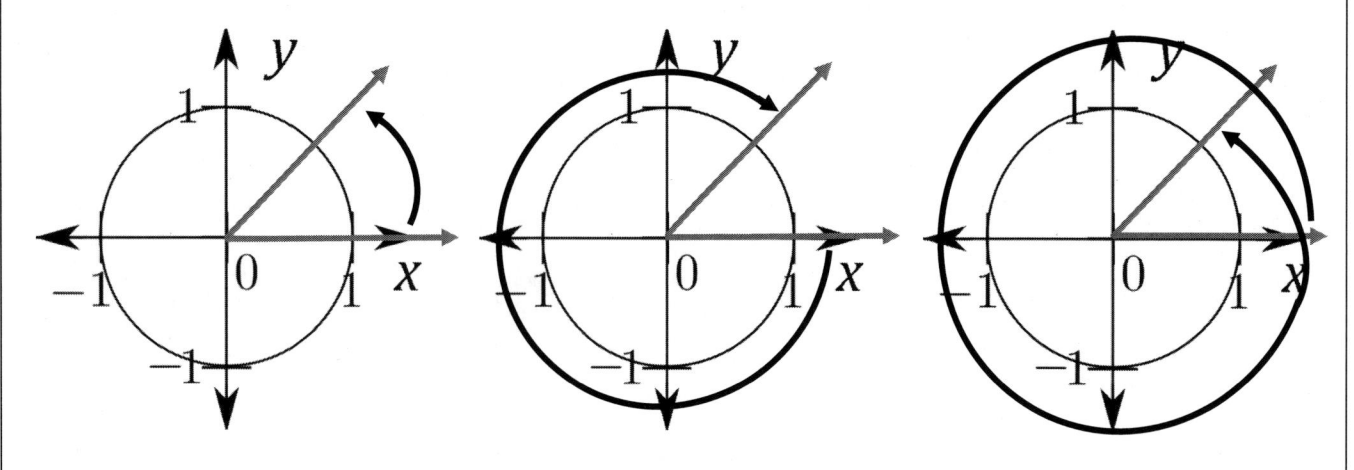

Observe that in the leftmost picture, the arc intercepted by the angle has a length that is one-eighth of the circumference of the circle. Since the circumference of the unit circle is 2π and the motion is in the counterclockwise direction, the angle has a radian measure of $\frac{2\pi}{8} = \frac{\pi}{4}$.

Similarly, in the center picture, the arc intercepted by the angle has a length that is seven-eighths of the circumference of the circle. This time the motion is in the clockwise direction, and so, the radian measure of the angle is $-\frac{7}{8} \cdot 2\pi = -\frac{7\pi}{4}$.

In the rightmost picture, the angle consists of a complete rotation, tracing out the entire circumference of the circle, followed by tracing out an additional length that is one-eighth the circumference of the circle. Since the motion is in the counterclockwise direction, the radian measure of the angle is $2\pi + \frac{2\pi}{8} = \frac{8\pi}{4} + \frac{\pi}{4} = \frac{9\pi}{4}$.

Let's find the point of intersection of the unit circle with the terminal ray of the angle $\frac{\pi}{4}$ that lies along the line with equation $y = x$ (as shown in the leftmost figure from Example 5.22 above). If we call this point (a, b), then we have $b = a$ (because (a, b) is on the line $y = x$) and $a^2 + b^2 = 1$ (because (a, b) is on the unit circle). Replacing b by a in the second equation gives us $a^2 + a^2 = 1$, or equivalently, $2a^2 = 1$. So, $a^2 = \frac{1}{2}$. The two solutions to this equation are $a = \pm\sqrt{\frac{1}{2}} = \pm\frac{\sqrt{1}}{\sqrt{2}} = \pm\frac{1}{\sqrt{2}}$. From the picture, it should be clear that we are looking for the positive solution, so that $a = \frac{1}{\sqrt{2}}$. Since $b = a$, we also have $b = \frac{1}{\sqrt{2}}$. Therefore, the point of intersection is $\left(\frac{1}{\sqrt{2}}, \frac{1}{\sqrt{2}}\right)$.

Notes: (1) The number $\frac{1}{\sqrt{2}}$ can also be written in the form $\frac{\sqrt{2}}{2}$. To see that these two numbers are equal, observe that we have $\frac{1}{\sqrt{2}} = \frac{1}{\sqrt{2}} \cdot 1 = \frac{1}{\sqrt{2}} \cdot \frac{\sqrt{2}}{\sqrt{2}} = \frac{1 \cdot \sqrt{2}}{\sqrt{2} \cdot \sqrt{2}} = \frac{\sqrt{2}}{2}$.

(2) In the figure below on the left, we see a visual representation of the circle, the given angle, and the desired point of intersection.

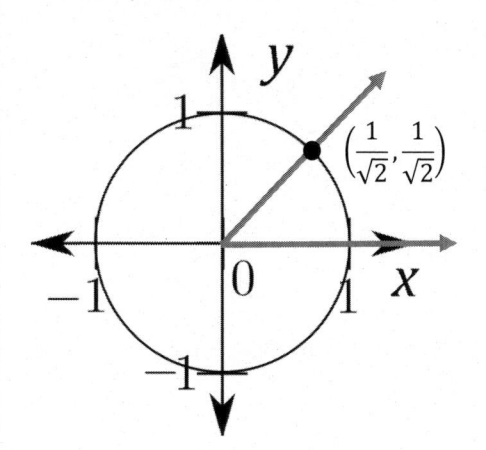

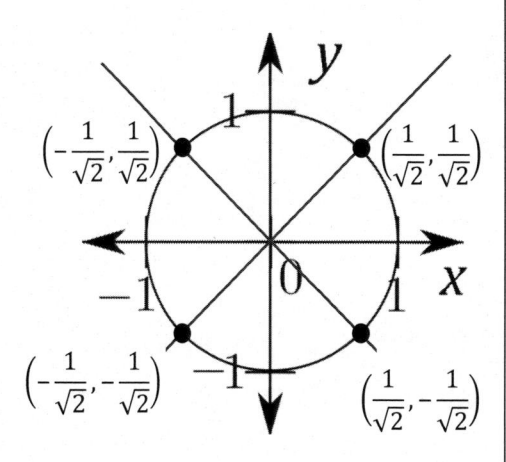

(3) In the figure above on the right, we have divided the Complex Plane into eight regions using the lines with equations $y = x$ and $y = -x$ (together with the x- and y-axes). We then used the symmetry of the circle to label the four points of intersection of the unit circle with each of these two lines.

If θ (pronounced "theta") is the radian measure of an angle in standard position such that the terminal ray intersects the unit circle at the point (x, y), then we will say that $W(\theta) = (x, y)$. This expression defines a function $W: \mathbb{R} \to \mathbb{R} \times \mathbb{R}$ called the **wrapping function**. Observe that the inputs of the wrapping function are real numbers, which we think of as the radian measure of angles in standard position. The outputs of the wrapping function are pairs of real numbers, which we think of as points in the Complex Plane. Also, observe that the range of the wrapping function is the unit circle.

We now define the cosine and sine of the angle θ by $\cos\theta = x$ and $\sin\theta = y$, where $W(\theta) = (x, y)$. For convenience, we also define the tangent of the angle by $\tan\theta = \frac{\sin\theta}{\cos\theta} = \frac{y}{x}$.

Note: The wrapping function is **not** one to one. For example, $W\left(\frac{\pi}{2}\right) = (0, 1)$ and $W\left(\frac{5\pi}{2}\right) = (0, 1)$. However, $\frac{\pi}{2} \neq \frac{5\pi}{2}$. There are actually infinitely many real numbers that map to $(0, 1)$ under the wrapping function. Specifically, $W\left(\frac{\pi}{2} + 2k\pi\right) = (0, 1)$ for every $k \in \mathbb{Z}$.

In general, each point on the unit circle is the image of infinitely many real numbers. Indeed, if $W(\theta) = (a, b)$, then $W(\theta + 2k\pi) = (a, b)$ for all $k \in \mathbb{Z}$.

Example 5.23: Using the rightmost figure above, we can make the following computations:

$$W\left(\frac{\pi}{4}\right) = \left(\frac{1}{\sqrt{2}}, \frac{1}{\sqrt{2}}\right) \quad W\left(\frac{3\pi}{4}\right) = \left(-\frac{1}{\sqrt{2}}, \frac{1}{\sqrt{2}}\right) \quad W\left(\frac{5\pi}{4}\right) = \left(-\frac{1}{\sqrt{2}}, -\frac{1}{\sqrt{2}}\right) \quad W\left(\frac{7\pi}{4}\right) = \left(\frac{1}{\sqrt{2}}, -\frac{1}{\sqrt{2}}\right)$$

$$\cos\frac{\pi}{4} = \frac{1}{\sqrt{2}} \qquad \sin\frac{\pi}{4} = \frac{1}{\sqrt{2}} \qquad \cos\frac{3\pi}{4} = -\frac{1}{\sqrt{2}} \qquad \sin\frac{3\pi}{4} = \frac{1}{\sqrt{2}}$$

$$\cos\frac{5\pi}{4} = -\frac{1}{\sqrt{2}} \qquad \sin\frac{5\pi}{4} = -\frac{1}{\sqrt{2}} \qquad \cos\frac{7\pi}{4} = \frac{1}{\sqrt{2}} \qquad \sin\frac{7\pi}{4} = -\frac{1}{\sqrt{2}}$$

It's also easy to compute the cosine and sine of the four **quadrantal angles** $0, \frac{\pi}{2}, \pi$, and $\frac{3\pi}{2}$. Here we use the fact that the points $(1, 0), (0, 1), (-1, 0)$, and $(0, -1)$ lie on the unit circle.

$$W(0) = (1, 0) \quad W\left(\frac{\pi}{2}\right) = (0, 1) \quad W(\pi) = (-1, 0) \quad W\left(\frac{3\pi}{2}\right) = (0, -1)$$

$$\cos 0 = 1 \qquad \sin 0 = 0 \qquad \cos\frac{\pi}{2} = 0 \qquad \sin\frac{\pi}{2} = 1$$

$$\cos \pi = -1 \qquad \sin \pi = 0 \qquad \cos\frac{3\pi}{2} = 0 \qquad \sin\frac{3\pi}{2} = -1$$

Also, if we add any integer multiple of 2π to an angle, the cosine and sine of the new angle have the same values as the old angle. For example, $\cos\frac{9\pi}{4} = \cos\left(\frac{\pi}{4} + \frac{8\pi}{4}\right) = \cos\left(\frac{\pi}{4} + 2\pi\right) = \cos\frac{\pi}{4} = \frac{1}{\sqrt{2}}$. This is a direct consequence of the fact that $W(\theta + 2k\pi) = W(\theta)$ for all $k \in \mathbb{Z}$.

We can also compute the tangent of each angle by dividing the sine of the angle by the cosine of the angle. For example, we have $\tan\frac{\pi}{4} = \frac{\sin\frac{\pi}{4}}{\cos\frac{\pi}{4}} = \frac{\frac{1}{\sqrt{2}}}{\frac{1}{\sqrt{2}}} = 1$.

Similarly, we have the following: $\tan\frac{3\pi}{4} = -1 \quad \tan\frac{5\pi}{4} = 1 \quad \tan\frac{7\pi}{4} = -1 \quad \tan 0 = 0 \quad \tan \pi = 0$

When $\theta = \frac{\pi}{2}$ or $\frac{3\pi}{2}$, $\tan \theta$ is **undefined**.

Notes: (1) If $z = x + yi$ is any complex number, then the point (x, y) lies on a circle of radius r centered at the origin, where $r = |z| = \sqrt{x^2 + y^2}$. If θ is the radian measure of an angle in standard position such that the terminal ray intersects this circle at the point (x, y), then it can be proved that the cosine and sine of the angle are equal to $\cos \theta = \frac{x}{r}$ and $\sin \theta = \frac{y}{r}$.

(2) It is standard to use the abbreviations $\cos^2 \theta$ and $\sin^2 \theta$ for $(\cos \theta)^2$ and $(\sin \theta)^2$, respectively.

From the definition of cosine and sine, we have the following formula called the **Pythagorean Identity**:

$$\cos^2 \theta + \sin^2 \theta = 1$$

(3) Also, from the definition of cosine and sine, we have the following two formulas called the **Negative Identities**:

$$\cos(-\theta) = \cos \theta \qquad \sin(-\theta) = -\sin \theta.$$

Theorem 5.24: Let θ and ϕ be the radian measures of angles A and B, respectively. Then we have

$$\cos(\theta + \phi) = \cos \theta \cos \phi - \sin \theta \sin \phi$$
$$\sin(\theta + \phi) = \sin \theta \cos \phi + \cos \theta \sin \phi.$$

Notes: (1) The two formulas appearing in Theorem 5.24 are called the **Sum Identities**. You will be asked to prove Theorem 5.24 in Problem 23 below (parts (i) and (v)).

(2) Theorem 5.24 will be used to prove De Moivre's Theorem (Theorem 5.26) below. De Moivre's Theorem provides a fast method for performing exponentiation of complex numbers.

(3) θ and ϕ are Greek letters pronounced "theta" and "phi," respectively. These letters are often used to represent angle measures. We may sometimes also use the capital versions of these letters, Θ and Φ, especially when insisting that the radian measures of the given angles are between $-\pi$ and π.

The **standard form** (or **rectangular form**) of a complex number z is $z = x + yi$, where x and y are real numbers. Recall from part 3 of Example 1.6 that we can visualize the complex number $z = x + yi$ as the point (x, y) in the Complex Plane.

If for $z \neq 0$, we let $r = |z| = |x + yi| = \sqrt{x^2 + y^2}$ and we let θ be the radian measure of an angle in standard position such that the terminal ray passes through the point (x, y), then we see that r and θ determine this point. So, we can also write this point as (r, θ).

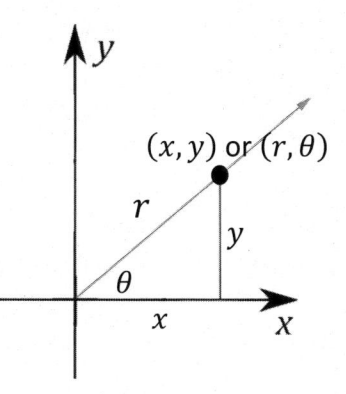

In Note 1 following Example 5.23, we saw that $\cos \theta = \frac{x}{r}$ and $\sin \theta = \frac{y}{r}$. By multiplying each side of these last two equations by r, we get $x = r \cos \theta$ and $y = r \sin \theta$. These equations allow us to rewrite the complex number $z = x + yi$ in the **polar form** $z = r \cos \theta + (r \sin \theta)i = r(\cos \theta + i \sin \theta)$.

If we also make the definition $e^{i\theta} = \cos\theta + i\sin\theta$, we can write the complex number $z = x + yi$ in the **exponential form** $z = re^{i\theta}$.

$r = |z|$ is called the **absolute value** or **modulus** of the complex number. We will call the angle θ an **argument** of the complex number and we may sometimes write $\theta = \arg z$.

Note that although $r = |z|$ and $\theta = \arg z$ uniquely determine a point (r, θ), there are infinitely many other values for $\arg z$ that represent the same point. Indeed, $(r, \theta + 2k\pi)$ represents the same point for each $k \in \mathbb{Z}$. However, there is a unique such value Θ for $\arg z$ such that $-\pi < \Theta \leq \pi$. We call this value Θ the **principal argument** of z, and we write $\Theta = \text{Arg } z$.

Notes: (1) The definition $e^{i\theta} = \cos\theta + i\sin\theta$ is known as **Euler's formula**.

(2) When written in exponential form, two complex numbers $z = re^{i\theta}$ and $w = se^{i\phi}$ are equal if and only if $r = s$ and $\phi = \theta + 2k\pi$ for some $k \in \mathbb{Z}$.

Example 5.25: Let's convert the complex number $z = 1 + i$ to exponential form. To do this, we need to find r and θ. We have $r = |z| = \sqrt{1^2 + 1^2} = \sqrt{1 + 1} = \sqrt{2}$. Next, we have $\tan\theta = \frac{1}{1} = 1$. It follows that $\theta = \frac{\pi}{4}$. So, in exponential form, we have $z = \sqrt{2}e^{\frac{\pi}{4}i}$.

Note: $\frac{\pi}{4}$ is the principal argument of $z = 1 + i$ because $-\pi < \frac{\pi}{4} \leq \pi$. When we write a complex number in exponential form, we will usually use the principle argument.

If $z \in \mathbb{C}$, we define z^2 to be the complex number $z \cdot z$. Similarly, $z^3 = z \cdot z \cdot z = z^2 \cdot z$. More generally, for $z \in \mathbb{C}$ and $n \in \mathbb{Z}$ we define z^n as follows:

- For $n = 0$, $z^n = z^0 = 1$.
- For $n \in \mathbb{Z}^+$, $z^{n+1} = z^n \cdot z$.
- For $n \in \mathbb{Z}^-$, $z^n = (z^{-n})^{-1} = \frac{1}{z^{-n}}$.

Notice that these definitions agree with the definition of powers of an element from an arbitrary group.

Due to the following theorem, it's often easier to compute z^n when z is written in exponential form.

Theorem 5.26 (De Moivre's Theorem): For all $n \in \mathbb{Z}$, $\left(e^{i\theta}\right)^n = e^{i(n\theta)}$.

Proof: For $n = 0$, we have $\left(e^{i\theta}\right)^0 = (\cos\theta + i\sin\theta)^0 = 1 = e^0 = e^{i(0\theta)}$.

We prove De Moivre's Theorem for $n \in \mathbb{Z}^+$ by induction on n.

Base Case ($k = 1$): $\left(e^{i\theta}\right)^1 = e^{i\theta} = e^{i(1\theta)}$.

Inductive Step: Assume that $k \geq 1$ and $\left(e^{i\theta}\right)^k = e^{i(k\theta)}$. We then have

$$(e^{i\theta})^{k+1} = (\cos\theta + i\sin\theta)^{k+1} = (\cos\theta + i\sin\theta)^k(\cos\theta + i\sin\theta) = (e^{i\theta})^k(\cos\theta + i\sin\theta)$$
$$= e^{i(k\theta)}(\cos\theta + i\sin\theta) = (\cos k\theta + i\sin k\theta)(\cos\theta + i\sin\theta)$$
$$= [(\cos k\theta)(\cos\theta) - (\sin k\theta)(\sin\theta)] + [(\sin k\theta)(\cos\theta) + (\cos k\theta)(\sin\theta)]i.$$
$$= \cos((k+1)\theta) + \sin((k+1)\theta)\,i \text{ (by Theorem 5.23)} = e^{i((k+1)\theta)}.$$

By the Principle of Mathematical Induction, $(e^{i\theta})^n = e^{i(n\theta)}$ for all $n \in \mathbb{Z}^+$.

If $n < 0$, then

$$(e^{i\theta})^n = \frac{1}{(e^{i\theta})^{-n}} = \frac{1}{e^{i(-n\theta)}} = \frac{1}{\cos(-n\theta) + i\sin(-n\theta)}$$
$$= \frac{1}{\cos(n\theta) - i\sin(n\theta)} \text{ (by the Negative Identities)}$$
$$= \frac{1}{\cos(n\theta) - i\sin(n\theta)} \cdot \frac{\cos(n\theta) + i\sin(n\theta)}{\cos(n\theta) + i\sin(n\theta)} = \frac{\cos(n\theta) + i\sin(n\theta)}{\cos^2(n\theta) + \sin^2(n\theta)}$$
$$= \cos(n\theta) + i\sin(n\theta) \text{ (by the Pythagorean Identity)} = e^{i(n\theta)}. \qquad \square$$

Note: De Moivre's Theorem generalizes to all $n \in \mathbb{C}$ with a small "twist." In general, the expression $(e^{i\theta})^n$ may have multiple values, whereas $e^{i(n\theta)}$ takes on just one value. However, for all $n \in \mathbb{C}$, $(e^{i\theta})^n = e^{i(n\theta)}$ in the sense that $e^{i(n\theta)}$ is equal to one of the possible values of $(e^{i\theta})^n$.

As a very simple example, let $\theta = 0$ and $n = \frac{1}{2}$. Then $e^{i(n\theta)} = e^0 = 1$ and $(e^{i\theta})^n = 1^{\frac{1}{2}}$, which has two values: 1 and -1 (because $1^2 = 1$ and $(-1)^2 = 1$). Observe that $e^{i(n\theta)}$ is equal to one of the two possible values of $(e^{i\theta})^n$.

We will not prove this more general result here.

Example 5.27: Let's compute $(2 - 2i)^6$. If we let $z = 2 - 2i$, we have $\tan\theta = \frac{-2}{2} = -1$, so that $\theta = \frac{7\pi}{4}$ (Why?). Also, $r = |z| = \sqrt{2^2 + (-2)^2} = \sqrt{2^2(1+1)} = \sqrt{2^2 \cdot 2} = \sqrt{2^2} \cdot \sqrt{2} = 2\sqrt{2}$. So, in exponential form, $z = 2\sqrt{2}e^{\frac{7\pi}{4}i}$, and therefore,

$$z^6 = \left(2\sqrt{2}e^{\frac{7\pi}{4}i}\right)^6 = 2^6\sqrt{2}^6\left(e^{\frac{7\pi}{4}i}\right)^6 = 64 \cdot 8 e^{6\left(\frac{7\pi}{4}\right)i} = 512 e^{\frac{21\pi}{2}i} = 512 e^{\left(\frac{\pi}{2} + 10\pi\right)i}$$
$$= 512 e^{\frac{\pi}{2}i} = 512\left(\cos\frac{\pi}{2} + i\sin\frac{\pi}{2}\right) = 512(0 + i \cdot 1) = \mathbf{512i}.$$

If x and y are real or complex numbers such that $y = x^2$, then we call x a **square root** of y. If x is a positive real number, then we say that x is the **positive square root** of y and we write $x = \sqrt{y}$.

For positive real numbers, we will use the square root symbol only for the positive square root of the number. For complex numbers, we will use the square root symbol for the **principal square root** of the number. The concept of principal square root will be explained shortly.

Example 5.28:

1. Since $2^2 = 4$, $2 \in \mathbb{R}$, and $2 > 0$, we see that 2 is the positive square root of 4 and we write $2 = \sqrt{4}$.

2. We have $(-2)^2 = 4$, but $-2 < 0$, and so we **do not** write $-2 = \sqrt{4}$. However, -2 is still a square root of 4, and we can write $-2 = -\sqrt{4}$.

3. Since $i^2 = -1$, we see that i is a square root of -1.

4. Since $(-i)^2 = (-i)(-i) = (-1)(-1)i^2 = 1(-1) = -1$, we see that $-i$ is also a square root of -1.

5. $(1+i)^2 = (1+i)(1+i) = (1-1) + (1+1)i = 0 + 2i = 2i$. So, $1+i$ is a square root of $2i$.

More generally, if $z \in \mathbb{C}$ and $n \in \mathbb{Z}^+$, we say that $w \in \mathbb{C}$ is an **nth root** of z if $z = w^n$.

Suppose that $z = re^{i\theta}$ and $w = se^{i\phi}$ are exponential forms of $z, w \in \mathbb{C}$ and that w is an nth root of z. Let's derive a formula for w in terms of r and θ.

We have $w^n = s^n(e^{i\phi})^n = s^n e^{i(n\phi)}$. Since $z = w^n$, $re^{i\theta} = s^n e^{i(n\phi)}$. So, $s^n = r$ and $n\phi = \theta + 2k\pi$, where $k \in \mathbb{Z}$. Therefore, $s = \sqrt[n]{r}$ and $\phi = \frac{\theta + 2k\pi}{n} = \frac{\theta}{n} + \frac{2k\pi}{n}$ for $k \in \mathbb{Z}$. Thus, $w = \sqrt[n]{r}\, e^{i\left(\frac{\theta}{n} + \frac{2k\pi}{n}\right)}$, $k \in \mathbb{Z}$.

If $k \geq n$, then $\frac{\theta}{n} + \frac{2k\pi}{n} = \frac{\theta}{n} + \frac{2(n+k-n)\pi}{n} = \frac{\theta}{n} + \frac{2n\pi + 2(k-n)\pi}{n} = \frac{\theta}{n} + \frac{2n\pi}{n} + \frac{2(k-n)\pi}{n} = \frac{\theta}{n} + \frac{2(k-n)\pi}{n} + 2\pi$, and therefore, $e^{i\left(\frac{\theta}{n} + \frac{2k\pi}{n}\right)} = e^{i\left(\frac{\theta}{n} + \frac{2(k-n)\pi}{n}\right)}$.

It follows that there are exactly n distinct n^{th} roots of z given by $w = \sqrt[n]{r}\, e^{i\left(\frac{\theta}{n} + \frac{2k\pi}{n}\right)}$, $k = 0, 1, \ldots, n-1$. The **principal n^{th} root** of z, written $\sqrt[n]{z}$, is $\sqrt[n]{r}\, e^{i\frac{\Theta}{n}}$, where $-\pi < \Theta \leq \pi$.

Example 5.29: Let's compute all the eighth roots of 1 (also called the **8th roots of unity**). If $1 = w^8$, then $w = \sqrt[8]{1}\, e^{i\left(\frac{0}{8} + \frac{2k\pi}{8}\right)} = e^{\frac{k\pi}{4}i}$ for $k = 0, 1, 2, 3, 4, 5, 6, 7$. Substituting each of these values for k into the expression $e^{\frac{k\pi}{4}i}$ gives us the following 8 eighth roots of unity.

$$1, \frac{1}{\sqrt{2}} + \frac{1}{\sqrt{2}}i, i, -\frac{1}{\sqrt{2}} + \frac{1}{\sqrt{2}}i, -1, -\frac{1}{\sqrt{2}} - \frac{1}{\sqrt{2}}i, -i, \frac{1}{\sqrt{2}} - \frac{1}{\sqrt{2}}i$$

Note: Notice how the eight 8th roots of unity are uniformly distributed on the unit circle.

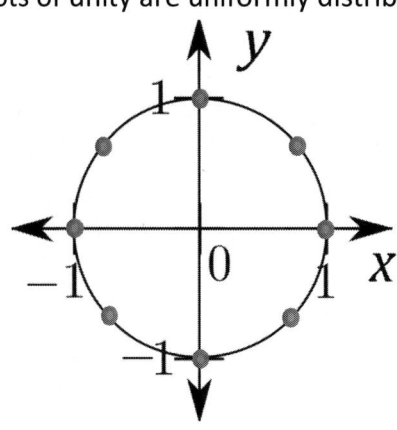

In general, an **nth root of unity** is a complex number w such that $w^n = 1$. Since for $z = 1$, we have $r = |1| = 1$ and $\Theta = \text{Arg } 1 = 0$, from the discussion above, we see that an nth root of unity has the form

$$w = \sqrt[n]{r}e^{i\left(\frac{\theta}{n}+\frac{2k\pi}{n}\right)} = \sqrt[n]{1}e^{i\left(\frac{0}{n}+\frac{2k\pi}{n}\right)} = e^{i\left(\frac{2k\pi}{n}\right)} = e^{\frac{2k\pi}{n}i}, \quad k = 0, 1, 2, 3, \ldots, n-1$$

It turns out that for each $n \in \mathbb{Z}^+$, the set of nth roots of unity together with multiplication forms a cyclic group of order n. We will show this in Example 5.30 below.

Example 5.30: Let $W_n = \{w \mid w \text{ is an } n\text{th root of unity}\}$. To see that $(W_n, \cdot)$ is a cyclic group of order n, first note from our remarks above that $W_n = \left\{e^{\frac{2k\pi}{n}i} \mid k = 0, 1, \ldots, n-1\right\}$ and therefore, $|W_n| = n$.

To see that W_n is closed under multiplication, observe that if $k_1, k_2 \in \mathbb{Z}$, then we have $e^{\frac{2k_1\pi}{n}i} \cdot e^{\frac{2k_2\pi}{n}i} = e^{\frac{2(k_1+k_2)\pi}{n}i}$ (Check this!) and $k_1 + k_2 \in \mathbb{Z}$. By our remarks above, we have $e^{\frac{2(k_1+k_2)\pi}{n}i} = e^{\frac{2k\pi}{n}i}$ for some $k \in \mathbb{Z}$ with $0 \leq k < n$.

Since $W_n \subseteq \mathbb{C}$ and multiplication is associative in $\mathbb{C}$, it follows that multiplication is associative in W_n (see Note 2 following Example 2.3).

The identity of $(W_n, \cdot)$ is $1 = e^0 = e^{\frac{2 \cdot 0 \cdot \pi}{n}i}$.

The inverse of $e^{\frac{2k\pi}{n}i}$ is $e^{\frac{2(n-k)\pi}{n}i}$ because $e^{\frac{2k\pi}{n}i} \cdot e^{\frac{2(n-k)\pi}{n}i} = e^{\frac{2(k+n-k)\pi}{n}i} = e^{\frac{2n\pi}{n}i} = e^{2\pi i} = 1$.

So, $(W_n, \cdot)$ is a group of order n.

To see that W_n is cyclic, note that by DeMoivre's Theorem, we have $e^{\frac{2k\pi}{n}i} = \left(e^{\frac{2\pi}{n}i}\right)^k$, and therefore, $W_n = \left\langle e^{\frac{2\pi}{n}i}\right\rangle$. □

Example 5.31: Let $\mathbb{S}$ be the unit circle in the Complex Plane. So, $\mathbb{S} = \{z \in \mathbb{C} \mid |z| = 1\}$. When written in exponential form, each element of $\mathbb{S}$ has the form $e^{i\theta}$. Since $e^{i(\theta+2k\pi)} = e^{i\theta}$ for all $k \in \mathbb{Z}$, we see that $\mathbb{S} = \{e^{i\theta} \in \mathbb{C} \mid 0 \leq \theta < 2\pi\}$.

Let's show that $(\mathbb{S}, \cdot)$ is a group. For closure, let $e^{i\theta}, e^{i\phi} \in \mathbb{S}$. Then we have $e^{i\theta}e^{i\phi} = e^{i(\theta+\phi)}$ (Check this!), and by our previous remark, we have $e^{i(\theta+\phi)} = e^{i\tau}$ for some τ with $0 \leq \tau < 2\pi$. Since $\mathbb{S} \subseteq \mathbb{C}$ and multiplication is associative in $\mathbb{C}$, it follows that multiplication is associative in $\mathbb{S}$ (see Note 2 following Example 2.3). The identity of $(\mathbb{S}, \cdot)$ is $1 = e^0 = e^{i \cdot 0}$. The inverse of $e^{i\theta}$ is $e^{i(2\pi-\theta)}$ because $e^{i\theta} \cdot e^{i(2\pi-\theta)} = e^{i(\theta+2\pi-\theta)} = e^{2\pi i} = 1$. So, $(\mathbb{S}, \cdot)$ is a group. Note that this is an infinite group. In fact, it is uncountable. Therefore, it **cannot** be cyclic.

Problem Set 5

Full solutions to these problems are available for free download here:
www.SATPrepGet800.com/AAFBTDW

LEVEL 1

1. Use the Principle of Mathematical Induction to prove each of the following:

    (i) $2^n > n$ for all natural numbers $n \geq 1$.

    (ii) $0 + 1 + 2 + \cdots + n = \frac{n(n+1)}{2}$ for all natural numbers.

    (iii) $n! > 2^n$ for all natural numbers $n \geq 4$ (where $n! = 1 \cdot 2 \cdots n$ for all natural numbers $n \geq 1$).

    (iv) $2^n \geq n^2$ for all natural numbers $n \geq 4$.

2. A natural number n is **divisible** by a natural number k, written $k|n$, if there is another natural number b such that $n = kb$. Prove that $n^3 - n$ is divisible by 3 for all natural numbers n.

3. Let $z = -4 - i$ and $w = 3 - 5i$. Compute each of the following:

    (i) $z + w$

    (ii) zw

    (iii) $\text{Im } w$

LEVEL 2

4. Prove each of the following. (You may assume that $<$ is a strict linear ordering of $\mathbb{N}$.)

    (i)   Addition is commutative in $\mathbb{N}$.

    (ii)   The set of natural numbers is closed under multiplication.

    (iii) 1 is a multiplicative identity in $\mathbb{N}$.

    (iv)   Multiplication is distributive over addition in $\mathbb{N}$.

    (v)   Multiplication is associative in $\mathbb{N}$.

    (vi)   Multiplication is commutative in $\mathbb{N}$.

    (vii) For all natural numbers m, n, and k, if $m + k = n + k$, then $m = n$.

    (viii) For all natural numbers m, n, and k, if $mk = nk$, then $m = n$.

    (ix)   For all natural numbers m and n, $m < n$ if and only if there is a natural number $k > 0$ such that $n = m + k$.

    (x)   For all natural numbers m, n, and k, $m < n$ if and only if $m + k < n + k$.

    (xi)   For all natural numbers m and n, if $m > 0$ and $n > 0$, then $mn > 0$.

5. A set A is **transitive** if $\forall x(x \in A \to x \subseteq A)$ (in words, every element of A is also a subset of A). Prove that every natural number is transitive.

6. Determine if each of the following sequences are Cauchy sequences. Are any of the Cauchy sequences equivalent?

    (i) $(x_n) = \left(1 + \frac{1}{n+1}\right)$

    (ii) $(y_n) = (2^n)$

    (iii) $(z_n) = \left(1 - \frac{1}{2n+1}\right)$

7. Each of the following complex numbers is written in exponential form. Rewrite each complex number in standard form:

    (i) $e^{\pi i}$

    (ii) $e^{-\frac{5\pi}{2}i}$

    (iii) $3e^{\frac{\pi}{4}i}$

    (iv) $2e^{\frac{\pi}{3}i}$

    (v) $\sqrt{2}e^{\frac{7\pi}{6}i}$

    (vi) $\pi e^{-\frac{5\pi}{4}i}$

    (vii) $e^{\frac{19\pi}{12}}$

8. Each of the following complex numbers is written in standard form. Rewrite each complex number in exponential form:

    (i) $-1 - i$

    (ii) $\sqrt{3} + i$

    (iii) $1 - \sqrt{3}i$

    (iv) $\left(\frac{\sqrt{6} + \sqrt{2}}{4}\right) + \left(\frac{\sqrt{6} - \sqrt{2}}{4}\right)i$

9. Write the following complex numbers in standard form:

    (i) $\left(\frac{\sqrt{2}}{2} + \frac{\sqrt{2}}{2}i\right)^4$

    (ii) $(1 + \sqrt{3}i)^5$

LEVEL 3

10. Prove that if $n \in \mathbb{N}$ and A is a nonempty subset of n, then A has a least element.

11. Prove POMI $\to$ WOP.

12. Prove that $<_\mathbb{Z}$ is a well-defined strict linear ordering on $\mathbb{Z}$. You may use the fact that $<_\mathbb{N}$ is a well-defined strict linear ordering on $\mathbb{N}$.

LEVEL 4

13. Prove that $3^n - 1$ is even for all natural numbers n.

14. Prove that the Principle of Mathematical Induction is equivalent to the following statement:

    ($\star$) Let $P(n)$ be a statement and suppose that (i) $P(0)$ is true and (ii) for all $k \in \mathbb{N}$, $P(k) \to P(k+1)$. Then $P(n)$ is true for all $n \in \mathbb{N}$.

15. Let $r \in \mathbb{N}$. Prove that the Principle of Mathematical Induction implies the following statement:

    ($\star_r$) Let $P(n)$ be a statement and suppose that (i) $P(r)$ is true and (ii) for all $k \in \mathbb{N}$ with $k \geq r$, $P(k) \to P(k+1)$. Then $P(n)$ is true for all $n \in \mathbb{N}$ with $n \geq r$.

16. Prove that addition of integers is well-defined.

17. Prove that addition and multiplication of rational numbers are well-defined.

18. Let $A = \{(x_n) \mid (x_n) \text{ is a Cauchy sequence of rational numbers}\}$ and define the relation R on A by $(x_n)R(y_n)$ if and only if for every $k \in \mathbb{N}^+$, there is $K \in \mathbb{N}$ such that $n > K$ implies $|x_n - y_n| < \frac{1}{k}$. Prove that R is an equivalence relation on A.

19. Prove that $\{A \in \mathcal{P}(\mathbb{N}) \mid A \text{ is finite}\}$ is countable and $\{A \in \mathcal{P}(\mathbb{N}) \mid A \text{ is infinite}\}$ is uncountable.

20. Consider triangle AOP, where $O = (0,0)$, $A = (1,0)$, and P is the point on the unit circle so that angle POA has radian measure $\frac{\pi}{3}$. Prove that triangle AOP is equilateral, and then use this to prove that $W\left(\frac{\pi}{3}\right) = \left(\frac{1}{2}, \frac{\sqrt{3}}{2}\right)$. You may use the following facts about triangles: (i) The interior angle measures of a triangle sum to π radians; (ii) Two sides of a triangle have the same length if and only if the interior angles of the triangle opposite these sides have the same measure; (iii) If two sides of a triangle have the same length, then the line segment beginning at the point of intersection of those two sides and terminating on the opposite base midway between the endpoints of that base is perpendicular to that base.

21. Prove that $W\left(\frac{\pi}{6}\right) = \left(\frac{\sqrt{3}}{2}, \frac{1}{2}\right)$. You can use facts (i), (ii), and (iii) described in Problem 20.

22. Let θ and ϕ be the radian measures of angles A and B, respectively. Prove the following identity:
$$\cos(\theta - \phi) = \cos\theta \cos\phi + \sin\theta \sin\phi$$

23. Let θ and ϕ be the radian measure of angles A and B, respectively. Prove the following identities:
    (i) $\cos(\theta + \phi) = \cos\theta \cos\phi - \sin\theta \sin\phi$
    (ii) $\cos(\pi - \theta) = -\cos\theta$
    (iii) $\cos\left(\frac{\pi}{2} - \theta\right) = \sin\theta$

(iv) $\sin\left(\frac{\pi}{2} - \theta\right) = \cos\theta$

(v) $\sin(\theta + \phi) = \sin\theta\cos\phi + \cos\theta\sin\phi$

(vi) $\sin(\pi - \theta) = -\sin\theta$

LEVEL 5

24. The Principle of Strong Induction is the following statement:

    (★★) Let $P(n)$ be a statement and suppose that (i) $P(0)$ is true and (ii) for all $k \in \mathbb{N}$, $\forall j \leq k \, (P(j)) \to P(k+1)$. Then $P(n)$ is true for all $n \in \mathbb{N}$.

    Use the Principle of Mathematical Induction to prove the Principle of Strong Induction.

25. Use the Principle of Mathematical Induction to prove that for every $n \in \mathbb{N}$, if S is a set with $|S| = n$, then S has 2^n subsets. (Hint: Use Problem 28 from Problem Set 1.)

26. Provide a formal definition of generalized commutativity. Then prove that if $\star$ is a binary operation that is commutative in a set S, then $\star$ satisfies generalized commutativity in S.

27. Prove that addition of real numbers is well-defined and that the sum of two real numbers is a real number.

28. Using the formal definitions of $\mathbb{N}$, $\mathbb{Z}$, $\mathbb{Q}$, and $\mathbb{R}$ provided in this lesson, prove each of the following:

    (i) $(\mathbb{N}, +, \cdot)$ is a commutative semiring, but not a ring.

    (ii) $(\mathbb{Z}, +, \cdot)$ is a commutative ring, but not a field.

    (iii) $(\mathbb{R}, +, \cdot)$ is a field.

CHALLENGE PROBLEMS

29. Prove that multiplication of integers is well-defined.

30. Prove that multiplication of real numbers is well-defined and that the product of two real numbers is a real number.

31. Prove that $<_\mathbb{N}$ is a strict linear ordering on $\mathbb{N}$, that $<_\mathbb{Q}$ is a well-defined strict linear ordering on $\mathbb{Q}$, and that $<_\mathbb{R}$ is a well-defined strict linear ordering on $\mathbb{R}$.

32. Define a set to be **selfish** if the number of elements it has is in the set. For example, the set $K_5 = \{1, 2, 3, 4, 5\}$ is selfish because it has 5 elements and 5 is in the set. A selfish set is **minimal** if none of its proper subsets is also selfish. For example, the set K_5 is not a minimal selfish set because $\{1\}$ is a selfish subset. Let $K_n = \{1, 2, 3, \ldots n\}$. Determine with proof how many minimal selfish subsets K_n has in terms of n.

LESSON 6
SUBSTRUCTURES

Structures and Substructures

Recall that an ***n*-ary relation** on a set S is a subset of S^n. We usually use the expressions **unary**, **binary**, and **ternary** in place of 1-ary, 2-ary, and 3-ary. Note that a unary relation on S is simply a subset of S. We do not define a 0-ary relation.

Example 6.1: Let $\mathbb{Z} = \{\ldots, -3, -2, -1, 0, 1, 2, 3, \ldots\}$ be the set of integers. The set $\mathbb{N} = \{0, 1, 2, 3, \ldots\}$ of natural numbers is a unary relation on $\mathbb{Z}$. In other words, $\mathbb{N} \subseteq \mathbb{Z}$. Some examples of binary relations on $\mathbb{Z}$ are the linear orderings $<, \leq, >,$ and $\geq$ (see Example 3.1 (part 2)) and the equivalence relations $\equiv_n = \{(a,b) \in \mathbb{Z}^2 \mid n \mid b - a\}$ (see Example 3.12 (part 4)). $R = \{(x,y,z) \in \mathbb{Z}^3 \mid x + y = z\}$ is an example of a ternary relation on $\mathbb{Z}$ (see Example 3.4 (part 2)).

An ***n*-ary operation** on a set S is a function from S^n to S. We also define a 0-ary operation to simply be an element of S. We will usually call a 0-ary operation a **constant** in S.

Example 6.2: Let $\mathbb{R}$ be the set of real numbers. Negation is an example of a unary operation on $\mathbb{R}$. This is the operation that maps each $x \in \mathbb{R}$ to $-x$. Addition, subtraction, and multiplication are examples of binary operations on $\mathbb{R}$. 0 is an example of a 0-ary operation on $\mathbb{R}$ or a constant in $\mathbb{R}$.

A **finitary relation** is an n-ary relation for some $n \in \mathbb{N}^*$ (recall that $\mathbb{N}^* = \mathbb{N} \setminus \{0\}$ is the set of nonzero natural numbers). A **finitary operation** is an n-ary operation for some $n \in \mathbb{N}$.

A **structure** is a set together with a collection of finitary operations and relations defined on the set. The set is called the **domain** of the structure.

Example 6.3:

1. Semigroups, monoids, and groups are structures of the form $(S, \star)$, where S is a set and $\star$ is a binary operation on S. This agrees with the notation we introduced for these structures in Lesson 2.

    We may want to view a monoid as a structure of the form $(S, \star, e)$ and a group as a structure of the form $(S, \star, ^{-1}, e)$, where e is a constant called the identity element of the monoid or group and $^{-1}$ is the unary inverse operator. One reason for including additional symbols might be to "force" additional conditions on substructures (to be defined shortly) without having to spell out each condition explicitly. This will be made clear in Example 6.5 below.

2. Rings and fields are structures of the form $(S, +, \cdot)$, where S is a set, and $+$ and $\cdot$ are binary operations on S. Again, we may want to include additional operations (see part 4 of Example 6.5 and also part 5 of Example 6.5 below).

3. Partially ordered sets and linearly ordered sets are structures of the form $(S, \leq)$, where S is a set and $\leq$ is a binary relation on S. Strict partially ordered sets and strict linearly ordered sets have the same form (although we would normally use the symbol $<$ instead of the symbol $\leq$).

4. Every set without any operations and relations is a structure. For example, $\mathbb{N}, \mathbb{Z}, \mathbb{Q}, \mathbb{R}$, and $\mathbb{C}$ are structures. (Notice that we abbreviate the structure (S) as S.)

5. We can view a vector space $(V, \oplus)$ over the field $(F, +, \cdot)$ as $(V \cup F, V, F, R_\oplus, R_+, R_\cdot, R_\star)$, where V and F are unary relations and $R_\oplus, R_+, R_\cdot, R_\star$ are the following ternary relations:

$$R_\oplus = \{(x, y, z) \in V^3 \mid x \oplus y = z\} \quad R_+ = \{(x, y, z) \in F^3 \mid x + y = z\}$$
$$R_\cdot = \{(x, y, z) \in F^3 \mid x \cdot y = z\} \quad R_\star = \{(x, y, z) \in F \times V \times V \mid xy = z\}$$

Notice that we had to use ternary relations instead of binary functions for the four operations because the definition of a structure demands that functions be defined on $(V \cup F)^2$. However, none of the functions are defined on $(V \cup F)^2$. Indeed, $\oplus$ is defined only on V^2, $+$ and $\cdot$ are defined only on F^2, and scalar multiplication is defined on $F \times V$.

6. We can view a left-module $(M, \oplus)$ over the ring $(S, +, \cdot)$ as $(M \cup S, M, S, R_\oplus, R_+, R_\cdot, R_\star)$, in exactly the same way we view a vector space over a field (see part 5 above). Specifically, M and S are unary relations, and $R_\oplus, R_+, R_\cdot, R_\star$ are the following ternary relations:

$$R_\oplus = \{(x, y, z) \in M^3 \mid x \oplus y = z\} \quad R_+ = \{(x, y, z) \in S^3 \mid x + y = z\}$$
$$R_\cdot = \{(x, y, z) \in S^3 \mid x \cdot y = z\} \quad R_\star = \{(x, y, z) \in S \times M \times M \mid xy = z\}$$

We can view a right-module in the same way, except we replace the relation $R_\star$ by the following:

$$R_\star = \{(x, y, z) \in M \times S \times M \mid xy = z\}$$

We will sometimes use a fraktur letter (such as $\mathfrak{A}, \mathfrak{B}, \mathfrak{C}$) for the name of a structure if we want to be clear that we are talking about the whole structure and not just the underlying set. For example, we might write $\mathfrak{G} = (G, \star)$ for a group $\mathfrak{G}$ with underlying set G and group operation $\star$.

At the other extreme, we will often abuse notation and use the same name for the structure and its domain. For example, if $(G, \star)$ is a group, we may talk about the group G under the operation $\star$. This is certainly less formal language and we need to be careful when we do this that it is clear what type of structure we are referring to. We have already used this less formal language extensively when talking about vector spaces is Lesson 2. A quick glance back at part 5 of Example 6.3 should make it clear why we would usually choose to be less formal when discussing vector spaces.

Notes: (1) A finitary operation on a set S is a function $f: S^n \to S$ for some $n \in \mathbb{N}$. There are two important facts implied by this definition:

1. The operation f is defined for every n-tuple $(a_1, a_2, \ldots, a_n) \in S^n$.
2. The set S is closed under f.

(2) A finitary relation on a set S is a subset R of S^n for some $n \in \mathbb{N}$. We have more flexibility with relations than we do with operations. For example, an $(n + 1)$-ary relation can be used to define a *partial n-ary function*. Suppose we want a structure that consists of the set of integers $\mathbb{Z}$ together with the partial function defined on only the even integers that divides each even integer by 2. We can define a relation $R = \{(2k, k) \mid k \in \mathbb{Z}\}$. The structure $(\mathbb{Z}, R)$ consists of the set of integers together with the function $f: 2\mathbb{Z} \to \mathbb{Z}$ defined by $f(n) = \frac{n}{2}$ (recall that $2\mathbb{Z}$ is the set of even integers). Notice that we defined a *unary* partial function on $\mathbb{Z}$ by using a *binary* relation. Also, see parts 5 and 6 of Example 6.3 above to see how we use ternary relations to define partial binary functions on a given domain.

We say that structures $\mathfrak{A}$ and $\mathfrak{B}$ have the same **type** if they have the same number of n-ary operations for each $n \in \mathbb{N}$, and the same number of n-ary relations for each $n \in \mathbb{N}^*$ (recall once again that $\mathbb{N}^* = \mathbb{N} \setminus \{0\}$ is the set of nonzero natural numbers).

Example 6.4:

1. $(\mathbb{Q}, \leq)$, $(\mathcal{P}(\mathbb{N}), \subseteq)$, and for each $n \in \mathbb{N}^*$, $(\mathbb{Z}, \equiv_n)$ all have the same type because they each have exactly one binary relation.

2. $(\mathbb{Z}, +)$ and $(\mathbb{Z}, +, 0)$ have different types. The first structure has one binary operation and nothing else. The second structure has a binary operation and a constant (or a 0-ary operation). Both of these are different ways of describing the group of integers under addition. The second way is specifically mentioning the identity element, while the first is not. Another structure (of yet another type) that describes the same group is $(\mathbb{Z}, +, -, 0)$, where $-$ is the unary additive inverse operator.

Note: For structures with only finitely many operations and relations, the definition we gave of being of the same type is adequate. However, for structures with infinitely many operations and/or relations, we should be a little more careful with what we mean by "the same number." A better definition in this case is that for each $n \in \mathbb{N}$, the set of n-ary operations in $\mathfrak{A}$ is equinumerous with the set of n-ary operations in $\mathfrak{B}$, and for each $n \in \mathbb{N}^*$, the set of n-ary relations in $\mathfrak{A}$ is equinumerous with the set of n-ary relations in $\mathfrak{B}$. See Lesson 4 for more information on equinumerosity.

In this book, we will be working with structures with finitely many operations and relations. Therefore, the simpler definition of type is satisfactory for our purposes.

$\mathfrak{A}$ is a **substructure** of $\mathfrak{B}$, written $\mathfrak{A} \subseteq \mathfrak{B}$ if

1. $\mathfrak{A}$ and $\mathfrak{B}$ have the same type.
2. $A \subseteq B$.
3. If f is an n-ary operation, and $(a_1, a_2, \ldots, a_n) \in A^n$, then $f_A(a_1, a_2, \ldots, a_n) = f_B(a_1, a_2, \ldots, a_n)$.
4. If R is an n-ary relation, and $(a_1, a_2, \ldots, a_n) \in A^n$, then $R_A(a_1, a_2, \ldots, a_n)$ if and only if $R_B(a_1, a_2, \ldots, a_n)$

Notes: (1) Part 1 of the definition says that in order for $\mathfrak{A}$ to be a substructure of $\mathfrak{B}$, the two structures must have the same number of n-ary operations and n-ary relations for each n. For example, $(\mathbb{N}, +)$ is a substructure of $(\mathbb{Z}, +)$, written $(\mathbb{N}, +) \subseteq (\mathbb{Z}, +)$, but $(\mathbb{N}, +)$ is **not** a substructure of $(\mathbb{Z}, +, 0)$.

(2) The notation in 3 and 4 might look confusing at first. Let's clarify with an example of each. Suppose that f is addition, so that $f(a_1, a_2) = a_1 + a_2$. Then 3 says that if $\mathfrak{A} \subseteq \mathfrak{B}$ and we choose a_1 and a_2 from A, then we get the same result whether we add a_1 and a_2 in A or B. We might write this as $a_1 +_A a_2 = a_1 +_B a_2$. Now suppose that R is $<$, so that $R(a_1, a_2)$ means $a_1 < a_2$. Then 4 says that if $\mathfrak{A} \subseteq \mathfrak{B}$ and we choose a_1 and a_2 from A, then $a_1 <_A a_2$ if and only if $a_1 <_B a_2$.

Example 6.5:

1. Let $(S,\star)$ be a semigroup. A substructure $(T,\star)$ of $(S,\star)$ is called a **subsemigroup**. Notice that $T \subseteq S$ and the operation $\star$ must be the same for both structures. Also, $\star$ is a binary operation on T, which means that T is closed under $\star$. Is $\star$ associative in T? Recall from Note 2 following Example 2.3 in Lesson 2 that associativity is closed downwards. In other words, since $\star$ is associative in S and $T \subseteq S$, it follows that $\star$ is associative in T. We just showed that a subsemigroup of a semigroup is itself a semigroup.

   Also note that if $\star$ is commutative in S, then $\star$ is automatically commutative in T. Commutativity is closed downwards for the same reason that associativity is closed downwards (once again, see Note 2 following Example 2.3 in Lesson 2).

   For example, let $\mathfrak{A} = (\mathbb{N}, +)$ and let $\mathfrak{B} = (2\mathbb{N}, +)$ (recall that $2\mathbb{N} = \{2k \mid k \in \mathbb{N}\}$ is the set of even natural numbers). Then $\mathfrak{B} \subseteq \mathfrak{A}$. That is, $\mathfrak{B}$ is a subsemigroup of $\mathfrak{A}$. Also, since $\mathfrak{A}$ is a commutative semigroup, $\mathfrak{B}$ is a commutative subsemigroup of $\mathfrak{A}$.

   On the other hand, $(2\mathbb{N} + 1, +)$ is not even a structure because $+$ is **not** a binary operation on $2\mathbb{N} + 1$ (recall that $2\mathbb{N} + 1 = \{2k + 1 \mid k \in \mathbb{N}\}$ is the set of odd natural numbers). For example, $3, 5 \in 2\mathbb{N} + 1$, but $3 + 5 \notin 2\mathbb{N} + 1$.

2. Let $(M, \star, e)$ be a monoid, where e is the identity of M. A substructure $(N, \star, e)$ of $(M, \star, e)$ is called a **submonoid**. Notice that the operation $\star$ and the identity e must be the same for both structures. As we saw in part 1 above, N is closed under $\star$ and $\star$ is associative in N. We just showed that a submonoid of a monoid is itself a monoid.

   Note that a substructure $(N, \star)$ of a monoid $(M, \star)$ is a subsemigroup of $(M, \star)$, but may or may not be a submonoid of $(M, \star)$. For example, let $A = \mathbb{N} \setminus \{0, 1\} = \{2, 3, 4, \dots\}$ be the set of natural numbers with 0 and 1 removed. Then $(A, \cdot)$ is a subsemigroup of the monoid $(\mathbb{N}, \cdot)$, but $(A, \cdot)$ is not a submonoid of $(\mathbb{N}, \cdot)$ because A is missing the multiplicative identity 1.

   If $(M, \star)$ is a monoid with identity e, we can also define a submonoid to be a substructure $(N, \star)$ of $(M, \star)$ such that N contains e. In other words, if we wish to leave the identity out of the structure, we need to explicitly mention that the domain of the substructure contains the identity in order to guarantee that we get a submonoid. For example, if we let $\mathfrak{A} = (\mathbb{N}, +)$ and $\mathfrak{B} = (2\mathbb{N}, +)$, we see that $\mathfrak{B}$ is a submonoid of $\mathfrak{A}$ because $2\mathbb{N} \subseteq \mathbb{N}$ is closed under $+$ and $0 \in 2\mathbb{N}$.

3. Let $(G, \star, {}^{-1}, e)$ be a group, where ${}^{-1}$ is the unary inverse operator and e is the identity of G. A substructure $(H, \star, {}^{-1}, e)$ of $(G, \star, {}^{-1}, e)$ is called a **subgroup**. Notice that the operations $\star$ and ${}^{-1}$, and the identity e must be the same for both structures. As we saw in parts 1 and 2 above, H is closed under $\star$ and $\star$ is associative in N. By making the unary inverse operator part of the structure, we have guaranteed that the inverse property holds for the substructure. So, a subgroup of a group is itself a group.

   For example, let $\mathfrak{A} = (\mathbb{Z}, +, -, 0)$ and let $\mathfrak{B} = (2\mathbb{Z}, +, -, 0)$. Then $\mathfrak{B}$ is a subgroup of $\mathfrak{A}$. More generally, for any positive integer n, the structure $(n\mathbb{Z}, +, -, 0)$ is a subgroup of the group $(\mathbb{Z}, +, -, 0)$ (recall that $n\mathbb{Z} = \{nk \mid k \in \mathbb{Z}\}$). Also, since $(\mathbb{Z}, +, -, 0)$ is a commutative group, $(n\mathbb{Z}, +, -, 0)$ is a commutative subgroup of $\mathfrak{A}$ (once again, commutativity is closed downwards – see part 1 above).

Note that a substructure $(H,\star)$ of a group $(G,\star)$ is a subsemigroup of $(G,\star)$, but may or may not be a subgroup of $(G,\star)$, as we saw in 2 above. Furthermore, a substructure $(H,\star,e)$ of a group $(G,\star,e)$ is a submonoid of $(G,\star,e)$ but still may not be a subgroup of $(G,\star,e)$. For example, $(\mathbb{N},+,0)$ is a substructure of the group $(\mathbb{Z},+,0)$ that is **not** a subgroup of $(\mathbb{Z},+,0)$ (it is a submonoid though). We need to include the unary inverse operator in the structure to guarantee that a substructure of a subgroup will be a subgroup.

If $(G,\star)$ is a group with identity e, we can define a subgroup to be a substructure $(H,\star)$ of $(G,\star)$ such that H contains e and for all $x \in H$, $x^{-1} \in H$ (in other words, we need to insist that H is closed under taking inverses). These conditions can be used in place of including symbols for inverse and identity in the structure itself. For example, if we let $\mathfrak{A} = (\mathbb{R}^*, \cdot)$ and $\mathfrak{B} = (\mathbb{Q}^*, \cdot)$, we see that $\mathfrak{B}$ is a subgroup of $\mathfrak{A}$ because $\mathbb{Q}^* \subseteq \mathbb{R}^*$, $1 \in \mathbb{Q}^*$, and $\mathbb{Q}^*$ is closed under taking multiplicative inverses. Don't forget that the definition of substructure also implies closure under the given operation. So, it is equally important that the product of two nonzero rational numbers is a nonzero rational number.

If the operation is understood, we can simplify notation even further. We may write $H \leq G$ and say that H is a subgroup of G. What we mean by this is $(H,\star,{}^{-1},e)$ is a substructure of $(G,\star,{}^{-1},e)$, or equivalently, $(H,\star)$ is a substructure of $(G,\star)$ (and so, H is closed under $\star$) such that the identity of G is in H and H is closed under taking inverses.

We use the same notation for other structures as well. Just be careful about one thing. When we write $A \leq B$, we don't just mean that the structure $\mathfrak{A}$ is a substructure of the structure $\mathfrak{B}$. We also mean that the structure $\mathfrak{A}$ has all the properties we need for the type of structure under discussion. For example, if we are talking about groups under addition, then we would **not** write $\mathbb{N} \leq \mathbb{Z}$. However, if we are talking about monoids under addition, then we could write $\mathbb{N} \leq \mathbb{Z}$.

Let's practice using the less formal notation for a moment with an important example of a subgroup. Let G be a group under $\star$ and let $Z(G) = \{x \in G \mid xz = zx \text{ for all } z \in G\}$. $Z(G)$ is called the **center** of G. It consists of the elements of G that commute with every other element of G. Let's prove that $Z(G) \leq G$. First, suppose that $x, y \in Z(G)$ and let $z \in G$. Then we have

$$(xy)z = x(yz) = x(zy) = (xz)y = (zx)y = z(xy).$$

It follows that $xy \in Z(G)$. Since $x, y \in Z(G)$ were arbitrary, we see that $Z(G)$ is closed under $\star$. Next, let $x \in Z(G)$ and let $z \in G$. Then $x^{-1}z = (z^{-1}x)^{-1} = (xz^{-1})^{-1} = zx^{-1}$. So, $x^{-1} \in Z(G)$. Since $x \in Z(G)$ was arbitrary, $Z(G)$ is closed under taking inverses. Finally, $ez = z = ze$ for any $z \in G$, and so, $e \in Z(G)$. This completes the argument.

Note that if G is a commutative group, then $Z(G) = G$.

It should be clear that every group G is a subgroup of itself. That is, $G \leq G$. It is equally obvious that $\{e\}$ consisting of the identity of G alone is a subgroup of G. That is, $\{e\} \leq G$. The subgroup $\{e\}$ is known as the **trivial** subgroup of G.

4. Let $(R, +, \cdot, -, 1)$ be a ring, where $-$ is the unary additive inverse operator and 1 is the multiplicative identity of R. A substructure $(S, +, \cdot, -, 1)$ of $(R, +, \cdot, -, 1)$ is called a **subring**. Notice that the operations $+$, $\cdot$, and $-$, and the multiplicative identity 1 must be the same for both structures. By the definition of a structure, S is closed under $+$, $\cdot$ and $-$.

You may be wondering why we didn't put a constant for 0 in the structure. The reason is simple: we don't need to. Since $1 \in S$ and S is closed under taking additive inverses, we have $0 = 1 + (-1) \in S$. Associativity of addition and multiplication, commutativity of addition, and distributivity all hold in S because these operations are closed downwards (see Note 2 following Example 2.3 in Lesson 2). It follows that a subring is itself a ring.

Alternatively, we can say that $(S, +, \cdot)$ is a subring of $(R, +, \cdot)$ if $(S, +, \cdot)$ is a substructure of $(R, +, \cdot)$ (and so, S is closed under $+$ and $\cdot$) such that S contains 1 and for all $x \in S$, $-x \in S$ (in other words, we need to insist that S is closed under taking additive inverses).

As we discussed above, we may write $S \leq R$ for S is a subring of R if it is clear that we are talking about the ring structures of S and R.

For example, $(\mathbb{Z}, +, \cdot)$ is a subring of the fields $(\mathbb{Q}, +, \cdot)$, $(\mathbb{R}, +, \cdot)$, and $(\mathbb{C}, +, \cdot)$. So, we might write $\mathbb{Z} \leq \mathbb{Q}$, $\mathbb{Z} \leq \mathbb{R}$, and $\mathbb{Z} \leq \mathbb{C}$.

$\mathbb{Z}$ has no subring other than itself. To see this, let $A \leq \mathbb{Z}$. First note that the multiplicative identity $1 \in A$. Using closure of addition and the principle of mathematical induction, we can then show that each positive integer is in A (for example, $2 = 1 + 1$). Since A is closed under the additive inverse of $\mathbb{Z}$, for each positive integer n, $-n \in A$. It follows that $A = \mathbb{Z}$. (Note that we know that $0 \in A$ because we have already shown above that 0 is in any subring of a ring.)

We can define the **center** $Z(R)$ of a ring R to be $Z(R) = \{x \in R \mid xz = zx \text{ for all } z \in R\}$ just like we did for a group. It consists of the elements of R that commute with every other element of R under the ring multiplication (note that every element commutes with every other element under addition because R is commutative for addition). Let's prove that $Z(R) \leq R$. First, suppose that $x, y \in Z(R)$ and let $z \in R$. Then we have

$$(x + y)z = xz + yz = zx + zy = z(x + y).$$
$$(xy)z = x(yz) = x(zy) = (xz)y = (zx)y = z(xy).$$

It follows that $x + y \in Z(R)$ and $xy \in Z(R)$. Since $x, y \in Z(R)$ were arbitrary, we see that $Z(R)$ is closed under $+$ and $\cdot$. Next, let $x \in Z(R)$ and let $z \in R$. Then we have

$$(-x)z = (-1x)z = (-1)(xz) = -1(zx) = (-1z)x = (z(-1))x = z(-1x) = z(-x).$$

(Note that we used part (v) of Problem 7 from Problem Set 2 here.) So, $x^{-1} \in Z(G)$. Since $x \in Z(G)$ was arbitrary, $Z(G)$ is closed under taking inverses. Finally, $1z = z = z \cdot 1$ for any $z \in G$, and so, $1 \in Z(G)$. This completes the argument.

It should be clear that every ring R is a subring of itself. That is, $R \leq R$.

Note that $\{0\}$ and $\{1\}$ are **not** subrings of any ring R with two or more elements. $\{0\}$ is not a subring because it does not contain 1 and $\{1\}$ is not a subring because $1 + (-1) = 0$ and $0 \notin \{1\}$.

5. Let $(F, +, \cdot, -, ^{-1}, 0, 1)$ be a field, where $-$ and $^{-1}$ are the unary additive inverse and multiplicative inverse operators, respectively, and 0 and 1 are the additive and multiplicative identities of R, respectively. Note that technically speaking, $^{-1}$ must be expressed as the binary relation $^{-1} = \{(x, y) \mid y = x^{-1}\}$ because $^{-1}$ isn't defined for $x = 0$.

A substructure $(K, +, \cdot, -, ^{-1}, 0, 1)$ of $(F, +, \cdot, -, ^{-1}, 0, 1)$ is a **subfield** provided that the domain and range of the multiplicative inverse relation $^{-1}$ are both K^*. Notice that the operations $+, \cdot, -$, the relation $^{-1}$, and the identities 0 and 1 must be the same for both structures. By the definition of a structure, K is closed under $+, \cdot,$ and $-$. Associativity and commutativity of addition and multiplication, and distributivity all hold in K because these operations are closed downwards (see Note 2 following Example 2.3). It follows that a subfield is itself a field.

Alternatively, we can say that $(K, +, \cdot)$ is a subfield of $(F, +, \cdot)$ if $(K, +, \cdot)$ is a substructure of $(F, +, \cdot)$ such that K contains 0 and 1, for all $x \in K$, $-x \in K$ and for all nonzero $x \in K$, $x^{-1} \in K$ (in other words, we need to insist that K is closed under taking additive inverses and K^* is closed under taking multiplicative inverses). In the case of fields, this alternate approach is quite a bit nicer. We will write $K \leq F$ when K is a subfield of F and it is clear we are talking about the field structures of K and F.

For example, $(\mathbb{Q}, +, \cdot)$ is a subfield of both $(\mathbb{R}, +, \cdot)$ and $(\mathbb{C}, +, \cdot)$, and $(\mathbb{R}, +, \cdot)$ is a subfield of $(\mathbb{C}, +, \cdot)$. So, we might write $\mathbb{Q} \leq \mathbb{R}$, $\mathbb{Q} \leq \mathbb{C}$, and $\mathbb{R} \leq \mathbb{C}$.

7. Let $(V \cup F, V, F, R_\oplus, R_+, R_\cdot, R_\star)$ be a vector space (in other words, $(V, \oplus)$ is a vector space over the field $(F, +, \cdot)$). Recall from part 5 of Example 6.3 above that V and F are unary relations, and $R_\oplus, R_+, R_\cdot, R_\star$ are the following ternary relations:

$$R_\oplus = \{(x, y, z) \in V^3 \mid x \oplus y = z\} \quad R_+ = \{(x, y, z) \in F^3 \mid x + y = z\}$$
$$R_\cdot = \{(x, y, z) \in F^3 \mid x \cdot y = z\} \quad R_\star = \{(x, y, z) \in F \times V \times V \mid xy = z\}$$

A substructure $(W \cup F, W, F, R_\oplus, R_+, R_\cdot, R_\star)$ of $(V \cup F, V, F, R_\oplus, R_+, R_\cdot, R_\star)$ is a **subspace** provided that (i) $0 \in W$ (the zero vector is in W), (ii) whenever $x, y \in W$ and $(x, y, z) \in R_\oplus$ (where $R_\oplus$ is being considered as part of the structure with domain $V \cup F$), then $z \in W$ (W is closed under addition), and (iii) whenever $x \in F, y \in W$, and $(x, y, z) \in R_\star$ (where $R_\star$ is being considered as part of the structure with domain $V \cup F$), then $z \in W$ (W is closed under scalar multiplication). Notice that the field F must be the same for both structures. We will see in Theorem 6.6 below that a subspace of a vector space is itself a vector space.

In practice, we will dispense with the formalities above and say that a subset W of a vector space V (written $W \subseteq V$) is a subspace of V (written $W \leq V$) if W is also a vector space with respect to the same operations of addition and scalar multiplication as they were defined in V. Once again, in Theorem 6.6 below, we will prove that $W \leq V$ is equivalent to W satisfying the following three conditions: (i) $0 \in W$, (ii) W is closed under addition, and (iii) W is closed under scalar multiplication. We will see examples of subspaces in Example 6.7 below.

8. A **left submodule** of a module over a ring R is defined in exactly the same way as a subspace of a vector space over a field F. Everything mentioned in 7 above about subspaces applies to left submodules as well. A **right submodule** of a module over R is almost the same. As we already discussed in part 6 of Example 6.3 above, the only difference is in the definition of $R_\star$. In the case of a right submodule W of a module V, $R_\star = \{(x, y, z) \in V \times F \times V \mid xy = z\}$ and we require that whenever $x \in W, y \in F$, and $(x, y, z) \in R_\star$ (where $R_\star$ is being considered as part of the structure with domain $V \cup F$), then $z \in W$.

9. If $(P, \leq)$ is a partially ordered set, then a substructure $(Q, \leq)$ of $(P, \leq)$ will also be a partially ordered set. This is because reflexivity, antisymmetry, and transitivity are all closed downwards. Once again, see Note 2 following Example 2.3 for an explanation of this. Similarly, any substructure of a linearly ordered set is linearly ordered, and similar results hold for strict partial and linear orders.

   For example, we have $(\mathbb{N}, \leq) \subseteq (\mathbb{Z}, \leq) \subseteq (\mathbb{Q}, \leq) \subseteq (\mathbb{R}, \leq)$, and each of these structures are linearly ordered sets. Similarly, we have $(\mathbb{N}, <) \subseteq (\mathbb{Z}, <) \subseteq (\mathbb{Q}, <) \subseteq (\mathbb{R}, <)$, and each of these structures are strict linearly ordered sets.

Subspaces of Vector Spaces

Let V be a vector space over a field F. A subset W of V is called a **subspace** of V, written $W \leq V$, if it is also a vector space with respect to the same operations of addition and scalar multiplication as they were defined in V.

Notes: (1) Recall from Note 2 following Example 2.3 that a **universal statement** is a statement that describes a property that is true for all elements without mentioning the existence of any new elements. A universal statement begins with the quantifier ∀ ("For all") and never includes the quantifier ∃ ("There exists" or "There is").

Recall that properties defined by universal statements are **closed downwards**. This means that if a property defined by a universal statement is true in V and W is a subset of V, then the property is true in W as well.

For example, the statement for commutativity is $\forall v, w\ (v + w = w + v)$. This is read "For all v and w, $v + w = w + v$." The quantifier ∀ is referring to whichever set we are considering. If we are thinking about the set V, then we mean "For all v and w in V, $v + w = w + v$." If we are thinking about the set W, then we mean "For all v and w in W, $v + w = w + v$."

If we assume that $+$ is commutative in V and $W \subseteq V$, we can easily show that $+$ is also commutative in W. To see this, let $v, w \in W$. Since $W \subseteq V$, we have $v, w \in V$. Since $+$ is commutative in V, we have $v + w = w + v$. Since v and w were arbitrary elements in W, we see that $+$ is commutative in W.

(2) Associativity, commutativity, and distributivity are all defined by universal statements, and therefore, when checking if W is a subspace of V, we **do not** need to check any of these properties—they will always be satisfied in the subset W.

(3) The identity property for addition is **not** defined by a universal statement. It begins with the existential quantifier ∃ "There is." Therefore, we **do** need to check that the identity 0 is in a subset W of V when determining if W is a subspace of V. However, once we have checked that 0 is there, we **do not** need to check that it satisfies the property of being an identity. As long as $0 \in W$ (the same 0 from V), then it will behave as an identity because the defining property of 0 contains only the quantifier ∀.

(4) The inverse property for addition will always be true in a subset W of a vector space V that is closed under scalar multiplication. To see this, we use the fact that $-1v = -v$ for all v in a vector space (see part (iv) of Problem 10 from Problem Set 2).

(5) Since the multiplicative identity 1 comes from the field F and not the vector space V, and we are using the same field for the subset W, we **do not** need to check the scalar multiplication identity when verifying that W is a subspace of V.

(6) The main issue when checking if a subset W of V is a subspace of V is closure. For example, we need to make sure that whenever we add 2 vectors in W, we get a vector that is also in W. If we were to take an arbitrary subset of V, then there is no reason this should happen. For example, let's consider the vector space $\mathbb{C}$ over the field $\mathbb{R}$. Let $A = \{2 + bi \mid b \in \mathbb{R}\}$. A is a subset of $\mathbb{C}$, but A is not a subspace of $\mathbb{C}$. To see this, we just need a single counterexample. $2 + i \in A$, but $(2 + i) + (2 + i) = 4 + 2i \notin A$ (because the real part is 4 and not 2).

(7) Notes 1 through 6 above tell us that to determine if a subset W of a vector space V is a subspace of V, we need only check that $0 \in W$ and that W is closed under addition and scalar multiplication.

(8) The statements for closure, as we have written them do look a lot like universal statements. For example, the statement for closure under addition is "For all $v, w \in V$, $v + w \in V$." The issue here is that the set V is not allowed to be explicitly mentioned in the formula. It needs to be understood.

For example, we saw in Note 1 that the statement for commutativity can be written as "$\forall v, w(v + w = w + v)$." The quantifier $\forall$ (for all) can be applied to any set for which there is a notion of addition defined. We also saw that if the statement is true in V, and W is a subset of V, then the statement will be true in W.

With the statement of closure, to eliminate the set V from the formula, we would need to say something like, "For all x and y, $x + y$ exists." However, there is no way to say "exists" using just logical notation without talking about the set we wish to exist inside of.

(9) The definitions for (left or right) submodules of a (left or right) module are similar to the definition of a subspace of a vector space. For example, if M is a left module over a ring R, then a subset W of M is a **left submodule** of M, written $W \leq M$, if it is also a left module with respect to the same operations of addition and scalar multiplication as they were defined in M.

We summarize Notes 1 through 8 above in the following theorem.

Theorem 6.6: Let V be a vector space over a field F and let $W \subseteq V$. Then $W \leq V$ if and only if (i) $0 \in W$, (ii) for all $v, w \in W$, $v + w \in W$, and (iii) for all $v \in W$ and $k \in F$, $kv \in W$.

Proof: Let V be a vector space over a field F, and $W \subseteq V$.

If W is a subspace of V, then by definition of W being a vector space, (i), (ii), and (iii) hold.

Now suppose that (i), (ii), and (iii) hold.

By (ii), $+$ is a binary operation on W.

Associativity and commutativity of $+$ are defined by universal statements, and therefore, since they hold in V and $W \subseteq V$, they hold in W.

We are given that $0 \in W$. If $v \in W$, then since $W \subseteq V$, $v \in V$. Since 0 is the additive identity for V, $0 + v = v + 0 = v$. Since $v \in W$ was arbitrary, the additive identity property holds in W.

Let $v \in W$. Since $W \subseteq V$, $v \in V$. Therefore, there is $-v \in V$ such that $v + (-v) = (-v) + v = 0$. By (iii), $-1v \in W$ and by part (iv) of Problem 10 from Problem Set 2, $-1v = -v$. Since $v \in W$ was arbitrary, the additive inverse property holds in W.

So, $(W, +)$ is a commutative group.

By (iii), W is closed under scalar multiplication.

Associativity of scalar multiplication and both types of distributivity are defined by universal statements, and therefore, since they hold in V and $W \subseteq V$, they hold in W.

Finally, if $v \in W$, then since $W \subseteq V$, $v \in V$. So, $1v = v$, and the scalar multiplication identity property holds in W.

Therefore, $W \leq V$. □

Example 6.7:

1. Let $V = \mathbb{R}^2 = \{(a,b) \mid a, b \in \mathbb{R}\}$ be the vector space over $\mathbb{R}$ with the usual definitions of addition and scalar multiplication and let $W = \{(a, 0) \mid a \in \mathbb{R}\}$. If $(a, 0) \in W$, then $a, 0 \in \mathbb{R}$, and so $(a, 0) \in V$. Thus, $W \subseteq V$. The 0 vector of V is $(0, 0)$, which is in W. If $(a, 0), (b, 0) \in W$ and $k \in \mathbb{R}$, then $(a, 0) + (b, 0) = (a + b, 0) \in W$ and $k(a, 0) = (ka, 0) \in W$. It follows from Theorem 6.6 that $W \leq V$.

   This subspace W of $\mathbb{R}^2$ looks and behaves just like $\mathbb{R}$, the set of real numbers. More specifically, we say that W is **isomorphic** to $\mathbb{R}$. Most mathematicians identify this subspace W of $\mathbb{R}^2$ with $\mathbb{R}$, and just call it $\mathbb{R}$. We will provide a precise definition of "isomorphic" in Lesson 7.

   In general, it is common practice for mathematicians to call various isomorphic copies of certain structures by the same name. As a generalization of this example, if $m < n$, then we can say $\mathbb{R}^m \leq \mathbb{R}^n$ by identifying $(a_1, a_2, \ldots, a_m) \in \mathbb{R}^m$ with the vector $(a_1, a_2, \ldots, a_m, 0, 0, \ldots, 0) \in \mathbb{R}^n$ that has a tail end of $n - m$ zeros. For example, we may say that $(2, \sqrt{2}, 7, -\frac{1}{2}, 0, 0, 0)$ is in $\mathbb{R}^4$, even though it is technically in $\mathbb{R}^7$. With this type of identification, we have $\mathbb{R}^4 \leq \mathbb{R}^7$.

2. Let $V = \mathbb{Q}^3 = \{(a, b, c) \mid a, b, c \in \mathbb{Q}\}$ be the vector space over $\mathbb{Q}$ with the usual definitions of addition and scalar multiplication and let $W = \{(a, b, c) \in \mathbb{Q}^3 \mid c = a + 2b\}$. Let's check that $W \leq V$.

   It's clear that $W \subseteq V$. Since $0 = 0 + 2 \cdot 0$, we see that the zero vector $(0, 0, 0)$ is in W. Let $(a, b, c), (d, e, f) \in U$ and $k \in \mathbb{Q}$. Then we have

   $$(a, b, c) + (d, e, f) = (a, b, a + 2b) + (d, e, d + 2e) = \big(a + d, b + e, (a + d) + 2(b + e)\big).$$

   $$k(a, b, c) = k(a, b, a + 2b) = (ka, kb, ka + 2kb).$$

   These vectors are both in W, and so, by Theorem 6.6, $W \leq V$.

3. Consider $V = \mathbb{C}$ as a vector space over $\mathbb{R}$ in the usual way and let $W = \{z \in \mathbb{C} \mid \text{Re } z = 1\}$. Then $W \subseteq V$, but $W \not\leq V$ because the zero vector is not in W. After all, $0 = 0 + 0i$, and so, $\text{Re } 0 = 0$.

4. Let $M_{2,3}^{\mathbb{Z}_5}$ be the set of all 2×3 matrices with entries in the field $\mathbb{Z}_5$ (see part 2 of Example 2.19). $M_{2,3}^{\mathbb{Z}_5}$ is a vector space over $\mathbb{Z}_5$. For example, the matrices $A = \begin{bmatrix} 2 & 1 & 0 \\ 3 & 2 & 4 \end{bmatrix}$ and $B = \begin{bmatrix} 4 & 2 & 0 \\ 4 & 4 & 3 \end{bmatrix}$ are in $M_{2,3}^{\mathbb{Z}_5}$ (note that, as usual, we are abbreviating $[k]_5 \in \mathbb{Z}_5$ by k). The sum of A and B is $A + B = \begin{bmatrix} 1 & 3 & 0 \\ 2 & 1 & 2 \end{bmatrix}$ and when we multiply A by the scalar $4 \in \mathbb{Z}_5$, we get $4A = \begin{bmatrix} 3 & 4 & 0 \\ 2 & 3 & 1 \end{bmatrix}$. Let $W = \left\{\begin{bmatrix} a & b & c \\ d & e & f \end{bmatrix} \in M_{2,3}^{\mathbb{Z}_5} \mid a + b + c = d + e + f \right\}$. Let's check that $W \leq M_{2,3}^{\mathbb{Z}_5}$. We have $\begin{bmatrix} 0 & 0 & 0 \\ 0 & 0 & 0 \end{bmatrix} \in W$ because $0 + 0 + 0 = 0 + 0 + 0$. Next, suppose that $\begin{bmatrix} a & b & c \\ d & e & f \end{bmatrix}, \begin{bmatrix} g & h & i \\ j & k & l \end{bmatrix} \in W$. Then we have

$$(a + g) + (b + h) + (c + i) = (a + b + c) + (g + h + i)$$
$$= (d + e + f) + (j + k + l) = (d + j) + (e + k) + (f + l)$$

Therefore, $\begin{bmatrix} a & b & c \\ d & e & f \end{bmatrix} + \begin{bmatrix} g & h & i \\ j & k & l \end{bmatrix} = \begin{bmatrix} a+g & b+h & c+i \\ d+j & e+k & f+l \end{bmatrix} \in W$.

I leave it to the reader to show that W is closed under scalar multiplication.

5. Let $V = \{ax^2 + bx + c \mid a, b, c \in \mathbb{R}\}$ be the set of polynomials of degree at most 2 with real coefficients over $\mathbb{R}$, and let $W = \{p(x) \in V \mid p(5) = 0\}$. Let's check that $W \leq V$ (note that if $p(x) = ax^2 + bx + c$, then $p(5) = 25a + 5b + c$).

It's clear that $W \subseteq V$. The zero polynomial $p(x) = 0$ satisfies $p(5) = 0$, and so, the zero vector is in W. Let $p(x), q(x) \in W$ and $k \in \mathbb{R}$. Then we have $p(5) + q(5) = 0 + 0 = 0$, so that $p(x) + q(x) \in W$, and we have $kp(5) = k \cdot 0 = 0$, so that $kp(x) \in W$. By Theorem 6.6, $W \leq V$.

6. Let $\mathbb{Q}[x] = \{a_k x^k + a_{k-1} x^{k-1} + \cdots + a_1 x + a_0 \mid k \in \mathbb{N} \wedge a_0, a_1, \ldots, a_k \in \mathbb{Q}\}$ be the set of all polynomials with rational coefficients and let W be the subset of $\mathbb{Q}[x]$ consisting of polynomials with coefficients in $\mathbb{Z}$: $W = \{a_k x^k + a_{k-1} x^{k-1} + \cdots + a_1 x + a_0 \mid k \in \mathbb{N} \wedge a_0, a_1, \ldots, a_k \in \mathbb{Z}\}$. If we are considering $\mathbb{Q}[x]$ as a vector space over $\mathbb{Q}$, then W is **not** a subspace of $\mathbb{Q}[x]$. For example, $x + 1 \in W$, but $\frac{1}{2}(x + 1) = \frac{1}{2}x + \frac{1}{2} \notin W$.

However, if we consider $\mathbb{Q}[x]$ as a left module over $\mathbb{Z}$, then W is a submodule of $\mathbb{Q}[x]$. I leave it to the reader to check the details.

7. Every vector space V is a subspace of itself, and the vector space consisting of just the 0 vector from the vector space V is a subspace of V. In other words, for any vector space V, $V \leq V$ and $\{0\} \leq V$. The empty set, however, can never be a subspace of a vector space because it doesn't contain a zero vector.

Substructures Generated by a Set

Let G be a group and let $S \subseteq G$. The subgroup of G **generated** by S, written $\langle S \rangle$, is the intersection of all subgroups of G containing S. You will be asked to prove in Problem 11 below that an arbitrary intersection of subgroups of a group G is also a subgroup of G. It follows that $\langle S \rangle$ is the "smallest" subgroup of G containing S. Stated more formally, $\langle S \rangle$ is a subgroup of G such that if H is any subgroup of G containing S, then $\langle S \rangle \subseteq H$ (Check this!).

If $S \neq \emptyset$, then $\langle S \rangle$ consists of all finite products (or sums when the operation is addition) of elements of S and their inverses. The dedicated reader should verify that for $S \neq \emptyset$, the two definitions of $\langle S \rangle$ just presented are equivalent.

The elements of S are called the **generators** of $\langle S \rangle$.

If S consists of a single element x, then we abbreviate $\langle S \rangle = \langle \{x\} \rangle$ as $\langle x \rangle$, and we call $\langle x \rangle$ the **cyclic subgroup generated by** x. In this case, $\langle x \rangle$ is generated by the single element x. If $G = \langle x \rangle$, then G is a **cyclic group**. More generally, if $G = \langle S \rangle$ and S is finite, then we say that G is **finitely generated**.

If G is a group and $x \in G$, we define the **order of the element** x, written $|x|$, to be the order of the cyclic subgroup $|\langle x \rangle|$. Equivalently, $|x|$ is the least positive integer n such that $x^n = e$ (or $nx = e$ if we are using additive notation), if such an n exists (otherwise x has infinite order). The equivalence of these two definitions is not completely obvious (although not too difficult) and requires the Division Algorithm. We will learn all about the Division Algorithm in Lesson 8 (see Theorem 8.13) and we will prove the equivalence of these two definitions in Lesson 9 (Theorems 9.1 and 9.2).

Example 6.8:

1. Let $G = \mathbb{Z}_6 = \{0, 1, 2, 3, 4, 5\}$ and consider the group $(G, +)$. Note that G is a cyclic group generated by 1. So, $\langle 1 \rangle = \mathbb{Z}_6$. We also have $\langle 2 \rangle = \{0, 2, 4\}$ and $\langle 3 \rangle = \{0, 3\}$. These are the cyclic subgroups of $\mathbb{Z}_6$ generated by 2 and 3, respectively. Note that 4 is also a generator for $\langle 2 \rangle$ because $4 + 4 = 2$ and $2 + 4 = 0$. Similarly, we see that 5 is a generator for $\mathbb{Z}_6$ (Check this!). If we let $S = \{2, 3\}$, then $\langle S \rangle = \mathbb{Z}_6$. To see this, simply observe that $5 = 2 + 3$, $1 = 5 + 2$, $4 = 3 + 1$, and $0 = 3 + 3$. The order of each element of $\mathbb{Z}_6$ is as follows: $|1| = |5| = 6$, $|2| = |4| = 3$, and $|3| = 2$ (note that the notation $|n|$ is being used here for the **order** of n and **not** the absolute value of n).

2. Let $G = \mathbb{Z}$. We saw in part 2 of Example 2.11 that $(\mathbb{Z}, +) = \langle 1 \rangle = \langle -1 \rangle$. We also have $\mathbb{Z} = \langle \{2, 3\} \rangle$. To see this, simply note that $1 = 3 + (-2)$ and $\mathbb{Z}$ is generated by 1. The other cyclic subgroups of $(\mathbb{Z}, +)$ have the form $\langle n \rangle = n\mathbb{Z}$. Each element of $\mathbb{Z}$, with the exception of 0, has infinite order.

3. Let $D_3 = \{e, r, r^2, s, rs, r^2s\}$, where $r^3 = e$, $s^2 = e$, and $sr = r^2s$. D_3 is called the **Dihedral group of degree 3**. It is defined by the **generators** r and s, together with the **relations** $r^3 = e, s^2 = e$, and $sr = r^2s$. We can write $D_3 = \langle r, s \mid r^3 = e, s^2 = e, sr = r^2s \rangle$. Defining a group using generators and relations this way provides us with an alternative to writing out the entire multiplication table for the group. The relations provide us with a method for determining any entry in the multiplication table. For example, let's compute $(r^2s)(r^2)$. Well, we have $(r^2s)(r^2) = r^2(sr)r = r^2(r^2s)r = (r^3)r(sr) = er(r^2s) = r^3s = es = s$.

Geometrically, we can think of D_3 as all the symmetries of an equilateral triangle. Specifically, we can think of r as a rotation and s as a reflection as follows.

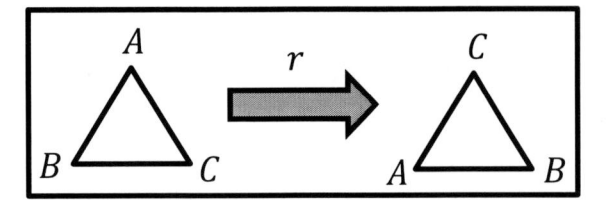

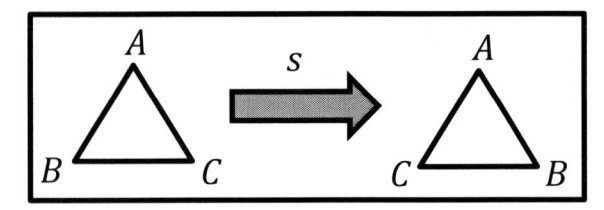

Notice how r is the group element that rotates the equilateral triangle 120° counterclockwise and s is the group element that reflects the equilateral triangle in the vertical line that passes through A and is perpendicular to the base BC. The group element r^2 rotates the equilateral triangle twice for a total rotation of 240° counterclockwise. The group element rs first reflects the equilateral triangle in the vertical line just mentioned above and then rotates the equilateral triangle 120° counterclockwise with the following result:

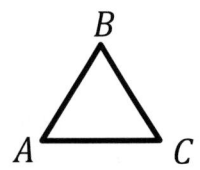

We can now see that rs is the group element that reflects the original equilateral triangle in the line that passes through C and is perpendicular to the base AB. The dedicated reader should draw similar pictures for each element of D_3 and write out the group's multiplication table.

Clearly, D_3 is generated by $S = \{r, s\}$. D_3 is **not** cyclic. r and r^2 each generate the cyclic subgroup $\{e, r, r^2\}$ and s, rs, and r^2s each generate a cyclic subgroup of order 2. For example, we have $(r^2s)(r^2s) = r^2(sr)(rs) = r^2(r^2s)(rs) = r^3r(sr)s = er(r^2s)s = r^3s^2 = ee = e$.

4. In general, for $n \in \mathbb{Z}^+$, $D_n = \langle r, s \mid r^n = e, s^2 = e, sr = r^{n-1}s \rangle$ is the **Dihedral group of degree n**. Geometrically, we can think of D_n as all the symmetries of a regular n-sided polygon (regular means that all sides have equal length and all angles have equal measure). For example, when $n = 4$, D_4 consists of all the symmetries of a square, r can be thought of as a counterclockwise rotation by 90°, and s can be thought of as a reflection. For $n > 2$, D_n is **not** cyclic, but it is finitely generated: it is generated by the 2-element set $S = \{r, s\}$. The rotation r generates a cyclic subgroup of order n and the reflection s generates a cyclic subgroup of order 2. There may be additional cyclic subgroups. For example, for $n = 4$, $\langle r^2 \rangle = \{e, r^2\}$ is a cyclic subgroup that is not equal to $\langle r \rangle$ or $\langle s \rangle$.

We can easily generalize the definition of a subgroup generated by a set to arbitrary structures such as semigroups, monoids, rings, vector spaces, and modules. If S is a subset of a structure X, we let $\langle S \rangle$ be the intersection of all substructures of X (with the appropriate properties) containing S.

For example, let M be a semigroup and let $S \subseteq M$. The subsemigroup of M **generated** by S, written $\langle S \rangle$, is the intersection of all subsemigroups of M containing S. If $S \neq \emptyset$, then $\langle S \rangle$ consists of all finite products (or sums when the operation is addition) of elements of S (Check this!).

If M is a monoid, the definition is similar. In this case, if $\langle S \rangle$ is the submonoid of M generated by S, then $\langle S \rangle$ must contain the identity of M.

As another example, let R be a ring and let $S \subseteq R$. The subring of R **generated** by S, written $\langle S \rangle$, is the intersection of all subrings of R containing S. If $S \neq \emptyset$, then $\langle S \rangle$ consists of all finite sums of finite products of elements of S, their additive inverses, and the multiplicative identity (Check this!).

Example 6.9:

1. $\mathbb{Z}^+ = \{1, 2, 3, \ldots\}$ is a semigroup under addition. The subsemigroup of $\mathbb{Z}^+$ generated by 3 is $\langle 3 \rangle = \{3, 6, 9, 12, \ldots\} = 3\mathbb{N} \setminus \{0\}$.

2. Let $R = \mathbb{Z}_6 = \{0, 1, 2, 3, 4, 5\}$ and consider the ring $(R, +, \cdot)$. Since the additive group structure is cyclic (generated by 1), we call $\mathbb{Z}_6$ a **cyclic ring**. Every subgroup of $\mathbb{Z}_6$ is closed under the ring multiplication, and so, as in part 1 of Example 6.8, we have the two proper cyclic subrings $\langle 2 \rangle = \{0, 2, 4\}$ and $\langle 3 \rangle = \{0, 3\}$.

3. Let $R = \mathbb{C}$ and consider the subring $\langle i \rangle$ generated by i. We call this subring the **ring of Gaussian integers** and usually denote it by $\mathbb{Z}[i]$. Since $\mathbb{Z}[i]$ is closed under finite sums and their inverses, we must have $bi \in \mathbb{Z}[i]$ for every integer b. Since $1 \in \mathbb{Z}[i]$ (by definition of $\langle i \rangle$), it follows that $a + bi \in \mathbb{Z}[i]$ for all integers a and b. In Problem 7 below, you will be asked to prove that $(\{a + bi \mid a, b \in \mathbb{Z}\}, +, \cdot)$ is a ring. So, $\mathbb{Z}[i] = \{a + bi \mid a, b \in \mathbb{Z}\}$.

   Note that since $i^4 = 1$, it follows that 1 is generated from i. So, even considering $\mathbb{C}$ as a rng (a ring without the multiplicative identity property), we would still have $1 = i^4 \in \mathbb{Z}[i]$.

4. More generally, if $c \in \mathbb{C}$ we define $\mathbb{Z}(c)$ to be the subring of $\mathbb{C}$ generated by c ($\mathbb{Z}(c) = \langle c \rangle$). It is not hard to see that $\mathbb{Z}(c) = \{a_k c^k + a_{k-1} c^{k-1} + \cdots + a_1 c + a_0 \mid k \in \mathbb{N} \wedge a_0, a_1, \ldots, a_k \in \mathbb{Z}\}$. We can describe $\mathbb{Z}(c)$ as the set of all polynomials "evaluated at c" with integer coefficients. The verification that $\mathbb{Z}(c)$ is a ring is Problem 21 below.

   We saw in part 3 above that if $c = i$, then $\mathbb{Z}(c)$ simplifies to $\mathbb{Z}(i) = \{a + bi \mid a, b \in \mathbb{Z}\}$. As another example, if $c = \frac{1}{2}$, then $\mathbb{Z}(c)$ simplifies to $\mathbb{Z}\left(\frac{1}{2}\right) = \left\{\frac{m}{2^n} \mid m, n \in \mathbb{Z}\right\}$. Given an arbitrary element $c \in \mathbb{C}$, such a simplification may or may not be possible. There will be such a simplification if there is a polynomial $p(z)$ with integer coefficients such that $p(c) = 0$. For i, such a polynomial is $z^2 + 1$ and for $\frac{1}{2}$, such a polynomial is $2z - 1$. We say that such a complex number c is **algebraic over** $\mathbb{Z}$. If there is no such polynomial, then c is called **transcendental over** $\mathbb{Z}$. When c is transcendental, two distinct polynomials in c represent distinct complex numbers, whereas if c is algebraic, there are infinitely many polynomials in c that represent each complex number. For example, since $i^2 + 1 = 0$, we see that 0 can be represented by $q(i)(i^2 + 1)$ for any polynomial $q(z)$, and more generally, the complex number d can be represented by $q(i)(i^2 + 1) + d$. As a specific example, $5 + i$ can be represented by $3i(i^2 + 1) + (5 + i)$, or equivalently, $3i^3 + 4i + 5$.

As one more example of a substructure generated by a set, let V be a vector space over a field F and let $S \subseteq V$. The subspace of V **generated** by S (or **spanned** by S), written $\langle S \rangle$ (or span S), is the intersection of all subspaces of V containing S. If $S \neq \emptyset$, then $\langle S \rangle = $ span S consists of all finite sums of scalar multiples of elements of S (these are called **linear combinations**). This same definition works for a (left or right) module over a ring.

147

Example 6.10: Consider the vector space $\mathbb{R}^3 = \{(a,b,c) \mid a,b,c \in \mathbb{R}\}$. The subspace of $\mathbb{R}^3$ generated by $\{(1,1,1)\}$ consists of all scalar multiples of $(1,1,1)$. So, span $\{(1,1,1)\} = \{(a,a,a) \mid a \in \mathbb{R}\}$. The subspace generated by $\{(1,0,0), (0,1,0)\}$ is

$$\text{span}\{(1,0,0), (0,1,0)\} = \{a(1,0,0) + b(0,1,0) \mid a,b \in \mathbb{R}\}$$
$$= \{(a,0,0) + (0,b,0) \mid a,b \in \mathbb{R}\} = \{(a,b,0) \mid a,b \in \mathbb{R}\}.$$

If $v, w \in V$ and $j, k \in F$, then $jv + kw$ is called a **linear combination** of the vectors v and w. We call the scalars j and k **weights**.

Example 6.11: Let $V = \mathbb{R}^2 = \{(a,b) \mid a,b \in \mathbb{R}\}$ be the vector space over $\mathbb{R}$ with the usual definitions of addition and scalar multiplication. Let $v = (1,0)$, $w = (0,1)$, $j = 4$, and $k = -2$. We have

$$jv + kw = 4(1,0) - 2(0,1) = (4,0) + (0,-2) = (4,-2).$$

It follows that the vector $(4,-2)$ is a linear combination of the vectors $(1,0)$ and $(0,1)$ with weights 4 and -2, respectively.

We will now extend the notion of linear combination to more than two vectors.

Let V be a vector space over a field F, let $v_1, v_2, \ldots, v_n \in V$, and $k_1, k_2, \ldots, k_n \in F$. The expression $k_1 v_1 + k_2 v_2 + \cdots + k_n v_n$ is called a **linear combination** of the vectors $v_1, v_2, \ldots, v_n$. Once again, we call the scalars $k_1, k_2, \ldots, k_n$ **weights**.

Example 6.12: Let $V = \mathbb{R}^3 = \{(a,b,c) \mid a,b,c \in \mathbb{R}\}$ be the vector space over $\mathbb{R}$ with the usual definitions of addition and scalar multiplication. Let $v_1 = (1,0,0)$, $v_2 = (0,1,0)$, $v_3 = (0,0,1)$, $k_1 = 3$, $k_2 = -5$, $k_3 = 6$ We have

$$k_1 v_1 + k_2 v_2 + k_3 v_3 = 3(1,0,0) - 5(0,1,0) + 6(0,0,1).$$
$$= (3,0,0) + (0,-5,0) + (0,0,6) = (3,-5,6).$$

It follows that the vector $(3,-5,6)$ is a linear combination of the vectors $(1,0,0)$, $(0,1,0)$, and $(0,0,1)$ with weights $3, -5$, and 6, respectively.

Once again, if $v, w \in V$, where V is a vector space over a field F, then the set of all linear combinations of v and w is called the **span** of v and w. Symbolically, we have span$\{v,w\} = \{jv + kw \mid j,k \in F\}$.

Example 6.13: in Example 6.11, we saw that $(4,-2)$ can be written as a linear combination of the vectors $(1,0)$ and $(0,1)$. It follows that $(4,-2) \in \text{span}\{(1,0), (0,1)\}$.

Theorem 6.14: Let $V = \mathbb{R}^2 = \{(a,b) \mid a,b \in \mathbb{R}\}$ be the vector space over $\mathbb{R}$ with the usual definitions of addition and scalar multiplication. Then span$\{(1,0), (0,1)\} = \mathbb{R}^2$.

Proof: Let $v \in \text{span}\{(1,0), (0,1)\}$. Then there are weights $j, k \in \mathbb{R}$ with $v = j(1,0) + k(0,1)$. So, we have $v = j(1,0) + k(0,1) = (j,0) + (0,k) = (j,k)$. Since $j, k \in \mathbb{R}$, we have $v = (j,k) \in \mathbb{R}^2$. Since $v \in \text{span}\{(1,0), (0,1)\}$ was arbitrary, span$\{(1,0), (0,1)\} \subseteq \mathbb{R}^2$.

Now, let $v \in \mathbb{R}^2$. Then there are $a, b \in \mathbb{R}$ with $v = (a, b) = (a, 0) + (0, b) = a(1, 0) + b(0, 1)$. Since we have expressed v as a linear combination of $(1, 0)$ and $(0, 1)$, we see that $v \in \text{span}\{(1, 0), (0, 1)\}$. Since $v \in \mathbb{R}^2$ was arbitrary, $\mathbb{R}^2 \subseteq \text{span}\{(1, 0), (0, 1)\}$.

Since $\text{span}\{(1, 0), (0, 1)\} \subseteq \mathbb{R}^2$ and $\mathbb{R}^2 \subseteq \text{span}\{(1, 0), (0, 1)\}$, we have $\text{span}\{(1, 0), (0, 1)\} = \mathbb{R}^2$. □

If $v, w \in V$, where V is a a vector space over a field F, then we say that v and w are **linearly independent** if neither vector is a scalar multiple of the other one. Otherwise, we say that v and w are **linearly dependent**.

Example 6.15:

1. The vectors $(1, 0)$ and $(0, 1)$ are linearly independent in $\mathbb{R}^2$ because for any $k \in \mathbb{R}$, we have $k(1, 0) = (k, 0) \neq (0, 1)$ and $k(0, 1) = (0, k) \neq (1, 0)$.

2. The vectors $(1, 2)$ and $(-3, -6)$ are linearly dependent in $\mathbb{R}^2$ because $(-3, -6) = -3(1, 2)$.

If $v, w \in V$, where V is a vector space over a field F, then we say that $\{v, w\}$ is a **basis** of V if v and w are linearly independent and $\text{span}\{v, w\} = V$.

Example 6.16:

1. In part 1 of Example 6.15, we saw that the vectors $(1, 0)$ and $(0, 1)$ are linearly independent in $\mathbb{R}^2$. By Theorem 6.14, $\text{span}\{(1, 0), (0, 1)\} = \mathbb{R}^2$. It follows that $\{(1, 0), (0, 1)\}$ is a basis of $\mathbb{R}^2$.

2. In part 2 of Example 6.15, we saw that the vectors $(1, 2)$ and $(-3, -6)$ are linearly dependent in $\mathbb{R}^2$. It follows that $\{(1, 2), (-3, -6)\}$ is **not** a basis of $\mathbb{R}^2$.

We would like to generalize the notion of linear dependence to more than two vectors. The definition of one vector being a scalar multiple of the other isn't quite good enough to do that. The following theorem gives us an alternative definition of linear dependence that generalizes nicely.

Theorem 6.17: Let V be a vector space over a field F and let $v, w \in V$. Then v and w are linearly dependent if and only if there are $j, k \in F$, not both 0, such that $jv + kw = 0$.

Proof: Let $v, w \in V$ and suppose that v and w are linearly dependent. Then one vector is a scalar multiple of the other. Without loss of generality, we may assume that there is $c \in F$ with $v = cw$. Then we have $1v + (-c)w = 0$. So, if we let $j = 1$ and $k = -c$, then $jv + kw = 0$, and $j = 1 \neq 0$.

Now suppose that there are $j, k \in F$, not both 0, such that $jv + kw = 0$. Without loss of generality, assume that $j \neq 0$. Then we have $jv = -kw$, and so, $v = -\frac{k}{j}w$. So, v is a scalar multiple of w. Therefore, v and w are linearly dependent. □

Note: See Note 1 following Example 4.4 in Lesson 4 for an explanation of the expression "Without loss of generality," and how to properly use it in a proof.

We will now extend the notions of span, linear dependence, linear independence, and basis to more than two vectors.

If $v_1, v_2, \ldots, v_n \in V$, where V is a vector space over a field F, then the set of all linear combinations of $v_1, v_2, \ldots, v_n \in V$ is called the **span** of $v_1, v_2, \ldots, v_n$. Symbolically, we have

$$\text{span}\{v_1, v_2, \ldots, v_n\} = \{k_1 v_1 + k_2 v_2 + \cdots + k_n v_n \mid k_1, k_2, \ldots, k_n \in \mathbb{F}\}.$$

Example 6.18: In Example 6.12, we saw that $(3, -5, 6)$ can be written as a linear combination of the vectors $(1, 0, 0)$, $(0, 1, 0)$, and $(0, 0, 1)$. It follows that $(3, -5, 6) \in \text{span}\{(1, 0, 0), (0, 1, 0), (0, 0, 1)\}$.

Theorem 6.19: Let $V = \mathbb{R}^n = \{(k_1, k_2, \ldots, k_n) \mid k_1, k_2, \ldots, k_n \in \mathbb{R}\}$ be the vector space over $\mathbb{R}$ with the usual definitions of addition and scalar multiplication. Then

$$\text{span}\{(1, 0, 0, \ldots, 0), (0, 1, 0, \ldots, 0), \ldots, (0, 0, 0, \ldots, 1)\} = \mathbb{R}^n.$$

Proof: Let $v \in \text{span}\{(1, 0, 0, \ldots, 0), (0, 1, 0, \ldots, 0), \ldots, (0, 0, 0, \ldots, 1)\}$. Then there are weights $k_1, k_2, \ldots, k_n \in \mathbb{R}$ with $v = k_1(1, 0, 0, \ldots, 0) + k_2(0, 1, 0, \ldots, 0) + \cdots + k_n(0, 0, 0, \ldots, 1)$. So, we have $v = (k_1, 0, 0, \ldots, 0) + (0, k_2, 0, \ldots, 0) + \cdots + (0, 0, 0, \ldots, k_n) = (k_1, k_2, \ldots, k_n)$. Since $k_1, k_2, \ldots, k_n \in \mathbb{R}$, we have $v = (k_1, k_2, \ldots, k_n) \in \mathbb{R}^n$. Since $v \in \text{span}\{(1, 0, 0, \ldots, 0), (0, 1, 0, \ldots, 0), \ldots, (0, 0, 0, \ldots, 1)\}$ was arbitrary, $\text{span}\{(1, 0, 0, \ldots, 0), (0, 1, 0, \ldots, 0), \ldots, (0, 0, 0, \ldots, 1)\} \subseteq \mathbb{R}^n$.

Now, let $v \in \mathbb{R}^n$. Then there are $k_1, k_2, \ldots, k_n \in \mathbb{R}$ with

$$v = (k_1, k_2, \ldots, k_n) = (k_1, 0, 0, \ldots, 0) + (0, k_2, 0, \ldots, 0) + \cdots + (0, 0, 0, \ldots, k_n)$$
$$= k_1(1, 0, 0, \ldots, 0) + k_2(0, 1, 0, \ldots, 0) + \cdots + k_n(0, 0, 0, \ldots, 1).$$

Since we have expressed v as a linear combination of $(1, 0, 0, \ldots, 0), (0, 1, 0, \ldots, 0), \ldots, (0, 0, 0, \ldots, 1)$, we see that $v \in \text{span}\{(1, 0, 0, \ldots, 0), (0, 1, 0, \ldots, 0), \ldots, (0, 0, 0, \ldots, 1)\}$. Since $v \in \mathbb{R}^n$ was arbitrary, we have $\mathbb{R}^n \subseteq \text{span}\{(1, 0, 0, \ldots, 0), (0, 1, 0, \ldots, 0), \ldots, (0, 0, 0, \ldots, 1)\}$.

Therefore, $\text{span}\{(1, 0, 0, \ldots, 0), (0, 1, 0, \ldots, 0), \ldots, (0, 0, 0, \ldots, 1)\} = \mathbb{R}^n$. □

If $v_1, v_2, \ldots, v_n \in V$, where V is a vector space over a field F, then we say that $v_1, v_2, \ldots, v_n$ are **linearly dependent** if there exist weights $k_1, k_2, \ldots, k_n \in F$, with at least one weight nonzero, such that $k_1 v_1 + k_2 v_2 + \cdots + k_n v_n = 0$. Otherwise, we say that $v_1, v_2, \ldots, v_n$ are **linearly independent**.

Notes: (1) $v_1, v_2, \ldots, v_n$ are **linearly independent** if whenever we write $k_1 v_1 + k_2 v_2 + \cdots + k_n v_n = 0$, it follows that all the weights $k_1, k_2, \ldots, k_n$ are 0.

(2) We will sometimes call the expression $k_1 v_1 + k_2 v_2 + \cdots + k_n v_n = 0$ a **dependence relation**. If any of the weights $k_1, k_2, \ldots, k_n$ are nonzero, then we say that the dependence relation is **nontrivial**.

Example 6.20:

1. The three vectors $(1, 0, 0)$, $(0, 1, 0)$, and $(0, 0, 1)$ are linearly independent in $\mathbb{R}^3$. To see this, note that we have

    $$k_1(1, 0, 0) + k_2(0, 1, 0) + k_3(0, 0, 1) = (k_1, 0, 0) + (0, k_2, 0) + (0, 0, k_3) = (k_1, k_2, k_3).$$

    So, $k_1(1, 0, 0) + k_2(0, 1, 0) + k_3(0, 0, 1) = (0, 0, 0)$ if and only if $(k_1, k_2, k_3) = (0, 0, 0)$ if and only if $k_1 = 0$, $k_2 = 0$, and $k_3 = 0$.

2. A similar computation shows that the n vectors $(1, 0, 0, \ldots, 0), (0, 1, 0, \ldots, 0), \ldots, (0, 0, 0, \ldots, 1)$ are linearly independent in $\mathbb{R}^n$.

3. The vectors $(1, 2, 3)$, $(-2, 4, 3)$, and $(1, 10, 12)$ are linearly dependent in $\mathbb{R}^3$. To see this, note that $3(1, 2, 3) + (-2, 4, 3) = (3, 6, 9) + (-2, 4, 3) = (1, 10, 12)$, and therefore,

$$3(1, 2, 3) + (-2, 4, 3) - (1, 10, 12) = 0.$$

This gives us a nontrivial dependence relation because we have at least one nonzero weight (in fact, all three weights are nonzero). The weights are 3, 1, and −1.

If $v_1, v_2, \ldots, v_n \in V$, where V is a vector space over a field $\mathbb{F}$, then we say that $\{v_1, v_2, \ldots, v_n\}$ is a **basis** of V if $v_1, v_2, \ldots, v_n$ are linearly independent and span$\{v_1, v_2, \ldots, v_n\} = V$.

Example 6.21:

1. In part 1 of Example 6.20, we saw that the vectors $(1, 0, 0)$, $(0, 1, 0)$, and $(0, 0, 1)$ are linearly independent in $\mathbb{R}^3$. By Theorem 6.19, span$\{(1, 0, 0), (0, 1, 0), (0, 0, 1)\} = \mathbb{R}^3$. It follows that $\{(1, 0, 0), (0, 1, 0), (0, 0, 1)\}$ is a basis of $\mathbb{R}^3$.

   Similarly, $\{(1, 0, 0, \ldots, 0), (0, 1, 0, \ldots, 0), \ldots, (0, 0, 0, \ldots, 1)\}$ is a basis of $\mathbb{R}^n$.

2. In part 3 of Example 6.20, we saw that the vectors $(1, 2, 3)$, $(-2, 4, 3)$, and $(1, 10, 12)$ are linearly dependent in $\mathbb{R}^3$. It follows that $\{(1, 2, 3), (-2, 4, 3), (1, 10, 12)\}$ is **not** a basis of $\mathbb{R}^3$.

In Problem 20 below, you will be asked to prove that every vector space has a basis. In general, this result is not true for modules, as the next example shows.

First note that the definitions of span, linear dependence, linear independence, and basis all make sense for left R-modules exactly as defined.

Example 6.22: Consider $\mathbb{Q}$ as a left $\mathbb{Z}$-module (see part 5 of Example 2.20). Any two rational numbers $\frac{a}{b}$ and $\frac{c}{d}$ are linearly independent because $(bc) \cdot \frac{a}{b} + (-ad) \cdot \frac{c}{d} = ca - ac = 0$. Therefore, if a basis exists, it must consist of just one nonzero rational number $\frac{a}{b}$. But span$\left\{\frac{a}{b}\right\} \neq \mathbb{Q}$ (Check this!). It follows that $\mathbb{Q}$ as a left $\mathbb{Z}$-module has no basis.

A left R-module with a basis is called a **free left R-module**. For example, every vector space V over a field F is a free left F-module. Let's look at a non-vector space example.

Example 6.23: $\mathbb{Z}^2$ is a free left $\mathbb{Z}$-module with basis $\{(1, 0), (0, 1)\}$. More generally, $\mathbb{Z}^n$ is a free left $\mathbb{Z}$-module with basis $\{(1, 0, 0, \ldots, 0), (0, 1, 0, \ldots, 0), \ldots, (0, 0, 0, \ldots, 1)\}$.

Since every commutative group is a left $\mathbb{Z}$-module, the definitions of span, linear dependence, linear independence, and basis all make sense for commutative groups as well. A commutative group with a basis is called a **free commutative group** (or **free abelian group**).

$\mathbb{Z}^n$ is a free commutative group with basis $\{(1, 0, 0, \ldots, 0), (0, 1, 0, \ldots, 0), \ldots, (0, 0, 0, \ldots, 1)\}$ by Example 6.23. In particular, $\mathbb{Z}$ is a free commutative group with basis $\{1\}$. By Example 6.22, $\mathbb{Q}$ is **not** a free commutative group.

Problem Set 6

Full solutions to these problems are available for free download here:
www.SATPrepGet800.com/AAFBTDW

LEVEL 1

1. Let H and K be the subsets of S_3 defined by $H = \{(1), (123), (132)\}$ and $K = \{(1), (12)\}$. Prove that H and K are subgroups of S_3.

2. Determine if each of the following subsets of $\mathbb{R}^2$ is a subspace of $\mathbb{R}^2$:

    (i) $A = \{(x,y) \mid x + y = 0\}$

    (ii) $B = \{(x,y) \mid xy = 0\}$

    (iii) $C = \{(x,y) \mid 2x = 3y\}$

    (iv) $D = \{(x,y) \mid x \in \mathbb{Q}\}$

3. For each of the following, determine if the given pair of vectors v and w are linearly independent or linearly dependent in the given vector space V:

    (i) $V = \mathbb{Q}^4$, $v = (3, 2, 2, -1)$, $w = \left(-1, -\frac{2}{3}, -\frac{2}{3}, -\frac{1}{3}\right)$

    (ii) $V = \mathbb{R}^3$, $v = (1, \sqrt{2}, 1)$, $w = (\sqrt{2}, 2, \sqrt{2})$

    (iii) $V = \mathbb{C}^5$, $v = (1, i, 2-i, 0, 3i)$, $w = (-i, 1, -1 - 2i, 0, 3)$

    (iv) $V = M_{22}^{\mathbb{Q}}$, $v = \begin{bmatrix} a & b \\ \frac{a}{2} & 3b \end{bmatrix}$, $w = \begin{bmatrix} 1 & \frac{b}{a} \\ \frac{1}{2} & 3 \end{bmatrix}$ $(a \neq 0, a \neq b)$

    (v) $V = \{ax^2 + bx + c \mid a, b, c \in \mathbb{R}\}$, $v = x$, $w = x^2$

LEVEL 2

4. Let G be a group and let $a \in G$. The **centralizer** of a in G is $C_a(G) = \{x \in G \mid ax = xa\}$. Prove that $C_a(G)$ is a subgroup of G. Then define the centralizer $C_a(R)$ of an element a in a ring R analogously and prove that $C_a(R)$ is a subring of R.

5. Let G be a group with H a nonempty subset of G. Prove that H is a subgroup of G if and only if for all $g, h \in H$, $gh^{-1} \in H$.

6. Let G be a group with H and K subgroups of G. Prove that if G is commutative, then $HK = \{hk \in G \mid h \in H \text{ and } k \in K\}$ is a subgroup of G. Is the result still true if G is not commutative?

7. Let $X = \{a + bi \mid a, b \in \mathbb{Z}\}$ be the set of Gaussian integers. Prove that X is a subring of $\mathbb{C}$.

8. Let G be a commutative group with identity e. Prove that each of the following subsets of G is a subgroup of G:

    (i) $H = \{x \in G \mid x^2 = e\}$

    (ii) $K = \{x^2 \mid x \in G\}$

    (iii) $L = \{x \in G \mid x \text{ has finite order}\}$

LEVEL 3

9. Let G be a group with H and K subgroups of G, and let $G = H \cup K$. Prove that $H = G$ or $K = G$.

10. Let $\mathbb{Z}[x]_n = \{a_n x^n + a_{n-1} x^{n-1} + \cdots + a_1 x + a_0 \mid a_0, a_1, \ldots, a_n \in \mathbb{Z}\}$. By Problem 17 from Problem Set 2, $(\mathbb{Z}_n[x], +)$ is a commutative group for $n = 0, 1,$ and 2, where addition is defined in the "usual way." Prove that $\mathbb{Z}[x]_0$ is a subgroup of $\mathbb{Z}[x]_1$ and $\mathbb{Z}[x]_1$ is a subgroup of $\mathbb{Z}[x]_2$.

11. Let V be a vector space over a field F and let X be a set of subspaces of V. Prove that $\bigcap X$ is a subspace of V. State and prove the analogous theorems for groups, rings, and fields.

12. Prove that a finite set with at least two vectors is linearly dependent if and only if one of the vectors in the set can be written as a linear combination of the other vectors in the set.

LEVEL 4

13. Let G be a group and let H be a nonempty finite subset of G that is closed under the operation of G. Prove that H is a subgroup of G.

14. Let U and W be subspaces of a vector space V. Determine necessary and sufficient conditions for $U \cup W$ to be a subspace of V.

15. Give an example of vector spaces U and V with $U \subseteq V$ such that U is closed under scalar multiplication, but U is not a subspace of V.

LEVEL 5

16. Let S be a set of two or more linearly dependent vectors in a vector space V. Prove that there is a vector v in the set so that $\text{span } S = \text{span } S \setminus \{v\}$.

17. Prove that a finite set of vectors S in a vector space V is a basis of V if and only if every vector in V can be written uniquely as a linear combination of the vectors in S.

18. Let $S = \{v_1, v_2, \ldots, v_m\}$ be a set of linearly independent vectors in a vector space V and let $T = \{w_1, w_2, \ldots, w_n\}$ be a set of vectors in V such that $\text{span } T = V$. Prove that $m \leq n$.

19. Let B be a basis of a vector space V with n vectors. Prove that any other basis of V also has n vectors.

CHALLENGE PROBLEMS

20. Prove that every vector space has a basis. (Note that the proof of this result will require the Axiom of Choice. There are many equivalent versions of the Axiom of Choice. The most useful version for proving this result is called Zorn's Lemma. The statement of Zorn's Lemma is provided below right after some preliminary definitions).

    If $(P, \leq)$ is a partially ordered set, then $(C, \leq)$ is a **chain** in P if $C \subseteq P$ and $(C, \leq)$ is linearly ordered. An **upper bound** of the chain $(C, \leq)$ in P is an element $s \in P$ such that for all $y \in C$, $y \leq s$. An element $m \in P$ is a **maximal element** of $(P, \leq)$ if there is no $y \in P$ such that $m < y$.

    Zorn's Lemma (ZL): Let $(P, \leq)$ be a partially ordered set such that each chain in P has an upper bound in P. Then P contains at least one maximal element.

21. Let $c \in \mathbb{C}$ and let $\mathbb{Z}(c) = \{a_k c^k + a_{k-1} c^{k-1} + \cdots + a_1 c + a_0 \mid k \in \mathbb{N} \wedge a_0, a_1, \ldots, a_k \in \mathbb{Z}\}$. Prove that $\mathbb{Z}(c)$ is the subring of $\mathbb{C}$ generated by c. More generally, let R be a subring of a ring A, let $c \in A$, and let $R(c) = \{a_k c^k + a_{k-1} c^{k-1} + \cdots + a_1 c + a_0 \mid k \in \mathbb{N} \wedge a_0, a_1, \ldots, a_k \in R\}$. Is $R(c)$ the subring of A generated by c? If R is a domain, is $R(c)$ necessarily a domain? If R is a field, is $R(c)$ necessarily a field?

22. Let E be a field.

    (i) Given that $S \subseteq E$, provide two equivalent definitions for $\langle S \rangle$, the subfield of E generated by S, and prove that these two definitions are equivalent.

    (ii) Suppose that F is a subfield of E and $c \in E$. Prove that $F(c) = \langle F \cup \{c\} \rangle$.

    (See Problem 21 above for the definition of $F(c)$.)

LESSON 7
HOMOMORPHISMS AND ISOMORPHISMS

Homomorphisms

Recall from Lesson 6 that structures $\mathfrak{A}$ and $\mathfrak{B}$ have the same **type** if they have the same number of n-ary operations for each $n \in \mathbb{N}$, and the same number of n-ary relations for each $n \in \mathbb{N}^*$.

A **homomorphism** is a function from one structure to another structure of the same type that preserves all the relations and functions of the structure. We will see a more formal definition shortly, but let's look at the types of homomorphisms we are most interested in first.

Example 7.1:

1. Let $(S,\star)$ and $(T,\circ)$ be semigroups. A **semigroup homomorphism** is a function $f: S \to T$ such that for all $a, b \in S$, $f(a \star b) = f(a) \circ f(b)$.

   For example, let $\mathfrak{A} = (\mathbb{Z}^+, +)$, let $\mathfrak{B} = (2\mathbb{N}, \cdot)$, and let $f: \mathbb{Z}^+ \to 2\mathbb{N}$ be defined by $f(n) = 2^n$. For all $n, m \in \mathbb{Z}^+$, we have $f(n + m) = 2^{n+m} = 2^n \cdot 2^m = f(n) \cdot f(m)$. Therefore, f is a semigroup homomorphism.

   As another example, let $\mathfrak{A} = (\mathbb{N}, +)$, $\mathfrak{B} = (\{T, F\}, \vee)$ (here $\vee$ is the disjunction operation—it's multiplication table can be drawn by relabeling it's truth table found in Note 2 before Example 1.16), and let $g: \mathbb{N} \to \{T, F\}$ be defined by $g(n) = T$. For all $n, m \in \mathbb{N}$, we have $g(n + m) = T = T \vee T = g(n) \vee g(m)$. Therefore, g is a semigroup homomorphism.

2. Let $(M, \star, e_M)$ and $(N, \circ, e_N)$ be monoids, where e_M and e_N are the identities of M and N, respectively. A **monoid homomorphism** is a function $f: M \to N$ such that for all $a, b \in M$, $f(a \star b) = f(a) \circ f(b)$ and $f(e_M) = e_N$.

   Note that we need to include the identity element of a monoid as part of the structure for a homomorphism to be a monoid homomorphism. Otherwise we get only a semigroup homomorphism. The second example in part 1 above is a semigroup homomorphism, but **not** a monoid homomorphism. Indeed, the identity of $(\mathbb{N}, +)$ is 0 and the identity of $(\{T, F\}, \vee)$ is F, but $g(0) = T \neq F$.

   On the other hand, if we change the domains of the structures in the first example from part 1 above slightly, we **do** get a monoid homomorphism. Let $\mathfrak{A} = (\mathbb{N}, +, 0)$, $\mathfrak{B} = (\mathbb{N}, \cdot, 1)$, and let $f: \mathbb{N} \to \mathbb{N}$ be defined by $f(n) = 2^n$. For all $n, m \in \mathbb{N}$, $f(n + m) = f(n) \cdot f(m)$, as we saw above, and $f(0) = 2^0 = 1$. Therefore, f is a monoid homomorphism.

3. Let $(G, \star)$ and $(H, \circ)$ be groups. A **group homomorphism** is a function $f: G \to H$ such that for all $a, b \in G$, $f(a \star b) = f(a) \circ f(b)$.

   You may be asking why we are not including constant symbols for the identity like we did for monoids. After all, we certainly want f to take the identity of G to the identity of H. And you may also be asking why we are not including a unary operator symbol for taking the inverse, as we certainly want $f(a^{-1}) = (f(a))^{-1}$.

For structures $(G, \star, ^{-1_G}, e_G)$ and $(H, \circ, ^{-1_H}, e_H)$, we can define a group homomorphism to be a function $f: G \to H$ such that for all $a, b \in G$, $f(a \star b) = f(a) \circ f(b)$, for all $a \in G$, $f(a^{-1}) = (f(a))^{-1}$, and $f(e_G) = e_H$. However, it turns out that this more complicated definition is equivalent to our first simpler one. In other words, if $f: G \to H$ is a group homomorphism using the simpler definition, then f already maps the identity of G to the identity of H, and f already preserves inverses. We will prove these facts in Theorems 7.2 and 7.3 below.

As an example, let $\mathfrak{A} = (\mathbb{Z}, +)$, $\mathfrak{B} = (\{1, -1\}, \cdot)$, and let $f: \mathbb{Z} \to \{1, -1\}$ be defined by $f(n) = \begin{cases} 1 & \text{if } n \text{ is even} \\ -1 & \text{if } n \text{ is odd} \end{cases}$. There are four cases to consider. If n and m are both even, then $n + m$ is even, and so, $f(n + m) = 1$ and $f(n) \cdot f(m) = 1 \cdot 1 = 1$. If n and m are both odd, then $n + m$ is even, and so, $f(n + m) = 1$ and $f(n) \cdot f(m) = (-1) \cdot (-1) = 1$. If n is even and m is odd, then $n + m$ is odd, and so, $f(n + m) = -1$ and $f(n) \cdot f(m) = 1 \cdot (-1) = -1$. Finally, if n is odd and m is even, then $n + m$ is odd, and so, we have $f(n + m) = -1$ and $f(n) \cdot f(m) = -1 \cdot 1 = -1$. Therefore, f is a group homomorphism.

Let's look at another example. Let $\mathfrak{A} = (\mathbb{R}, +)$, let $\mathfrak{B} = (\mathbb{R}, +)$, and let $g: \mathbb{R} \to \mathbb{R}$ be defined by $g(x) = x^2$. Then g is **not** a group homomorphism. To see this, we just need a single counterexample. We have $g(1) = 1^2 = 1$, $g(2) = 2^2 = 4$, $g(1 + 2) = g(3) = 3^2 = 9$, and $g(1) + g(2) = 1 + 4 = 5$. Since $g(1 + 2) \neq g(1) + g(2)$, g fails to be a homomorphism.

However, if we let $\mathfrak{A} = (\mathbb{R}^*, \cdot)$ and $\mathfrak{B} = (\mathbb{R}^*, \cdot)$, then $g: \mathbb{R}^* \to \mathbb{R}^*$ defined by $g(x) = x^2$ is a group homomorphism. Indeed, for $x, y \in \mathbb{R}^*$, $f(xy) = (xy)^2 = x^2 y^2 = f(x)f(y)$.

4. Let $(R, +_R, \cdot_R, 1_R)$ and $(S, +_S, \cdot_S, 1_S)$ be rings, where 1_R and 1_S are the multiplicative identities of R and S, respectively. A **ring homomorphism** is a function $f: R \to S$ such that for all $a, b \in R$, $f(a +_R b) = f(a) +_S f(b)$, $f(a \cdot_R b) = f(a) \cdot_S f(b)$, and $f(1_R) = 1_S$.

Notice that we did not include constant symbols for the additive identities of the rings and we did not include unary operator symbols for taking the additive inverses of elements in the rings. Once again, we will see in Theorems 7.2 and 7.3 below that with f defined as above, it follows that for all $a \in R$, $f(-a) = -f(a)$, and $f(0_R) = 0_S$.

Let's look at an example. If R is a ring, then by Problem 13 in Problem Set 2, $R \times R$ with addition and multiplication defined componentwise is also a ring. That is, for $a, b, c, d \in R$, we define addition and multiplication by $(a, b) + (c, d) = (a + c, b + d)$ and $(a, b)(c, d) = (ac, bd)$. Let $\mathfrak{A} = (\mathbb{Z} \times \mathbb{Z}, +, \cdot, (1,1))$, $\mathfrak{B} = (\mathbb{Z}, +, \cdot, 1)$, and let $f: \mathbb{Z} \times \mathbb{Z} \to \mathbb{Z}$ be defined by $f((n, m)) = n$. Then if $n, m, j, k \in \mathbb{Z}$, then we have $f((n, m) + (j, k)) = f((n + j, m + k)) = n + j$ and we have $f((n, m)) + f((j, k)) = n + j$. We also have $f((n, m) \cdot (j, k)) = f((nj, mk)) = nj$ and $f((n, m)) \cdot f((j, k)) = nj$. Finally, $f((1,1)) = 1$. Therefore, f is a ring homomorphism.

As another example, let $\mathfrak{A} = \mathfrak{B} = (\mathbb{Z}, +, \cdot, 1)$, and let $g: \mathbb{Z} \to \mathbb{Z}$ be defined by $g(n) = 2n$. Then g is **not** a ring homomorphism. To see this, we just need a single counterexample. We have $g(3) = 2 \cdot 3 = 6$, $g(5) = 2 \cdot 5 = 10$, $g(3 \cdot 5) = g(15) = 2 \cdot 15 = 30$, and we have $g(3) \cdot g(5) = 6 \cdot 10 = 60$. Since $g(3 \cdot 5) \neq g(3) \cdot g(5)$, g fails to be a ring homomorphism. Note, however, that g **is** a group homomorphism from $(\mathbb{Z}, +)$ to itself. Indeed, if $n, m \in \mathbb{Z}$, then $g(n + m) = 2(n + m) = 2n + 2m = g(n) + g(m)$.

As one more example of a ring homomorphism, let $n \in \mathbb{Z}^+$, let $\mathfrak{A} = (\mathbb{Z}, +, \cdot, 1)$, let $\mathfrak{B} = (\mathbb{Z}_n, +, \cdot, [1]_n)$, and let $h: \mathbb{Z} \to \mathbb{Z}_n$ be defined by $h(k) = [k]_n$. Then h is a ring homomorphism. To see this, just observe that $h(1) = [1]_n$ and

$$h(k_1 + k_2) = [k_1 + k_2]_n = [k_1]_n + [k_2]_n = h(k_1) + f(k_2),$$
$$h(k_1 k_2) = [k_1 k_2]_n = [k_1]_n \cdot [k_2]_n = h(k_1) h(k_2).$$

5. A **field homomorphism** is the same as a ring homomorphism. The multiplicative inverse is automatically preserved (see Theorem 7.3 below), and so, nothing additional needs to be added to the definition.

6. A homomorphism of vector spaces over the same field F (and more generally a homomorphism of left modules over the same ring R) is called a **linear transformation**. Linear transformations will be discussed in detail below.

7. Let $(A, \leq_A)$ and $(B, \leq_B)$ be partially ordered sets. An **order homomorphism** (also known as a **monotonic function**) is a function $f: A \to B$ such that for all $x, y \in A$, $x \leq_A y$ if and only if $f(x) \leq_B f(y)$.

   For example, let $\mathfrak{A} = \mathfrak{B} = (\mathbb{N}, \leq)$ and let $f: \mathbb{N} \to \mathbb{N}$ be defined by $f(n) = n + 3$. For all $n, m \in \mathbb{N}$, we have $n \leq m$ if and only if $n + 3 \leq m + 3$ if and only if $f(n) \leq f(m)$. Therefore, f is an order homomorphism.

   As another example, let $\mathfrak{A} = (\mathbb{Z}, \geq)$, $\mathfrak{B} = (\mathcal{P}(\mathbb{Z}), \subseteq)$, and let $g: \mathbb{Z} \to \mathcal{P}(\mathbb{Z})$ be defined by $g(n) = \{k \in \mathbb{Z} \mid n \leq k\}$. Let $m, n \in \mathbb{Z}$. We will show that $m \geq n$ if and only if the relationship $\{k \in \mathbb{Z} \mid m \leq k\} \subseteq \{k \in \mathbb{Z} \mid n \leq k\}$ holds. Suppose that $m \geq n$ and let $j \in \{k \in \mathbb{Z} \mid m \leq k\}$. Then $j \geq m$. Since $m \geq n$, $j \geq n$, and so, $j \in \{k \in \mathbb{Z} \mid n \leq k\}$. Now, let $\{k \in \mathbb{Z} \mid m \leq k\} \subseteq \{k \in \mathbb{Z} \mid n \leq k\}$. Since $m \leq m$, we have $m \in \{k \in \mathbb{Z} \mid m \leq k\}$. So, $m \in \{k \in \mathbb{Z} \mid n \leq k\}$. Thus, $n \leq m$, or equivalently, $m \geq n$. Therefore, g is an order homomorphism.

Note: For completeness, here is a more rigorous definition of a homomorphism for arbitrary structures $\mathfrak{A}$ and $\mathfrak{B}$ (however, in this book, it suffices to understand the specific cases presented in Example 7.1).

If $\mathfrak{A}$ and $\mathfrak{B}$ are structures of the same type with underlying domains A and B, then a homomorphism is a function $f: A \to B$ such that for each $n \in \mathbb{N}$,

1. if R is an n-ary relation, then $R_A(a_1, a_2, \ldots, a_n)$ if and only if $R_B(f(a_1), f(a_2), \ldots, f(a_n))$.
2. If F is an n-ary function, then $f(F_A(a_1, a_2, \ldots, a_n)) = F_B(f(a_1), f(a_2), \ldots, f(a_n))$.

In particular, 2 implies that if c is a constant, then $f(c_A) = c_B$.

Theorem 7.2: Let $(G, \star)$ and $(H, \circ)$ be groups with identities e_G and e_H, respectively, and let $f: G \to H$ be a group homomorphism. Then $f(e_G) = e_H$.

Proof: Since $e_G = e_G \star e_G$, we have $f(e_G) = f(e_G \star e_G) = f(e_G) \circ f(e_G)$. So,

$$f(e_G) = f(e_G) \circ e_H = f(e_G) \circ \left(f(e_G) \circ (f(e_G))^{-1}\right)$$
$$= (f(e_G) \circ f(e_G)) \circ (f(e_G))^{-1} = f(e_G) \circ (f(e_G))^{-1} = e_H. \qquad \square$$

Notes: (1) The computations in the proof take place in the group $(H, \circ)$. In particular, $f(e_G) \in H$ and $e_H \in H$. If the proof seems confusing because $f(e_G)$ appears so often, try making the substitutions $h = f(e_G)$ and $e = e_H$. Notice that $h, e \in H$ and by the first line of the proof, $h = h \circ h$. The rest of the proof then looks like this:
$$h = h \circ e = h \circ (h \circ h^{-1}) = (h \circ h) \circ h^{-1} = h \circ h^{-1} = e.$$
Remember that $h = f(e_G)$ and $e = e_H$. So, we have $f(e_G) = e_H$, as desired.

(2) $h = h \circ e$ because e is the identity for H.

(3) $e = h \circ h^{-1}$ by the definition of inverse and because e is the identity for H. From this equation, it follows that $h \circ e = h \circ (h \circ h^{-1})$.

(4) $h \circ (h \circ h^{-1}) = (h \circ h) \circ h^{-1}$ because $\circ$ is associative in H.

(5) $h \circ h = h$ from the first line of the proof (this is equivalent to $f(e_G) \circ f(e_G) = f(e_G)$). It follows that $(h \circ h) \circ h^{-1} = h \circ h^{-1}$.

(6) Finally, $h \circ h^{-1} = e$, again by the definition of inverse and because e is the identity for H.

(7) If the group operation is addition, then we usually use the symbols 0_G and 0_H for the identities.

Theorem 7.3: Let $(G, \star)$ and $(H, \circ)$ be groups and let $f: G \to H$ be a group homomorphism. Then for all $g \in G$, $f(g^{-1}) = (f(g))^{-1}$.

Proof: By Theorem 7.2, we have $f(e_G) = e_H$. So, for $g \in G$, we have
$$e_H = f(e_G) = f(g \star g^{-1}) = f(g) \circ f(g^{-1}).$$
Since $f(g) \circ f(g^{-1}) = e_H$, $f(g^{-1}) = (f(g))^{-1}$. □

Notes: (1) $e_G = g \star g^{-1}$ by the definition of inverse and because e_G is the identity for G. From this equation, it follows that $f(e_G) = f(g \star g^{-1})$.

(2) $f(g \star g^{-1}) = f(g) \circ f(g^{-1})$ because f is a homomorphism.

(3) In a group with identity e, if $xy = e$ and $yx = e$, then $y = x^{-1}$. We actually need to verify only one of the equations $xy = e$ or $yx = e$ to determine that $y = x^{-1}$ (see Note 4 after the proof of Theorem 2.9). Letting $x = f(g)$, $y = f(g^{-1})$, and $e = e_H$, we showed in the proof that $xy = e$. It follows that $y = x^{-1}$. That is, $f(g^{-1}) = (f(g))^{-1}$.

(4) If $(R, +_R, \cdot_R)$ and $(S, +_S, \cdot_S)$ are rings, and $f: R \to S$ is a ring homomorphism, then Theorem 7.3 implies that for all $x \in R$, $f(-x) = -f(x)$ (notice how we needed to switch to additive notation here). Additionally, if $(R, +_R, \cdot_R)$ and $(S, +_S, \cdot_S)$ are fields, then Theorem 7.3 also gives us for $x \neq 0$ that $f(x^{-1}) = (f(x))^{-1}$, or equivalently, $f\left(\frac{1}{x}\right) = \frac{1}{f(x)}$.

Isomorphisms

An **isomorphism** is a bijective homomorphism. If there is an isomorphism from a structure $\mathfrak{A}$ to a structure $\mathfrak{B}$, then we say that $\mathfrak{A}$ and $\mathfrak{B}$ are **isomorphic**, and we write $\mathfrak{A} \cong \mathfrak{B}$. Mathematicians generally consider isomorphic structures to be the same. Indeed, they behave identically. The only difference between them is the "names" of the elements.

Example 7.4:

1. For $n \in \mathbb{Z}^+$, the function $f: \mathbb{Z} \to n\mathbb{Z}$ defined by $f(k) = nk$ is an isomorphism between the groups $(\mathbb{Z}, +)$ and $(n\mathbb{Z}, +)$. It's easy to see that f is injective ($j \neq k \to nj \neq nk$) and surjective (if $nk \in n\mathbb{Z}$, then $f(k) = nk$). If $j, k \in \mathbb{Z}$, then $f(j + k) = n(j + k) = nj + nk = f(j) + f(k)$. It follows that $(\mathbb{Z}, +) \cong (n\mathbb{Z}, +)$.

   Note that this map is **not** a ring isomorphism for $n > 1$. First, $(n\mathbb{Z}, +, \cdot)$ is technically not even a ring for $n > 1$ because $1 \notin n\mathbb{Z}$. But it is "almost a ring." In fact, the multiplicative identity property is the only property that fails. See Note 6 following the definition of a ring in Lesson 2.

   Let's show that for $n > 1$, f is **not** an isomorphism between the "almost rings" $(\mathbb{Z}, +, \cdot)$ and $(n\mathbb{Z}, +, \cdot)$. Let's use $2, 3 \in \mathbb{Z}$ to provide a counterexample: $f(2 \cdot 3) = f(6) = n \cdot 6 = 6n$ and $f(2) \cdot f(3) = (n \cdot 2)(n \cdot 3) = 6n^2$. If $f(2 \cdot 3) = f(2) \cdot f(3)$, then $6n = 6n^2$, so that $n = n^2$. This equation is equivalent to $n^2 - n = 0$, or $n(n - 1) = 0$. So, $n = 0$ or $n = 1$.

   In fact, as "almost rings," $(\mathbb{Z}, +, \cdot)$ is **not** isomorphic to $(n\mathbb{Z}, +, \cdot)$ at all for $n > 1$. If $f: \mathbb{Z} \to n\mathbb{Z}$ were an isomorphism, then $f(1) = nm$ for some $m \in \mathbb{Z}$. But also, since f is a homomorphism, $f(1) = f(1 \cdot 1) = f(1)f(1) = (nm)(nm) = n^2m^2$. So, $nm = n^2m^2$, and thus, $m = 0, n = 0$, or $1 = nm$. If $m = 0$, then $f(1) = 0$, and so, $f(2) = f(1 + 1) = f(1) + f(1) = 0 + 0 = 0$. So, f is not injective. Since $n > 1$, $n \neq 0$ and $1 \neq nm$.

2. If $z = a + bi$ is a complex number, then the conjugate of z is the complex number $\bar{z} = a - bi$. Let's check that the function $g: \mathbb{C} \to \mathbb{C}$ defined by $g(z) = \bar{z}$ is an isomorphism between the field $(\mathbb{C}, +, \cdot)$ and itself. We have $g(1) = \bar{1} = 1$ and by Problem 5 below,

$$g(z + w) = \overline{z + w} = \bar{z} + \bar{w} = g(z) + g(w),$$
$$g(zw) = \overline{zw} = \bar{z} \cdot \bar{w} = g(z)g(w).$$

   Thus, g is a homomorphism. Since for all $z \in \mathbb{C}$, $g(\bar{z}) = z$, f is surjective. Since $z \neq w$ implies that $\bar{z} \neq \bar{w}$, g is injective. Therefore, g is a bijective homomorphism, and so, g is an isomorphism.

   An isomorphism from a structure to itself is called an **automorphism**. The identity function is always an automorphism from any structure to itself. The function g just described is an example of a nontrivial automorphism from $\mathbb{C}$ to $\mathbb{C}$.

Note: There are a few more definitions that come up in the context of homomorphisms. A **monomorphism** is an injective homomorphism. An **epimorphism** is a surjective homomorphism. Therefore, an isomorphism is both a monomorphism and an epimorphism. Finally, an **endomorphism** is a homomorphism from a structure to itself. So, an automorphism is both an isomorphism and an endomorphism.

Theorem 7.5 (Cayley's Theorem): Every group is isomorphic to a group of permutations.

Analysis: Recall from Note 4 following Example 2.10 that for a finite group, each element appears exactly once in every row and every column of the group's multiplication table. The following multiplication table is for a group known as the **Klein four group**. Let's look carefully at the second row:

$(G,\star)$	e	a	b	c
e	e	a	b	c
a	a	e	c	b
b	b	c	e	a
c	c	b	a	e

The second row of the table describes what happens when we multiply each element of G on the left by the element a. If we define $\phi_a: G \to G$ by $\phi_a(x) = ax$, we see from the table that $\phi_a(e) = a$, $\phi_a(a) = e$, $\phi_a(b) = c$, and $\phi_a(c) = b$. Notice that ϕ_a is a permutation of G. Similarly, $\phi_b: G \to G$ is the permutation of G satisfying $\phi_b(e) = b$, $\phi_b(a) = c$, $\phi_b(b) = e$, and $\phi_b(c) = a$.

The composition $\phi_a \circ \phi_b: G \to G$ satisfies $(\phi_a \circ \phi_b)(e) = \phi_a(\phi_b(e)) = \phi_a(be) = \phi_a(b) = ab = c$, and similarly, $(\phi_a \circ \phi_b)(a) = b$, $(\phi_a \circ \phi_b)(b) = a$, and $(\phi_a \circ \phi_b)(c) = e$. It follows that $\phi_a \circ \phi_b = \phi_c$. Look at that...from the multiplication table, we see that $a \star b = c$ and by computation, $\phi_a \circ \phi_b = \phi_c$. In other words, composing the permutation ϕ_a with the permutation ϕ_b behaves just like multiplying a with b. In fact, if we write out the rest of the multiplication table for these permutations, it looks just like the multiplication table for $(G,\star)$.

$\circ$	ϕ_e	ϕ_a	ϕ_b	ϕ_c
ϕ_e	ϕ_e	ϕ_a	ϕ_b	ϕ_c
ϕ_a	ϕ_a	ϕ_e	ϕ_c	ϕ_b
ϕ_b	ϕ_b	ϕ_c	ϕ_e	ϕ_a
ϕ_c	ϕ_c	ϕ_b	ϕ_a	ϕ_e

So, if we let $H = \{\phi_x \mid x \in G\}$, we see that $f: G \to H$ defined by $f(x) = \phi_x$ is an isomorphism from G to a group of permutations of G. Let's use the idea presented here to write out the details of the proof.

Proof of Theorem 7.5: Let G be a group. For each $x \in G$, we define a function $\phi_x: G \to G$ by $\phi_x(y) = xy$. To see that ϕ_x is injective, suppose that $\phi_x(y) = \phi_x(z)$. Then $xy = xz$. Since the left cancellation law holds in a group (see Note 1 after Example 2.10), $y = z$. To see that ϕ_x is surjective, let $z \in G$. Then we have $\phi_x(x^{-1}z) = x(x^{-1}z) = (xx^{-1})z = ez = z$. It follows that ϕ_x is a permutation of G.

We let $H = \{\phi_x \mid x \in G\}$ and consider the structure $(H, \circ)$, where $\circ$ is composition of functions. Note that H is a subset of $S(G)$, the group of all permutations of G. Also, since ϕ_e is the identity permutation on G and $\phi_e \in H$, we see that H contains an identity. If $x_1, x_2 \in G$, then

$$(\phi_{x_1} \circ \phi_{x_2})(y) = \phi_{x_1}(\phi_{x_2}(y)) = \phi_{x_1}(x_2 y) = x_1(x_2 y) = (x_1 x_2)y = \phi_{x_1 x_2}(y).$$

So, $\phi_{x_1} \circ \phi_{x_2} = \phi_{x_1 x_2}$. It follows that H is closed under $\circ$. It also follows that $\phi_x \circ \phi_{x^{-1}} = \phi_{xx^{-1}} = \phi_e$, and therefore, $\phi_{x^{-1}} = (\phi_x)^{-1}$. So, H is a subgroup of $S(G)$. In particular, H is a group of permutations.

We define $f: G \to H$ by $f(x) = \phi_x$. To see that f is injective, suppose that $f(x_1) = f(x_2)$. Then $\phi_{x_1} = \phi_{x_2}$. So, for all $y \in G$, $\phi_{x_1}(y) = \phi_{x_2}(y)$, or equivalently, $x_1 y = x_2 y$. In particular, $x_1 e = x_2 e$, and so, $x_1 = x_2$. To see that f is surjective, let $\phi_x \in H$. Then $x \in G$ and $f(x) = \phi_x$. Finally, to see that f is a homomorphism, observe that $f(x_1 x_2) = \phi_{x_1 x_2} = \phi_{x_1} \circ \phi_{x_2} = f(x_1) \circ f(x_2)$. Therefore, f is an isomorphism. □

Linear Transformations

Let V and W be vector spaces over a field F, and let $T: V \to W$ be a function from V to W.

We say that T is **additive** if for all $u, v \in V$, $T(u + v) = T(u) + T(v)$.

We say that T is **homogenous** if for all $k \in F$ and all $v \in V$, $T(kv) = kT(v)$.

T is a **linear transformation** (or **vector space homomorphism**) if it is additive and homogeneous.

Note: The definitions of additive, homogenous, and linear transformation are exactly the same if we use left modules over a ring instead of vector spaces over a field. For right modules over a ring, the definitions are all the same, except that we modify the definition of homogeneous, to say: for all $k \in F$ and all $v \in V$, $T(vk) = T(v) \cdot k$.

Example 7.6:

1. Let $V = W = \mathbb{C}$ be vector spaces over $\mathbb{R}$ and define $T: \mathbb{C} \to \mathbb{C}$ by $T(z) = 5z$. We see that $T(z + w) = 5(z + w) = 5z + 5w = T(z) + T(w)$. So, T is additive. Furthermore, we have $T(kz) = 5(kz) = k(5z) = kT(z)$. So, T is homogenous. Therefore, T is a linear transformation.

    More generally, for any vector space V over $\mathbb{R}$ and any $m \in \mathbb{R}$, the function $S: V \to V$ defined by $S(v) = mv$ is a linear transformation. The verification is nearly identical to what we did in the last paragraph. This type of linear transformation is called a **dilation**.

    Note that if $m, b \in \mathbb{R}$ with $b \neq 0$, then the function $R: V \to V$ defined by $R(v) = mv + b$ is **not** a linear transformation. To see this, observe that $R(2v) = m(2v) + b = 2mv + b$ and $2R(v) = 2(mv + b) = 2mv + 2b$. If $R(2v) = 2R(v)$, then $2mv + b = 2mv + 2b$, or equivalently, $b = 2b$. Subtracting b from each side of this equation yields $b = 0$, contrary to our assumption that $b \neq 0$. So, the linear functions that we learned about in high school are usually not linear transformations. The only linear functions that **are** linear transformations are the ones that pass through the origin (in other words, b must be 0).

2. Let $V = \mathbb{R}^4$ and $W = \mathbb{R}^3$ be vector spaces over $\mathbb{R}$ and define $T: \mathbb{R}^4 \to \mathbb{R}^3$ by
$$T(x, y, z, w) = (x + z, 2x - 3y, 5y - 2w).$$
We have
$$T\big((x, y, z, w) + (s, t, u, v)\big) = T(x + s, y + t, z + u, w + v)$$
$$= \big((x + s) + (z + u), 2(x + s) - 3(y + t), 5(y + t) - 2(w + v)\big)$$
$$= \big((x + z) + (s + u), (2x - 3y) + (2s - 3t), (5y - 2w) + (5t - 2v)\big)$$
$$= (x + z, 2x - 3y, 5y - 2w) + (s + u, 2s - 3t, 5t - 2v)$$
$$= T(x, y, z, w) + T(s, t, u, v).$$

So, T is additive. Also, we have

$$T(k(x,y,z,w)) = T(kx, ky, kz, kw)$$
$$= (kx + kz, 2(kx) - 3(ky), 5(ky) - 2(kw))$$
$$= (k(x+z), k(2x-3y), k(5y-2w))$$
$$= k(x+z, 2x-3y, 5y-2w) = kT(x,y,z,w).$$

So, T is homogenous. Therefore, T is a linear transformation.

Note that in certain cases, we removed a set of parentheses for easier readability. For example, we abbreviated $T((kx, ky, kz, kw))$ by $T(kx, ky, kz, kw)$. We will continue to do this without further mention.

3. Let $V = \mathbb{R}^2$ and $W = \mathbb{R}$ be vector spaces over $\mathbb{R}$ and define $T: \mathbb{R}^2 \to \mathbb{R}$ by $T(x,y) = xy$. Then T is **not** a linear transformation. Indeed, consider $(1,0), (0,1) \in \mathbb{R}^2$. We have

$$T((1,0) + (0,1)) = T(1,1) = 1 \cdot 1 = 1.$$
$$T(1,0) + T(0,1) = 1 \cdot 0 + 0 \cdot 1 = 0 + 0 = 0.$$

So, $T((1,0) + (0,1)) \neq T(1,0) + T(0,1)$. This shows that T is **not** additive, and therefore, T is not a linear transformation.

Observe that T is also **not** homogeneous. To see this, consider $(1,1) \in \mathbb{R}^2$ and $2 \in \mathbb{R}$. We have $T(2(1,1)) = T(2,2) = 2 \cdot 2 = 4$, but $2T(1,1) = 2(1 \cdot 1) = 2 \cdot 1 = 2$.

In Problem 6 below, you will be asked to show that neither additivity nor homogeneity alone is enough to guarantee that a function is a linear transformation.

Recall from Lesson 6 that if $v, w \in V$ and $j, k \in F$, then $jv + kw$ is called a **linear combination** of the vectors v and w with weights j and k. The next theorem says that a function is a linear transformation if and only if it "behaves well" with respect to linear combinations.

Theorem 7.7: Let V and W be vector spaces over a field F. A function $T: V \to W$ is a linear transformation if and only if for all $v, w \in V$ and all $a, b \in F$, $T(av + bw) = aT(v) + bT(w)$.

Proof: Suppose that $T: V \to W$ is a linear transformation, let $v, w \in V$, and let $a, b \in F$. Since T is additive, $T(av + bw) = T(av) + T(bw)$. Since T is homogenous, $T(av) = aT(v)$ and $T(bw) = bT(w)$. Therefore, $T(av + bw) = T(av) + T(bw) = aT(v) + bT(w)$, as desired.

Conversely, suppose that for all $a, b \in F$, $T(av + bw) = aT(v) + bT(w)$. Let $v, w \in V$ and let $a = b = 1$. Then $T(v + w) = T(1v + 1w) = 1T(v) + 1T(w) = T(v) + T(w)$. Therefore, T is additive. Now, let $v \in V$ and $k \in \mathbb{F}$. Then $T(kv) = T(kv + 0v) = kT(v) + 0T(v) = kT(v)$. Therefore, T is homogenous. It follows that T is a linear transformation. □

We can use induction to extend Theorem 7.7 to arbitrary linear combinations. If $v \in V$ can be written as a linear combination of vectors $v_1, v_2, \ldots, v_n \in V$, then $T(v)$ is determined by $T(v_1), T(v_2), \ldots, T(v_n)$. Specifically, if $v = c_1 v_1 + c_2 v_2 + \cdots + c_n v_n$, then we have

$$T(v) = T(c_1 v_1 + c_2 v_2 + \cdots + c_n v_n) = c_1 T(v_1) + c_2 T(v_2) + \cdots + c_n T(v_n).$$

In particular, if $B = \{v_1, v_2, \ldots, v_n\}$ is a basis of V, then T is completely determined by the values of $T(v_1), T(v_2), \ldots, T(v_n)$.

Notes: (1) Recall from Lesson 6 that the vectors $v_1, v_2, \ldots, v_n \in V$ are **linearly independent** if whenever $k_1 v_1 + k_2 v_2 + \cdots + k_n v_n = 0$, it follows that all the weights $k_1, k_2, \ldots, k_n$ are 0.

(2) Also, recall that the set of all linear combinations of $v_1, v_2, \ldots, v_n \in V$ is called the **span** of $v_1, v_2, \ldots, v_n$, written span$\{v_1, v_2, \ldots, v_n\}$.

(3) The set of vectors $\{v_1, v_2, \ldots, v_n\}$ is a **basis** of V if $v_1, v_2, \ldots, v_n$ are linearly independent and span$\{v_1, v_2, \ldots, v_n\} = V$.

In particular, if $\{v_1, v_2, \ldots, v_n\}$ is a basis of V, then every vector in V can be written as a linear combination of $v_1, v_2, \ldots, v_n$.

So, if we know the values of $T(v_1), T(v_2), \ldots, T(v_n)$, then we know the value of $T(v)$ for any $v \in V$, as shown above.

In other words, given a basis B of V, any function $f: B \to W$ extends uniquely to a linear transformation $T: V \to W$.

Let V and W be vector spaces over a field F. We define $\mathcal{L}(V, W)$ to be the set of all linear transformations from V to W. Symbolically, $\mathcal{L}(V, W) = \{T: V \to W \mid T \text{ is a linear transformation}\}$.

Theorem 7.8: Let V and W be vector spaces over a field F. Then $\mathcal{L}(V, W)$ is a vector space over F, where addition and scalar multiplication are defined as follows:

$S + T \in \mathcal{L}(V, W)$ is defined by $(S + T)(v) = S(v) + T(v)$ for $S, T \in \mathcal{L}(V, W)$.

$kT \in \mathcal{L}(V, W)$ is defined by $(kT)(v) = kT(v)$ for $T \in \mathcal{L}(V, W)$ and $k \in F$.

The reader will be asked to prove Theorem 7.8 in Problem 16 below.

If V, W, and U are vector spaces over F, and $T: V \to W$, $S: W \to U$ are linear transformations, then the composition $S \circ T: V \to U$ is a linear transformation, where $S \circ T$ is defined by $(S \circ T)(v) = S(T(v))$ for all $v \in V$. To see this, let $v, w \in V$ and $a, b \in \mathbb{F}$. Then we have

$$(S \circ T)(av + bw) = S(T(av + bw)) = S(aT(v) + bT(w))$$
$$= a\big(S(T(v))\big) + b\big(S(T(w))\big) = a(S \circ T)(v) + b(S \circ T)(w).$$

Example 7.9: Let $T: \mathbb{R}^2 \to \mathbb{R}^3$ be the linear transformation defined by $T(x, y) = (x, x + y, y)$ and let $S: \mathbb{R}^3 \to \mathbb{R}^2$ be the linear transformation defined by $S(x, y, z) = (z - y, x - z)$. Then $S \circ T: \mathbb{R}^2 \to \mathbb{R}^2$ is a linear transformation and we have

$$(S \circ T)(x, y) = S(T(x, y)) = S(x, x + y, y) = (-x, x - y).$$

Notes: (1) In Example 7.9, the composition $T \circ S: \mathbb{R}^3 \to \mathbb{R}^3$ is also a linear transformation and we have

$$(T \circ S)(x, y, z) = T(S(x, y, z)) = T(z - y, x - z) = (z - y, x - y, x - z).$$

(2) In general, if $T: V \to W, S: X \to U$ are linear transformations, then $S \circ T$ is defined if and only if $W = X$. So, just because $S \circ T$ is defined, it does not mean that $T \circ S$ is also defined. For example, if $T: \mathbb{R} \to \mathbb{R}^2$ and $S: \mathbb{R}^2 \to \mathbb{R}^3$, then $S \circ T$ is defined and $S \circ T: \mathbb{R} \to \mathbb{R}^3$. However, $T \circ S$ is not defined. The "outputs" of the linear transformation S are ordered triples of real numbers, while the "inputs" of the linear transformation T are real numbers. They just don't "match up."

(3) If S and T are both linear transformations from a vector space V to itself (that is $S, T: V \to V$), then the compositions $S \circ T$ and $T \circ S$ will both also be linear transformations from V to itself.

By Note 3 above, in the vector space $\mathcal{L}(V, V)$, we can define a multiplication by $ST = S \circ T$. This definition of multiplication gives $\mathcal{L}(V, V)$ a ring structure. In fact, with addition, scalar multiplication, and composition as previously defined, $\mathcal{L}(V, V)$ is a structure called a **linear algebra**. From now on, we will use the abbreviation $\mathcal{L}(V)$ in place of $\mathcal{L}(V, V)$.

A **linear algebra** (sometimes just called an **algebra**) over a field F is a triple $(A, +, \cdot)$, where $(A, +)$ is a vector space over F, $(A, +, \cdot)$ is a ring, and for all $u, v \in A$ and $k \in F$, $k(uv) = (ku)v = u(kv)$.

We will call the last property "**compatibility of scalar and vector multiplication.**"

Notes: (1) There are two multiplications defined in a linear algebra. As for a vector space, we have **scalar multiplication**. We will refer to the ring multiplication as **vector multiplication**.

(2) Recall from Lesson 2 that a ring $(A, +, \cdot)$ satisfies the first property of a vector space (that $(A, +)$ is a commutative group) together with the following three additional properties of vector multiplication:

- **(Closure)** For all $u, v \in A$, $u \cdot v \in A$.
- **(Associativity)** For all $u, v, w \in A$, $(u \cdot v) \cdot w = u \cdot (v \cdot w)$.
- **(Identity)** There exists an element $1 \in A$ such that for all $v \in A$, $1 \cdot v = v \cdot 1 = v$.

Example 7.10:

1. $(\mathbb{R}, +, \cdot)$ is a linear algebra over $\mathbb{R}$, where addition and multiplication are defined in the usual way. In this example, scalar and vector multiplication are the same.

2. Similarly, $(\mathbb{C}, +, \cdot)$ is a linear algebra over $\mathbb{C}$, where addition and multiplication are defined in the usual way (see Lesson 2). Again, in this example, scalar and vector multiplication are the same.

3. If V is a vector space over a field $\mathbb{F}$, then $\mathcal{L}(V)$ is a linear algebra over F, where addition and scalar multiplication are defined as in Theorem 7.8, and vector multiplication is given by composition of linear transformations. You will be asked to verify this in Problem 17 below.

Recall from Lesson 4 that a function $f: A \to B$ is **injective** if $a, b \in A$ and $a \neq b$ implies $f(a) \neq f(b)$. Also, f is **surjective** if for all $b \in B$, there is $a \in A$ with $f(a) = b$. A **bijective** function is one that is both injective and surjective.

Also recall that a bijective function f is **invertible**. The **inverse** of f is then the function $f^{-1}: B \to A$ defined by $f^{-1}(b) =$ "the unique $a \in A$ such that $f(a) = b$."

By Theorem 4.9, $f^{-1} \circ f = i_A$ and $f \circ f^{-1} = i_B$, where i_A and i_B are the identity functions on A and B, respectively. Furthermore, f^{-1} is the **only** function that satisfies these two equations. Indeed, if $h: B \to A$ also satisfies $h \circ f = i_A$ and $f \circ h = i_B$, then

$$h = h \circ i_B = h \circ (f \circ f^{-1}) = (h \circ f) \circ f^{-1} = i_A \circ f^{-1} = f^{-1}.$$

A bijection $T: V \to W$ that is also a linear transformation is called an **isomorphism** (just like for any bijective homomorphism). If an isomorphism $T: V \to W$ exists, we say that V and W are isomorphic. As is always the case with algebraic structures, isomorphic vector spaces are essentially identical. The only difference between them are the "names" of the elements.

If a bijective function happens to be a linear transformation between two vector spaces, it's nice to know that the inverse function is also a linear transformation. We prove this now.

Theorem 7.11: Let $T: V \to W$ be an invertible linear transformation. Then $T^{-1}: W \to V$ is also a linear transformation.

Proof: Let $T: V \to W$ be an invertible linear transformation, let $u, v \in W$, and let $a, b \in \mathbb{F}$. Then by the linearity of T, we have

$$T(aT^{-1}(u) + bT^{-1}(v)) = aT(T^{-1}(u)) + bT(T^{-1}(v)) = au + bv.$$

Since T is injective, $aT^{-1}(u) + bT^{-1}(v)$ is the unique element of V whose image under T is $au + bv$. By the definition of T^{-1}, $T^{-1}(au + bv) = aT^{-1}(u) + bT^{-1}(v)$. □

Note: This theorem can be generalized to invertible homomorphisms between any two structures of the same type. See Problem 22 below.

Example 7.12:

1. Let $V = W = \mathbb{C}$ be vector spaces over $\mathbb{R}$ and define $T: \mathbb{C} \to \mathbb{C}$ by $T(z) = 5z$, as we did in part 1 of Example 7.6. if $z \neq w$, then $5z \neq 5w$, and so T is injective. Also, if $w \in \mathbb{C}$, then we have $T\left(\frac{1}{5}w\right) = 5\left(\frac{1}{5}w\right) = w$. So, T is surjective. It follows that T is invertible and that the inverse of T is defined by $T^{-1}(z) = \frac{1}{5}z$. By Theorem 7.11, $T^{-1}: \mathbb{C} \to \mathbb{C}$ is also a linear transformation. T is actually an automorphism. In other words, T is an isomorphism from $\mathbb{C}$ to itself.

2. Let V be a vector space over a field F with basis $\{v_1, v_2, v_3\}$. Then let $T: V \to F^3$ be the unique linear transformation such that $T(v_1) = (1, 0, 0)$, $T(v_2) = (0, 1, 0)$, and $T(v_3) = (0, 0, 1)$. In other words, if $v \in V$, since $\{v_1, v_2, v_3\}$ is a basis of V, we can write $v = c_1 v_1 + c_2 v_2 + c_3 v_3$, and T is defined by $T(v) = c_1 T(v_1) + c_2 T(v_2) + c_3 T(v_3) = (c_1, c_2, c_3)$.

   To see that T is injective, suppose that $T(c_1 v_1 + c_2 v_2 + c_3 v_3) = T(d_1 v_1 + d_2 v_2 + d_3 v_3)$. Then $(c_1, c_2, c_3) = (d_1, d_2, d_3)$. It follows that $c_1 = d_1$, $c_2 = d_2$, and $c_3 = d_3$. Therefore, $c_1 v_1 + c_2 v_2 + c_3 v_3 = d_1 v_1 + d_2 v_2 + d_3 v_3$, and so, T is injective.

   Now, if $(a, b, c) \in F^3$, then $T(av_1 + bv_2 + cv_3) = (a, b, c)$ and so, T is surjective. From this computation, we also see that $T^{-1}: F^3 \to V$ is defined by $T^{-1}(a, b, c) = av_1 + bv_2 + cv_3$.

   It follows that $T: V \to F$ is an isomorphism, so that V is isomorphic to F^3.

Essentially the same argument as above can be used to show that if V is a vector space over a field F with a basis consisting of n vectors, then V is isomorphic to F^n.

Matrices

Recall from Example 2.19 in Lesson 2 that for $m, n \in \mathbb{Z}^+$, an $m \times n$ **matrix** over a field F is a rectangular array with m rows and n columns, and entries in F. For example, the matrix $H = \begin{bmatrix} i & 2 - 5i & \frac{1}{5} \\ -1 & \sqrt{3} & 7 + i \end{bmatrix}$ is a 2×3 matrix over $\mathbb{C}$. We will generally use a capital letter to represent a matrix, and the corresponding lowercase letter with double subscripts to represent the entries of the matrix. We use the first subscript for the row and the second subscript for the column. Using the matrix H above as an example, we see that $h_{11} = i$, $h_{12} = 2 - 5i$, $h_{13} = \frac{1}{5}$, $h_{21} = -1$, $h_{22} = \sqrt{3}$, and $h_{23} = 7 + i$.

If A is an $m \times n$ matrix, then we can visualize A as follows:

$$A = \begin{bmatrix} a_{11} & \cdots & a_{1n} \\ \vdots & & \vdots \\ a_{m1} & \cdots & a_{mn} \end{bmatrix}$$

We let $M_{m,n}^F$ be the set of all $m \times n$ matrices over the field F. Recall that we add two matrices $A, B \in M_{m,n}^F$ to get $A + B \in M_{m,n}^F$ using the rule $(a + b)_{ij} = a_{ij} + b_{ij}$. We multiply a matrix $A \in M_{m,n}^F$ by a scalar $k \in F$ using the rule $(ka)_{ij} = ka_{ij}$. We can visualize these computations as follows:

$$\begin{bmatrix} a_{11} & \cdots & a_{1n} \\ \vdots & & \vdots \\ a_{m1} & \cdots & a_{mn} \end{bmatrix} + \begin{bmatrix} b_{11} & \cdots & b_{1n} \\ \vdots & & \vdots \\ b_{m1} & \cdots & b_{mn} \end{bmatrix} = \begin{bmatrix} a_{11} + b_{11} & \cdots & a_{1n} + b_{1n} \\ \vdots & & \vdots \\ a_{m1} + b_{m1} & \cdots & a_{mn} + b_{mn} \end{bmatrix}$$

$$k \begin{bmatrix} a_{11} & \cdots & a_{1n} \\ \vdots & & \vdots \\ a_{m1} & \cdots & a_{mn} \end{bmatrix} = \begin{bmatrix} ka_{11} & \cdots & ka_{1n} \\ \vdots & & \vdots \\ ka_{m1} & \cdots & ka_{mn} \end{bmatrix}$$

With these operations of addition and scalar multiplication, $M_{m,n}^F$ is a vector space over F.

Given $n \in \mathbb{Z}^+$, we would now like to turn $M_{n,n}^F$ into a linear algebra over F by defining a vector multiplication in F. Notice that we will not be turning all vector spaces $M_{m,n}^F$ into linear algebras. We will be able to do this only when $m = n$. That is, the linear algebra will consist only of **square matrices** of a specific size.

We first define the product of an $m \times n$ matrix with an $n \times p$ matrix, where m, n, p are positive integers. Note that to take the product AB we first insist that the number of columns of A be equal to the number of rows of B (these are the "inner" two numbers in the expressions "$m \times n$" and "$n \times p$").

So, how do we actually multiply two matrices? This is a bit complicated and requires just a little practice. Let's begin by walking through an example while informally describing the procedure, so that we can get a feel for how matrix multiplication works before getting caught up in the "messy looking" definition.

Let $A = \begin{bmatrix} 0 & 1 \\ 3 & 2 \end{bmatrix}$ and $B = \begin{bmatrix} 1 & 2 & 0 \\ 0 & 3 & 6 \end{bmatrix}$. Notice that A is a 2×2 matrix and B is a 2×3 matrix. Since A has 2 columns and B has 2 rows, we will be able to multiply the two matrices.

For each row of the first matrix and each column of the second matrix, we add up the products entry by entry. Let's compute the product AB as an example.

$$AB = \begin{bmatrix} 0 & 1 \\ 3 & 2 \end{bmatrix} \cdot \begin{bmatrix} 1 & 2 & 0 \\ 0 & 3 & 6 \end{bmatrix} = \begin{bmatrix} x & y & z \\ u & v & w \end{bmatrix}$$

Since x is in the first row and first column, we use the first row of A and the first column of B to get $x = \begin{bmatrix} 0 & 1 \end{bmatrix} \begin{bmatrix} 1 \\ 0 \end{bmatrix} = 0 \cdot 1 + 1 \cdot 0 = 0 + 0 = 0$.

Since u is in the second row and first column, we use the second row of A and the first column of B to get $u = \begin{bmatrix} 3 & 2 \end{bmatrix} \begin{bmatrix} 1 \\ 0 \end{bmatrix} = 3 \cdot 1 + 2 \cdot 0 = 3$.

The reader should attempt to follow this procedure to compute the values of the remaining entries. The final product is

$$AB = \begin{bmatrix} 0 & 3 & 6 \\ 3 & 12 & 12 \end{bmatrix}$$

Notes: (1) The product of a **2 × 2** matrix and a **2 × 3** matrix is a **2 × 3** matrix.

(2) More generally, the product of an $m \times n$ matrix and an $n \times p$ matrix is an $m \times p$ matrix. Observe that the inner most numbers (both n) must agree, and the resulting product has dimensions given by the outermost numbers (m and p).

We formally define matrix multiplication as follows. Let A be the $m \times n$ matrix $A = \begin{bmatrix} a_{11} & \cdots & a_{1n} \\ \vdots & & \vdots \\ a_{m1} & \cdots & a_{mn} \end{bmatrix}$

and let B be the $n \times p$ matrix $B = \begin{bmatrix} b_{11} & \cdots & b_{1p} \\ \vdots & & \vdots \\ b_{n1} & \cdots & b_{np} \end{bmatrix}$. We define the product AB to be the $m \times p$ matrix

$C = \begin{bmatrix} c_{11} & \cdots & c_{1p} \\ \vdots & & \vdots \\ c_{m1} & \cdots & c_{mp} \end{bmatrix}$ such that

$$c_{ij} = a_{i1}b_{1j} + a_{i2}b_{2j} + \cdots + a_{in}b_{nj} = \sum_{k=1}^{n} a_{ik}b_{kj}.$$

Notes: (1) The symbol Σ is the Greek letter Sigma. In mathematics, this symbol is often used to denote a sum. Σ is generally used to abbreviate a very large sum or a sum of unknown length by specifying what a typical term of the sum looks like. Let's look at a simpler example first before we analyze the more complicated one above:

$$\sum_{k=1}^{5} k^2 = 1^2 + 2^2 + 3^2 + 4^2 + 5^2 = 1 + 4 + 9 + 16 + 25 = 55.$$

The expression "$k = 1$" written underneath the symbol indicates that we get the first term of the sum by replacing k by 1 in the given expression. When we replace k by 1 in the expression k^2, we get 1^2.

For the second term, we simply increase k by 1 to get $k = 2$. So, we replace k by 2 to get $k^2 = 2^2$.

We continue in this fashion, increasing k by 1 each time until we reach the number written above the symbol. In this case, that is $k = 5$.

(2) Let's now get back to the expression that we're interested in.

$$c_{ij} = \sum_{k=1}^{n} a_{ik}b_{kj} = a_{i1}b_{1j} + a_{i2}b_{2j} + \cdots + a_{in}b_{nj}$$

Once again, the expression "$k = 1$" written underneath the symbol indicates that we get the first term of the sum by replacing k by 1 in the given expression. When we replace k by 1 in the expression $a_{ik}b_{kj}$, we get $a_{i1}b_{1j}$. Notice that this is the first term of c_{ij}.

For the second term, we simply increase k by 1 to get $k = 2$. So, we replace k by 2 to get $a_{i2}b_{2j}$.

We continue in this fashion, increasing k by 1 each time until we reach the number written above the symbol. In this case, that is $k = n$. So, the last term is $a_{in}b_{nj}$.

(3) In general, we get the entry c_{ij} in the ith row and jth column of $C = AB$ by "multiplying" the ith row of A with the jth column of B. We can think of the computation like this:

$$[a_{i1} \ a_{i2} \cdots a_{in}] \begin{bmatrix} b_{1j} \\ b_{2j} \\ \vdots \\ b_{nj} \end{bmatrix} = a_{i1}b_{1j} + a_{i2}b_{2j} + \cdots + a_{in}b_{nj}$$

Notice how we multiply the leftmost entry a_{i1} by the topmost entry b_{1j}. Then we move one step to the right to a_{i2} and one step down to b_{2j} to form the next product, ... and so on.

Example 7.13: For this example, we are considering the matrices to be over the field $\mathbb{Q}$.

1. $[1 \ 2 \ 3 \ 4] \cdot \begin{bmatrix} 5 \\ 1 \\ -2 \\ 3 \end{bmatrix} = [1 \cdot 5 + 2 \cdot 1 + 3(-2) + 4 \cdot 3] = [5 + 2 - 6 + 12] = [13]$.

    We generally identify a 1×1 matrix with its only entry. So, $[1 \ 2 \ 3 \ 4] \cdot \begin{bmatrix} 5 \\ 1 \\ -2 \\ 3 \end{bmatrix} = \mathbf{13}$.

2. $\begin{bmatrix} 5 \\ 1 \\ -2 \\ 3 \end{bmatrix} \cdot [1 \ 2 \ 3 \ 4] = \begin{bmatrix} 5 & 10 & 15 & 20 \\ 1 & 2 & 3 & 4 \\ -2 & -4 & -6 & -8 \\ 3 & 6 & 9 & 12 \end{bmatrix}$.

Notice that $[1 \ 2 \ 3 \ 4] \cdot \begin{bmatrix} 5 \\ 1 \\ -2 \\ 3 \end{bmatrix} \neq \begin{bmatrix} 5 \\ 1 \\ -2 \\ 3 \end{bmatrix} \cdot [1 \ 2 \ 3 \ 4]$, and in fact, the two products do not even have the same size. This shows that if AB and BA are both defined, then they **do not** need to be equal.

3. $\begin{bmatrix} 1 & 2 \\ 0 & 1 \end{bmatrix} \cdot \begin{bmatrix} 0 & 2 \\ 3 & 2 \end{bmatrix} = \begin{bmatrix} 0+6 & 2+4 \\ 0+3 & 0+2 \end{bmatrix} = \begin{bmatrix} 6 & 6 \\ 3 & 2 \end{bmatrix}$.

$\begin{bmatrix} 0 & 2 \\ 3 & 2 \end{bmatrix} \cdot \begin{bmatrix} 1 & 2 \\ 0 & 1 \end{bmatrix} = \begin{bmatrix} 0+0 & 0+2 \\ 3+0 & 6+2 \end{bmatrix} = \begin{bmatrix} 0 & 2 \\ 3 & 8 \end{bmatrix}$.

Notice that $\begin{bmatrix} 1 & 2 \\ 0 & 1 \end{bmatrix} \cdot \begin{bmatrix} 0 & 2 \\ 3 & 2 \end{bmatrix} \neq \begin{bmatrix} 0 & 2 \\ 3 & 2 \end{bmatrix} \cdot \begin{bmatrix} 1 & 2 \\ 0 & 1 \end{bmatrix}$.

This shows that even if A and B are square matrices of the same size, in general $AB \neq BA$. So, matrix multiplication is **not** commutative. M_{nn}^F is a **noncommutative linear algebra**.

Example 7.14: For this example, we are considering the matrices to be over the field $\mathbb{Z}_7$.

1. $[1 \ 2 \ 3 \ 4] \cdot \begin{bmatrix} 5 \\ 1 \\ 5 \\ 3 \end{bmatrix} = [1 \cdot 5 + 2 \cdot 1 + 3 \cdot 5 + 4 \cdot 3] = [5 + 2 + 1 + 5] = [6] = \mathbf{6}$.

2. $\begin{bmatrix} 5 \\ 1 \\ 2 \\ 3 \end{bmatrix} \cdot [1 \ 2 \ 3 \ 4] = \begin{bmatrix} 5 & 3 & 1 & 6 \\ 1 & 2 & 3 & 4 \\ 2 & 4 & 6 & 1 \\ 3 & 6 & 2 & 5 \end{bmatrix}$.

3. $\begin{bmatrix} 1 & 2 \\ 3 & 4 \end{bmatrix} \cdot \begin{bmatrix} 5 & 6 \\ 0 & 1 \end{bmatrix} = \begin{bmatrix} 5+0 & 6+2 \\ 1+0 & 4+4 \end{bmatrix} = \begin{bmatrix} 5 & 1 \\ 1 & 1 \end{bmatrix}$.

$\begin{bmatrix} 5 & 6 \\ 0 & 1 \end{bmatrix} \cdot \begin{bmatrix} 1 & 2 \\ 3 & 4 \end{bmatrix} = \begin{bmatrix} 5+4 & 3+3 \\ 0+3 & 0+4 \end{bmatrix} = \begin{bmatrix} 2 & 6 \\ 3 & 4 \end{bmatrix}$.

Once again, notice that $\begin{bmatrix} 1 & 2 \\ 3 & 4 \end{bmatrix} \cdot \begin{bmatrix} 5 & 6 \\ 0 & 1 \end{bmatrix} \neq \begin{bmatrix} 5 & 6 \\ 0 & 1 \end{bmatrix} \cdot \begin{bmatrix} 1 & 2 \\ 3 & 4 \end{bmatrix}$.

Notes: (1) We have seen in part 2 of Example 2.19 that if F is a field and $m, n \in \mathbb{Z}^+$ then M_{mn}^F (the set of $m \times n$ matrices with entries in F) is a vector space over F.

(2) If R is a ring and $n \in \mathbb{Z}^+$, then $M_{n,n}^R$ (the set of $n \times n$ matrices with entries in R) is a ring. I leave the verification of this to the reader. Note that it is important here that the number of rows and columns of our matrices are the same. Otherwise, the matrix products will not be defined. This ring is **not** commutative, as we saw in the examples above. Also, this ring is **not** a domain. For example, $\begin{bmatrix} 1 & 0 \\ 0 & 0 \end{bmatrix}, \begin{bmatrix} 0 & 0 \\ 0 & 1 \end{bmatrix} \in M_{2,2}^R$ are zero divisors because

$$\begin{bmatrix} 1 & 0 \\ 0 & 0 \end{bmatrix} \cdot \begin{bmatrix} 0 & 0 \\ 0 & 1 \end{bmatrix} = \begin{bmatrix} 0 & 0 \\ 0 & 0 \end{bmatrix}.$$

(3) If F is a field and $n \in \mathbb{Z}^+$, then $M_{n,n}^F$ is a linear algebra over F. The reader may want to verify that $M_{n,n}^F$ satisfies compatibility of scalar and vector multiplication (as well as the ring axioms).

The Matrix of a Linear Transformation

Let $T \in \mathcal{L}(V, W)$ and let $B = \{v_1, v_2, \ldots, v_n\}$ and $C = \{w_1, w_2, \ldots, w_m\}$ be bases of V and W, respectively. Recall that T is completely determined by the values of $T(v_1), T(v_2), \ldots, T(v_n)$. Furthermore, since $T(v_1), T(v_2), \ldots, T(v_n) \in W$ and C is a basis of W, each of $T(v_1), T(v_2), \ldots, T(v_n)$ can be written as a linear combination of the vectors in C. So, we have

$$T(v_1) = a_{11}w_1 + a_{21}w_2 + \cdots + a_{m1}w_m$$
$$T(v_2) = a_{12}w_1 + a_{22}w_2 + \cdots + a_{m2}w_m$$
$$\vdots$$
$$T(v_j) = a_{1j}w_1 + a_{2j}w_2 + \cdots + a_{mj}w_m$$
$$\vdots$$
$$T(v_n) = a_{1n}w_1 + a_{2n}w_2 + \cdots + a_{mn}w_m$$

Here we have $a_{ij} \in F$ for each $i = 1, 2, \ldots, m$ and $j = 1, 2, \ldots, n$. We form the following matrix:

$$\mathcal{M}_T(B, C) = \begin{bmatrix} a_{11} & \cdots & a_{1n} \\ \vdots & & \vdots \\ a_{m1} & \cdots & a_{mn} \end{bmatrix}$$

$\mathcal{M}_T(B, C)$ is called the **matrix of the linear transformation T with respect to the bases B and C**.

Note: The coefficients in the expression $T(v_j) = a_{1j}w_1 + a_{2j}w_2 + \cdots + a_{mj}w_m$ become the jth **column** of $\mathcal{M}_T(B, C)$. Your first instinct might be to form the row $[a_{1j} \; a_{2j} \cdots a_{mj}]$, but this is incorrect. Pay careful attention to how we form $\mathcal{M}_T(B, C)$ in part 2 of Example 7.15 below to make sure that you avoid this error.

Example 7.15:

1. Consider the linear transformation $T: \mathbb{C} \to \mathbb{C}$ from part 1 of Example 7.6. We are considering $\mathbb{C}$ as a vector space over $\mathbb{R}$ and T is defined by $T(z) = 5z$. Let's use the standard basis of $\mathbb{C}$, so that $B = C = \{1 + 0i, 0 + 1i\} = \{1, i\}$. We have

$$T(1) = 5 = 5 \cdot 1 + 0 \cdot i$$
$$T(i) = 5i = 0 \cdot 1 + 5 \cdot i$$

The matrix of T with respect to the standard basis is $\mathcal{M}_T(\{1, i\}, \{1, i\}) = \begin{bmatrix} 5 & 0 \\ 0 & 5 \end{bmatrix}$.

In this case, since T is being mapped from a vector space to itself and we are using the same basis for both "copies" of $\mathbb{C}$, we can abbreviate $\mathcal{M}_T(\{1, i\}, \{1, i\})$ as $\mathcal{M}_T(\{1, i\})$. Furthermore, since we are using the standard basis, we can abbreviate $\mathcal{M}_T(\{1, i\}, \{1, i\})$ even further as $\mathcal{M}_T$. So, we can simply write $\mathcal{M}_T = \begin{bmatrix} 5 & 0 \\ 0 & 5 \end{bmatrix}$.

Now, let $z = a + bi \in \mathbb{C}$ and write z as the column vector $z = \begin{bmatrix} a \\ b \end{bmatrix}$. We have

$$\mathcal{M}_T \cdot z = \begin{bmatrix} 5 & 0 \\ 0 & 5 \end{bmatrix} \begin{bmatrix} a \\ b \end{bmatrix} = \begin{bmatrix} 5a \\ 5b \end{bmatrix} = 5 \begin{bmatrix} a \\ b \end{bmatrix} = 5z = T(z).$$

So, multiplication on the left by $\mathcal{M}_T$ gives the same result as applying the transformation T.

2. Consider the linear transformation $T: \mathbb{R}^4 \to \mathbb{R}^3$ from part 2 of Example 7.6. We are considering $\mathbb{R}^4$ and $\mathbb{R}^3$ as vector spaces over $\mathbb{R}$ and T is defined by

$$T((x, y, z, w)) = (x + z, 2x - 3y, 5y - 2w).$$

Let's use the standard bases for $\mathbb{R}^4$ and $\mathbb{R}^3$, so that

$B = \{(1,0,0,0), (0,1,0,0), (0,0,1,0), (0,0,0,1)\}$ and $C = \{(1,0,0), (0,1,0), (0,0,1)\}$.

We have

$$T(1,0,0,0) = (1, 2, 0)$$
$$T(0,1,0,0) = (0, -3, 5)$$
$$T(0,0,1,0) = (1, 0, 0)$$
$$T(0,0,0,1) = (0, 0, -2)$$

The matrix of T with respect to the standard bases is $\mathcal{M}_T = \begin{bmatrix} 1 & 0 & 1 & 0 \\ 2 & -3 & 0 & 0 \\ 0 & 5 & 0 & -2 \end{bmatrix}$

Once again, we abbreviate $\mathcal{M}_T(B, C)$ as $\mathcal{M}_T$ because we are using the standard bases.

Now, let $v = (x, y, z, w) \in \mathbb{R}^4$ and write v as the column vector $v = \begin{bmatrix} x \\ y \\ z \\ w \end{bmatrix}$. We have

$$\mathcal{M}_T \cdot v = \begin{bmatrix} 1 & 0 & 1 & 0 \\ 2 & -3 & 0 & 0 \\ 0 & 5 & 0 & -2 \end{bmatrix} \begin{bmatrix} x \\ y \\ z \\ w \end{bmatrix} = \begin{bmatrix} x + z \\ 2x - 3y \\ 5y - 2w \end{bmatrix} = T(v).$$

So, once again, multiplication on the left by $\mathcal{M}_T$ gives the same result as applying the transformation T.

Let V be a vector space over F with a finite basis. Then we say that V is **finite-dimensional**. If $B = \{v_1, v_2, \ldots, v_n\}$, then by Problem 19 in Problem Set 6, all bases of V have n elements. In this case, we say that V is **n-dimensional**, and we write $\dim V = n$.

Theorem 7.16: Let V be an n-dimensional vector space over a field F. Then there is a linear algebra isomorphism $F: \mathcal{L}(V) \to M_{nn}^F$.

You will be asked to prove Theorem 7.16 in Problem 21 below.

Images and Kernels

Let $f: A \to B$ be a homomorphism. The **image** of f is the set $f[A] = \{f(x) \mid x \in A\}$ and the **kernel** of f is the set $\ker(f) = \{x \in A \mid f(x) = e_B\}$. In the case where B has both an additive and multiplicative identity, then e_B will always be the additive identity (in other words, if $0, 1 \in B$, then the kernel of f is the set of all elements of A that map to 0).

Note: The definition of image given above makes sense for homomorphisms between any two structures of the same type. For the definition of kernel to make sense, the codomain of the homomorphism must have an identity with respect to one of the operations.

Example 7.17:

1. Consider the monoid $\mathfrak{A} = (\mathbb{R}, \cdot)$, and let $f \colon \mathbb{R} \to \mathbb{R}$ be defined by $f(x) = |x|$. Then given $x, y \in \mathbb{R}$, we have $f(xy) = |xy| = |x| \cdot |y| = f(x) \cdot f(y)$ (Check this!). Therefore, f is a homomorphism.

   Now, $x \in \ker(f)$ if and only if $f(x) = 1$ if and only if $|x| = 1$ if and only if $x = \pm 1$. So, $\ker(f) = \{-1, 1\}$.

   Also, it is easy to see that $f[\mathbb{R}] = \mathbb{R}^+ \cup \{0\} = [0, \infty)$.

2. Consider the rings $\mathfrak{A} = (\mathbb{Z}, +, \cdot)$, $\mathfrak{B} = (\mathbb{Z}_8, +, \cdot)$, and let $h \colon \mathbb{Z} \to \mathbb{Z}_8$ be defined by $h(k) = [k]_8$. In part 4 of Example 7.1, we saw that h is a ring homomorphism.

   Now, $k \in \ker(h)$ if and only if $h(k) = [0]_8$ if and only if $[k]_8 = [0]_8$ if and only if $k \equiv_8 0$ if and only if $8 | k - 0$ if and only if $8 | k$ if and only if there is an integer j such that $k = 8j$ if and only if $k \in 8\mathbb{Z}$. So, $\ker(h) = 8\mathbb{Z}$.

   Also, it is easy to see that $h[\mathbb{Z}] = \mathbb{Z}_8$, so that f is surjective.

   Note that there is nothing special about $\mathbb{Z}_8$ here. If $n \in \mathbb{Z}^+$, then we can define $h \colon \mathbb{Z} \to \mathbb{Z}_n$ by $h(k) = [k]_n$ and the same argument above shows that $\ker(h) = n\mathbb{Z}$ and $h[\mathbb{Z}] = \mathbb{Z}_n$.

3. Consider $\mathbb{R}^3$ and $\mathbb{R}^4$ as vector spaces over $\mathbb{R}$ and let $T \colon \mathbb{R}^4 \to \mathbb{R}^3$ be the linear transformation defined by $T(x, y, z, w) = (x + y, x - z, x + 2w)$. Let's compute $T[\mathbb{R}^4]$ and $\ker(T)$. First, $T[\mathbb{R}^4]$ consists of all vectors of the form

   $$(x + y, x - z, x + 2w) = (x + y)(1, 0, 0) + (x - z)(0, 1, 0) + (x + 2w)(0, 0, 1)$$

   So, if $(v_1, v_2, v_3) \in \mathbb{R}^3$, let $x = 0, y = v_1, z = -v_2$, and $w = \frac{1}{2}v_3$. Then we see that

   $$(x + y)(1, 0, 0) + (x - z)(0, 1, 0) + (x + 2w)(0, 0, 1)$$
   $$= v_1(1, 0, 0) + v_2(0, 1, 0) + v_3(0, 0, 1) = (v_1, v_2, v_3).$$

   Therefore, $\mathbb{R}^3 \subseteq T[\mathbb{R}^4]$. Since it is clear that $T[\mathbb{R}^4] \subseteq \mathbb{R}^3$, we have $T[\mathbb{R}^4] = \mathbb{R}^3$ (T is surjective).

   Now, $(x, y, z, w) \in \ker(T)$ if and only if $(x + y, x - z, x + 2w) = (0, 0, 0)$ if and only if $x + y = 0$, $x - z = 0$, and $x + 2w = 0$ if and only if $y = -x$, $z = x$, and $w = -\frac{x}{2}$ if and only if $(x, y, z, w) = \left(x, -x, x, -\frac{x}{2}\right) = x\left(1, -1, 1, -\frac{1}{2}\right)$.

   So, every element of $\ker(T)$ is a scalar multiple of $\left(1, -1, 1, -\frac{1}{2}\right)$. Therefore, $\ker(T) \subseteq \text{span}\left\{\left(1, -1, 1, -\frac{1}{2}\right)\right\}$.

   Conversely, an element of $\text{span}\left\{\left(1, -1, 1, -\frac{1}{2}\right)\right\}$ has the form $\left(v, -v, v, -\frac{1}{2}v\right)$, and we have $T\left(v, -v, v, -\frac{1}{2}v\right) = \left(v - v, v - v, v + 2\left(-\frac{1}{2}v\right)\right) = (0, 0, 0)$. Therefore, it follows that $\text{span}\left\{\left(1, -1, 1, -\frac{1}{2}\right)\right\} \subseteq \ker(T)$. So, $\ker(T) = \text{span}\left\{\left(1, -1, 1, -\frac{1}{2}\right)\right\}$.

Notice that $T[\mathbb{R}^4]$ is a subspace of $\mathbb{R}^3$ (in fact, $T[\mathbb{R}^4] = \mathbb{R}^3$) and $\ker(T)$ is a subspace of $\mathbb{R}^4$. Also, the sum of the dimensions of $T[\mathbb{R}^4]$ and $\ker(T)$ is $3 + 1 = 4$, which is the dimension of $\mathbb{R}^4$. None of this is a coincidence, as we will see in the next few theorems and Problem 24 below.

Theorem 7.18: Let $f: R \to S$ be a ring homomorphism. Then $f[R]$ is a subring of S.

Proof: Since $f(x) + f(y) = f(x+y)$ and $f(x)f(y) = f(xy)$, we see that $f[R]$ is closed under addition and multiplication. Since $1_S = f(1_R)$, $1_S \in f[R]$. By Theorem 7.3, $-f(x) = f(-x)$ (this is the conclusion of Theorem 7.3 when additive notation is used). So, for each element $f(x) \in f[R]$, $-f(x) \in f[R]$. It follows that $f[R]$ is a subring of S. □

Note: The same result holds if we replace "ring" by semigroup, monoid, group, or field. If $(S,\star)$ and $(T,\circ)$ are semigroups, and $f: S \to T$ is a semigroup homomorphism, then $f(x) \circ f(y) = f(x \star y)$ shows that $f[S]$ is closed under $\circ$, and therefore, $f[S]$ is a subsemigroup of T.

Furthermore, if $(M,\star)$ and $(N,\circ)$ are monoids, and $f: M \to N$ is a monoid homomorphism, then by definition, $f(e_M) = e_N$, and therefore, $f[M]$ is a submonoid of N.

If $(G,\star)$ and $(H,\circ)$ are groups, and $f: G \to H$ is a group homomorphism, then $f(e_G) = e_H$ by Theorem 7.2, and for all $g \in G$, $(f(g))^{-1} = f(g^{-1})$ by Theorem 7.3. Therefore, $f[G]$ is a subgroup of H.

If $(F, +, \cdot)$ and $(K, +, \cdot)$ are fields, and $f: F \to K$ is a field homomorphism, then for all $x \in F^*$, $(f(x))^{-1} = f(x^{-1})$ by Theorem 7.3 again. Therefore, $f[F]$ is a subfield of K.

If V and W are vector spaces over a field F, and $f: V \to W$ is a linear transformation, then $f(0_V) = 0_W$ by Theorem 7.2, $f(x) + f(y) = f(x+y)$, and $kf(x) = f(kx)$. By Theorem 6.6, $f[V]$ is a subspace of W. The same argument works if V and W are left modules over a ring R to show that $f[V]$ is a left submodule of W. If V and W are right modules over R, we replace the last computation by $f(x) \cdot k = f(xk)$ to see that $f[V]$ is a right submodule of W.

Theorem 7.19: Let $(G,\star)$ and $(H,\circ)$ be groups and let $f: G \to H$ be a group homomorphism. Then $\ker(f)$ is a subgroup of G.

Proof: Let $x, y \in \ker(f)$. Then $f(x) = e_H$ and $f(y) = e_H$. So $f(x \star y) = f(x) \circ f(y) = e_H \circ e_H = e_H$. Thus, $x \star y \in \ker(f)$. Since $f(e_G) = e_H$ (by Theorem 7.2), $e_G \in \ker(f)$. Suppose $x \in \ker(f)$. By Theorem 7.3, we have $f(x^{-1}) = (f(x))^{-1} = e_H^{-1} = e_H$. So $x^{-1} \in \ker(f)$. Therefore, $\ker(f)$ is a subgroup of G. □

Notes: (1) The same result holds for semigroups and monoids (however, H must have an identity for the theorem to make sense). This should be clear from the proof.

(2) Recall that we say that $(R, +, \cdot)$ is **almost a ring** (or a **rng**) if all the ring properties hold **except** the existence of a multiplicative identity. Similarly, we will say that $(S, +, \cdot)$ is **almost a subring** (or a **subrng**) of the ring $(R, +, \cdot)$ if all the properties of being a subring hold **except** S does not contain the multiplicative identity.

(3) If $f: R \to S$ is a ring homomorphism, then unless S is the trivial ring $\{0\}$, $\ker(f)$ is not a ring because $f(1_R) = 1_S \neq 0_S$. So, $1_R \notin \ker(f)$. However, every other property holds and so $\ker(f)$ is almost a subring of R. Indeed, if $x, y \in \ker(f)$, then

$$f(x+y) = f(x) + f(y) = 0_S + 0_S = 0_S \text{ and } f(xy) = f(x)f(y) = 0_S \cdot 0_S = 0_S.$$

Also, $f(0_R) = 0_S$ by Theorem 7.2, and if $x \in \ker(f)$, then $f(-x) = -f(x) = -0_S = 0_S$ by Theorem 7.3 (this is the conclusion of Theorem 7.3 when additive notation is used).

(4) Some authors exclude the existence of a multiplicative identity from the definition of a ring. Note 3 above gives a good justification for doing so. However, removing a property creates other complexities. So, there is no right or wrong answer here. In this book, rings always include a multiplicative identity. If we wish to exclude the multiplicative identity, we call the structure "almost a ring."

(5) If V and W are vector spaces over a field F, and $f: V \to W$ is a linear transformation, then if $k \in F$ and $x \in \ker(f)$, we have $f(kx) = kf(x) = k \cdot 0_W = 0_W$ by part (iii) of Problem 10 from Problem Set 2. This computation together with the computations given in Note 2 above, show that $\ker(f)$ is a subspace of V. If V and W are left modules over a ring R, an identical argument shows that $\ker(f)$ is a left submodule of V. As usual, the argument can be adjusted slightly for right modules over R.

Theorem 7.20: Let $f: G \to H$ be a group homomorphism. Then $\ker(f) = \{e_G\}$ if and only f is injective.

Proof: Suppose that $\ker(f) = \{e_G\}$, let $x, y \in G$, and let $f(x) = f(y)$. Then $f(x)(f(y))^{-1} = e_H$. It follows from Theorem 7.3 that $f(xy^{-1}) = f(x)f(y^{-1}) = f(x)(f(y))^{-1} = e_H$. So, $xy^{-1} \in \ker(f)$. Since $\ker(f) = \{e_G\}$, $xy^{-1} = e_G$. Therefore, $x = y$. Since $x, y \in G$ were arbitrary, f is injective.

Conversely, suppose that f is injective, and let $x \in \ker(f)$. Then $f(x) = e_H$. But also, by Theorem 7.2, $f(e_G) = e_H$. So, $f(x) = f(e_G)$. Since f is injective, $x = e_G$. Since $x \in G$ was arbitrary, $\ker(f) \subseteq \{e_G\}$. By Theorem 7.2, $f(e_G) = e_H$, so that $e_G \in \ker(f)$, and therefore, $\{e_G\} \subseteq \ker(f)$. It follows that $\ker(f) = \{e_G\}$. □

Note: (1) The theorem also holds for ring homomorphisms. Specifically, if $f: R \to S$ is a ring homomorphism, then $\ker(f) = \{0_R\}$ if and only if f is injective. The proof is the same, except additive notation should be used. Here is a sketch of the proof using additive notation:

If $\ker(f) = \{0_R\}$ and $f(x) = f(y)$, then $f(x + (-y)) = f(x) + f(-y) = f(x) - f(y) = 0_S$, so that $x + (-y) \in \ker(f)$, and thus, $x + (-y) = 0_R$, and so, $x = y$.

Conversely, if f is injective and $x \in \ker(f)$, then $f(x) = 0_S$. Since $f(0_R) = 0_S$ and f is injective, we have $x = 0_R$. So, $\ker(f) \subseteq \{0_R\}$. Also, $f(0_R) = 0_S$. So, $0_R \in \ker(f)$, and therefore, $\{0\}_R \subseteq \ker(f)$.

(2) The theorem holds for linear transformations as well. The details are covered in Note 1 above.

Normal Subgroups and Ring Ideals

Let $(G, \star)$ be a group and $h, k \in G$. We say that k is a **conjugate** of h if there is a $g \in G$ such that $k = ghg^{-1}$ (as usual, we abbreviate $g \star h \star g^{-1}$ as ghg^{-1}).

If $(G,\star)$ is a group, we say that a subgroup N of G is **normal**, and write $N \triangleleft G$, if whenever $h \in N$ and $k \in G$ is a conjugate of h, then $k \in N$. (In this case, we may say that N is **closed under conjugation**.)

Example 7.21:

1. If G is a commutative group, then every subgroup H of G is normal. Indeed, if $h \in H$ and $g \in G$, then $ghg^{-1} = hgg^{-1} = he = h \in H$.

2. If $f: G \to H$ is a group homomorphism, then $\ker(f)$ is a normal subgroup of G. We already showed in Theorem 7.19 that $\ker(f)$ is a subgroup of G. To see that $\ker(f) \triangleleft G$, let $h \in \ker(f)$ and let $g \in G$. Then $f(ghg^{-1}) = f(g)f(h)f(g^{-1}) = f(g)e\big(f(g)\big)^{-1} = f(g)\big(f(g)\big)^{-1} = e$.

3. Any group is a normal subgroup of itself. Indeed, if $h \in G$ and $g \in G$, then clearly $ghg^{-1} \in G$.

4. The trivial subgroup of a group G consisting of just the identity e is a normal subgroup of G. Indeed, if $h \in \{e\}$ and $g \in G$, then $ghg^{-1} = geg^{-1} = gg^{-1} = e \in \{e\}$.

5. If $f: G \to H$ is a group homomorphism and N is a normal subgroup of G, then $f[N]$ is a normal subgroup of $f[G]$. To see this, first observe that by the Note following Theorem 7.18, $f[G]$ is a subgroup of H. To see that $f[N]$ is a subgroup of $f[G]$, first observe that $e_G \in N$ (because N is a subgroup of G), and so, $e_H = f(e_G) \in f[N]$. Next, if $x, y \in N$, then $xy \in N$ and $x^{-1} \in N$ (again, because N is a subgroup of G), and so, $f(x)f(y) = f(xy) \in f[N]$ and $f(x)^{-1} = f(x^{-1}) \in f[N]$. So, $f[N]$ is closed under the group operation and taking inverses, and therefore, is a subgroup of $f[G]$. To see that $f[N]$ is normal in $f[G]$, let $x \in f[N]$ and $y \in f[G]$. Then there are $n \in N$ and $g \in G$ with $f(n) = x$ and $f(g) = y$. Since N is normal in G, we have $gng^{-1} \in N$ and therefore, $yxy^{-1} = f(g)f(n)f(g)^{-1} = f(gng^{-1}) \in f[N]$, as desired.

Let $(R, +, \cdot)$ be a ring and let $A \subseteq R$. We say that A **absorbs** R if for every $a \in A$ and $x \in R$, $ax \in A$ and $xa \in A$.

Note: Since in a ring, multiplication is not necessarily commutative, both conditions $ax \in A$ and $xa \in A$ may be necessary. In a commutative ring, either condition follows from the other.

If $(R, +, \cdot)$ is a ring, we say that a subset I of R is an **ideal** of R, and write $I \triangleleft R$, if $(I, +)$ is a subgroup of $(R, +)$ and I absorbs R.

Example 7.22:

1. Consider the ring $(\mathbb{Z}, +, \cdot)$. Then $(2\mathbb{Z}, +, \cdot)$ is an ideal of $\mathbb{Z}$ because $(2\mathbb{Z}, +)$ is a subgroup of $(\mathbb{Z}, +)$ (see part 3 of Example 6.5) and when we multiply an even integer by **any** other integer, we get an even integer (so, $2\mathbb{Z}$ absorbs $\mathbb{Z}$).

    More generally, for each $n \in \mathbb{Z}^+$, $(n\mathbb{Z}, +, \cdot)$ is an ideal of $(\mathbb{Z}, +, \cdot)$.

2. If $f: R \to S$ is a ring homomorphism, then $\ker(f)$ is an ideal of R. We already showed in Note 3 following Theorem 7.19 that $(\ker(f), +)$ is a subgroup of $(R, +)$. To see that $\ker(f)$ absorbs R, let $a \in \ker(f)$ and let $x \in R$. Then $f(ax) = f(a)f(x) = 0_S \cdot f(x) = 0_S$ (by Theorem 2.14), so that $ax \in \ker(f)$. Also, $f(xa) = f(x)f(a) = f(x) \cdot 0_S = 0_S$ (again by Theorem 2.14), so that $xa \in \ker(f)$.

3. If R is a commutative ring and $a \in R$, let $\langle a \rangle = \{ar \mid r \in R\}$. Let's show that $\langle a \rangle$ is an ideal of R. Since $0 = a \cdot 0$ (by Theorem 2.14), $0 \in \langle a \rangle$. If $x, y \in \langle a \rangle$, then there are $r, s \in R$ with $x = ar$ and $y = as$. Since R is a ring, $r - s \in R$, and so, $x - y = ar - as = a(r - s) \in \langle a \rangle$. By Problem 5 in Problem Set 6, $(\langle a \rangle, +)$ is a subgroup of $(R, +)$. Now, let $x \in \langle a \rangle$ and $s \in R$. Since $x \in \langle a \rangle$, there is $r \in R$ with $x = ar$. Since R is a ring, $rs \in R$ and $sr \in R$, and therefore, $xs = (ar)s = a(rs) \in \langle a \rangle$ and $sx = s(ar) = (sa)r = (as)r = a(sr) \in \langle a \rangle$. Note that for the third equality in this last sequence of equations, we needed for R to be commutative.

   An ideal of the form $\langle a \rangle$ is called a **principal ideal**. As a specific example, the ideal $(2\mathbb{Z}, +, \cdot)$ of $(\mathbb{Z}, +, \cdot)$ is a principal ideal. More generally, for each $n \in \mathbb{Z}^+$, the ideal $(n\mathbb{Z}, +, \cdot)$ of $(\mathbb{Z}, +, \cdot)$ is a principal ideal.

4. If R is a commutative ring and $a_1, a_2, \ldots, a_n \in R$, let $S = \{a_1, a_2, \ldots, a_n\}$, and let
$$\langle S \rangle = \{a_1 r_1 + a_2 r_2 + \cdots + a_n r_n \mid r_1, r_2, \ldots, r_n \in R\}.$$

   Then $\langle S \rangle$ is an ideal of R. The verification is similar to what we did in part 3 above, and so, I leave it to the reader. The ideal $\langle S \rangle$ is called the **ideal generated by S**. We may also write $\langle S \rangle$ as $\langle a_1, a_2, \ldots, a_n \rangle$ and we may call it the ideal generated by $a_1, a_2, \ldots, a_n$.

5. Any ring is an ideal of itself. Indeed, if $a \in R$ and $x \in R$, then clearly $ax \in R$ and $xa \in R$.

6. $\{0_R\}$ is an ideal of R because for all $x \in R$, $0_R \cdot x = 0_R$ and $x \cdot 0_R = 0_R$ (by Theorem 2.14).

7. Every field F has exactly two ideals: $\{0\}$ and F. To see this, let I be an ideal of F such that $I \neq \{0\}$. Then there is $a \in I$ with $a \neq 0$. Since F is a field and $a \neq 0$, a^{-1} exists. Since I is an ideal, we have $1 = aa^{-1} \in I$. Now, let $x \in F$ be arbitrary. Then $x = x \cdot 1 \in I$. Therefore, $F \subseteq I$. By the definition of an ideal, $I \subseteq F$. So, $I = F$. Since I was an arbitrary ideal not equal to $\{0\}$, the only ideals of F are $\{0\}$ and F.

Theorem 7.23: Let F and K be fields and let $f: F \to K$ be a field homomorphism. Then f is injective.

Proof: Since a field homomorphism is the same as a ring homomorphism, by part 2 of Example 7.22, $\ker(f)$ is an ideal of F. Since F is a field, by part 7 of Example 7.22, $\ker(f) = \{0_F\}$ or $\ker(f) = F$. Since $f(1_F) = 1_K \neq 0_K$, we see that $1_F \notin \ker(f)$, and so, $\ker(f) \neq F$. Therefore, $\ker(f) = \{0_F\}$. By Note 1 following Theorem 20, f is injective. $\square$

If $f: F \to K$ is a field homomorphism, by the Note following Theorem 7.18, $f[F]$ is a subfield of K. Clearly $f: F \to f[F]$ is surjective. This observation together with Theorem 7.23 gives us the following corollary.

Corollary 7.24: Let F and K be fields and let $f: F \to K$ be a field homomorphism. Then F is isomorphic to $f[F]$.

Problem Set 7

Full solutions to these problems are available for free download here:
www.SATPrepGet800.com/AAFBTDW

LEVEL 1

1. Let H and K be the subgroups of S_3 defined by $H = \{(1), (123), (132)\}$ and $K = \{(1), (12)\}$ (see Problem 1 from Problem Set 6). Determine which of these is a normal subgroup of S_3.

2. Let V and W be vector spaces over $\mathbb{R}$. Determine if each of the following functions is a linear transformation:

    (i) $f: \mathbb{R} \to \mathbb{R}$ defined by $f(x) = 2x + 1$

    (ii) $g: \mathbb{R} \to \mathbb{R}^2$ defined by $g(x) = (2x, 3x)$

    (iii) $h: \mathbb{R}^3 \to \mathbb{R}^3$ defined by $h\big((x, y, z)\big) = (x + y, x + z, z - y)$

3. Compute each of the following:

    (i) $\begin{bmatrix} 2 & 0 & -3 \\ 0 & 1 & 4 \end{bmatrix} \cdot \begin{bmatrix} 1 & 1 & 3 & 0 \\ 1 & -4 & 2 & 0 \\ 2 & 0 & 1 & -4 \end{bmatrix}$

    (ii) $\begin{bmatrix} 3 & -1 & 5 \end{bmatrix} \cdot \begin{bmatrix} -4 \\ -7 \\ 2 \end{bmatrix}$

    (iii) $\begin{bmatrix} -4 \\ -7 \\ 2 \end{bmatrix} \cdot \begin{bmatrix} 3 & -1 & 5 \end{bmatrix}$

    (iv) $\begin{bmatrix} a & b & c \\ d & e & f \\ g & h & i \end{bmatrix} \cdot \begin{bmatrix} 1 & 0 & 1 \\ 0 & 2 & 0 \\ 3 & 1 & 4 \end{bmatrix}$.

LEVEL 2

4. Let $(R, +, \cdot)$ be a ring and define addition and multiplication on $R \times R$ componentwise, as was done in Problem 13 from Problem Set 2. Prove that $(R, +, \cdot)$ is isomorphic to a subring of $(R \times R, +, \cdot)$.

5. Let z and w be complex numbers. Prove the following:

    (i) $\overline{z + w} = \bar{z} + \bar{w}$

    (ii) $\overline{zw} = \bar{z} \cdot \bar{w}$

    (iii) $\overline{\left(\frac{z}{w}\right)} = \frac{\bar{z}}{\bar{w}}$

6. Consider $\mathbb{C}$ as a vector space over itself. Give an example of a function $f: \mathbb{C} \to \mathbb{C}$ such that f is additive, but **not** a linear transformation. Then give an example of vector spaces V and W and a homogenous function $g: V \to W$ that is **not** a linear transformation.

LEVEL 3

7. Prove that there are exactly two ring homomorphisms from $\mathbb{Z}$ to itself.

8. Prove each of the following:

    (i) Ring isomorphism is an equivalence relation.

    (ii) If we let $\text{Aut}(R)$ be the set of automorphisms of a ring R, then $(\text{Aut}(R), \circ)$ is a group, where $\circ$ is composition. $\text{Aut}(R)$ is called the **automorphism group** of R.

    (iii) Let R be a ring and for each $x \in R$, define the function $\phi_x: R \to R$ by $\phi_x(y) = xyx^{-1}$. Let $\text{Inn}(R) = \{\phi_x \mid x \in R\}$. Prove that $\text{Inn}(R)$ is a normal subgroup of $\text{Aut}(R)$. $\text{Inn}(R)$ is called the **inner automorphism group** of R.

9. Prove that a commutative ring R is a field if and only if the only ideals of R are $\{0\}$ and R.

10. Prove that if X is a nonempty set of normal subgroups of a group G then $\bigcap X$ is a normal subgroup of G. Similarly, prove that if X is a nonempty set of ideals of a ring R, then $\bigcap X$ is an ideal of R. Is the union of normal subgroups always a normal subgroup? Is the union of ideals always an ideal?

11. Let $P = \{ax^2 + bx + c \mid a, b, c \in \mathbb{R}\}$ be the vector space of polynomials of degree at most 2 with real coefficients (see part 3 of Example 2.19 from Lesson 2). Define the linear transformation $D: P \to P$ by $D(ax^2 + bx + c) = 2ax + b$. Find the matrix of T with respect to each of the following bases:

    (i) The standard basis $B = \{1, x, x^2\}$

    (ii) $C = \{x + 1, x^2 + 1, x^2 + x\}$

12. Let V and W be vector spaces with V finite-dimensional, let $U \leq V$, and let $T \in \mathcal{L}(U, W)$. Prove that there is an $S \in \mathcal{L}(V, W)$ such that $S(v) = T(v)$ for all $v \in U$.

LEVEL 4

13. Let G and H be groups and let $f: G \to H$ be a homomorphism. Prove each of the following:

    (i) If G is commutative, then $f[G]$ is commutative.

    (ii) If $n \in \mathbb{Z}$ and $x \in G$, then $f(x^n) = (f(x))^n$.

    (iii) If G is cyclic, then $f[G]$ is cyclic.

    (iv) If f is an isomorphism, then $f[Z(G)] = Z(H)$, where $Z(G)$ is the center of G (in other words, $Z(G) = \{x \in G \mid xz = zx \text{ for all } z \in G\}$).

14. Prove that each of the following pairs of groups are **not** isomorphic.

    (i) $(\mathbb{Q}, +)$ and $(\mathbb{Z}, +)$.

    (ii) $(\mathbb{Q}, +)$ and $(\mathbb{Q}^+, \cdot)$.

15. Let $T: V \to W$ be a linear transformation and let $v_1, v_2, \ldots, v_n \in V$. Prove the following:

    (i) If T is injective and $v_1, v_2, \ldots, v_n$ are linearly independent in V, then $T(v_1), T(v_2), \ldots, T(v_n)$ are linearly independent in W.

    (ii) If T is surjective and $\text{span}\{v_1, v_2, \ldots, v_n\} = V$, then $\text{span}\{T(v_1), T(v_2), \ldots, T(v_n)\} = W$.

16. Let V and W be vector spaces over a field F. Prove that $\mathcal{L}(V, W)$ is a vector space over F, where addition and scalar multiplication are defined as in Theorem 7.8.

17. Let V be a vector space over a field F. Prove that $\mathcal{L}(V)$ is a linear algebra over F, where addition and scalar multiplication are defined as in Theorem 7.8 and vector multiplication is given by composition of linear transformations.

18. Let $T: V \to W$ and $S: W \to V$ be linear transformations such that $ST = i_V$ and $TS = i_W$. Prove that S and T are bijections and that $S = T^{-1}$.

LEVEL 5

19. Prove that $(^{\mathbb{R}}\mathbb{R}, +, \cdot)$ is a ring, where addition and multiplication are defined pointwise. Then prove that for each $x \in \mathbb{R}$, $I_x = \{f \in {}^{\mathbb{R}}\mathbb{R} \mid f(x) = 0\}$ is an ideal of $^{\mathbb{R}}\mathbb{R}$ and the only ideal of $^{\mathbb{R}}\mathbb{R}$ containing I_x and not equal to I_x is $^{\mathbb{R}}\mathbb{R}$.

20. Let V be a vector space with $\dim V > 1$. Show that $\{T \in \mathcal{L}(V) \mid T \text{ is not invertible}\} \not\leq \mathcal{L}(V)$.

21. Let V be an n-dimensional vector space over a field F. Prove that there is a linear algebra isomorphism $G: \mathcal{L}(V) \to M_{nn}^F$.

CHALLENGE PROBLEMS

22. Let $\mathfrak{A}$ and $\mathfrak{B}$ be structures of the same type and let $f: A \to B$ be an invertible homomorphism. Prove that $f^{-1}: B \to A$ is also an invertible homomorphism.

23. Let R be an integral domain. Prove that there exists a field F that contains a subring isomorphic to R. This field is called the **field of quotients of R**. (Hint: Mimic the procedure of constructing the rational numbers from the integers, as was done in Lesson 5.)

24. Let V and W be vector spaces over a field F with $\dim V = n$ and let $T: V \to W$ be a linear transformation. Prove that $\dim(T[V]) + \dim(\ker(T)) = n$.

25. Let V be a vector space over a field F and let U and W be subspaces of V. Prove that
$$\dim(U + W) = \dim U + \dim W - \dim(U \cap W).$$

LESSON 8
NUMBER THEORY

Divisibility

Recall: An integer a is called **even** if there is another integer b such that $a = 2b$.

Example 8.1:

1. 6 is even because $6 = 2 \cdot 3$.
2. -14 is even because $-14 = 2 \cdot (-7)$.
3. We can write $1 = 2 \cdot \frac{1}{2}$, but this does **not** show that 1 is even (and as we all know, it is not). In the definition of even, it is very important that b is an integer. The problem here is that $\frac{1}{2}$ is not an integer, and so, it cannot be used as a value for b in the definition of even.

We define the **sum** of integers a and b to be $a + b$. We define the **product** of a and b to be $a \cdot b$.

Theorem 8.2: The sum of two even integers is even.

Strategy: Before writing the proof, let's think about our strategy. We need to start with two arbitrary but specific even integers. Let's call them m and n. Notice that we need to give them different names because there is no reason that they need to have the same value.

When we try to add m and n, we get $m + n$. Hmmm...I see no reason yet why the expression $m + n$ should represent an even integer.

The problem is that we haven't yet used the definition of even. If we invoke the definition, we get integers j and k such that $m = 2j$ and $n = 2k$.

Now, when we add m and n, we get $m + n = 2j + 2k$.

Is it clear that $2j + 2k$ represents an even integer? Nope...not yet. To be even, our final expression needs to have the form $2b$, where b is an integer.

Here is where we use the fact that $\mathbb{Z}$ is a ring. Specifically, we use the distributive property to rewrite $2j + 2k$ as $2(j + k)$.

It looks like we've done it. We just need to verify one more thing: is $j + k$ an integer? Once again, we can use the fact that $\mathbb{Z}$ is a ring to verify this. Specifically, we use the fact that $+$ is a binary operation on $\mathbb{Z}$.

I think we're now ready to write the proof.

Proof of Theorem 8.2: Let m and n be even integers. Then there are integers j and k such that $m = 2j$ and $n = 2k$. So, $m + n = 2j + 2k = 2(j + k)$ because multiplication is distributive over addition in $\mathbb{Z}$. Since $\mathbb{Z}$ is closed under addition, $j + k \in \mathbb{Z}$. Therefore, $m + n$ is even. □

The property of being even is a special case of the more general notion of divisibility.

An integer a is **divisible** by an integer k, written $k|a$, if there is another integer b such that $a = kb$. We also say that k is a **factor** of a, k is a **divisor** of a, k **divides** a, or a is a **multiple** of k.

Example 8.3:
1. Note that being divisible by 2 is the same as being even.
2. 18 is divisible by 3 because $18 = 3 \cdot 6$.
3. -56 is divisible by 7 because $-56 = 7 \cdot (-8)$.

Theorem 8.4: The product of two integers that are each divisible by k is also divisible by k.

Proof: Let m and n be integers that are divisible by k. Then there are integers b and c such that $m = kb$ and $n = kc$. So, $m \cdot n = (k \cdot b) \cdot (k \cdot c) = k \cdot (b \cdot (k \cdot c))$ because multiplication is associative in $\mathbb{Z}$. Since $\mathbb{Z}$ is closed under multiplication, $b \cdot (k \cdot c) \in \mathbb{Z}$. Thus, $m \cdot n$ is divisible by k. □

Notes: (1) If you're confused about how associativity was used here, it might help to make the substitution $u = (k \cdot c)$. Then we have $(k \cdot b) \cdot (k \cdot c) = (k \cdot b) \cdot u = k \cdot (b \cdot u) = k(b \cdot (k \cdot c))$.

(2) Although it may seem tempting to simplify $k \cdot (b \cdot (k \cdot c))$ further, it is unnecessary. The definition of divisibility by k requires us only to generate an expression of the form k times some integer, and that's what we have done.

(3) If the generality of the proof confuses you, try replacing k by a specific integer. For example, if we let $k = 2$, we have $m = 2b$, $n = 2c$, and therefore $m \cdot n = (2b) \cdot (2c) = 2(b \cdot (2c))$. Is it clear that this final expression is even (divisible by 2)?

(4) It's worth noting that the product $m \cdot n$ is actually divisible by k^2. Indeed, we have

$$m \cdot n = k \cdot (b \cdot (k \cdot c)) = k \cdot ((b \cdot k) \cdot c) = k \cdot ((k \cdot b) \cdot c) = k \cdot (k \cdot (b \cdot c)) = k^2(b \cdot c)$$

Prime Numbers

Before defining a prime number, let's make note of a few basic facts.

Notes: (1) Every integer is divisible by 1. Indeed, if $n \in \mathbb{Z}$, then $n = 1 \cdot n$.

(2) Every integer is divisible by itself. Indeed, if $n \in \mathbb{Z}$, then $n = n \cdot 1$.

(3) It follows from Notes 1 and 2 above that every integer greater than 1 has at least 2 factors.

A **prime number** is a natural number with **exactly** two positive integer factors.

Notes: (1) An equivalent definition of a prime number is the following: A prime number is an integer greater than 1 that is divisible only by 1 and itself.

(2) An integer greater than 1 that is not prime is called **composite**.

Example 8.5:

1. 0 is **not** prime because every positive integer is a factor of 0. Indeed, if $n \in \mathbb{Z}^+$, then $0 = n \cdot 0$, so that $n|0$.

2. 1 is **not** prime because it has only one positive integer factor: if $1 = kb$ with $k > 0$, then $k = 1$ and $b = 1$.

3. The first ten prime numbers are 2, 3, 5, 7, 11, 13, 17, 19, 23, and 29.

4. 4 is not prime because $4 = 2 \cdot 2$. In fact, the only even prime number is 2 because by definition, an even integer has 2 as a factor.

5. 9 is the first odd integer greater than 1 that is not prime. Indeed, 3, 5, and 7 are prime, but 9 is not because $9 = 3 \cdot 3$.

6. The first ten composite numbers are 4, 6, 8, 9, 10, 12, 14, 15, 16, and 18.

Two very important facts about prime numbers (that we will prove in this Lesson) are the following.

1. There are infinitely many prime numbers.

2. Every integer greater than 1 can be written uniquely as a product of prime numbers, up to the order in which the factors are written.

The second fact is known as **The Fundamental Theorem of Arithmetic**. It is used often in many branches of mathematics.

When we write an integer n as a product of other integers, we call that product a **factorization** of n. If all the factors in the product are prime, we call the product a **prime factorization** of n.

Example 8.6:

1. $20 = 4 \cdot 5$ is a factorization of 20. This is **not** a prime factorization of 20 because 4 is **not** prime. $20 = 2 \cdot 10$ is another factorization of 20. This example shows that factorizations in general are **not** unique.

2. An example of a prime factorization of 20 is $20 = 2 \cdot 2 \cdot 5$. We can also write this prime factorization as $2 \cdot 5 \cdot 2$ or $5 \cdot 2 \cdot 2$. So, you can see that if we consider different orderings of the factors as different factorizations, then prime factorizations are **not** unique. This is why we say that prime factorizations are unique, **up to the order in which the factors are written**.

3. A prime number is equal to its own prime factorization. In other words, we consider a prime number to be a product of primes with just one factor in the product. For example, the prime factorization of 2 is 2.

Recall from Lesson 5 that the **Well Ordering Principle** says that every nonempty subset of natural numbers has a least element.

We will now use the Well Ordering Principle to prove half of the Fundamental Theorem of Arithmetic.

Theorem 8.7: Every integer greater than 1 can be written as a product of prime numbers.

Note that we left out the word "uniquely" here. The uniqueness is the second half of the Fundamental Theorem of Arithmetic, which we will prove later in this lesson.

Analysis: We will prove this theorem by contradiction using the Well Ordering Principle. The idea is simple. If an integer n greater than 1 is not prime, then it can be factored as kr with $1 < k < n$ and $1 < r < n$. If k and r can be written as a product of primes, then so can n because n is simply the product of all the factors of k and r. For example, $6 = 2 \cdot 3$ and $20 = 2 \cdot 2 \cdot 5$. Therefore, we have $120 = 6 \cdot 20 = (2 \cdot 3) \cdot (2 \cdot 2 \cdot 5)$. Let's write the proof.

Proof of Theorem 8.7: Suppose toward contradiction that there exists an integer greater than 1 that cannot be written as a product of prime numbers. By the Well Ordering Principle, there is a least such integer, let's call it n. Since n cannot be written as a product of prime numbers, then in particular, n is not prime. So, we can write $n = kr$ with $k, r \in \mathbb{N}$ and $1 < k < n$ and $1 < r < n$. Since n is the least integer greater than 1 that cannot be written as a product of prime numbers, k and r can both be written as products of prime numbers. But then $n = kr$ is also a product of prime numbers, contradicting our choice of n. This contradiction shows that every integer greater than 1 can be written as a product of prime numbers. □

Notes: (1) Recall that a proof by contradiction works as follows:

1. We assume the negation of what we are trying to prove.

2. We use a logically valid argument to derive a statement which is false.

3. Since the argument was logically valid, the only possible error is our original assumption. Therefore, the negation of our original assumption must be true.

The negation of the statement "Every integer greater than 1 can be written as a product of prime numbers" is "There is an integer greater than 1 that **cannot** be written as a product of prime numbers." If we let $S = \{k \in \mathbb{N} \mid k > 1 \land k$ cannot be written as a product of prime numbers$\}$, then by our assumption, $S \neq \emptyset$. It follows from the Well Ordering Principle that S has a least element, which in the proof above, we name n.

The argument then proceeds to factor n as kr, where k and r are both greater than 1 and less than n. We can factor n this way because n in not prime.

Since n is the least element of S, it follows that k and r are not in S. Therefore, k and r **can** be written as a product of prime numbers. But this immediately gives us a prime factorization of n, contradicting our original assumption.

Since every step of our argument was logically valid, the only thing that could have been wrong was our original assumption. So, **every** integer greater than 1 **can** be written as a product of prime numbers.

(2) In general, if $P(x)$ is a property, then the negation of $\forall x(P(x))$ is $\exists x(\neg P(x))$. In other words, when we pass a negation symbol through a universal quantifier, the quantifier changes to an existential quantifier. So, $\neg \forall x(P(x)) \equiv \exists x(\neg P(x))$, where $\equiv$ is pronounced "is logically equivalent to." For Theorem 8.7, the property $P(x)$ is $q(x) \rightarrow r(x)$, where $q(x)$ is "$x > 1$" and $r(x)$ is "x can be written as a product of prime numbers." It is not too hard to show that $\neg(q(x) \rightarrow r(x))$ is logically equivalent to $q(x) \wedge \neg r(x)$ (just check that each truth assignment of $q(x)$ and $r(x)$ results in the same truth value for both statements). So $\exists x(\neg P(x))$ says, "There is an integer x such that $x > 1$ and x cannot be written as a product of prime numbers."

In general (although not needed here), we also have $\neg \exists x(P(x)) \equiv \forall x(\neg P(x))$.

Corollary 8.8: Every integer greater than 1 has a prime factor.

Proof: Let n be an integer greater than 1. By Theorem 8.7, n can be written as a product of prime numbers. Let p be any of the prime numbers in that product. Then p is a prime factor of n. □

Theorem 8.9: There are infinitely many primes.

Analysis: Starting with a prime number $p > 1$, we want to find a prime number greater than p. This will prove that there infinitely many prime numbers, because if P is a finite set of prime numbers, then the previous statement implies that we can find a prime number greater than the biggest number in the set P.

Now recall that if n is a positive integer, then the number $n!$ (pronounced "**n factorial**") is defined by $n! = 1 \cdot 2 \cdots n$. For example, $3! = 1 \cdot 2 \cdot 3 = 6$ and $4! = 1 \cdot 2 \cdot 3 \cdot 4 = 24$.

If $n > 2$, then $n!$ is a number larger than n that is divisible by every positive integer less than or equal to n. For example, $3! = 6$ is divisible by $1, 2$, and 3, and $4! = 24$ is divisible by $1, 2, 3,$ and 4.

Now, $n!$ Is certainly **not** prime. In fact, it has lots of factors! For example, $4! = 24$ has 8 factors (what are they?). Therefore, $n!$ itself won't work for us. So, we add 1 to this number to get the number $M = n! + 1$.

By adding 1 to $n!$ to produce M, we have destroyed almost all the divisibility that we had. Specifically, M is **not** divisible by any integer k with $1 < k \leq n$. To see this, let k be an integer satisfying $1 < k \leq n$. We know that there is an integer r such that $n! = kr$ (because $n!$ Is divisible by k). If M were divisible by k, then there would be an integer s such that $M = ks$. But then, by subtracting $n!$ from each side of the equation $M = n! + 1$, we get $1 = M - n! = ks - kr = k(s - r)$. Since $k > 1$ and $s - r$ is an integer, this is impossible! Therefore, M is not divisible by k.

It would be nice if we could prove that M is prime. Then M would be a prime number greater than n, thus completing the proof. Sometimes M does turn out to be prime. For example, if $n = 2$, then $M = 2! + 1 = 2 + 1 = 3$, which is prime. However, it is unfortunate for us that M is not always prime. In Problem 6 below you will find values for n for which M is not prime.

However, even if M is not prime, all is not lost. By Corollary 8.8, we know that M has a prime factor, let's call it p. We also know that M is **not** divisible by any integer k with $1 < k \leq n$. It follows that p is a prime number greater than n.

I think we're ready to write out the proof.

Proof of Theorem 8.9: Let P be a finite set of prime numbers with greatest member q and let $M = q! + 1$. By Corollary 8.8, M has a prime factor p. So, there is an integer k such that $M = pk$.

We show that $p > q$.

Suppose toward contradiction that $p \leq q$. Then $p|q!$. So, there is an integer r such that $q! = pr$. It follows that $1 = M - q! = pk - pr = p(k - r)$. So, $p = 1$, which contradicts that p is prime.

It follows that $p > q$ and so, p is greater than every prime number in P. Since P was an arbitrary finite set of prime numbers, we have shown that there are infinitely many prime numbers. □

The Division Algorithm

Before we state the Division Algorithm in its full generality, let's first look at a relatively simple special case.

Recall from the beginning of this lesson that an integer a is **even** if there is another integer b such that $a = 2b$. Additionally, we say that an integer a is **odd** if there is another integer b such that $a = 2b + 1$.

Theorem 8.10 (The Baby Division Algorithm): Every integer is even or odd, but not both.

Proof: We already proved in Theorem 5.10 that every natural number is even or odd. If $n < 0$ is an integer, then $-n > 0$, and so, there is a natural number j such that $-n = 2j$ or $-n = 2j + 1$. If $-n = 2j$, then $n = 2(-j)$ (and since $j \in \mathbb{N}$, $-j \in \mathbb{Z}$). If $-n = 2j + 1$, then

$$n = -(2j + 1) = -2j - 1 = -2j - 1 - 1 + 1 \text{ (SACT)} = -2j - 2 + 1 = 2(-j - 1) + 1.$$

Here we used the fact that $\mathbb{Z}$ is a ring. Since $\mathbb{Z}$ is closed under addition, it follows that $-j - 1 = -j + (-1) \in \mathbb{Z}$.

Now, if $n = 2j$ and $n = 2k + 1$, then $2j = 2k + 1$. So, we have

$$2(j - k) = 2j - 2k = (2k + 1) - 2k = 2k + (1 - 2k) = 2k + (-2k + 1)$$
$$= (2k - 2k) + 1 = 0 + 1 = 1.$$

So, $2(j - k) = 1$. But 2 does not have a multiplicative inverse in $\mathbb{Z}$, and so, this is a contradiction. □

Let's take a moment to restate the Baby Division Algorithm in a way that will more easily generalize to the full Division Algorithm

The Baby Division Algorithm: if $n \in \mathbb{Z}$, there are unique integers k and r such that $n = 2k + r$, where $r = 0$ or $r = 1$.

Take a moment to convince yourself that this restatement is equivalent to Theorem 8.10. We sometimes say, "When n is divided by 2, k is the **quotient** and r is the **remainder**." Observe that when an integer n is divided by 2, the quotient can be any integer, but the remainder can be only 0 or 1.

Example 8.11:

1. When 11 is divided by 2, the quotient is 5 and the remainder is 1. That is, $11 = 2 \cdot 5 + 1$.

2. When 20 is divided by 2, the quotient is 10 and the remainder is 0. That is, $20 = 2 \cdot 10 + 0$, or equivalently, $20 = 2 \cdot 10$. Notice that in this case, 20 is divisible by 2.

3. When -11 is divided by 2, the quotient is -6 and the remainder is 1. That is $-11 = 2(-6) + 1$. Compare this to the first example. Based on that example, most students would probably guess that the quotient here would turn out to be -5. But as you can see, that is not the case.

The **Division Algorithm** generalizes the notion of an integer n being "even or odd" ($2k$ or $2k + 1$) to n being equal to $mk + r$, where $0 \leq r < m$.

For example, for $m = 3$, the Division Algorithm will tell us that every integer can be written uniquely in one of the three forms $3k$, $3k + 1$, or $3k + 2$. Observe that when an integer n is divided by 3, the quotient can be any integer, but the remainder can be only 0, 1, or 2.

As one more example, for $m = 4$, the Division Algorithm will tell us that every integer can be written uniquely in one of the four forms $4k$, $4k + 1$, $4k + 2$, or $4k + 3$. Observe that when an integer n is divided by 4, the quotient can be any integer, but the remainder can be only 0, 1, 2, or 3.

Example 8.12:

1. When 14 is divided by 3, the quotient is 4 and the remainder is 2. That is, $14 = 3 \cdot 4 + 2$.

2. When 36 is divided by 4, the quotient is 9 and the remainder is 0. That is, $36 = 4 \cdot 9 + 0$, or equivalently, $36 = 4 \cdot 9$. Notice that in this case, 36 is divisible by 4.

3. When 17 is divided by 5, the quotient is 3 and the remainder is 2. That is, $17 = 5 \cdot 3 + 2$.

4. When -17 is divided by 5, the quotient is -4 and the remainder is 3. That is $-17 = 5(-4) + 3$.

Theorem 8.13 (The Division Algorithm): Let n and m be integers with $m > 0$. Then there are unique integers k and r such that $n = mk + r$ with $0 \leq r < m$.

Many students find the standard proof of the Division Algorithm to be quite hard to follow. I know that when I read the proof for the first time, I found it quite confusing. To better understand the argument, let's first run a couple of simulations using specific examples that mimic the proof.

Simulation 1: Let's let $n = 7$ and $m = 2$. With these choices for n and m, the Division Algorithm says that there are unique integers k and r such that $7 = 2k + r$ and $0 \leq r < 2$ (in other words, $r = 0$ or $r = 1$).

Let's look at the equation $7 = 2k + r$ in the form $7 - 2k = r$. In particular, let's look at the possible values of $7 - 2k$ as k ranges over all possible integers. Let's do this by matching up each integer k with the corresponding value of $7 - 2k$:

k	$\cdots$	-4	-3	-2	-1	0	1	2	3	4	$\cdots$
$r?$ $\quad 7-2k$	$\cdots$	15	13	11	9	7	5	3	**1**	-1	$\cdots$

Observe that the top row is simply "listing" all the integers. The "$\cdots$" to the left of -4 and to the right of 4 are there to indicate that this list keeps going infinitely in each direction. However, I did make sure to include the most important values in the visible part of our list.

We get each value in the bottom row by substituting the value above it for k in the expression $7 - 2k$. For example, for $k = -4$, we have $7 - 2k = 7 - 2(-4) = 7 + 8 = 15$.

Notice that the values in the bottom row decrease by 2 units for each 1 unit increase in k. This is because $m = 2$.

We highlighted the column where $k = 3$ and $r = 7 - 2k = 1$. This is the column where the smallest nonnegative number appears in the bottom row. In other words, we let r be the least positive value of $7 - 2t$, as t ranges over all the integers, and we let k be the corresponding t-value.

In general, how do we know that these values exist?

Well, since $n \geq 0$ ($n = 7$ in this example), the expression $n - mt \geq 0$ when $t = 0$. It follows that the set $\{n - mt \mid t \in \mathbb{Z} \wedge n - mt \geq 0\} = \{7 - 2t \mid t \in \mathbb{Z} \wedge 7 - 2t \geq 0\}$ is not empty ($7 - 2 \cdot 0 = 7$ is in this set). So, we can invoke the Well Ordering Principle to get a least element r. In this simulation, r will turn out to be 1 with a corresponding k-value of 3. (We will see what happens if $n < 0$ in the next simulation).

By taking r to be the least element from a set of natural numbers, we know that r will be nonnegative. But how do we know that r will be less than 2? We use the fact that the bottom row decreases by 2 units for each 1 unit increase in the top row.

Suppose we accidentally chose $r = 3$. Then we have $7 - 2k = 3$. If we subtract 2 from each side of this equation, we get $7 - 2k - 2 = 1$. Using distributivity, we have that $7 - 2k - 2$ is equal to $7 - 2(k + 1)$. So, $7 - 2(k + 1) = 1$. Looks like we chose the wrong value for r. What we just showed is that if we increase k by 1 (from 2 to 3), we decrease r by 2 (from 3 to 1).

In general, if $r \geq 2$, then we have $n - 2k \geq 2$, so that $n - 2k - 2 \geq 0$. Thus, $n - 2(k + 1) \geq 0$. But $n - 2(k + 1) = n - 2k - 2 < n - 2k$. This contradicts that r was the least possible value of $n - 2t$ with $n - 2t \geq 0$. It follows that $r < 2$.

Now let's check uniqueness. So, we have $7 = 2 \cdot 3 + 1$. How do we know that there aren't two other numbers k' and r' with $0 \leq r' < 2$ such that $7 = 2k' + r'$?

Well, if there were, then we would have $2 \cdot 3 + 1 = 2k' + r'$. Subtracting $2k'$ from each side of the equation and subtracting 1 from each side of the equation gives us $2 \cdot 3 - 2k' = r' - 1$. We now use the distributive property on the left to get $2(3 - k') = r' - 1$. This equation shows that 2 is a factor of $r' - 1$. r' can't be 0 because 2 is not a factor of -1. Therefore, $r' = 1$ (remember that 0 and 1 are the only two choices for r'). So, $2(3 - k') = 0$, and therefore, $3 - k' = 0$. So, $k' = 3$. Oh, look at that! r' and k' are the same as r and k.

So, we just proved that there is exactly one way to write 7 in the form $2k + r$ with k and r integers and $0 \leq r < 2$. We showed that $7 = 2 \cdot 3 + 1$ is the only way to do it.

Simulation 2: This time, let's let $n = -4$ and $m = 3$. With these choices for n and m, the Division Algorithm says that there are unique integers k and r such that $-4 = 3k + r$ and $0 \leq r < 3$ (in other words, $r = 0$, $r = 1$, or $r = 2$).

Let's look at the equation $-4 = 3k + r$ in the form $-4 - 3k = r$, and as we did in Simulation 1, let's match up each integer k with the corresponding value of $-4 - 3k$:

	k	$\cdots$	-4	-3	**-2**	-1	0	1	2	3	4	$\cdots$
$r?$	$-4 - 3k$	$\cdots$	8	5	**2**	-1	-4	-7	-10	-13	-16	$\cdots$

This time, since $m = 3$, the values in the bottom row decrease by 3 units for each 1 unit increase in k.

We highlighted the column where $k = -2$ and $r = -4 - 3(-2) = -4 + 6 = 2$ because it is the column where the smallest nonnegative number appears in the bottom row. This time 2 is the smallest possible value of r, and this r-value corresponds to a k-value of -2.

Since $n < 0$ this time ($n = -4$ in this example), setting $t = 0$ in the expression $n - mt$ does **not** produce a nonnegative value. This time, we let $t = n$ to get $n - m \cdot n$ (specifically, for this simulation we set $t = -4$ to get $-4 - 3(-4) = -4 + 12 = 8$, which is greater than 0). It follows that the set $\{n - mt \mid t \in \mathbb{Z} \wedge n - mt \geq 0\} = \{-4 - 3t \mid t \in \mathbb{Z} \wedge -4 - 3t \geq 0\}$ is not empty. So, once again, we can invoke the Well Ordering Principle to get a least element r. In this simulation, r will turn out to be 2 with a corresponding k-value of -2.

As in Simulation 1, it is clear that $r \geq 0$, and we use the fact that the bottom row decreases by 3 units for each 1 unit increase in the top row to show that $r < 3$.

Suppose we accidentally chose $r = 5$. Then we have $-4 - 3k = 5$. If we subtract 3 from each side of this equation, we get $-4 - 3k - 3 = 2$. But using distributivity, we have that $-4 - 3k - 3$ is equal to $-4 - 3(k + 1)$. So, $-4 - 3(k + 1) = 2$. We just showed is that if we increase k by 1 (from -3 to -2), we decrease r by 3 (from 5 to 2).

In general, if $r \geq 3$, then we have $n - 3k \geq 3$, so that $n - 3k - 3 \geq 0$. Thus, $n - 3(k + 1) \geq 0$. But $n - 3(k + 1) = n - 3k - 3 < n - 3k$. This contradicts that r was the least possible value of $n - 3t$ with $n - 3t \geq 0$. It follows that $r < 3$.

I leave it as an exercise for the reader to check uniqueness for this special case.

Let's move on to the proof of the Theorem.

Proof of Theorem 8.13: Let $n, m \in \mathbb{Z}$ with $m > 0$, and let $S = \{n - mt \mid t \in \mathbb{Z} \wedge n - mt \geq 0\}$. To see that $S \neq \emptyset$, we consider two cases. If $n \geq 0$, then let $t = 0$, and we have $n - mt = n \in S$. If $n < 0$, then let $t = n$, so that we have $n - mt = n - mn = n(1 - m)$. Since $m \geq 1$, we have $1 - m \leq 0$. It follows that $n(1 - m) \geq 0$, and so, $n - mt \in S$. In both cases, we have shown that $S \neq \emptyset$.

Since S is a nonempty subset of natural numbers, by the Well Ordering Principle, S has a least element $r = n - mk$, where $k \in \mathbb{Z}$. Since $S \subseteq \mathbb{N}$, $r \geq 0$. By adding mk to each side of the equation, we have $n = mk + r$.

We need to show that $r < m$. Suppose toward contradiction that $r \geq m$. Substituting $n - mk$ for r gives us $n - mk \geq m$. Subtracting m from each side of this last equation gives $(n - mk) - m \geq 0$. Now, since $m > 0$, $r > r - m = (n - mk) - m$. But $(n - mk) - m = n - mk - m = n - m(k + 1)$, and so, $(n - mk) - m$ is an element of S smaller than r, contradicting r being the least element of S. This contradiction tells us that we must have $r < m$.

We still need to prove that k and r are unique. Suppose that $n = mk_1 + r_1$ and $n = mk_2 + r_2$ with both $0 \leq r_1 < m$ and $0 \leq r_2 < m$. Without loss of generality, we may assume that $r_2 \geq r_1$.

By a simple substitution, $mk_1 + r_1 = mk_2 + r_2$. Subtracting mk_2 from each side of the equation and simultaneously subtracting r_1 from each side of the equation, we get $mk_1 - mk_2 = r_2 - r_1$. Factoring m on the left gives $m(k_1 - k_2) = r_2 - r_1$, and we see that $m | r_2 - r_1$.

Since $r_2 \geq r_1$, we have $r_2 - r_1 \geq 0$. Since we have $r_1 \geq 0$ and $r_2 < m$, we have $r_2 - r_1 < m - 0 = m$. So, $m | r_2 - r_1$ and $0 \leq r_2 - r_1 < m$. It follows that $r_2 - r_1 = 0$. So, $r_2 = r_1$. Finally, $r_2 = r_1$ and $mk_1 + r_1 = mk_2 + r_2$ together imply that $mk_1 = mk_2$, and so, $k_1 = k_2$. □

GCD and LCM

Let a and b be two integers. An integer j is a **common divisor** (or **common factor**) of a and b if j is a factor of both a and b. An integer k is a **common multiple** of a and b if k is a multiple of both a and b.

Example 8.14: Let $a = 6$ and $b = 15$. The positive divisors of a are **1**, 2, **3**, and 6. The positive divisors of b are **1**, **3**, 5, and 15. Therefore, the positive common divisors of a and b are **1 and 3**.

For each positive divisor there is a corresponding negative divisor. So, a complete list of the divisors of a are $1, 2, 3, 6, -1, -2, -3,$ and -6 and a complete list of the divisors of b are $1, 3, 5, 15, -1, -3, -5,$ and -15. Therefore, a complete list of the common divisors of a and b are **1, 3, −1, and −3**.

If both a and $-a$ are in a list, we will sometimes use the notation $\pm a$ instead of listing a and $-a$ separately. In this example, we can say that the complete list of common divisors of a and b is $\pm 1, \pm 3$.

The multiples of a are $\pm 6, \pm 12, \pm 18, \pm 24, \pm 30, \pm 36, \ldots$ and so on. The multiples of 15 are $\pm 15, \pm 30, \pm 45, \pm 60, \ldots$ and so on. Therefore, the common multiples of a and b are $\pm 30, \pm 60, \pm 90, \pm 120, \ldots$ **and so on**.

Again, let a and b be distinct integers. The **greatest common divisor** (or **greatest common factor**) of a and b, written $\gcd(a, b)$, is the largest common divisor of a and b. The **least common multiple** of a and b, written $\text{lcm}(a, b)$, is the smallest positive common multiple of a and b.

Example 8.15:

1. From Example 8.14, it's easy to see that $\gcd(6, 15) = 3$ and $\text{lcm}(6, 15) = 30$.

2. $\gcd(2,3) = 1$ and $\text{lcm}(2,3) = 6$. More generally, if p and q are prime numbers with $p \neq q$, then $\gcd(p,q) = 1$ and $\text{lcm}(p,q) = pq$.

3. $\gcd(4,15) = 1$ and $\text{lcm}(4,15) = 60$. Observe that neither 4 nor 15 is prime, and yet their gcd is 1 and their lcm is the product of 4 and 15. This is because 4 and 15 have no common factors except for 1 and −1. We say that 4 and 15 are **relatively prime**.

   Note that if p and q are prime numbers with $p \neq q$, then p and q are relatively prime.

   We have the following more general result: if a and b are relatively prime integers, then $\gcd(a,b) = 1$ and $\text{lcm}(a,b) = ab$ (see Theorem 8.26 below).

We can extend all these ideas to larger sets of numbers. Specifically, let X be a finite set of integers containing at least one nonzero integer. Then the **greatest common divisor** of the integers in X, written $\gcd(X)$ (or $\gcd(a_1, a_2, \ldots, a_n)$, where $X = \{a_1, a_2, \ldots, a_n\}$) is the largest integer that divides every integer in the set X, and the **least common multiple** of the integers in X, written $\text{lcm}(X)$ (or $\text{lcm}(a_1, a_2, \ldots, a_n)$) is the smallest positive integer that each integer in the set X divides.

For convenience, if X contains only 0, we define $\gcd(X) = 0$ and $\text{lcm}(X) = 0$.

Also, the integers in the set X are said to be **mutually relatively prime** if $\gcd(X) = 1$. The integers in the set X are said to be **pairwise relatively prime** if for each pair $a, b \in X$ with $a \neq b$, $\gcd(a,b) = 1$.

Example 8.16:

1. $\gcd(10, 15, 35) = 5$ and $\text{lcm}(10, 15, 35) = 210$.

2. $\gcd(2, 3, 12) = 1$ and $\text{lcm}(2, 3, 12) = 12$. Notice that here 2, 3, and 12 are mutually relatively prime, but **not** pairwise relatively prime because for example, $\gcd(2, 12) = 2 \neq 1$.

3. $\gcd(10, 21, 143) = 1$ and $\text{lcm}(10, 21, 143) = 30{,}030$. In this case, we have 10, 21, and 143 are pairwise relatively prime.

   We have the following result: if $X = \{a_1, a_2, \ldots, a_n\}$ is a set of pairwise relatively prime integers, then $\gcd(X) = 1$ and $\text{lcm}(X) = a_1 a_2 \cdots a_n$. The proof of this is left as an optional exercise for the reader. Also note that pairwise relatively prime implies mutually relatively prime.

4. For a set X with just one element a, $\gcd(a) = a$ and $\text{lcm}(a) = a$. In particular, $\gcd(0) = 0$ and $\text{lcm}(0) = 0$.

Let $a, b \in \mathbb{Z}$. A **linear combination** of a and b is an expression of the form $ma + nb$ with $m, n \in \mathbb{Z}$. We call the integers m and n **weights**.

Example 8.17:

1. Since $5 \cdot 10 - 2 \cdot 15 = 50 - 30 = 20$, we see that 20 is a linear combination of 10 and 15. When we write 20 as $5 \cdot 10 - 2 \cdot 15$, the weights are 5 and −2.

   This is not the only way to write 20 as a linear combination of 10 and 15. For example, we also have $-1 \cdot 10 + 2 \cdot 15 = -10 + 30 = 20$. When we write 20 as $-1 \cdot 10 + 2 \cdot 15$, the weights are −1 and 2.

2. Any number that is a multiple of either 10 or 15 is a linear combination of 10 and 15 because we can allow weights to be 0. For example, 80 is a linear combination of 10 and 15 because $80 = 8 \cdot 10 + 0 \cdot 15$.

   Also, 45 is a linear combination of 10 and 15 because $45 = 0 \cdot 10 + 3 \cdot 15$.

3. We will see in Theorem 8.18 below that $\gcd(a, b)$ can always be written as a linear combination of a and b. For example, $\gcd(10, 15) = 5$, and we have $5 = -1 \cdot 10 + 1 \cdot 15$.

4. Using the same theorem mentioned in 3, if a and b are relatively prime, then 1 can be written as a linear combination of a and b. For example, 4 and 15 are relatively prime and we have $1 = 4 \cdot 4 - 1 \cdot 15$.

Theorem 8.18: Let a and b be integers, at least one of which is not 0. Then $\gcd(a, b)$ is the least positive integer k such that there exist $m, n \in \mathbb{Z}$ with $k = ma + nb$.

This theorem says two things. First, it says that $\gcd(a, b)$ can be written as a linear combination of a and b. Second, it says that any positive integer smaller than $\gcd(a, b)$ **cannot** be written as a linear combination of a and b.

Proof: We first prove the theorem for $a, b \in \mathbb{Z}^+$. So, let a, b be positive integers and let S be the set of all positive linear combinations of a and b with weights in $\mathbb{Z}$.

$$S = \{ma + nb \mid m, n \in \mathbb{Z} \land ma + nb > 0\}$$

Notice that $a, b \in S$ because $a = 1a + 0b$ and $b = 0a + 1b$. In particular, $S \neq \emptyset$. By the Well Ordering Principle, S has a least element k. By the definition of S, there exist $m, n \in \mathbb{Z}$ with $k = ma + nb$.

By the Division Algorithm, there are $s, r \in \mathbb{Z}$ with $a = ks + r$ and $0 \leq r < k$.

So, $r = a - ks = a - (ma + nb)s = a - mas - nbs = (1 - ms)a - (ns)b$. We see that r is a linear combination of a and b. Since $r < k$ and r is a linear combination of a and b, r cannot be in S (because k is the least element of S). So, r must be 0. It follows that $a = ks$. Therefore, $k | a$.

Replacing a by b in the last two paragraphs shows that $k | b$ as well. So, k is a common divisor of a and b. Now, if c is another common divisor of a and b, then by Problem 16 below, c is a divisor of any linear combination of a and b. Since k is a linear combination of a and b, c is a divisor of k. Since every common divisor of a and b is also a divisor of k, it follows that $k = \gcd(a, b)$.

Since $ma = (-m)(-a)$ and $nb = (-n)(-b)$, the result holds whenever a and b are both nonzero.

Finally, suppose $a = 0$ or $b = 0$. Without loss of generality, let $a = 0$. Then $b \neq 0$. So, $\gcd(a, b) = b$ (or $-b$ if $b < 0$). We also have for any $m, n \in \mathbb{Z}$, $ma + nb = m \cdot 0 + nb = nb$. The least positive integer of the form nb is $1 \cdot b = b$ (or $-1 \cdot b$ if $b < 0$). So, the result holds in this case as well. □

We're almost ready to finish proving the Fundamental Theorem of Arithmetic. We will first prove two preliminary results that will make the proof easier.

Theorem 8.19 (Euclid's Principle): Let $a, b, c \in \mathbb{Z}^+$ with a and b relatively prime and $a | bc$. Then $a | c$.

Proof: Let $a, b, c \in \mathbb{Z}^+$ with a and b relatively prime and let $a|bc$. Since $\gcd(a,b) = 1$, by Theorem 8.18, there are integers m and n with $1 = ma + nb$. Since $a|bc$, there is an integer k such that $bc = ak$. Multiplying each side of the equation $1 = ma + nb$ by c and using the distributive property, $c = c(ma + nb) = cma + cnb = cma + nbc = cma + nak = a(cm + nk)$. Since $c, m, n, k \in \mathbb{Z}$ and $\mathbb{Z}$ is closed under addition and multiplication, $cm + nk \in \mathbb{Z}$. Therefore, $a|c$. □

Theorem 8.20: Let p be prime and let $a_1, a_2, \ldots, a_n$ be positive integers such that $p|a_1 a_2 \cdots a_n$. Then there is an integer j with $1 \leq j \leq n$ such that $p|a_j$.

Proof: We will prove this theorem by induction on $n \geq 1$.

Base Case ($n = 1$): We are given that p is prime, $a_1 \in \mathbb{Z}^+$, and $p|a_1$. Wait a sec... $p|a_1$ is the conclusion we were looking for. So, the theorem holds for $n = 1$.

Inductive Step: Let $k \in \mathbb{N}$ and assume that the result holds for $n = k$.

Let p be prime and let $a_1, a_2, \ldots a_k, a_{k+1}$ be positive integers such that $p|a_1 a_2 \cdots a_k a_{k+1}$. Since p is prime, its only positive factors are 1 and p. Therefore, $\gcd(p, a_1 a_2 \cdots a_k)$ is either 1 or p.

If $\gcd(p, a_1 a_2 \cdots a_k) = 1$, then by Theorem 8.19, $p|a_{k+1}$. If $\gcd(p, a_1 a_2 \cdots a_k) = p$, then $p|a_1 a_2 \cdots a_k$, and by our inductive assumption, there is an integer j with $1 \leq j \leq k$ such that $p|a_j$.

Therefore, the result holds for $n = k + 1$.

By the Principle of Mathematical Induction, the result holds for all $n \in \mathbb{N}$ with $n \geq 1$. □

We are finally ready to finish the proof of the Fundamental Theorem of Arithmetic.

Theorem 8.21 (The Fundamental Theorem of Arithmetic): Every integer greater than 1 can be written uniquely as a product of prime numbers, up to the order in which the factors are written.

Proof: By Theorem 8.7, every integer greater than 1 can be written as a product of prime numbers. We need to show that any two such prime factorizations are equal. Assume toward contradiction that n can be written in the following two different ways: $n = p_1 p_2 \cdots p_k = q_1 q_2 \cdots q_r$, where $p_1, p_2, \ldots, p_k, q_1, q_2, \ldots, q_r$ are prime numbers. Without loss of generality, assume $p_1 \leq p_2 \leq \cdots \leq p_k$ and $q_1 \leq q_2 \leq \cdots \leq q_r$. Also, by cancelling common primes on the left with common primes on the right, we may assume that for all $i \leq k$ and $j \leq r$, $p_i \neq q_j$. Suppose $1 \leq i \leq k$. Then $p_i|p_1 p_2 \cdots p_k$. Since $p_1 p_2 \cdots p_k = q_1 q_2 \cdots q_r$, we have $p_i|q_1 q_2 \cdots q_r$. By Theorem 8.20, there is j with $1 \leq j \leq r$ such that $p_i|q_j$. This is a contradiction. So, there cannot exist two different prime factorizations of n. □

Since prime factorizations are unique only up to the order in which the factors are written, there can be many ways to write a prime factorization. For example, 10 can be written as $2 \cdot 5$ or $5 \cdot 2$. To make things as simple as possible we always agree to use the **canonical representation** (or **canonical form**). The word "canonical" is just a fancy name for "natural," and the most natural way to write a prime factorization is in increasing order of primes. So, the canonical representation of 10 is $2 \cdot 5$.

As another example, the canonical representation of 18 is $2 \cdot 3 \cdot 3$. We can tidy this up a bit by rewriting $3 \cdot 3$ as 3^2. So, the canonical representation of 18 is $2 \cdot 3^2$.

If you are new to factoring, you may find it helpful to draw a factor tree.

For example, here is a factor tree for 18:

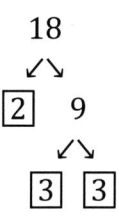

```
      18
      ╱╲
     2   9
         ╱╲
        3  3
```

To draw this tree, we started by writing 18 as the product $2 \cdot 9$. We put a box around 2 because 2 is prime and does not need to be factored any more. We then proceeded to factor 9 as $3 \cdot 3$. We put a box around each 3 because 3 is prime. We now see that we are done, and the prime factorization can be found by multiplying all the boxed numbers together. Remember that we will usually want the canonical representation, and so, we write the final product in increasing order of primes.

By the Fundamental Theorem of Arithmetic above it does not matter how we factor the number—we will always get the same canonical form. For example, here is a different factor tree for 18:

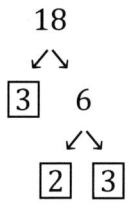

```
      18
      ╱╲
     3   6
         ╱╲
        2  3
```

Now, to prove that a positive integer n is composite, we simply need to produce a factor of n that is different from 1 and n itself. This may sound easy, but in practice, as we look at larger and larger values of n it can become very difficult to find factors of n. For example, the largest prime number that we are currently aware of (at the time I am writing this book) is $2^{82,589,933} - 1$. This is an enormous number with 24,862,048 digits. By Theorem 8.9, we know that there are prime numbers larger than this, but we have not yet found one.

The following theorem provides a couple of tricks to help us (or a computer) determine if a positive integer is prime more quickly.

Theorem 8.22: If n is composite, then n has a prime factor $p \leq \sqrt{n}$.

Proof: Let n be composite, so that there are integers a, b with $1 < a, b < n$ and $n = ab$. If both a and b are greater than $\sqrt{n}$, then we would have $n = ab > \sqrt{n} \cdot \sqrt{n} = n$, a contradiction. So, either $a \leq \sqrt{n}$ or $b \leq \sqrt{n}$. Without loss of generality, suppose that $a \leq \sqrt{n}$. By Corollary 8.8, a has a prime factor p. Since p is a factor of a and a is a factor n, it follows that p is a factor of n. Also, since p is a factor of a and $a \leq \sqrt{n}$, we have $p \leq \sqrt{n}$. □

Example 8.23:

1. Let's determine if 187 is prime or composite. Since $\sqrt{187} < \sqrt{196} = 14$, by Theorem 8.22, we need only check to see if 187 is divisible by 2, 3, 5, 7, 11, and 13. Checking each of these, we see that $187 = 11 \cdot 17$. So, 187 is composite.

2. Let's determine if 359 is prime or composite. Since $\sqrt{359} < \sqrt{361} = 19$, by Theorem 8.22, we need only check to see if 359 is divisible by 2, 3, 5, 7, 11 13, and 17. A quick check shows that 359 is **not** divisible by any of these numbers, and so, 359 is prime.

Sometimes in a prime factorization we will want to make sure that we do not "skip" any primes, and that each prime has a power.

For example, the canonical representation of 50 is $2 \cdot 5^2$. Note that we "skipped over" the prime 3 and there is no exponent written for 2. We can easily give 2 an exponent by rewriting it as 2^1, and since $x^0 = 1$ for any nonzero x (by definition), we can write $1 = 3^0$. Therefore, the prime factorization of 50 can be written as $2^1 \cdot 3^0 \cdot 5^2$.

This convention can be especially useful when comparing two or more positive integers or performing an operation on two or more integers. We will say that $p_0^{a_0} p_1^{a_1} \cdots p_n^{a_n}$ is a **complete prime factorization** if $p_0, p_1, \ldots, p_n$ are the first $n+1$ primes ($p_0 = 2, p_1 = 3$, and so on) and $a_0, a_1, \ldots, a_n \in \mathbb{N}$.

Example 8.24:

1. The prime factorization of 364 in canonical form is $2^2 \cdot 7 \cdot 13$. However, this is **not** a complete factorization.

    A complete factorization of 364 is $2^2 \cdot 3^0 \cdot 5^0 \cdot 7^1 \cdot 11^0 \cdot 13^1$. This is not the only complete factorization of 364. Another one is $2^2 \cdot 3^0 \cdot 5^0 \cdot 7^1 \cdot 11^0 \cdot 13^1 \cdot 17^0$.

    Given a complete factorization $p_0^{a_0} p_1^{a_1} \cdots p_n^{a_n}$ of a positive integer, $p_0^{a_0} p_1^{a_1} \cdots p_n^{a_n} p_{n+1}^0$ is another complete factorization, and in fact, for any $k \in \mathbb{N}$, $p_0^{a_0} p_1^{a_1} \cdots p_n^{a_n} p_{n+1}^0 p_{n+2}^0 \cdots p_{n+k}^0$ is also a complete factorization of that same positive integer. In words, we can include finitely many additional prime factors at the tail end of the original factorization all with exponent 0. Just be careful not to skip any primes!

2. $2^0 \cdot 3^5 \cdot 5^0 \cdot 7^2 \cdot 11^0 \cdot 13^0 \cdot 17^2$ and $2^3 \cdot 3^1 \cdot 5^0 \cdot 7^0 \cdot 11^6$ are complete prime factorizations. In many cases, it is useful to rewrite the second factorization as $2^3 \cdot 3^1 \cdot 5^0 \cdot 7^0 \cdot 11^6 \cdot 13^0 \cdot 17^0$. This is also a complete prime factorization. However, this one has all the same prime factors as the first number given.

Complete prime factorizations give us an easy way to compute greatest common divisors and least common multiples of positive integers.

Suppose that $a = p_0^{a_0} p_1^{a_1} \cdots p_n^{a_n}$ and $b = p_0^{b_0} p_1^{b_1} \cdots p_n^{b_n}$ are complete prime factorizations of a and b. Then we have

$$\gcd(a,b) = p_0^{\min\{a_0,b_0\}} p_1^{\min\{a_1,b_1\}} \cdots p_n^{\min\{a_n,b_n\}} \qquad \text{lcm}(a,b) = p_0^{\max\{a_0,b_0\}} p_1^{\max\{a_1,b_1\}} \cdots p_n^{\max\{a_n,b_n\}}.$$

Example 8.25: Let $a = 2 \cdot 5^2 \cdot 7$ and $b = 3 \cdot 5 \cdot 11^2$. We can rewrite a and b with the following complete prime factorizations: $a = 2^1 \cdot 3^0 \cdot 5^2 \cdot 7^1 \cdot 11^0$ and $b = 2^0 \cdot 3^1 \cdot 5^1 \cdot 7^0 \cdot 11^2$. From these factorizations, it is easy to compute $\gcd(a,b)$ and $\text{lcm}(a,b)$.

$\gcd(a,b) = 2^0 \cdot 3^0 \cdot 5^1 \cdot 7^0 \cdot 11^0 = 5$ and $\text{lcm}(a,b) = 2^1 \cdot 3^1 \cdot 5^2 \cdot 7^1 \cdot 11^2 = 127{,}050$.

Observe that in this example, $ab = 350 \cdot 1815 = 635{,}250 = 5 \cdot 127{,}050 = \gcd(a,b) \cdot \text{lcm}(a,b)$.

We will now show that the equation $ab = \gcd(a,b) \cdot \text{lcm}(a,b)$ is true for all positive integers a and b.

Before we state and prove the theorem, note that $\min\{x,y\} + \max\{x,y\} = x + y$ (Check this!).

Theorem 8.26: Let $a, b \in \mathbb{Z}^+$. Then $\gcd(a,b) \cdot \text{lcm}(a,b) = ab$.

Proof: Let $a = p_0^{a_0} p_1^{a_1} \cdots p_n^{a_n}$ and $b = p_0^{b_0} p_1^{b_1} \cdots p_n^{b_n}$ be complete prime factorizations of a and b. Then

$$\gcd(a,b) \cdot \text{lcm}(a,b)$$
$$= p_0^{\min\{a_0,b_0\}} p_1^{\min\{a_1,b_1\}} \cdots p_n^{\min\{a_n,b_n\}} \cdot p_0^{\max\{a_0,b_0\}} p_1^{\max\{a_1,b_1\}} \cdots p_n^{\max\{a_n,b_n\}}$$
$$= p_0^{\min\{a_0,b_0\}} p_0^{\max\{a_0,b_0\}} p_1^{\min\{a_1,b_1\}} p_1^{\max\{a_1,b_1\}} \cdots p_n^{\min\{a_n,b_n\}} p_n^{\max\{a_n,b_n\}}$$
$$= p_0^{\min\{a_0,b_0\}+\max\{a_0,b_0\}} p_1^{\min\{a_1,b_1\}+\max\{a_1,b_1\}} \cdots p_n^{\min\{a_n,b_n\}+\max\{a_n,b_n\}}$$
$$= p_0^{a_0+b_0} p_1^{a_1+b_1} \cdots p_n^{a_n+b_n}$$
$$= p_0^{a_0} p_0^{b_0} p_1^{a_1} p_1^{b_1} \cdots p_n^{a_n} p_n^{b_n}$$
$$= p_0^{a_0} p_1^{a_1} \cdots p_n^{a_n} \cdot p_0^{b_0} p_1^{b_1} \cdots p_n^{b_n}$$
$$= ab \qquad \square$$

We will now look at a well-known algorithm for computing the gcd of two positive integers. It also provides a method for expressing the gcd as a linear combination of the two integers.

Theorem 8.27 (The Euclidean Algorithm): Let $a, b \in \mathbb{Z}^+$ with $a \geq b$. Let $r_0 = a$, $r_1 = b$. Apply the Division Algorithm to r_0 and r_1 to find $k_1, r_2 \in \mathbb{Z}^+$ such that $r_0 = r_1 k_1 + r_2$, where $0 \leq r_2 < r_1$. Iterate this process to get $r_j = r_{j+1} k_{j+1} + r_{j+2}$, where $0 \leq r_{j+2} < r_{j+1}$ for $j = 0, 1, \ldots, n-1$ so that $r_{n+1} = 0$. Then $\gcd(a,b) = r_n$.

You will be asked to prove the Euclidean Algorithm in Problem 18 below.

Example 8.28: Let's use the Euclidean Algorithm to find $\gcd(305, 1040)$.

$$1040 = 305 \cdot 3 + 125$$
$$305 = 125 \cdot 2 + 55$$
$$125 = 55 \cdot 2 + 15$$
$$55 = 15 \cdot 3 + 10$$
$$15 = 10 \cdot 1 + \mathbf{5}$$
$$10 = 5 \cdot 2 + \mathbf{0}$$

So, $\gcd(305, 1040) = \mathbf{5}$.

Notes: (1) In this example, we have $a = r_0 = 1040$ and $b = r_1 = 305$. By the Division Algorithm we can write $1040 = 305 k_1 + r_2$, where $0 < r_2 < 305$. To find k_1, we are simply looking for the largest integer k such that $305k \leq 1040$. Well, $305 \cdot 3 = 915$ and $305 \cdot 4 = 1220$. So, 4 is too big and therefore, we let $k_1 = 3$. It follows that $r_2 = 1040 - 305 \cdot 3 = 1040 - 915 = 125$.

We now repeat the procedure using $r_1 = 305$ and $r_2 = 125$ to get $305 = 125 \cdot 2 + 55$. Notice that $125 \cdot 3 = 375$, which is too big because $375 > 305$. This is why we let $k_2 = 2$. It follows that $r_3 = 305 - 125 \cdot 2 = 305 - 250 = 55$.

Continuing this process, we eventually wind up with $10 = 5 \cdot 2 + 0$, so that $r_7 = 0$. By Theorem 12.11, $\gcd(305, 1040) = r_6 = 5$.

(2) As we go through the algorithm, we get $r_0 = 1040, r_1 = 305, r_2 = 125, r_3 = 55, r_4 = 15, r_5 = 10, r_6 = 5$, and $r_7 = 0$.

We also get $k_1 = 3, k_2 = 2, k_3 = 2, k_4 = 3, k_5 = 1$, and $k_6 = 2$.

(3) We can now go backwards through the algorithm to express $\gcd(305, 1040)$ as a linear combination of 305 and 1040.

We start with the second to last line (line 5): $15 = 10 \cdot 1 + 5$. We solve this equation for 5 to get $5 = 15 - 1 \cdot 10$.

Working backwards, we next look at line 4: $55 = 15 \cdot 3 + 10$. We solve this equation for 10 and then substitute into the previous equation: $10 = 55 - 15 \cdot 3$. After substituting, we get

$$5 = 15 - 1 \cdot 10 = 15 - 1(55 - 15 \cdot 3)$$

We then distribute and group all the 15's together and all the 55's together. So, we have

$$5 = 15 - 1 \cdot 10 = 15 - 1(55 - 15 \cdot 3) = 15 - 1 \cdot 55 + 3 \cdot 15 = 4 \cdot 15 - 1 \cdot 55.$$

Line 3 is next: $125 = 55 \cdot 2 + 15$. We solve this equation for 15 to get $15 = 125 - 2 \cdot 55$. And once again we now substitute into the previous equation to get

$$5 = 4 \cdot 15 - 1 \cdot 55 = 4(125 - 2 \cdot 55) - 1 \cdot 55 = 4 \cdot 125 - 8 \cdot 55 - 1 \cdot 55 = 4 \cdot 125 - 9 \cdot 55.$$

Let's go to line 2: $305 = 125 \cdot 2 + 55$. We solve this equation for 55 to get $55 = 305 - 2 \cdot 125$. Substituting into the previous equation gives us

$$5 = 4 \cdot 125 - 9 \cdot 55 = 4 \cdot 125 - 9(305 - 2 \cdot 125)$$
$$= 4 \cdot 125 - 9 \cdot 305 + 18 \cdot 125 = 22 \cdot 125 - 9 \cdot 305.$$

And finally line 1: $1040 = 305 \cdot 3 + 125$. Solving this equation for 125 gives us $125 = 1040 - 3 \cdot 305$. Substituting into the previous equation gives

$$5 = 22 \cdot 125 - 9 \cdot 305 = 22(1040 - 3 \cdot 305) - 9 \cdot 305$$
$$= 22 \cdot 1040 - 66 \cdot 305 - 9 \cdot 305 = 22 \cdot 1040 - 75 \cdot 305.$$

So, we see that $\gcd(305, 1040) = 5 = 22 \cdot 1040 - 75 \cdot 305 = -75 \cdot 305 + 22 \cdot 1040$.

(4) With a little practice, the computations done in Note 3 can be done fairly quickly. Here is what the quicker computation might look like:

$$5 = 15 - 1 \cdot 10 = 15 - 1 \cdot (55 - 15 \cdot 3) = 4 \cdot 15 - 1 \cdot 55 = 4(125 - 55 \cdot 2) - 1 \cdot 55 = 4 \cdot 125 - 9 \cdot 55$$
$$= 4 \cdot 125 - 9(305 - 125 \cdot 2) = 22 \cdot 125 - 9 \cdot 305 = 22(1040 - 305 \cdot 3) - 9 \cdot 305 = 22 \cdot 1040 - 75 \cdot 305$$

So, $5 = \gcd(305, 1040) = -75 \cdot 305 + 22 \cdot 1040$.

Problem Set 8

Full solutions to these problems are available for free download here:
www.SATPrepGet800.com/AAFBTDW

LEVEL 1

1. Write each of the following positive integers as a product of prime factors in canonical form:

    (i) 9

    (ii) 13

    (iii) 21

    (iv) 30

    (v) 44

    (vi) 693

    (vii) 67,500

    (viii) 384,659

    (ix) 9,699,690

2. List all prime numbers less than 100.

3. Find the gcd and lcm of each of the following sets of numbers:

    (i) $\{4, 6\}$

    (ii) $\{12, 180\}$

    (iii) $\{2, 3, 5\}$

    (iv) $\{14, 21, 77\}$

    (v) $\{720, 2448, 5400\}$

    (vi) $\{2^{17} 5^4 11^9 23, \, 2^5 3^2 7^4 11^3 13\}$

LEVEL 2

4. Determine if each of the following numbers is prime:

    (i) 101

    (ii) 399

    (iii) 1829

    (iv) 1933

    (v) 8051

    (vi) 13,873

    (vii) 65,623

5. Use the Division Algorithm to find the quotient and remainder when 723 is divided by 17.

6. For $n \in \mathbb{Z}^+$, let $M_n = n! + 1$. Determine if M_n is prime for $n = 1, 2, 3, 4, 5, 6$, and 7.

7. Prove that the sum of three integers that are each divisible by 5 is divisible by 5.

LEVEL 3

8. Prove that if $a, b, c \in \mathbb{Z}$ with $a|b$ and $b|c$, then $a|c$.

9. Prove that $n^3 - n$ is divisible by 3 for all natural numbers n.

10. Use the Euclidean Algorithm to find $\gcd(825, 2205)$. Then express $\gcd(825, 2205)$ as a linear combination of 825 and 2205.

11. Prove that if $k \in \mathbb{Z}$ with $k > 1$, then $k^3 + 1$ is not prime.

12. Prove that $\gcd(a, b) \mid \text{lcm}(a, b)$.

13. Let $a, b, c \in \mathbb{Z}$. Prove that $\gcd(a, b) = \gcd(a + bc, b)$.

14. Let $a, b, k, r \in \mathbb{Z}$ with $a = bk + r$. Prove that $\gcd(a, b) = \gcd(r, b)$.

LEVEL 4

15. Prove that the product of two odd integers is odd.

16. Prove that if $a, b, c, d, e \in \mathbb{Z}$ with $a|b$ and $a|c$, then $a|(db + ec)$.

17. Prove that $3^n - 1$ is even for all natural numbers n.

18. Prove the Euclidean Algorithm: Let $a, b \in \mathbb{Z}^+$ with $a \geq b$. Let $r_0 = a$, $r_1 = b$. Apply the Division Algorithm to r_0 and r_1 to find $k_1, r_2 \in \mathbb{Z}^+$ such that $r_0 = r_1 k_1 + r_2$, where $0 \leq r_2 < r_1$. Iterate this process to get $r_j = r_{j+1} k_{j+1} + r_{j+2}$, where $0 \leq r_{j+2} < r_{j+1}$ for $j = 0, 1, ..., n-1$ so that $r_{n+1} = 0$. Then $\gcd(a, b) = r_n$.

19. Prove that if $a|c$ and $b|c$, then $\text{lcm}(a, b) \mid c$.

20. Suppose that $a, b \in \mathbb{Z}^+$, $\gcd(a, b) = 1$, and $c|ab$. Prove that there are integers d and e such that $c = de$, $d|a$, and $e|b$.

21. A **prime triple** is a sequence of three prime numbers of the form $p, p + 2$, and $p + 4$. For example, 3, 5, 7 is a prime triple. Prove that there are no other prime triples.

LEVEL 5

22. If $a, b \in \mathbb{Z}^+$ and $\gcd(a, b) = 1$, find the following:

   (i) $\gcd(a, a + 1)$

   (ii) $\gcd(a, a + 2)$

   (iii) $\gcd(3a + 2, 5a + 3)$

   (iv) $\gcd(a + b, a - b)$

   (v) $\gcd(a + 2b, 2a + b)$

23. Find the smallest ideal of $\mathbb{Z}$ containing 6 and 15. Find the smallest ideal of $\mathbb{Z}$ containing 2 and 3. In general, find the smallest ideal of $\mathbb{Z}$ containing j and k, where $j, k \in \mathbb{Z}$.

24. Find all subgroups of $(\mathbb{Z}, +)$ and all submonoids of $(\mathbb{Z}, +)$.

CHALLENGE PROBLEMS

25. Let $a, b \in \mathbb{Z}^+$. Prove that $a | b$ if and only if $(2^a - 1) | (2^b - 1)$.

26. Find all subsemigroups of $(\mathbb{Z}, +)$.

LESSON 9
NUMBER THEORETIC APPLICATIONS

Cyclic Groups

Let G be a group and let $x \in G$. In Lesson 6, we defined the order of x, written $|x|$, in two different ways:

1. $|x| = |\langle x \rangle|$, the order of the cyclic subgroup generated by x.
2. $|x|$ is the least positive integer n such that $x^n = e$, if such an n exists. Otherwise x has infinite order.

Now that we have the Division Algorithm, we are able to prove the equivalence of these two definitions. We will do this now in two steps—see Theorems 9.1 and 9.2 below.

Notes: (1) Recall that the cyclic group $\langle x \rangle = \{x^k \mid k \in \mathbb{Z}\}$ consists of all powers of x. There is nothing in this definition that says that these powers all need to be distinct. In fact, if $\langle x \rangle$ is finite, then the powers **cannot** be distinct. However, if $\langle x \rangle$ is infinite, we will see below that the powers are all distinct.

(2) If we are using additive notation for the group, then $\langle x \rangle = \{kx \mid k \in \mathbb{Z}\}$ consists of all multiples of x instead of powers of x.

(3) It would be worthwhile to take a few minutes to review Example 6.8 before proceeding.

Theorem 9.1: Let G be a group and let $x \in G$. $\langle x \rangle$ is infinite if and only if there does not exist $n \in \mathbb{Z}^+$ such that $x^n = e$.

Analysis: We will first prove that if $x^n \neq e$ for all $n \in \mathbb{Z}^+$, then the powers x^a with $a > 0$ are all distinct. This will prove that $\langle x \rangle$ is infinite.

We will prove the other direction by contrapositive (see the analysis after the statement of Theorem 1.28). Instead of proving "$\langle x \rangle$ is infinite implies there does not exist $n \in \mathbb{Z}^+$ such that $x^n = e$," we will prove "there is $n \in \mathbb{Z}^+$ with $x^n = e$ implies that $\langle x \rangle$ is finite." To prove this statement, we will need to use the Division Algorithm.

Proof: First suppose that $x^n \neq e$ for all $n \in \mathbb{Z}^+$. We will show that $0 < a < b$ implies $x^a \neq x^b$. Suppose toward contradiction that $x^a = x^b$. Then

$$x^{b-a} = x^{b+(-a)} = x^b x^{-a} = x^b (x^a)^{-1} = x^b (x^b)^{-1} = e.$$

For the first equality, we used the definition of subtraction of integers. For the second equality, we used Theorem 5.12. For the third equality, we used the definition of x^{-a}. For the fourth equality, we used a direct substitution. For the fifth equality, we used the inverse property of a group.

Now, $a < b$ implies that $b - a > 0$, contradicting our assumption that $x^n \neq e$ for all $n \in \mathbb{Z}^+$. Therefore, $0 < a < b$ implies $x^a \neq x^b$. It follows that $\langle x \rangle$ is infinite.

Conversely, suppose that there is $n \in \mathbb{Z}^+$ with $x^n = e$. We will show that $\langle x \rangle \subseteq \{e, x, x^2, \ldots, x^{n-1}\}$. For this, we will need the Division Algorithm. Let $x^j \in \langle x \rangle$ (so, $j \in \mathbb{Z}$). By the Division Algorithm, there are unique integers k and r such that $j = nk + r$ and $0 \leq r < n$. Then

$$x^j = x^{nk+r} = x^{nk}x^r = (x^n)^k x^r = e^k x^r = e x^r = x^r.$$

For the first and fourth equalities, we used direct substitutions. For the second equality, we used Theorem 5.12. For the third equality, we used Theorem 5.14. For the fifth and sixth equalities, we used the identity property of a group.

It follows that $x^j = x^r \in \{e, x, x^2, \ldots, x^{n-1}\}$. Since $x^j \in \langle x \rangle$ was arbitrary, $\langle x \rangle \subseteq \{e, x, x^2, \ldots, x^{n-1}\}$. Therefore, $\langle x \rangle$ is finite. □

Note: In the second part of the proof above, we did **not** prove that $\langle x \rangle$ has order n. We proved only that the order of $\langle x \rangle$ is less than or equal to n. We did **not** rule out the possibility that there are positive integers $c, d < n$ with $x^c = x^d$. In fact, this could certainly happen. In other words, the elements of $\{e, x, x^2, \ldots, x^{n-1}\}$ are not necessarily distinct. If we wanted to ensure that these elements are distinct, we would have chosen n to be the **least** positive integer such that $x^n = e$. We did not do this because it was not necessary to reach our desired conclusion. In the proof of the next result, however, we will need to be more careful.

Theorem 9.2: Let G be a group, let $x \in G$, and let $n \in \mathbb{Z}^+$. Then $\langle x \rangle$ has order n if and only if n is the least positive integer such that $x^n = e$.

Proof: Let n be the least positive integer such that $x^n = e$. Let's first show that $e = x^0, x^1, \ldots, x^{n-1}$ are all **distinct** elements of $\langle x \rangle$. Suppose toward contradiction that $0 \leq a < b < n$ and $x^a = x^b$. Then $x^{b-a} = x^{b+(-a)} = x^b x^{-a} = x^b(x^a)^{-1} = x^b(x^b)^{-1} = e$ (just like in the proof of Theorem 9.1). But we have $0 < b - a < n$, contradicting that n is the **least** positive integer such that $x^n = e$. It follows that the set $\{e, x, x^2, \ldots, x^{n-1}\}$ consists of **exactly** n elements.

We will now show that $\langle x \rangle = \{e, x, x^2, \ldots, x^{n-1}\}$. For this, we will need the Division Algorithm. Let $x^j \in \langle x \rangle$ (so, $j \in \mathbb{Z}$). By the Division Algorithm, there are unique integers k and r such that $j = nk + r$ and $0 \leq r < n$. Then $x^j = x^{nk+r} = x^{nk}x^r = (x^n)^k x^r = e^k x^r = e x^r = x^r$ (just like in the proof of Theorem 9.1). It follows that $x^j = x^r \in \{e, x, x^2, \ldots, x^{n-1}\}$. So, $\langle x \rangle \subseteq \{e, x, x^2, \ldots, x^{n-1}\}$. Since the reverse inclusion is clear, we have $\langle x \rangle = \{e, x, x^2, \ldots, x^{n-1}\}$, as desired.

Since $\langle x \rangle = \{e, x, x^2, \ldots, x^{n-1}\}$ and the elements of $\{e, x, x^2, \ldots, x^{n-1}\}$ are distinct, it follows that $|\langle x \rangle| = n$.

Conversely, assume that $\langle x \rangle$ has order n. If $0 < j < n$ and $x^j = e$, then $\langle x \rangle \subseteq \{e, x, x^2, \ldots, x^{j-1}\}$, as we showed above. So, $|\langle x \rangle| \leq j < n$. It follows that $x^j \neq e$ for $0 < j < n$. Now, if $x^n \neq e$, then $e, x, x^2, \ldots, x^n$ are distinct elements of $\langle x \rangle$, and so, $|\langle x \rangle| > n$. It follows that $x^n = e$ and in fact that n is the least such positive integer such that $x^n = e$. □

Note: Theorems 9.1 and 9.2 together imply that the two definitions of the order of an element are equivalent. In other words, $|\langle x \rangle|$ is equal to the least positive integer n if such a positive integer exists (by Theorem 9.2) and $|\langle x \rangle|$ is infinite if no such positive integer exists (by Theorem 9.1).

Example 9.3:

1. Let $G = (\mathbb{Z}_{18}, +)$. Let's find the order of the element 15 in $\mathbb{Z}_{18}$. Since the group operation is addition, we look at multiples of 15 instead of powers. We have $1 \cdot 15 = 15$, $2 \cdot 15 = 12$, $3 \cdot 15 = 9$, $4 \cdot 15 = 6$, $5 \cdot 15 = 3$, $6 \cdot 15 = 0$. Since 6 is the least positive integer such that $6 \cdot 15 = 0$, the order of 15 is $|15| = \mathbf{6}$.

   Alternatively, $\langle 15 \rangle = \{0, 3, 6, 9, 12, 15\}$. So, $|15| = |\langle 15 \rangle| = |\{0, 3, 6, 9, 12, 15\}| = \mathbf{6}$.

2. Let $G = S_5 = (S(\{0,1,2,3,4\}), \circ)$ be the symmetric group with 5 elements. Let's find the order of the element (1234) (see Example 4.12 for an explanation of the cycle notation being used here). $(1234)^1 = (1234)$, $(1234)^2 = (13)(24)$, $(1234)^3 = (1234)(13)(24) = (1432)$, and $(1234)^4 = (1234)(1432) = (1)$. Since 4 is the least positive integer such that $(1234)^4 = (1)$, the order of (1234) is $|(1234)| = 4$.

   Alternatively, $\langle (1234) \rangle = \{(1), (1234), (13)(24), (1432)\}$. So, $|(1432)| = |\langle (1432) \rangle| = \mathbf{4}$.

Notes: (1) There are several ways to compute $2 \cdot 15$ in $\mathbb{Z}_{18}$.

<u>Method 1:</u> Think of $2 \cdot 15$ as $15 + 15$ and then think of adding 15 and 15 on an 18 hour clock, so that 18 is the same as 0. In this way, $15 + 15 = 15 + (3 + 12) = (15 + 3) + 12 = 0 + 12 = 12$.

<u>Method 2:</u> We can add $15 + 15$ normally and then reduce modulo 18. So, we have $15 + 15 = 30$. We can reduce modulo 18 in two different ways: (1) we can subtract 18 to get $30 - 18 = 12$, or (2) we can take the remainder upon dividing 30 by 18. 18 goes into 30 one time with a remainder of 12.

<u>Method 3:</u> We can multiply $2 \cdot 15$ normally to get 30 and then reduce modulo 18 in either of the two ways we did in Method 2 to get 12.

(2) Here are a few ways to compute $3 \cdot 15$ in $\mathbb{Z}_{18}$.

<u>Method 1:</u> Think of $3 \cdot 15$ as $2 \cdot 15 + 15$. We already saw in Note 1 that in $\mathbb{Z}_{18}$, $2 \cdot 15 = 12$. So, we have $3 \cdot 15 = 2 \cdot 15 + 15 = 12 + 15 = 12 + (6 + 9) = (12 + 6) + 9 = 0 + 9 = 9$.

<u>Method 2:</u> We can add $3 \cdot 15 = 15 + 15 + 15$ normally and then reduce modulo 18. So, we have $15 + 15 + 15 = 45$. We can reduce modulo 18 in two different ways: (1) we can subtract 18 repeatedly until we get an integer between 0 and 18. In this case, we have $45 - 18 = 27$ and then we have $27 - 18 = 9$, or (2) we can take the remainder upon dividing 45 by 18. 18 goes into 45 two times with a remainder of 9.

<u>Method 3:</u> We can multiply $3 \cdot 15$ normally to get 45 and then reduce modulo 18 in either of the two ways we did in Method 2 to get 9.

(3) We can use any of the three methods given in Notes 1 and 2 above to compute $4 \cdot 15$, $5 \cdot 15$, and $6 \cdot 15$. I leave the details to the reader.

(4) In S_5, (1234) is the permutation that sends 1 to 2, 2 to 3, 3 to 4, 4 to 1, and 5 to itself. So, (1234) is the function $\pi: \{1,2,3,4,5\} \to \{1,2,3,4,5\}$ such that $\pi(1) = 2$, $\pi(2) = 3$, $\pi(3) = 4$, $\pi(4) = 1$, and $\pi(5) = 5$. Equivalently, $\pi = \{(1,2), (2,3), (3,4), (4,1), (5,5)\}$. Below are a few ways to visualize π.

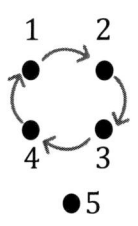

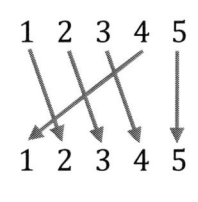

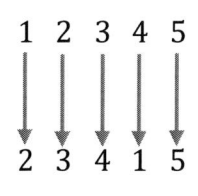

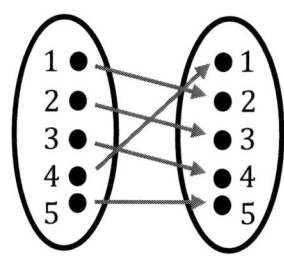

(5) $(1234)^2 = (1234)(1234)$. Recall from Example 4.12 that to compose these two permutations, we start with the one on the right. Starting with 1, we see that the rightmost cycle sends 1 to 2 and the leftmost cycle sends 2 to 3. Therefore, the composition sends 1 to 3. Let's do 3 next. The rightmost cycle sends 3 to 4 and then the leftmost cycle sends 4 to 1. So, the composition sends 3 to 1. So, we get the cycle (13). Similar reasoning shows that the composition sends 2 to 4 and 4 to 2, so that we get the cycle (24). It follows that the composition $(1234)(1234)$ simplifies to $(13)(24)$ or $(24)(13)$ (we will see in Lesson 12 that cycles with no elements in common—called disjoint cycles—commute). $(1234)^3$ and $(1234)^4$ can be computed similarly.

Recall: For integers a and b, we say that $a \equiv_n b$ if and only if $n|b - a$. In other words, $a \equiv_n b$ if and only if there is an integer k such that $b - a = nk$, ore equivalently, $b = nk + a$.

Theorem 9.4: Let G be a group, let $x \in G$ have order $n \in \mathbb{Z}^+$, and let $a, b \in \mathbb{Z}$. Then $x^a = x^b$ if and only if $a \equiv_n b$.

Proof: Assume that $x \in G$ has order n and let $a, b \in \mathbb{Z}$ with $x^a = x^b$. Then $x^{b-a} = e$ (see the beginning of the proof of Theorem 9.1 for details). By the Division Algorithm, there are unique integers k and r such that $b - a = nk + r$ with $0 \leq r < n$. Then, we have

$$e = x^{b-a} = x^{nk+r} = x^{nk}x^r = (x^n)^k x^r = e^k x^r = ex^r = x^r.$$

See the proof of Theorem 9.1 for a justification of each of these equalities.

Since n is the least positive integer such that $x^n = e$ and $r < n$, we must have $r = 0$. So, $b - a = nk$. Therefore, $n|b - a$, or equivalently, $a \equiv_n b$.

Conversely, assume that $a \equiv_n b$, or equivalently, $n|b - a$. Then there is $k \in \mathbb{Z}$ such that $b - a = nk$. So, $b = nk + a$. It follows that $x^b = x^{nk+a} = x^{nk}x^a = (x^n)^k x^a = e^k x^a = ex^a = x^a$, as desired. □

Note: If we let $b = 0$ in Theorem 9.4 (so that $x^b = x^0 = e$), we see that $x^a = e$ if and only if $n|a$ (n divides a), where $n = |x|$.

Theorem 9.5: A subgroup of a cyclic group is cyclic.

Proof: Let G be a cyclic group and let H be a subgroup of G. Since G is cyclic, there is $x \in G$ with $G = \langle x \rangle$. If $H = \{e\}$, then H is cyclic because $H = \langle e \rangle$. So, assume that $H \neq \{e\}$. Let n be the least positive integer such that $x^n \in H$. We will now show that $H = \langle x^n \rangle$. To see this, suppose that $y \in H$. Then there is $a \in \mathbb{Z}$ such that $y = x^a$. By the Division Algorithm, there are unique integers k and r such that $a = nk + r$ (or equivalently, $r = a - nk$) with $0 \leq r < n$. It follows that

$$x^r = x^{a-nk} = x^{a+(-nk)} = x^a x^{-nk} = x^a (x^n)^{-k}.$$

Since $x^a, x^n \in H$ and H is a group, it follows that $x^r = x^a(x^n)^{-k} \in H$. Since n is the least positive integer such that $x^n \in H$ and $r < n$, we must have $r = 0$. So, $a = nk$. Therefore, we have

$$y = x^a = x^{nk} = (x^n)^k \in \langle x^n \rangle.$$

Since $y \in H$ was arbitrary, we see that $H \subseteq \langle x^n \rangle$.

Conversely let $y \in \langle x^n \rangle$. Then there is $a \in \mathbb{Z}$ such that $y = (x^n)^a$. Since $x^n \in H$ and H is a group, $y = (x^n)^a \in H$. Since $y \in \langle x^n \rangle$ was arbitrary, we see that $\langle x^n \rangle \subseteq H$.

Since $H \subseteq \langle x^n \rangle$ and $\langle x^n \rangle \subseteq H$, we have $H = \langle x^n \rangle$. Therefore, H is cyclic. □

Note: The proof of Theorem 9.5 shows that if H is a subgroup of a cyclic group G, then H is a cyclic group that is generated by x^n, where n is the least positive integer such that $x^n \in H$.

Theorem 9.6: Let G be a finite cyclic group and suppose that k divides $|G|$. Then there is **exactly one** subgroup H of G such that $|H| = k$.

Proof: Suppose that $G = \langle x \rangle$ is a cyclic group of order n and suppose that $k | n$. Then there is an integer j such that $n = kj$. Let $H = \langle x^j \rangle$. Then $H \leq G$. Since $|G| = n$, we have $(x^j)^k = x^{jk} = x^n = e$. Therefore, $|H| \leq k$. If $0 < t < k$, then $tj < kj = n$, and so, $(x^j)^t = x^{jt} \neq e$ (because jt is less than the order of x). It follows that $|H| \geq k$. Since $|H| \leq k$ and $|H| \geq k$, we have $|H| = k$.

Now suppose that K is any subgroup of G such that $|K| = k$. By Theorem 9.5, K is cyclic. So, there is $a \in \mathbb{Z}$ with $K = \langle x^a \rangle$. Since $|\langle x^a \rangle| = k$, $x^{ak} = (x^a)^k = e = x^0$. By Theorem 9.4, $ak \equiv_n 0$. Therefore, $n | ak$. So, there is an integer b such that $ak = nb$. Since $n = kj$, we have $ak = kjb$, and so, $a = jb$. Thus, $x^a = x^{jb} = (x^j)^b \in \langle x^j \rangle = H$. Since H is a group, $K = \langle x^a \rangle \subseteq H$. Since $|K| = |H|$ (K and H both have the same finite order), we must have $K = H$, as desired. □

Note: If G is a finite cyclic group of order n and k divides n, then by Theorem 9.6, there is exactly one subgroup H of G of order k. This subgroup H consists precisely of all elements $y \in G$ such that $y^k = e$. Indeed, since k divides n, there is an integer j such that $n = kj$ and the proof above shows that $H = \langle x^j \rangle$. So, if $y \in H$, then $y = (x^j)^t = x^{jt}$ for some $t \in \mathbb{Z}$. So,

$$y^k = (x^{jt})^k = (x^{kj})^t = (x^n)^t = e^t = e.$$

Conversely, suppose $y \in G = \langle x \rangle$ with $y^k = e$. Since $y \in \langle x \rangle$, there is $a \in \mathbb{Z}$ with $y = x^a$. So, $x^{ak} = (x^a)^k = y^k = e$. Therefore, by the proof above, $x^a \in H$.

Example 9.7:

1. $\mathbb{Z}_6 = \{0, 1, 2, 3, 4, 5\}$ is a cyclic group of order 6 under addition. The divisors of 6 are 1, 2, 3, and 6. By Theorem 9.6, there is exactly one subgroup of $\mathbb{Z}_6$ of order 1, exactly one subgroup of order 2, exactly one subgroup of order 3, and exactly one subgroup of order 6. By Theorem 9.5, all these subgroups are cyclic. The cyclic subgroup of order 1 is $\langle 0 \rangle = \{0\}$. The cyclic subgroup of order 2 is $\langle 3 \rangle = \{0, 3\}$. The cyclic subgroup of order 3 is $\langle 2 \rangle = \{0, 2, 4\}$. The cyclic subgroup of order 6 is the group itself, $\langle 1 \rangle = \{0, 1, 2, 3, 4, 5\} = \mathbb{Z}_6$. There are no other subgroups of $\mathbb{Z}_6$.

2. $\mathbb{Z}_7 = \{0, 1, 2, 3, 4, 5, 6\}$ is a cyclic group of order 7. The divisors of 7 are just 1 and 7. The trivial group $\langle 0 \rangle = \{0\}$ is the only subgroup of order 1 and the group itself, $\langle 1 \rangle = \{1, 2, 3, 4, 5, 6\} = \mathbb{Z}_7$ is the only subgroup of order 7. There are no other subgroups of $\mathbb{Z}_7$.

Note: Theorem 9.6 tells us that if G is a cyclic group of finite order n, then there is exactly one subgroup of order k for each divisor k of n. Example 9.7 provides evidence that if j is a positive integer that is **not** a divisor of n, then there is no subgroup of G of order j. The next theorem will confirm this.

Theorem 9.8: Let G be a finite cyclic group and suppose that H is a subgroup of G. Then the order of H divides the order of G.

Proof: Suppose that $G = \langle x \rangle$, $|G| = n$, and H is a subgroup of G. By the Note following Theorem 9.5, $H = \langle x^j \rangle$, where j is the least positive integer such that $x^j \in H$. By The Division Algorithm, there are integers k and r such that $n = jk + r$ and $0 \leq r < j$. So, $r = n - jk$ and we have

$$x^r = x^{n-jk} = x^{n+(-jk)} = x^n x^{-jk} = x^n (x^j)^{-k} = e(x^j)^{-k} = (x^j)^{-k} \in H.$$

Since j is the least positive integer such that $x^j \in H$ and $r < j$, we must have $r = 0$. So, $n = jk$. Therefore, $k | n$. We will now finish the proof by showing that $|H| = k$. To see this, first observe that $(x^j)^k = x^{jk} = x^n = e$. So, $|H| \leq k$. If $t < k$, then $jt < jk = n$, and so, $(x^j)^t = x^{jt} \neq e$ because n is the least positive integer such that $x^n = e$. So, $|H| \geq k$. It follows that $|H| = k$. □

We will now show that all cyclic groups of the same order look and behave exactly the same way.

Theorem 9.9: Every infinite cyclic group is isomorphic to $(\mathbb{Z}, +)$. Every finite cyclic group of order $n \in \mathbb{Z}^+$ is isomorphic to $(\mathbb{Z}_n, +)$.

Proof: First, let $(G, \cdot)$ be an infinite cyclic group with generator x and define $f: \mathbb{Z} \to G$ by $f(k) = x^k$. To see that f is a homomorphism, we use Theorem 5.12. Let $k_1, k_2 \in \mathbb{Z}$. Then we have $f(k_1 + k_2) = x^{k_1 + k_2} = x^{k_1} \cdot x^{k_2} = f(k_1) \cdot f(k_2)$. To see that f is surjective, let $x^k \in G$. Then $k \in \mathbb{Z}$ and $f(k) = x^k$. Finally, to see that f is injective, assume that $f(k_1) = f(k_2)$. Then $x^{k_1} = x^{k_2}$. So, we have $x^{k_1 - k_2} = x^{k_1 + (-k_2)} = x^{k_1} x^{-k_2} = x^{k_1} (x^{k_2})^{-1} = x^{k_1} (x^{k_1})^{-1} = e$. Suppose toward contradiction that $k_1 \neq k_2$. Without loss of generality, assume that $k_1 > k_2$ and let $n = k_1 - k_2$. Then $n \in \mathbb{Z}^+$ and $x^n = e$. By Theorem 9.1, $G = \langle x \rangle$ is finite, contradicting our assumption that G is infinite. So, $k_1 = k_2$.

Now, let $(G, \cdot)$ be a finite cyclic group of order n and let x be a generator of G. Define $f: \mathbb{Z}_n \to G$ by $f([k]_n) = x^k$. To see that f is well-defined, suppose that $k_1 \equiv_n k_2$. Then by Theorem 9.4, $x^{k_1} = x^{k_2}$. To see that f is a homomorphism, we use Theorem 5.12. Let $k_1, k_2 \in \mathbb{Z}$. Then we have $f([k_1]_n + [k_2]_n) = f([k_1 + k_2]_n) = x^{k_1 + k_2} = x^{k_1} \cdot x^{k_2} = f([k_1]_n) \cdot f([k_2]_n)$. To see that f is surjective, let $x^k \in G$. Then $k \in \mathbb{Z}$ and $f([k]_n) = x^k$. Finally, to see that f is injective, assume that $f([k_1]_n) = f([k_2]_n)$. Then $x^{k_1} = x^{k_2}$. By Theorem 9.4, $k_1 \equiv_n k_2$. So, $[k_1]_n = [k_2]_n$. □

Theorem 9.10: Let G be a group, let $x \in G$ be an element of order n, let $a \in \mathbb{Z}^+$, and let $b = \gcd(a, n)$. Then $\langle x^a \rangle = \langle x^b \rangle$ and $n = b \cdot |x^a|$.

Note: In additive notation, the conclusion of Theorem 9.10 says $\langle ax \rangle = \langle bx \rangle$ and $n = b \cdot |ax|$.

Before proving Theorem 9.10, let's look at a few simple examples. Let $G = \mathbb{Z}_6$ under the operation of addition, let $x = 1$, and let $a = 5$. Then $n = |x| = 6$ and $b = \gcd(5,6) = 1$. Theorem 9.10 tells us that

$$\langle 5 \rangle = \langle 5 \cdot 1 \rangle = \langle 1 \cdot 1 \rangle = \langle 1 \rangle \quad \text{and} \quad 6 = 1 \cdot |5 \cdot 1|.$$

Important note: In this context, $|5 \cdot 1|$ means the order of $5 \cdot 1$ and **not** the absolute value of $5 \cdot 1$.

Let's check that these statements are true. First note that $\langle 1 \rangle = \{0,1,2,3,4,5\} = \mathbb{Z}_6$. Let's check that $\langle 5 \rangle$ is also $\mathbb{Z}_6$. Well, $2 \cdot 5 = 5 + 5 = 4$, $3 \cdot 5 = 4 + 5 = 3$, $4 \cdot 5 = 3 + 5 = 2$, $5 \cdot 5 = 2 + 5 = 1$. We also have $1 \cdot 5 = 5$ and $0 \cdot 5 = 0$. So, we see that $\langle 5 \rangle = \{0,1,2,3,4,5\} = \mathbb{Z}_6$. Also, we see that the order of 5 is 6 ($|5 \cdot 1| = |5| = 6$). So, $1 \cdot |5 \cdot 1| = 6$.

Let's looks at another example in $\mathbb{Z}_6$. Let $x = 1$ and $a = 4$. Then $n = |x| = 6$ and $b = \gcd(4,6) = 2$. Theorem 9.10 tells us that

$$\langle 4 \rangle = \langle 4 \cdot 1 \rangle = \langle 2 \cdot 1 \rangle = \langle 2 \rangle \quad \text{and} \quad 6 = 2 \cdot |4 \cdot 1|.$$

Let's check these statements. First note that $\langle 2 \rangle = \{0,2,4\}$. Let's check that $\langle 4 \rangle$ is also $\{0,2,4\}$. Well, $1 \cdot 4 = 4$, $2 \cdot 4 = 4 + 4 = 2$, $3 \cdot 4 = 2 + 4 = 0$, $4 \cdot 4 = 0 + 4 = 4$. Since 4 has already appeared, we see that $\langle 4 \rangle = \{0,2,4\}$. Also, we see that the order of 4 is 3 ($|4 \cdot 1| = |4| = 3$). So, $2 \cdot |4 \cdot 1| = 6$.

Let's do one last example in $\mathbb{Z}_6$. Let $x = 2$ and $a = 5$. Then $n = |x| = 3$ and $b = \gcd(5,3) = 1$. It follows that

$$\langle 4 \rangle = \langle 5 \cdot 2 \rangle = \langle 1 \cdot 2 \rangle = \langle 2 \rangle \quad \text{and} \quad 3 = 1 \cdot |5 \cdot 2|.$$

We have already shown that $\langle 4 \rangle = \{0,2,4\} = \langle 2 \rangle$ above. Now, the order of $5 \cdot 2$ is $|5 \cdot 2| = |4| = 3$. So, $1 \cdot |5 \cdot 2| = 1 \cdot 3 = 3$.

Proof of Theorem 9.10: First note that since $b|a$, there is an integer k such that $a = bk$. Now, $x^a = x^{bk} = (x^b)^k \in \langle x^b \rangle$. So, $\langle x^a \rangle \subseteq \langle x^b \rangle$. By Theorem 8.18, there exist $c, d \in \mathbb{Z}$ with $b = ca + dn$. It follows that $x^b = x^{ca+dn} = x^{ca}x^{dn} = (x^a)^c(x^n)^d = (x^a)^c e^d = (x^a)^c e = (x^a)^c \in \langle x^a \rangle$. Therefore, $\langle x^b \rangle \subseteq \langle x^a \rangle$. Since $\langle x^a \rangle \subseteq \langle x^b \rangle$ and $\langle x^b \rangle \subseteq \langle x^a \rangle$, we have $\langle x^a \rangle = \langle x^b \rangle$.

Now, since $b|n$, there is an integer j such that $n = bj$. We have $(x^b)^j = x^{bj} = x^n = e$. So, $|x^b| \leq j$. If t is a positive integer such that $t < j$, then $bt < bj = n$, and so, $(x^b)^t = x^{bt} \neq e$. So, $|x^b| \geq j$. Therefore, $|x^b| = j$. So, $|x^a| = |\langle x^a \rangle| = |\langle x^b \rangle| = |x^b| = j$. Thus, $n = bj = b \cdot |x^a|$. □

Modular Arithmetic

Recall once more that for integers a and b, we say that $a \equiv_n b$ if and only if $n|b - a$. In other words, $a \equiv_n b$ if and only if there is an integer k such that $b - a = nk$, or equivalently, $b = nk + a$.

In Problem 5 from Problem Set 3, you were asked to prove that $\equiv_n$ is an equivalence relation on $\mathbb{Z}$. If we let $[a]_n$ be the equivalence class of a under $\equiv_n$, then we have $b \in [a]_n$ if and only if $a \equiv_n b$ if and only if there is an integer k such that $b - a = nk$ if and only if there is an integer k such that $b = nk + a$ if and only if $b \in n\mathbb{Z} + a$. So, $[a]_n = n\mathbb{Z} + a$.

206

Since the equivalence classes form a partition of $\mathbb{Z}$, we see that $\{n\mathbb{Z} + a \mid a \in \mathbb{Z}\}$ is a partition of $\mathbb{Z}$. Over and over again throughout this book, we have used the fact that $\{n\mathbb{Z} + a \mid a \in \mathbb{Z}\}$ consists of the n distinct sets $n\mathbb{Z}, n\mathbb{Z} + 1, n\mathbb{Z} + 2, \ldots, n\mathbb{Z} + (n-1)$. However, we have never proved this fact. Now, with the Division Algorithm at our disposal, we are finally ready to do so.

Theorem 9.11: Let $n \in \mathbb{Z}^+$. Then $\{n\mathbb{Z} + a \mid a \in \mathbb{Z}\} = \{n\mathbb{Z}, n\mathbb{Z} + 1, n\mathbb{Z} + 2, \ldots, n\mathbb{Z} + (n-1)\}$. Also, if $b, c \in \mathbb{Z}$ with $0 \leq b < c < n$, then $n\mathbb{Z} + b \neq n\mathbb{Z} + c$.

Proof: Let $a \in \mathbb{Z}$. By the Division Algorithm, there are integers k and r with $a = nk + r$ and $0 \leq r < n$. We will show that $n\mathbb{Z} + a = n\mathbb{Z} + r$. To see this, first let $x \in n\mathbb{Z} + a$. Then there is $j \in \mathbb{Z}$ such that $x = nj + a = nj + nk + r = n(j + k) + r \in n\mathbb{Z} + r$. So, $n\mathbb{Z} + a \subseteq n\mathbb{Z} + r$. Now, let $x \in n\mathbb{Z} + r$. Then there is $j \in \mathbb{Z}$ such that $x = nj + r = nj + a - nk = n(j - k) + a \in n\mathbb{Z} + a$. Therefore, $n\mathbb{Z} + r \subseteq n\mathbb{Z} + a$. Since $n\mathbb{Z} + a \subseteq n\mathbb{Z} + r$ and $n\mathbb{Z} + r \subseteq n\mathbb{Z} + a$, we have $n\mathbb{Z} + a = n\mathbb{Z} + r$.

So, we have proved that $\{n\mathbb{Z} + a \mid a \in \mathbb{Z}\} \subseteq \{n\mathbb{Z}, n\mathbb{Z} + 1, n\mathbb{Z} + 2, \ldots, n\mathbb{Z} + (n-1)\}$. Since the reverse inclusion is obvious, we have $\{n\mathbb{Z} + a \mid a \in \mathbb{Z}\} = \{n\mathbb{Z}, n\mathbb{Z} + 1, n\mathbb{Z} + 2, \ldots, n\mathbb{Z} + (n-1)\}$.

Now, let $0 \leq b, c < n$ and suppose that $n\mathbb{Z} + b = n\mathbb{Z} + c$. Since $b \in n\mathbb{Z} + b$ (because $b = n \cdot 0 + b$), we have $b \in n\mathbb{Z} + c$. So, there is an integer k such that $b = nk + c$. Since we also have $b = n \cdot 0 + b$, by the Division Algorithm, we must have $b = c$. It follows that if $b, c \in \mathbb{Z}$ with $0 \leq b < c < n$, then $n\mathbb{Z} + b \neq n\mathbb{Z} + c$. □

We are also now ready to determine which of the rings $(\mathbb{Z}_n, +, \cdot)$ are fields. We first prove a preliminary result.

Theorem 9.12: Let $a \in \mathbb{Z}^+$. Then $[a]_n$ is a zero divisor in $\mathbb{Z}_n$ if and only if $\gcd(a, n) \neq 1$.

Proof: Let $d = \gcd(a, n)$.

First assume that $d > 1$. Since d is a divisor of a, there is an integer k such that $a = dk$. Since d is a divisor of n, there is an integer j such that $n = dj$. So, $aj = (dk)j = (dj)k = nk$. It follows that $[a]_n \cdot [j]_n = [0]_n$. So, $[a]_n$ is a zero divisor in $\mathbb{Z}_n$.

Next, assume that $d = 1$ and that $[a]_n \cdot [b]_n = [0]_n$. Then $[ab]_n = [0]_n$, and so, $ab \equiv_n 0$. Therefore, $n | ab$. By Euclid's Principle (Theorem 8.19), $n | b$. So, $b \equiv_n 0$ and therefore, $[b]_n = [0]_n$. It follows that $[a]_n$ is not a zero divisor in $\mathbb{Z}_n$. □

Example 9.13:

1. Let's find all the zero divisors in $(\mathbb{Z}_{12}, +, \cdot)$. Since $1, 5, 7,$ and 11 are relatively prime with 12, by Theorem 9.12, the elements $1, 5, 7,$ and 11 are **not** zero divisors in $\mathbb{Z}_{12}$. The remaining elements $2, 3, 4, 6, 8, 9,$ and 10 are zero divisors in $\mathbb{Z}_{12}$. Indeed, in $\mathbb{Z}_{12}$, we have $2 \cdot 6 = 0$, $3 \cdot 4 = 0, 8 \cdot 9 = 0$, and $10 \cdot 6 = 0$.

2. $(\mathbb{Z}_{13}, +, \cdot)$ has no zero divisors. Since 13 is prime, every positive integer less than 13 is relatively prime with 13. It follows that $(\mathbb{Z}_{13}, +, \cdot)$ is an integral domain. Similarly, $(\mathbb{Z}_2, +, \cdot), (\mathbb{Z}_3, +, \cdot),$ $(\mathbb{Z}_5, +, \cdot), (\mathbb{Z}_7, +, \cdot),$ and $(\mathbb{Z}_{11}, +, \cdot)$ are all integral domains. In general, if p is prime, then $(\mathbb{Z}_p, +, \cdot)$ is an integral domain.

Theorem 9.14: $(\mathbb{Z}_n, +, \cdot)$ is a field if and only if n is prime.

Proof: By Problem 10 in Problem Set 3, $(\mathbb{Z}_n, +, \cdot)$ is a commutative ring for all positive integers n. If n is not prime, then there are positive integers a and b with $0 < a, b < n$ such that $n = ab$. It follows that $[a]_n \cdot [b]_n = [ab]_n = [n]_n = [0]_n$. So, a and b are zero divisors. Therefore, $(\mathbb{Z}_n, +, \cdot)$ is not an integral domain. By part (iv) of Problem 7 from Problem Set 2, $(\mathbb{Z}_n, +, \cdot)$ is not a field.

Now, let n be prime and let $[a]_n \in \mathbb{Z}_n$ with $0 < a < n$. Since n is prime, $\gcd(a, n) = 1$. By Theorem 9.12, $[a]_n$ is not a zero divisor. Since $[a]_n \in \mathbb{Z}_n^*$ was arbitrary, $(\mathbb{Z}_n, +, \cdot)$ has no zero divisors, and therefore, it is an integral domain. By part 5 of Example 2.16, every finite integral domain is a field. So, $(\mathbb{Z}_n, +, \cdot)$ is a field. □

The next theorem is very helpful for performing calculations in the rings $(\mathbb{Z}_n, +, \cdot)$.

Theorem 9.15 (Fermat's Little Theorem): Let p be a prime number. Then for any integer a, $a^p \equiv_p a$.

Note that Fermat's Little Theorem is often stated in the following equivalent form: Let p be a prime number. If a is an integer not divisible by p, then $a^{p-1} \equiv_p 1$.

Proof: If $a \equiv_p 0$, then $a^p \equiv_p 0^p = 0$. Now, let a be an integer such that $a \not\equiv_p 0$. By Theorem 9.14, $(\mathbb{Z}_p, +, \cdot)$ is a field. So, $(\mathbb{Z}_p^*, \cdot)$ is a group of order $p - 1$. We have $\mathbb{Z}_p^* = \{[1]_p, [2]_p, \ldots, [p-1]_p\}$. If we multiply each element in $\mathbb{Z}_p^*$ by $[a]_p$, we get the same group: $\mathbb{Z}_p^* = \{[1a]_p, [2a]_p, \ldots, [(p-1)a]_p\}$. So, we have $[1]_p \cdot [2]_p \cdots [p-1]_p = [1a]_p \cdot [2a]_p \cdots [(p-1)a]_p$, or equivalently,

$$[1 \cdot 2 \cdots (p-1)]_p = [a \cdot 2a \cdots (p-1)a]_p = \left[(1 \cdot 2 \cdots (p-1))a^{p-1}\right]_p = [1 \cdot 2 \cdots (p-1)]_p \cdot [a^{p-1}]_p$$

So, we have $[1 \cdot 2 \cdots (p-1)]_p \cdot [1]_p = [1 \cdot 2 \cdots (p-1)]_p = [1 \cdot 2 \cdots (p-1)]_p \cdot [a^{p-1}]_p$. Since $(\mathbb{Z}_p^*, \cdot)$ is a group, the left cancellation law holds, and so, $[1]_p = [a^{p-1}]_p$. Multiplying each side of this last equation by $[a]_p$ yields $[a]_p = [a \cdot 1]_p = [a]_p \cdot [1]_p = [a]_p \cdot [a^{p-1}]_p = [a \cdot a^{p-1}]_p = [a^p]_p$. Therefore, $a^p \equiv_p a$. □

Note: There is actually a quicker proof of Fermat's Little Theorem. However, the tools necessary for that proof won't be developed until the next lesson. Nonetheless, I'll provide the argument now. It turns out that if G is a group of order n with identity e and $a \in G$, then $a^n = e$ (see Theorem 10.10). With this information, we can prove Fermat's Little Theorem as follows: If $a \not\equiv_p 0$ is an integer, then since $[a]_p \in \mathbb{Z}_p^*$ and $|\mathbb{Z}_p^*| = p - 1$, $[a^{p-1}]_p = [a]_p^{p-1} = [1]_p$. Therefore, as in the proof above, we have $[a^p]_p = [a]_p$, and so, $a^p \equiv_p a$.

Example 9.16: Let's use Fermat's Little Theorem to compute 2^{126} modulo 5. Well, Fermat's Little Theorem tells us that $2^5 \equiv_5 2$. Therefore,

$$2^{126} = 2^{125} \cdot 2^1 = 2^{5 \cdot 25} \cdot 2 = (2^5)^{25} \cdot 2 \equiv_5 2^{25} \cdot 2 \equiv_5 2^{5 \cdot 5} \cdot 2 \equiv_5 (2^5)^5 \cdot 2 \equiv_5 2^5 \cdot 2 \equiv_5 2 \cdot 2 \equiv_5 4.$$

Note: It's usually faster to apply Fermat's Little Theorem in the form $a^{p-1} \equiv_p 1$. Using this form, we have $2^4 \equiv_5 1$, and therefore, $2^{126} = 2^{124} \cdot 2^2 = 2^{4 \cdot 31} \cdot 4 = (2^4)^{31} \cdot 4 \equiv_5 1^{31} \cdot 4 = 1 \cdot 4 = 4$. The reason that this is preferable is simply because when we raise 1 to a power, we always get 1. In this way, tedious computations can often be avoided.

If $n \in \mathbb{Z}^+$ is not prime, then by Theorem 9.14, $(\mathbb{Z}_n, +, \cdot)$ is **not** a field. Since $(\mathbb{Z}_n, +, \cdot)$ is a commutative ring, it follows that $(\mathbb{Z}_n^*, \cdot)$ is **not** a group. We will now define a subset U_n of $\mathbb{Z}_n^*$ such that $(U_n, \cdot)$ always forms a group.

We define U_n to be the subset of $\mathbb{Z}_n^*$ consisting of all the elements in $\mathbb{Z}_n^*$ that have multiplicative inverses. For example, if n is prime, then $U_n = \mathbb{Z}_n^*$. In general, $(U_n, \cdot)$ is a group. Indeed, the product of two invertible elements is invertible $((rs)^{-1} = s^{-1}r^{-1})$, associativity follows from the associativity of multiplication in $\mathbb{Z}_n$, the identity $[1]_n$ is its own inverse, and the inverse of an invertible element is invertible $(a^{-1})^{-1} = a$).

Example 9.17:

1. $(\mathbb{Z}_5^*, \cdot)$ is already a group because 5 is prime. Therefore, $U_5 = \mathbb{Z}_5^*$.

2. The invertible elements of $(\mathbb{Z}_6^*, \cdot)$ are 1 and 5 (since $1 \cdot 1 = 1$ and $5 \cdot 5 = 1$, they are each their own inverses). To see that 2 and 4 are not invertible, simply observe that $2x$ and $4x$ are even for any integer x, while 1 is odd (because 1 more than a multiple of 6 is always odd). To see that 3 is not invertible, let's check all the possible products of 3 with elements of $\mathbb{Z}_6^*$. We have $3 \cdot 0 = 0$, $3 \cdot 1 = 3$, $3 \cdot 2 = 0$, $3 \cdot 3 = 3$, $3 \cdot 4 = 0$, and $3 \cdot 5 = 3$. In particular, there is no $x \in \mathbb{Z}_6^*$ such that $3x = 1$. It follows that $U_6 = \{1, 5\}$. Below is the multiplication table for U_6.

$\cdot$	1	5
1	1	5
5	5	1

In part 2 of Example 9.17, we did a lot more work than we had to in order to determine the invertible elements of $\mathbb{Z}_6^*$. You'll notice that the invertible elements turned out to be precisely the elements that are relatively prime with 6. This is not a coincidence, as we now show.

Theorem 9.18: $U_n = \{[a]_n \in \mathbb{Z}_n \mid \gcd(a, n) = 1\}$.

Proof: Suppose that $[a]_n \in U_n$, so that $[a]_n$ is invertible. Then there is $[b]_n \in U_n$ such that $[ab]_n = [a]_n \cdot [b]_n = [1]_n$. So, $ab \equiv_n 1$, or equivalently, $n | ab - 1$. Therefore, there is an integer k such that $ab - 1 = nk$, or equivalently, $ab - nk = 1$. By Theorem 8.18, $\gcd(a, n) = 1$.

Conversely, suppose that $\gcd(a, n) = 1$. By Theorem 8.18, there are integers m and k such that $ma + kn = 1$. So, $ma - 1 = (-k)n$. Therefore, $n | ma - 1$, and so, $ma \equiv_n 1$. So, $[m]_n \cdot [a]_n = [1]_n$. This shows that $[a]_n$ is invertible with inverse $[m]_n$. □

Example 9.19: Using Theorem 9.18, we can easily compute U_n for small values of n. For example, we have $U_9 = \{1, 2, 4, 5, 7, 8\}$ and $U_{18} = \{1, 5, 7, 11, 13, 17\}$. Note that U_9 and U_{18} have the same order, even though $9 \neq 18$.

For each $n \in \mathbb{Z}^+$, let $\Phi(n)$ be the number of positive integers less than n that are relatively prime to n. Then $\Phi: \mathbb{Z}^+ \to \mathbb{N}$ is a function known as **Euler's totient function** (or sometimes the **Euler phi function**). By Theorem 9.18, for each $n \in \mathbb{Z}^+$, the order of $(U_n, \cdot)$ is $\Phi(n)$.

Theorem 9.20: For any $a \in U_n$, $a^{\Phi(n)} \equiv_p 1$.

The proof of Theorem 9.20 is similar to the proof of Fermat's Little Theorem (Theorem 9.15), and so, I leave it to the reader. See Problem 17 below.

Example 9.21:

1. $(\mathbb{Z}_4^*, \cdot)$ is **not** a group. In fact, it's not even a semigroup because $2 \cdot 2 = 0$ in $\mathbb{Z}_4$ (and so, $\mathbb{Z}_4^*$ is not closed under multiplication). The positive integers less than 4 that are relatively prime with 4 are 1 and 3. It follows that $U_4 = \{1, 3\}$ and so, $(\{1, 3\}, \cdot)$ is a group, where the operation is multiplication modulo 4. In this group, we have $1 \cdot 1 = 1$ and $3 \cdot 3 = 1$, and so, 1 and 3 are each their own inverses. These computations also show that 1 and 3 are both elements of order 2. Below is the multiplication table for U_4.

$\cdot$	1	3
1	1	3
3	3	1

2. $U_{10} = \{1, 3, 7, 9\}$ and the multiplication table for U_{10} is shown below.

$\cdot$	1	3	7	9
1	1	3	7	9
3	3	9	1	7
7	7	1	9	3
9	9	7	3	1

    Let's compute the order of each element. Since $1^1 = 1$, $|1| = 1$. Now, $3^2 = 9$, $3^3 = 9 \cdot 3 = 7$, and $3^4 = 7 \cdot 3 = 1$. So, $|3| = 4$. We have $7^2 = 9$, $7^3 = 9 \cdot 7 = 3$, and $7^4 = 3 \cdot 7 = 1$. So, $|7| = 4$. Finally, $9^2 = 1$. So, $|9| = 2$. Observe that U_{10} is a cyclic group. Either 3 or 7 can be used as a generator. In this group, 1 and 9 are each their own inverse, whereas 3 and 7 are inverses of each other.

Solving Linear Congruences

We will now explore the question of how to solve the linear congruence $ax \equiv_n b$.

Example 9.22:

1. Let's solve the linear congruence $2x \equiv_3 1$. We have $2 \cdot 0 \equiv_3 0$, $2 \cdot 1 \equiv_3 2$, and $2 \cdot 2 \equiv_3 1$. It follows that 2 is the unique solution of this linear congruence.

    Perhaps it would be more correct to say that c is a solution of $2x \equiv_3 1$ if and only if $c \equiv_3 2$. For example, since $5 \equiv_3 2$, we see that $2 \cdot 5 \equiv_3 10 \equiv_3 1$, and so, 5 is also a solution of the linear congruence. As another example, since $-1 \equiv_3 2$, we see that $2(-1) \equiv_3 -2 \equiv_3 1$, and so, -1 is yet another solution of the linear congruence. In other words, the set of solutions of $2x \equiv_3 1$ is $[2]_3$. However, from now on, we will consider elements of the same equivalence class to be the same element. This identification allows us to say that the linear congruence $2x \equiv_3 1$ has the unique solution $x \equiv_3 2$.

    Note that saying that 2 is a solution of $2x \equiv_3 1$ is equivalent to saying that 2 is its own inverse in $U_3 = \mathbb{Z}_3^*$.

210

2. Let's solve the linear congruence $2x \equiv_4 1$. We have $2 \cdot 0 \equiv_4 0$, $2 \cdot 1 \equiv_4 2$, $2 \cdot 2 \equiv_4 0$, and $2 \cdot 3 \equiv_4 2$. It follows that $2x \equiv_4 1$ has **no solution**. This is equivalent to saying that 2 has no multiplicative inverse in $\mathbb{Z}_4^*$. The same computations show us that the linear congruence $2x \equiv_4 2$ has two solutions: 1 and 3. Similarly, the linear congruence $2x \equiv_4 0$ has the two solutions 0 and 2, whereas the linear congruence $2x \equiv_4 3$ has no solutions.

Based upon Example 9.22, it appears that some linear congruences have solutions and other do not. The next theorem tells us when we can expect a linear congruence to have a solution.

Theorem 9.23: Let $a, b \in \mathbb{Z}$ and let $n \in \mathbb{Z}^+$. The linear congruence $ax \equiv_n b$ has a solution for x if and only if $\gcd(a, n)$ is a divisor of b.

Proof: First assume that $ax \equiv_n b$ has a solution x. Then $n | ax - b$, and so, there is an integer k such that $ax - b = nk$, or equivalently, $ax - nk = b$. Since $\gcd(a, n)$ divides both a and n, this last equation shows that $\gcd(a, n)$ is a divisor of b.

Conversely, assume that $\gcd(a, n)$ is a divisor of b. By Theorem 8.18, there are integers j and k such that $\gcd(a, n) = ja + kn$. Since $\gcd(a, n) | b$, there is an integer t such that $b = \gcd(a, n) \cdot t$. So, $b = (ja + kn)t = a(tj) + n(tk)$. So, $a(tj) - b = n(-tk)$. It follows that $n | a(tj) - b$, and so, $x = tj$ is a solution of the linear congruence $ax \equiv_n b$. □

Note: If p is prime, and $a, b \in \mathbb{Z}^+$ are not multiples of p (or equivalently, $a, b \not\equiv_p 0$), then $\gcd(a, p) = 1$. It follows from Theorem 9.23 that for all $a, b \in U_p = \mathbb{Z}_p^*$, the linear congruence $ax \equiv_p b$ has a solution for x.

Example 9.24:

1. By Theorem 9.23 (and the Note following), the linear congruence $ax \equiv_3 b$ has a solution for all $a, b \not\equiv_3 0$, We saw in part 1 of Example 9.22 that $2x \equiv_3 1$ has the solution $x \equiv_3 2$. It's also easy to see that $2x \equiv_3 2$ has the solution $x \equiv_3 1$, $1x \equiv_3 1$ has the solution $x \equiv_3 1$, and $1x \equiv_3 2$ has the solution $x \equiv_3 2$.

2. Since $\gcd(2, 4) = 2$ and 2 is **not** a divisor of 1, by Theorem 9.23, the linear congruence $2x \equiv_4 1$ does **not** have a solution. We saw this in part 2 of Example 9.22 above. Similarly, the linear congruence $2x \equiv_4 3$ does not have a solution (because $\gcd(2, 4) = 2$ and 2 is not a divisor of 3).

   On the other hand, the linear congruence $3x \equiv_4 b$ has a solution for each of $b = 1, 2$, and 3 because $\gcd(3, 4) = 1$. What are these solutions?

3. Since $\gcd(6, 15) = 3$ and 3 is a divisor of 12, by Theorem 9.23, the linear congruence $6x \equiv_{15} 12$ has a solution. For example, $x \equiv_{15} 2$ is a solution because $6 \cdot 2 \equiv_{15} 12$.

   However, $x \equiv_{15} 7$ is another solution. Indeed, we have $6 \cdot 7 = 42 \equiv_{15} 12$. This solution is genuinely different from the solution $x \equiv_{15} 2$ (up to congruence modulo 15) because $2 \not\equiv_{15} 7$. There is also one more solution, namely $x \equiv_{15} 12$ (Check this!). Although $x \equiv_{15} 17$ is also a solution, it is not a "new" solution because $17 \equiv_{15} 2$. Up to congruence modulo 15, there are three distinct solutions: 2, 7, and 12. Note that $\gcd(6, 15) = 3$ and there are 3 distinct solutions, up to equivalence modulo 15. This is not a coincidence (see Problem 18 below).

Part 3 of Example 9.24 above (as well as part 2 of Example 9.22) shows that if a linear congruence has a solution, then it might **not** be unique.

Let me just clarify what we mean by unique here. We will say that c is a **unique solution** to the linear congruence $ax \equiv_n b$ if whenever d is another solution to the linear congruence, then $d \equiv_n c$. For example, we do not consider 2 and 5 to be different solutions to the linear congruence $2x \equiv_3 1$ because $2 \equiv_3 5$. However, 1 and 3 are different solutions to the linear congruence $2x \equiv_4 2$ because $1 \not\equiv_4 3$.

Theorem 9.23 gives us a simple condition for determining when a solution to a linear congruence exists. The next theorem gives us a simple condition that implies uniqueness.

Theorem 9.25: Let $a \in \mathbb{Z}$, let $n \in \mathbb{Z}^+$, and suppose that $\gcd(a, n) = 1$. Then for each $b \in \mathbb{Z}$, the linear congruence $ax \equiv_n b$ has a unique solution.

Proof: Suppose that $\gcd(a, n) = 1$. Then $\gcd(a, n)$ is a divisor of b (1 divides every integer), and so, by Theorem 9.23, $ax \equiv_n b$ has a solution. Now, suppose that c and d are both solutions of $ax \equiv_n b$. Then $ac \equiv_n ad$. By Theorem 9.18, $[a]_n$ is in the group U_n. It follows that $[a]_n$ has a multiplicative inverse in $\mathbb{Z}_n^*$. So, we have

$$[c]_n = [1]_n[c]_n = ([a]_n^{-1}[a]_n)[c]_n = [a]_n^{-1}([a]_n[c]_n) = [a]_n^{-1}([a]_n[d]_n)$$
$$= ([a]_n^{-1}[a]_n)[d]_n = [1]_n[d]_n = [d]_n.$$

By Theorem 3.14, $c \equiv_n d$, as desired. □

Notes: (1) Theorem 9.25 tells us that if a and n are relatively prime, then the linear congruence $ax \equiv_n b$ has a unique solution for all integers b.

(2) What if a and n are not relatively prime? There are two cases to consider. If $\gcd(a, n)$ is **not** a divisor of b, then by Theorem 9.23, the linear congruence $ax \equiv_n b$ has no solution.

Suppose $\gcd(a, n) \neq 1$ is a divisor of b. We know by Theorem 9.23 that the linear congruence $ax \equiv_n b$ has a solution. Let's call this solution c. Since $\gcd(a, n)$ is a divisor of both a and n, there are integers s and t such that $a = \gcd(a, n) \cdot s$ and $n = \gcd(a, n) \cdot t$. We will show that $c + t$ is another solution of $ax \equiv_n b$ that is **not** congruent to c modulo n.

First, let's show that $c + t$ is a solution. Since $a = \gcd(a, n) \cdot s$ and $n = \gcd(a, n) \cdot t$, we have $at = \gcd(a, n) \cdot st = sn$. So, $a(c + t) = ac + at = ac + sn \equiv_n ac \equiv_n b$, as desired.

Next, let's show that $c \not\equiv_n c + t$. Suppose toward contradiction that $c \equiv_n c + t$. Then $0 \equiv_n t$, and so, $n | t$. Therefore, $t = nr$ for some $r \in \mathbb{Z}$. So, we have $n = \gcd(a, n) \cdot t = \gcd(a, n) \cdot nr$. Therefore, $1 = \gcd(a, n) \cdot r$. So, $\gcd(a, n) = 1$, contrary to our original assumption.

(3) To summarize Notes 1 and 2 above, the linear congruence $ax \equiv_n b$ has a unique solution if and only if $\gcd(a, n) = 1$. If $\gcd(a, n) \neq 1$, then the linear congruence either has no solutions or more than one solution.

Example 9.26:

1. If $a \not\equiv_3 0$, then $\gcd(a, 3) = 1$. So, by Theorem 9.25, for each $b \in \mathbb{Z}$, the linear congruence $ax \equiv_n b$ has a unique solution. We saw this in part 1 of Example 9.24 for $b \not\equiv_3 0$.

   We also have that $ax \equiv_3 0$ has the unique solution $x \equiv_3 0$.

2. Since $\gcd(3, 4) = 1$, we see that $3x \equiv_4 b$ has a unique solution for all $b \in \mathbb{Z}$. The unique solution to $3x \equiv_4 0$ is $x \equiv_4 0$. The unique solution to $3x \equiv_4 1$ is $x \equiv_4 3$ The unique solution to $3x \equiv_4 2$ is $x \equiv_4 2$. Finally, The unique solution to $3x \equiv_4 3$ is $x \equiv_4 1$.

3. In part 2 of Example 9.22, we saw that the linear congruence $2x \equiv_4 2$ has two distinct solutions: 1 and 3. They are distinct in the sense that $1 \not\equiv_4 3$. However, we do have $1 \equiv_2 3$. So, although the solution 1 of the linear congruence $2x \equiv_4 2$ is **not** unique up to congruence modulo 4, it is unique up to congruence modulo 2. We get the 2 by dividing a, b, and n by $\gcd(a, n)$. In this case, since $\gcd(2, 4) = 2$ and $\gcd(2, 4)$ divides the right hand side (2 divides 2), we can "reduce" the linear congruence $2x \equiv_4 2$ by dividing each of 2, 2, and 4 by 2. This gives us the linear congruence $1x \equiv_2 1$, or equivalently, $x \equiv_2 1$. This linear congruence does have a unique solution, namely $x \equiv_2 1$, and this solution is also a solution of the original linear congruence $2x \equiv_4 2$. In Theorem 9.27 below, we will see that this procedure of "reducing" the linear congruence $ax \equiv_n b$ to one with a unique solution always works when $\gcd(a, n)$ divides b.

Theorem 9.27: Let $a, b \in \mathbb{Z}$ and let $n \in \mathbb{Z}^+$. If $ax \equiv_n b$ has a solution, then this solution is unique up to congruence modulo t, where $n = \gcd(a, n) \cdot t$.

Proof: Since $\gcd(a, n)$ divides a and n, there are integers s and t such that $a = \gcd(a, n) \cdot s$ and $n = \gcd(a, n) \cdot t$. Since $ax \equiv_n b$ has a solution, by Theorem 9.23, $\gcd(a, n)$ divides b, and so, there is an integer u such that $b = \gcd(a, n) \cdot u$. Now, $ax \equiv_n b \Leftrightarrow n | ax - b \Leftrightarrow$ there is $k \in \mathbb{Z}$ such that $ax - b = nk \Leftrightarrow$ there is $k \in \mathbb{Z}$ such that $\gcd(a, n) \cdot sx - \gcd(a, n) \cdot u = \gcd(a, n) \cdot tk \Leftrightarrow$ there is $k \in \mathbb{Z}$ such that $sx - u = tk \Leftrightarrow t | sx - u \Leftrightarrow sx \equiv_t u$. To summarize, we have

$$ax \equiv_n b \text{ if and only if } sx \equiv_t u.$$

Since $\gcd(s, t) = 1$ (see the Note below), by Theorem 9.25, $sx \equiv_t u$ has a unique solution. So, suppose that c and d are both solutions of $ax \equiv_n b$. Then c and d are both solutions of $sx \equiv_t u$. Since $sx \equiv_t u$ has a unique solution, $c \equiv_t d$. So, a solution of $ax \equiv_n b$ is unique up to congruence modulo t. □

Note: To see that $\gcd(s, t) = 1$, let d be a common divisor of s and t. Then there are integers v and w such that $s = dv$ and $t = dw$. Multiplying each side of each equation by $\gcd(a, n)$ gives us $\gcd(a, n) \cdot s = d \cdot \gcd(a, n) \cdot v$ and $\gcd(a, n) \cdot t = d \cdot \gcd(a, n) \cdot w$. By direct substitution, we have $a = \gcd(a, n) \cdot s = d \cdot \gcd(a, n) \cdot v$ and $n = \gcd(a, n) \cdot t = d \cdot \gcd(a, n) \cdot w$. It follows that $d \cdot \gcd(a, n)$ is a common divisor of a and n. Since $\gcd(a, n)$ is the greatest common divisor of a and n, we have $d = 1$ (if d were bigger than 1, then $d \cdot \gcd(a, n)$ would be a common divisor of a and n that is bigger than $\gcd(a, n)$, which of course is impossible). So, the only common divisor of s and t is 1. Thus, $\gcd(s, t) = 1$, as desired.

Example 9.28: Let's solve the linear congruence $15x \equiv_{21} 18$. First observe that $\gcd(15, 21) = 3$ and $3 | 18$. We have $15 = 3 \cdot 5$, $18 = 3 \cdot 6$, and $21 = 3 \cdot 7$. So, we solve $5x \equiv_7 6$. Since $\gcd(5, 7) = 1$, by Theorem 9.25, this last linear congruence has a unique solution. $5 \cdot 4 = 20 \equiv_7 6$. So, $x \equiv_7 4$ is the unique solution of $5x \equiv_7 6$. By Theorem 9.27, this is also the unique solution of $15x \equiv_{21} 18$ (up to congruence modulo 7).

Notes: (1) The conditions $\gcd(a, n) \neq 1$ and $\gcd(a, n) | b$ guarantee that we can "reduce" the linear congruence $ax \equiv_n b$ by "dividing out" $\gcd(a, n)$. In Example 9.28, since $\gcd(15, 21) = 3$ and $3|18$, we can divide each of 15, 18, and 21 by 3 to get a new linear congruence with a unique solution that is also a solution of the original linear congruence (and this solution is unique, up to the "new" modulus).

(2) In Example 9.28, 4 is still a solution of the original linear congruence if we consider it modulo 21. However, it is **not** the unique solution modulo 21. We can get the complete set of solutions up to the original modulus by adding multiples of the new modulus. Here, we have $4 + 7 = 11$ is also a solution. Let's check: $15 \cdot 11 = 165 \equiv_{21} 18$. We also have that $4 + 2 \cdot 7 = 4 + 14 = 18$ is another solution. Let's check again: $15 \cdot 18 = 270 \equiv_{21} 18$. Although $4 + 3 \cdot 7 = 4 + 21 = 25$ is also a solution, it is not a "new" solution because $25 \equiv_{21} 4$. So, the set of solutions of the linear congruence $15x \equiv_{21} 18$ up to congruence modulo 21 is $\{4, 11, 18\}$.

(3) Notice that in Example 9.28, $\gcd(15, 21) = 3$ and there are exactly 3 solutions to the linear congruence $15x \equiv_{21} 18$ up to congruence modulo 21. This is not a coincidence. See Problem 18 below.

We now turn our attention to systems of linear congruences. Let's begin with a small system consisting of just two linear congruences. Such a system looks like this:

$$ax \equiv_n b,$$
$$cx \equiv_m d.$$

A solution of this system of linear congruences is an integer x that satisfies both equations simultaneously. To solve this system, we first use Theorem 9.23 to determine if each of these linear congruences has a solution. If not, then the system has no solution as well. If so, then we can solve each linear congruence, as we did in Example 9.28. Let's say that $ax \equiv_n b$ has the unique solution $x \equiv_j c_1$ and $cx \equiv_m d$ has the unique solution $x \equiv_k c_2$. We have now reduced the problem to the following question. Is there a solution to the following system of "simple" linear congruences?

$$x \equiv_j c_1,$$
$$x \equiv_k c_2.$$

Note that j and k may be different from n and m.

Example 9.29:

1. Consider the following system of linear congruences.

$$x \equiv_6 3,$$
$$x \equiv_{10} 5.$$

Direct substitution shows that $x = 15$ is a solution to this system. It's also easy to see that $x = 15$ is the least positive solution (the only smaller positive solution of $x \equiv_{10} 5$ is $x = 5$, but this is **not** a solution of $x \equiv_6 3$). Note that if we add any multiple of 30 to the solution $x = 15$, we get another solution. So the complete list of positive solutions is 15, 45, 75, 105, ... Where did this number 30 come from? Well, 30 is the least common multiple of 6 and 10. In fact, the set of all solutions of this system of linear congruences is $[15]_{30}$.

2. Consider the following system of linear congruences.

$$x \equiv_6 1,$$
$$x \equiv_{10} 2.$$

This system has no solutions. The solutions of $x \equiv_{10} 2$ are $\{..., -18, -8, 2, 12, 22, 32\}$. In particular, every solution of this linear congruence is even. On the other hand, every solution of $x \equiv_6 1$ is odd.

3. Consider the following system of linear congruences.

$$3x \equiv_7 4,$$
$$2x \equiv_{70} 26.$$

We first solve each of these linear congruences individually. Since $\gcd(3, 7) = 1$, we know that $3x \equiv_7 4$ has a unique solution modulo 7. Direct substitution shows that $3 \cdot 6 = 18 \equiv_7 4$. So, the unique solution is $x \equiv_7 6$.

Since $\gcd(2, 70) = 2$ and 2 divides 26, we divide each of 2, 26, and 70 by 2 to transform the linear congruence $2x \equiv_{70} 26$ into the linear congruence $x \equiv_{35} 13$. Since $\gcd(1, 35) = 1$, this linear congruence has the unique solution $x \equiv_{35} 13$ and this is also the unique solution of $2x \equiv_{70} 26$ up to congruence modulo 35.

We have now reduced the original problem to solving the following system of linear congruences.

$$x \equiv_7 6,$$
$$x \equiv_{35} 13.$$

Direct substitution shows that $x = 13$ is a solution to this system. Since $\text{lcm}(7, 35) = 35$, we see that the complete solution set of this system is $[13]_{35}$.

We are now ready to precisely describe when a "simple" system of two linear congruences has a solution.

Theorem 9.30: Let $n_1, n_2 \in \mathbb{Z}^+$, let $c_1, c_2 \in \mathbb{Z}$, let $d = \gcd(n_1, n_2)$, and consider the following system of linear congruences.

$$x \equiv_{n_1} c_1,$$
$$x \equiv_{n_2} c_2.$$

This system has a solution if and only if $c_1 \equiv_d c_2$.

Proof: First assume that x is a solution of the system. Then $n_1 | x - c_1$ and $n_2 | x - c_2$. So, there are integers j and k such that $x - c_1 = n_1 j$ and $x - c_2 = n_2 k$. Mutiplying the first equation by -1 gives us $c_1 - x = -n_1 j$. We now add this equation to the second equation to get $c_1 - c_2 = n_2 k - n_1 j$. Since d divides both n_1 and n_2, $d | c_1 - c_2$. Therefore, $c_1 \equiv_d c_2$.

Conversely, assume that $c_1 \equiv_d c_2$. Then $d | c_1 - c_2$, and so, there is an integer k such that $c_1 - c_2 = dk$. By Theorem 8.18, there are integers a and b such that $d = an_1 + bn_2$. So, we have

$$c_1 - c_2 = dk = (an_1 + bn_2)k = akn_1 + bkn_2.$$

So, $c_1 - akn_1 = c_2 + bkn_2$. If we let $x = c_1 - akn_1 = c_2 + bkn_2$, then we have $x - c_1 = -akn_1$ and $x - c_2 = bkn_2$. So, $n_1 | x - c_1$ and $n_2 | x - c_2$. Therefore, $x \equiv_{n_1} c_1$ and $x \equiv_{n_2} c_2$. □

Example 9.31:

1. Consider the following system of linear congruences from part 1 of Example 9.29.

$$x \equiv_6 3,$$
$$x \equiv_{10} 5.$$

   If we let $d = \gcd(3, 5) = 1$, we see that $3 \equiv_1 5$ (this is trivial because 1 divides every integer). It follows from Theorem 9.30 that this system has a solution. We found that solution in part 1 of Example 9.29.

2. Consider the following system of linear congruences.

$$x \equiv_6 1,$$
$$x \equiv_{10} 2.$$

   In part 2 of Example 9.29, we saw that this system has no solution. We can also use Theorem 9.30 to get this result. If we let $d = \gcd(6, 10) = 2$, we see that $1 \not\equiv_2 2$. It follows from Theorem 9.30 that this system does **not** have a solution.

3. Consider the following system of linear congruences.

$$x \equiv_7 6,$$
$$x \equiv_{35} 13.$$

   If we let $d = \gcd(7, 35) = 7$, we see that $6 \equiv_7 13$ (because $13 - 6 = 7$). It follows from Theorem 9.30 that this system has a solution. We found that solution in part 3 of Example 9.29.

Example 9.29 hinted at how we can use a least common multiple to determine the entire solution set of a system of two linear congruences. We make this precise in the next theorem.

Theorem 9.32: Let $n_1, n_2 \in \mathbb{Z}^+$, let $c_1, c_2 \in \mathbb{Z}$, let $m = \text{lcm}(n_1, n_2)$, and consider the following system of linear congruences.

$$x \equiv_{n_1} c_1,$$
$$x \equiv_{n_2} c_2.$$

If this system has a solution, then it has a unique solution up to congruence modulo m.

Proof: Let x and y be solutions to the system of linear congruences. So, $x \equiv_{n_1} c_1$, $x \equiv_{n_2} c_2$, $y \equiv_{n_1} c_1$, and $y \equiv_{n_2} c_2$. Since $\equiv_{n_1}$ and $\equiv_{n_2}$ are equivalence relations, $x \equiv_{n_1} y$ and $x \equiv_{n_2} y$. So, $n_1 | x - y$ and $n_2 | x - y$. By Problem 19 in Problem Set 8, $m | x - y$. Therefore, $x \equiv_m y$. □

In the special case where n_1 and n_2 are relatively prime, then by Theorem 8.26, $\gcd(n_1, n_2) = 1$ and $\text{lcm}(n_1, n_2) = n_1 n_2$. In this case, by Theorem 9.32, the pair of linear congruences $x \equiv_{n_1} c_1$, $x \equiv_{n_2} c_2$ always has a unique solution modulo $n_1 n_2$. This is a special case of an important theorem called the Chinese Remainder Theorem.

Theorem 9.33 (Baby Chinese Remainder Theorem): Let $n_1, n_2 \in \mathbb{Z}^+$ with $\gcd(n_1, n_2) = 1$, let $c_1, c_2 \in \mathbb{Z}$, and consider the following system of linear congruences.

$$x \equiv_{n_1} c_1,$$
$$x \equiv_{n_2} c_2.$$

This system has a unique solution up to congruence modulo $n_1 n_2$.

Proof: Since $\gcd(n_1, n_2) = 1$ and $c_1 \equiv_1 c_2$ (because 1 divides **every** integer), by Theorem 9.30, the system has a solution. By Theorem 8.26, $\text{lcm}(n_1, n_2) = n_1 n_2$. So, by Theorem 9.32, the system has a unique solution up to congruence modulo $n_1 n_2$. □

We can easily extend Theorem 9.33 to any finite number of linear congruences by a simple application of the Principle of Mathematical Induction. This more general result is known as the **Chinese Remainder Theorem**. The proof is straightforward (all the hard work was already done) and therefore, I leave it to the reader. See Problem 13 below.

Theorem 9.34 (Chinese Remainder Theorem): Let $n_1, n_2, \ldots, n_k$ be pairwise relatively prime integers, let $c_1, c_2, \ldots, c_k \in \mathbb{Z}$, and consider the following system of linear congruences.

$$x \equiv_{n_1} c_1,$$
$$x \equiv_{n_2} c_2,$$
$$\vdots \quad \vdots \quad \vdots$$
$$x \equiv_{n_k} c_k.$$

This system has a unique solution up to congruence modulo $n_1 n_2 \cdots n_k$.

Problem Set 9

Full solutions to these problems are available for free download here:
www.SATPrepGet800.com/AAFBTDW

LEVEL 1

1. Find the order of 3 in $(\mathbb{Z}_4, +)$.

2. List the elements of U_{11}, U_{15}, and U_{20}.

3. Find the inverse of each element of U_9.

4. Draw the multiplication table for U_{28}.

LEVEL 2

5. Find the order of 32 in $(\mathbb{Z}_{40}, +)$.

6. Let $(H, +)$ be a subgroup of $(\mathbb{Z}, +)$. Prove that there is $n \in H$ such that $H = n\mathbb{Z}$.

7. Compute 128^{129} modulo 17.

8. Provide a counterexample for the following statement: For each $n \in \mathbb{Z}^+$, U_n is a cyclic group.

LEVEL 3

9. Find the order of $(123)(45)(6789)$ in S_{10}.

10. Find all subgroups of $(\mathbb{Z}_{30}, +)$.

11. Find all generators of $(\mathbb{Z}_{75}, +)$.

12. Let G be a group and let $a \in G$. Prove each of the following:
    (i) $|a^{-1}| = |a|$.
    (ii) $|a| = 1$ if and only if $a = e$.
    (iii) The order of any conjugate of a is equal to the order of a.
    (iv) If p is prime, $a \neq e$, and $a^p = e$, then $|a| = p$.

13. Prove the Chinese Remainder Theorem.

LEVEL 4

14. Let G be a group and let $a, b \in G$. Prove that $|ab| = |ba|$.

15. Find all subgroups of $\mathbb{Z}_5 \times \mathbb{Z}_3$.

16. Let p be a prime number, let $n \in \mathbb{Z}^+$, and let Φ be Euler's totient function. Prove that $\Phi(p^n) = p^n - p^{n-1}$. Use this result to compute $\Phi(1024)$ and $\Phi(625)$.

17. Let $n \in \mathbb{Z}^+$ and $[a]_n \in U_n$. Prove that $[a]_n^{\Phi(n)} \equiv_n [1]_n$, where Φ is Euler's totient function.

LEVEL 5

18. Let $a, b \in \mathbb{Z}$, let $n \in \mathbb{Z}^+$, let $d = \gcd(a, n)$ and suppose that $d \mid b$. Prove that the linear congruence $ax \equiv_n b$ has exactly d solutions, up to congruence modulo n.

19. Let $a, b \in \mathbb{Z}^+$ with $\gcd(a, b) = 1$. Prove each of the following:
    (i) $(\mathbb{Z}_{ab}, +, \cdot) \cong (\mathbb{Z}_a, +, \cdot) \times (\mathbb{Z}_b, +, \cdot)$.
    (ii) $U_{ab} \cong U_a \times U_b$.
    (iii) If $n \in \mathbb{Z}^+$ and $n = p_1^{m_1} \cdot p_2^{m_2} \cdots p_k^{m_k}$ is the prime factorization of n and Φ is Euler's totient function, then $\Phi(n) = (p_1^{m_1} - p_1^{m_1-1})(p_1^{m_2} - p_1^{m_2-1}) \cdots (p_1^{m_k} - p_1^{m_k-1})$.
    (iv) $\Phi(1{,}047{,}816) = 299{,}376$.

20. Let G be a group, let $m, n \in \mathbb{Z}^+$, and let $a \in G$ with $|a| = n$. Prove that $\gcd(m, n) = 1$ if and only if $|a^m| = n$.

21. Let G be a group, let $m, n \in \mathbb{Z}^+$, and let $a \in G$ with $|a| = n$. Prove that $m \cdot |a^m| = \mathrm{lcm}(m, n)$.

CHALLENGE PROBLEMS

22. Determine necessary and sufficient conditions on $n \in \mathbb{Z}^+$ for U_n to be a cyclic group.

23. Let n_1, n_2, n_3 be pairwise relatively prime integers and let Φ be Euler's totient function. Prove that $(n_1 n_2)^{\Phi(n_3)} + (n_1 n_3)^{\Phi(n_2)} + (n_2 n_3)^{\Phi(n_1)} \equiv_{n_1 n_2 n_3} 1$.

LESSON 10
QUOTIENTS

Cosets

Let G be a group, let H be a subgroup of G, and let $x \in G$. We define the **left coset** of H in G, written xH, and the **right coset** of H in G, written Hx, as follows:

$$xH = \{xy \mid y \in H\} \qquad Hx = \{yx \mid y \in H\}$$

Note: In the special case where the group operation is addition, we will use the notation $x + H$ and $H + x$ for left and right cosets of H in G, respectively. In this case, we have

$$x + H = \{x + y \mid y \in H\} \qquad H + x = \{y + x \mid y \in H\}$$

Example 10.1:

1. Let $G = \mathbb{Z}$ and $H = 2\mathbb{Z} = \{\ldots, -6, -4, -2, 0, 2, 4, 6, \ldots\}$. There are two left cosets of H in G.

   $$0 + H = \{0 + 2k \mid k \in \mathbb{Z}\} = \{2k \mid k \in \mathbb{Z}\} = \{\ldots, -6, -4, -2, 0, 2, 4, 6, \ldots\} = H = 2\mathbb{Z}$$
   $$1 + H = \{1 + 2k \mid k \in \mathbb{Z}\} = \{2k + 1 \mid k \in \mathbb{Z}\} = \{\ldots, -5, -3, -1, 1, 3, 5, 7, \ldots\} = 2\mathbb{Z} + 1$$

   If m is even, then $m + H = 2\mathbb{Z} = 0 + H$ and if m is odd, then $m + H = 2\mathbb{Z} + 1 = 1 + H$.

   For this example, the right cosets are the same as the left cosets: $H + 0 = H = 2\mathbb{Z}$ and $H + 1 = 2\mathbb{Z} + 1 = H + 1$. This will happen whenever the group is commutative.

2. Let $G = \mathbb{Z}$, let $n \in \mathbb{Z}^+$, and let $H = n\mathbb{Z} = \{\ldots, -3n, -2n, -n, 0, n, 2n, 3n, ,\ldots\}$. There are n left (or equivalently, right) cosets of H in G.

   $$0 + H = \{0 + nk \mid k \in \mathbb{Z}\} = \{nk \mid k \in \mathbb{Z}\} = \{\ldots, -3n, -2n, -n, 0, n, 2n, 3n, ,\ldots\} = H = n\mathbb{Z}$$
   $$1 + H = \{1 + nk \mid k \in \mathbb{Z}\} = \{nk + 1 \mid k \in \mathbb{Z}\} = n\mathbb{Z} + 1$$
   $$2 + H = \{2 + nk \mid k \in \mathbb{Z}\} = \{nk + 2 \mid k \in \mathbb{Z}\} = n\mathbb{Z} + 2$$
   $$\vdots \qquad \vdots \qquad \vdots$$
   $$(n-1) + H = \{(n-1) + nk \mid k \in \mathbb{Z}\} = \{nk + (n-1) \mid k \in \mathbb{Z}\} = n\mathbb{Z} + (n-1)$$

   Note that we have

   $$n + H = \{n + nk \mid k \in \mathbb{Z}\} = \{\ldots, -3n, -2n, -n, 0, n, 2n, 3n, ,\ldots\} = \{nk \mid k \in \mathbb{Z}\} = 0 + H.$$

   So, the coset $n + H$ is equal to the coset $0 + H$. Similarly, the coset $(n + 1) + H$ is equal to the coset $1 + H$, and so on.

3. Let $G = S_3 = \{(1), (12), (13), (23), (123), (132)\}$ (see Example 4.12) and let $H = \{(1), (12)\}$. By Problem 1 in Problem Set 6, $H \leq G$. The three left cosets of H in G are

   $$H = \{(1), (12)\} \qquad (13)H = \{(13), (123)\} \qquad (23)H = \{(23), (132)\}$$

   The three right cosets of H in G are

   $$H = \{(1), (12)\} \qquad H(13) = \{(13), (132)\} \qquad H(23) = \{(23), (123)\}$$

Notice that in this case the left cosets are **not** the same as the right cosets. However, it is worth noting that (i) each coset has the same number of elements and (ii) there are the same number of left and right cosets. See Theorems 10.6 and 10.7 below

Also notice that each coset can be represented by either element in the coset. For example, $H = (1)H = \{(1), (12)\}$ is the same as $(12)H$. Similarly, $(13)H = \{(13), (123)\}$ is the same as $(123)H$. The reader may want to check that this is true for all five cosets (there are three left cosets and three right cosets, but H is both a left and right coset).

4. Let $G = U_{28} = \{1, 3, 5, 9, 11, 13, 15, 17, 19, 23, 25, 27\}$ (recall that U_{28} is the group of positive integers relatively prime with 28; the group operation is multiplication) and let $H = \langle 9 \rangle = \{1, 9, 25\}$. There are three left (or equivalently, right) cosets of H in G.

$$H = \{1, 9, 25\} \qquad 3H = \{3, 19, 27\} \qquad 5H = \{5, 13, 17\} \qquad 11H = \{11, 15, 23\}$$

Once again, notice that each coset can be represented by either element in the coset. For example, $H = 1H = \{1, 9, 25\}$ is the same as $9H$. Indeed, $9 \cdot 1 = 9$, $9 \cdot 9 = 25$, and $9 \cdot 25 = 1$. The reader may want to check that $25H = H$ too and that each of the other cosets can be represented by each of its elements.

5. Let $G = \mathbb{R}$ and $H = \mathbb{Z}$. Then G is a group under addition and H is a subgroup of G. The cosets of $\mathbb{Z}$ in $\mathbb{R}$ have the form $x + \mathbb{Z}$ for each $x \in \mathbb{R}$. In this case, there are uncountably many cosets. Let's look at a few examples:

$$0 + \mathbb{Z} = \mathbb{Z} = \{\ldots, -4, -3, -2, -1, 0, 1, 2, 3, 4, \ldots\}$$

$$\tfrac{1}{2} + \mathbb{Z} = \left\{\ldots, -\tfrac{7}{2}, -\tfrac{5}{2}, -\tfrac{3}{2}, -\tfrac{1}{2}, \tfrac{1}{2}, \tfrac{3}{2}, \tfrac{5}{2}, \tfrac{7}{2}, \tfrac{9}{2}, \ldots\right\}$$

$$0.1 + \mathbb{Z} = \{\ldots, -3.9, -2.9, -1.9, -0.9, 0.1, 1.1, 2.1, 3.1, 4.1, \ldots\}$$

$$\tfrac{\sqrt{2}}{2} + \mathbb{Z} = \left\{\ldots, \tfrac{\sqrt{2}}{2} - 3, \tfrac{\sqrt{2}}{2} - 2, \tfrac{\sqrt{2}}{2} - 1, \tfrac{\sqrt{2}}{2}, \tfrac{\sqrt{2}}{2} + 1, \tfrac{\sqrt{2}}{2} + 2, \tfrac{\sqrt{2}}{2} + 3, \ldots\right\}$$

Notice that for each of these cosets, I chose a representative that was between 0 and 1. Every coset will have a representative x such that x is between any two consecutive integers we choose. So, we may as well always choose a representative x such that $0 \leq x < 1$. We see that the complete set of cosets of $\mathbb{Z}$ in $\mathbb{R}$ is $\{x + \mathbb{Z} \mid 0 \leq x < 1\}$.

Notes: (1) We usually use additive notation when a group is commutative. In this case, each left coset is equal to the corresponding right coset. To see this, note that $z \in x + H$ if and only if there is $y \in H$ with $z = x + y$ if and only if there is $y \in H$ with $z = y + x$ if and only if $z \in H + x$. Therefore, for commutative groups, we can simply talk about **cosets**, since it does not matter whether we think of them as left or right cosets.

(3) The definition of coset can be modified slightly to make sense for rings, fields, vector spaces and modules. If X is one of these structures and H is a subgroup of the additive group $(X, +)$, then we can define the coset $x + H$ (or $H + x$) by $x + H = \{x + y \mid y \in H\}$.

Theorem 10.2: Let G be a group, let H be a subgroup of G, and let $x, y \in G$. If $x \in yH$, then $xH = yH$.

Proof: Let G be a group, $H \leq G$, and let $x \in yH$. Then there is $w \in H$ such that $x = yw$.

First assume that $z \in xH$. Then there is $v \in H$ such that $z = xv$. Since $x = yw$, $z = (yw)v = y(wv)$. Since $w, v \in H$ and H is closed under the group operation (because H is itself a group), $wv \in H$. It follows that $z = y(wv) \in yH$. Since $z \in xH$ was arbitrary, we have $xH \subseteq yH$.

For the converse, first note that since $x = yw$, we have $y = ye = y(ww^{-1}) = (yw)w^{-1} = xw^{-1}$. Now, let $z \in yH$. Then there is $v \in H$ such that $z = yv = (xw^{-1})v = x(w^{-1}v) \in xH$ (once again, we see that $w^{-1}v \in H$ because H is itself a group). Since $z \in yH$ was arbitrary, we have $yH \subseteq xH$.

Since $xH \subseteq yH$ and $yH \subseteq xH$, we have $xH = yH$. $\square$

Notes: (1) If the group operation is addition, the last sentence of Theorem 10.2 becomes "If $x \in y + H$, then $x + H = y + H$."

(2) The right coset version of Theorem 10.2 is "If $x \in Hy$, then $Hx = Hy$." The argument is very similar to the proof given above. The dedicated reader should write out the details.

Example 10.3:

1. Let $G = \mathbb{Z}$ and $H = 5\mathbb{Z} = \{\ldots, -15, -10, -5, 0, 5, 10, 15, \ldots\}$. Let's first look at the coset $0 + H = H = 5\mathbb{Z}$. Since $5 \in H$, we have $5 + H = 0 + H = H$. Similarly, $10 + H = H$ and in general, for each $n \in \mathbb{Z}$, $5n + H = H$.

   Next, let's look at the coset $1 + H = 5\mathbb{Z} + 1 = \{\ldots, -14, -9, -4, 1, 6, 11, 16, \ldots\}$. Since $6 \in 1 + H$, we have $6 + H = 1 + H = 5\mathbb{Z} + 1$. Similarly, $-4 + H = 1 + H$ and in general, for each $n \in \mathbb{Z}$, $(5n + 1) + H = 1 + H$.

   Note once again that since $\mathbb{Z}$ is commutative, each right coset is the same as the corresponding left coset. For example, $H + 6 = 6 + H = 5\mathbb{Z} + 1$.

2. Let $G = S_3 = \{(1), (12), (13), (23), (123), (132)\}$ and $H = \{(1), (12)\}$, as in part 3 of Example 10.1. Let's look at the left coset $(13)H = \{(13), (123)\}$. Since $(123) \in (13)H$, by Theorem 10.2, $(123)H = (13)H = \{(13), (123)\}$.

We will now show that the left cosets of H in G form a partition of G (and the same can be said about the right cosets).

Theorem 10.4: Let G be a group and let H be a subgroup of G. Then $\{xH \mid x \in G\}$ is a partition of G.

Proof: We first show that each coset of H in G is nonempty. Let $x \in G$. Since G is a group, it has an identity e. Since H is a subgroup of G, $e \in H$. It follows that $x = xe \in xH$. So, $xH \neq \emptyset$.

We next show that each coset of H in G is a subset of G. Let $x \in G$ and let $y \in xH$. Then there is $z \in H$ with $y = xz$. Since H is a subgroup of G, $z \in G$. Since $x, z \in G$ and G is closed under the group operation, $y = xz \in G$. Since $y \in xH$ was arbitrary, $xH \subseteq G$.

Now, we show that any two cosets are either disjoint or equal. Let $x, y \in G$ and suppose that xH and yH are **not** disjoint. Then there is $z \in xH \cap yH$. Since $z \in xH$, there is $w \in H$ with $z = xw$. Since $z \in yH$, there is $v \in H$ with $z = yv$. So, $xw = yv$. It follows that

$$x = xe = x(ww^{-1}) = (xw)w^{-1} = (yv)w^{-1} = y(vw^{-1}).$$

Since $w \in H$ and H is has the inverse property (because H is a group), $w^{-1} \in H$. Since $v, w^{-1} \in H$ and H is closed under the group operation (again, because H is a group), $vw^{-1} \in H$. Therefore, we have $x = y(vw^{-1}) \in yH$. By Theorem 10.2, $xH = yH$.

Finally, we need to show that every element of G is in some coset of H in G. Let $x \in G$. Then $x = xe$. Since H is a subgroup of G, $e \in H$. So, $x = xe \in xH$. □

Notes: (1) By a similar argument, $\{Hx \mid x \in G\}$ is a partition of G.

(2) In the second to last paragraph of the proof, we could have used Problem 5 from Problem Set 6 to give a shorter argument that $vw^{-1} \in H$ as follows: since $v, w \in H$ and $H \leq G$, $vw^{-1} \in H$.

(3) You may want to take another look at Example 10.1. Observe that for each group and subgroup in that example, the cosets form a partition of the given group G.

If G is a group and H is a subgroup of G, then by Theorem 10.4 and Theorem 3.18 (and its proof), we can define an equivalence relation $\equiv_H$ on G by $x \equiv_H y$ if and only if x and y are in the same left coset of H in G (or x and y are in the same right coset of H in G).

Theorem 10.5: Let G be a group, let $H \leq G$, and let $x, y \in G$. The following are equivalent:

1. $x \equiv_H y$
2. $x^{-1}y \in H$
3. $y \in xH$
4. $xH = yH$

Proof: (1 → 2) Assume that $x \equiv_H y$. Then x and y are in the same left coset of H in G. So, let $z \in G$ with $x, y \in zH$. Then there are $w, v \in H$ with $x = zw$ and $y = zv$. Since $w \in H$ and H has the inverse property, $w^{-1} \in H$. Since $w^{-1}, v \in H$ and H is closed under the group operation, $w^{-1}v \in H$. It follows that $x^{-1}y = (zw)^{-1}(zv) = w^{-1}z^{-1}zv$ (by part (i) of Problem 15 in Problem Set 2) $= w^{-1}v \in H$.

(2 → 3) Assume that $x^{-1}y \in H$. Then there is $z \in H$ such that $x^{-1}y = z$. Therefore, we have $y = ey = (xx^{-1})y = x(x^{-1}y) = xz \in xH$.

(3 → 4) This is Theorem 10.2.

(4 → 1) Assume that $xH = yH$. Since $x = xe$ and $e \in H$ (because H is a subgroup of G), $x \in xH$. Similarly, $y \in yH$. Since $xH = yH$, $y \in xH$. So, x and y are in the same left coset of H in G. Therefore, $x \equiv_H y$. □

Notes: (1) As a simple example, consider $G = S_3 = \{(1), (12), (13), (23), (123), (132)\}$ and the subgroup $H = \{(1), (12)\}$, as defined in part 3 of Example 10.1. Let's consider the left coset $(13)H = \{(13), (123)\}$. If we let $x = (13)$ and $y = (123)$, then we have $x \equiv_H y$ because they are in the same left coset of H in G. We have $x^{-1}y = (13)^{-1}(123) = (13)(123) = (12)$, which is in H. We also see that $y = (123) \in (13)H = xH$. Finally, we have $(123)(1) = (123)$ and $(123)(12) = (13)$, and so, $yH = (123)H = \{(123), (13)\} = \{(13), (123)\} = (13)H = xH$.

(2) The reader is encouraged to state and prove the analogous result for right cosets. In this case, clause 2 should be replaced with $yx^{-1} \in H$.

(3) If the group G is commutative, then we would use additive notation and the statements in Theorem 10.5 would look as follows:

1. $x \equiv_H y$
2. $y - x \in H$
3. $y \in x + H$
4. $x + H = y + H$

(4) As an example of Note 3, consider $G = \mathbb{R}$ and $H = \mathbb{Z} = \{\ldots, -4, -3, -2, -1, 0, 1, 2, 3, 4, \ldots\}$, as defined in part 5 of Example 10.1. Let's consider the coset $\frac{1}{2} + \mathbb{Z} = \left\{\ldots, -\frac{7}{2}, -\frac{5}{2}, -\frac{3}{2}, -\frac{1}{2}, \frac{1}{2}, \frac{3}{2}, \frac{5}{2}, \frac{7}{2}, \frac{9}{2}, \ldots\right\}$. If we let $x = -\frac{3}{2}$ and $y = \frac{9}{2}$, then we have $x \equiv_H y$ because they are in the same coset of H in G. We have $y - x = \frac{9}{2} - \left(-\frac{3}{2}\right) = \frac{9}{2} + \frac{3}{2} = \frac{12}{2} = 6$, which is in $\mathbb{Z}$. The last two conditions are also easy to check.

We will now prove that if $H \leq G$, then all cosets of H in G (left or right) have the same cardinality. We will prove this by showing that any coset xH or Hx is equinumerous with H.

Theorem 10.6: Let G be a group, H a subgroup of G, and $x \in G$. Then $|xH| = |H|$ and $|Hx| = |H|$.

Proof: Define $f: H \to xH$ by $f(y) = xy$. If $f(y) = f(z)$, then $xy = xz$. So, we have
$$y = ey = (x^{-1}x)y = x^{-1}(xy) = x^{-1}(xz) = (x^{-1}x)z = ez = z.$$

So, f is injective.

Let $z \in xH$ and let $y \in H$ with $z = xy$. Then $f(y) = xy = z$. So, f is surjective.

Since f is injective and surjective, f is bijective, and so, $|xH| = |H|$.

To see that $|Hx| = |H|$, we define $g: H \to Hx$ by $g(y) = yx$ and use a similar argument. □

We will now prove that if $H \leq G$, then the number of left cosets of H in G is the same as the number of right cosets of H in G. In other words, we will prove that $|\{xH \mid x \in G\}| = |\{Hx \mid x \in G\}|$. We call this common value the **index** of H in G. We will write $[G:H]$ for the index of H in G.

Theorem 10.7: Let G be a group and let H be a subgroup of G. Then $|\{xH \mid x \in G\}| = |\{Hx \mid x \in G\}|$.

It is tempting to try to define a bijection $f: \{xH \mid x \in G\} \to \{Hx \mid x \in G\}$ by $f(xH) = Hx$. Unfortunately, with this definition, f may not be well-defined. For example, let $G = S_3$ and let $H = \{(1), (12)\}$. If we were to define f as above, then by part 3 of Example 10.1, we must have $f((13)H) = H(13) = \{(13), (132)\}$. Since $(123) \in (13)H$, by Theorem 10.5, $(123)H = (13)H$. Therefore, we must have $f((123)H) = \{(13), (132)\}$. But this is not the case. Indeed, we have $f((123)H) = H(123) = \{(23), (123)\} \neq \{(13), (132)\}$.

So, it seems that the most "natural" choice for f doesn't work. Luckily, there is a function that will work that is almost as simple. We show this now.

Proof of Theorem 10.7: Define $f: \{xH \mid x \in G\} \to \{Hx \mid x \in G\}$ by $f(xH) = Hx^{-1}$. We first show that f is well-defined. To see this, suppose that $xH = yH$. By Theorem 10.5, $x^{-1}y \in H$. Since H is a group, it has the inverse property, and therefore, $(x^{-1}y)^{-1} \in H$. Now, by part (i) of Problem 15 in Problem Set 2, we have $(x^{-1}y)^{-1} = y^{-1}(x^{-1})^{-1}$, and so, $y^{-1}(x^{-1})^{-1} \in H$. By Note 2 following Theorem 10.5, we have $Hx^{-1} = Hy^{-1}$.

To see that f is injective, suppose that $Hx^{-1} = Hy^{-1}$. By reversing the steps in the previous paragraph (and using both parts of Problem 15 in Problem Set 2), we get that $xH = yH$.

Finally, to see that f is surjective, let Hy be an arbitrary right coset. Then $y \in G$, and so, $y^{-1} \in G$. We have $f(y^{-1}H) = H(y^{-1})^{-1} = Hy$ (by part (ii) of Problem 15 in Problem Set 2). □

Theorem 10.8 (Lagrange's Theorem): Let G be a finite group and let H be a subgroup of G. Then $|G| = |H| \cdot [G:H]$.

Recall that $|G|$ is the order of G. Since G is finite, the order of G is just the number of elements in G. Similarly, $|H|$ is just the number of elements in H.

Also, recall that $[G:H]$ is the index of H in G. Again, since G is finite, this is just the number of cosets of H in G.

In words, Lagrange's Theorem says that the number of elements in G is equal to the product of the number of elements in H and the number of cosets of H in G (by Theorem 10.7, it does not matter if we use left cosets or right cosets).

Proof of Theorem 10.8: By Theorem 10.4, $\{xH \mid x \in G\}$ is a partition of G. In other words, we can write $G = x_1H \cup x_2H \cup \cdots \cup x_nH$, where this union is pairwise disjoint (note that this implies that $|\{xH \mid x \in G\}| = n$). By Theorem 10.6, we have $|x_1H| = |x_2H| = \cdots = |x_nH| = |H|$. So,

$$|G| = |x_1H \cup x_2H \cup \cdots \cup x_nH| = |x_1H| + |x_2H| + \cdots + |x_nH|$$
$$= |H| \cdot n = |H| \cdot |\{xH \mid x \in G\}| = |H| \cdot [G:H].$$ □

Notes: (1) Lagrange's Theorem is still true if we remove the word "finite" from the statement of the theorem. The proof in this case requires the **Axiom of Choice** (which we will abbreviate as **AC**). One version of AC says the following: given any collection $\mathcal{X}$ of pairwise disjoint nonempty sets, we can form a new set Y consisting of one element from each set in the collection. We call Y a **selector** for $\mathcal{X}$.

For example, let $\mathcal{X} = \{\{(1), (12)\}, \{(13), (123)\}, \{(23), (132)\}\}$. Notice that these are the left cosets of $H = \{(1), (12)\}$ in $G = S_3$ from part 3 of Example 9.1. If we let $Y = \{(1), (13), (23)\}$, then Y is a selector for $\mathcal{X}$. AC is **not** needed to provide a selector for this example. In fact, whenever $\mathcal{X}$ is a finite collection of sets, we can always find a selector without resorting to AC.

As a less trivial example, for each natural number n, let A_n be a set with two distinct elements such that the collection $\mathcal{X} = \{A_n \mid n \in \mathbb{N}\}$ is pairwise disjoint. Can we *explicitly define* a selector Y for $\mathcal{X}$?

The answer is no. Even though all the sets in $\mathcal{X}$ are finite, we do not have enough information about the sets to explicitly describe a selector. However, if we assume AC, then we can say that a selector exists. Furthermore, we can feel free to use this selector to prove other results. This example gives a little insight into why the axiom of choice was so controversial when it was first introduced. It's like magic. All of a sudden, we have this set that we can't describe. We weren't able to come up with the set ourselves, so we simply said, "No big deal—AC will take care of this for us. Here's your selector." Someone might ask, "Well, what does it look like?" And you would have to respond, "I have no idea, but here it is."

We can replace the natural numbers with any set here. For example, we can let A_r be a set with two distinct elements for each $r \in \mathbb{R}$. Even though $\mathbb{R}$ is uncountable, AC still gives us a selector. The axiom of choice is indeed very powerful.

(2) Let's use AC to show that we can eliminate the word "finite" in Lagrange's Theorem and the result is still true.

Given a subgroup H of a group G, let Y be a selector for the collection $\mathcal{X}$ of cosets of H in G. Note that $|Y| = |\mathcal{X}|$. By Theorem 10.6, for each $x \in Y$, there is a bijection $f_x: xH \to H$. We define a function $g: G \to H \times \{zH \mid z \in G\}$ by $g(y) = (f_x(y), xH)$, where $x \in Y$ and $y \in xH$. Then g is a bijection (see Problem 13 below). It follows that $|G| = |H| \cdot |\{xH \mid x \in G\}| = |H| \cdot [G:H]$.

In Lesson 9, we proved that if H is a subgroup of a finite cyclic group G, then the order of H divides the order of G (see Theorem 9.8). Lagrange's Theorem allows us to extend this result to all finite groups.

Theorem 10.9: Let G be a finite group and let H be a subgroup of G. Then the order of H divides the order of G.

Proof: By Lagrange's Theorem, $|G| = |H| \cdot [G:H]$. Since $[G:H]$ is a natural number, we see that $|H|$ divides $|G|$. □

Theorem 10.9 gives us a lot of information about the subgroups of a group G. For example, if H is a subgroup of a group G of order 21, then H has 1, 3, 7, or 21 elements. If H has 1 element, then H is the trivial subgroup, $H = \{e\}$. If H has 21 elements, then $H = G$. So, if we wish to find all the subgroups of G, we can focus our attention on finding the subgroups of orders 3 and 7.

In general, if H is a subgroup of a group G, $H \neq \{e\}$ and $H \neq G$, then we will call H a **proper subgroup** of G and we can write $H < G$. Notice that this terminology is slightly different than the definition of a proper subset (see Lesson 1). The reason is just a matter of convenience. Since for any group G, we have $\{e\} \leq G$ and $G \leq G$, it is convenient to be able to exclude these trivial cases by simply saying "H is a proper subgroup of G."

If the order of G is prime, then Theorem 10.9 further tells us that G has no proper subgroups.

Furthermore, if the order of G is prime, then G **must** be cyclic. To see this, let $x \in G$ with $x \neq e$. Then $\langle x \rangle$ (the subgroup generated by x) must have finite order n for some $n > 1$. By Theorem 10.9, $n|p$. Since p is prime, we have $n = p$. Therefore, $\langle x \rangle$ has p distinct elements, and so, $G = \langle x \rangle$, proving that G is cyclic.

The following theorem is another useful consequence of Lagrange's Theorem.

Theorem 10.10: Let G be a group of order n with identity e and let $a \in G$. Then $a^n = e$.

Proof: Let $a \in G$ and let $H = \langle a \rangle$, the cyclic subgroup generated by a. Suppose that the order of H is k. By Theorem 10.9, the order of H divides the order of G. So, there is an integer j such that $n = kj$. By Theorem 9.2, $x^k = e$. So, $x^n = x^{kj} = (x^k)^j = e^j = e$. □

In general, the converse of Theorem 10.9 is false. In other words, if k divides the order of G, then G may or may not have a subgroup of order k. Consider the following group:

$$A_4 = \{(1), (123), (124), (134), (234), (132), (142), (143), (243), (12)(34), (13)(24), (14)(23)\}.$$

This is a subgroup of S_4 called the **alternating group** on 4 elements. Note that the order of A_4 is 12. In Problem 6 below, you will be asked to show that A_4 has no subgroup of order 6.

In the special case where G is a cyclic group, the converse of Theorem 10.9 is true. This follows immediately from Theorem 9.6. For another special case, see Problem 24 below.

Quotient Groups

Let G be a group and let H be a subgroup of G. We define G/H to be the set of left cosets of H in G. Symbolically, we have

$$G/H = \{xH \mid x \in G\}.$$

Note: The choice of using left cosets here is arbitrary. We could have used right cosets instead. We will see shortly that we will be interested in G/H only in the case where each left coset is equal to the corresponding right coset anyway.

We would like to define an operation $\circ$ on G/H in the most naïve way possible. For $x, y \in G$, we define $(xH) \circ (yH) = (xy)H$. However, as it turns out, this operation is not always well-defined.

Note: Since the group operation is associative, for any $h \in H$, we have $(xy)h = x(yh)$. It follows that $(xy)H = x(yH)$. Indeed, $z \in (xy)H$ if and only if there is $h \in H$ with $z = (xy)h$ if and only if there is $h \in H$ with $z = x(yh)$ if and only if $z \in x(yH)$. So, we may write $(xH) \circ (yH) = xyH$.

Example 10.11. Let $G = S_3 = \{(1), (12), (13), (23), (123), (132)\}$ and $H = \{(1), (12)\}$, as in part 3 of Example 10.1. Then $G/H = \{H, (13)H, (23)H\}$. If we try to define $(xH)(yH) = xyH$, then we have the following:

$$\big((13)H\big) \circ \big((23)H\big) = (13)(23)H = (132)H = \{(23), (132)\}$$

$$\big((123)H\big) \circ \big((23)H\big) = (123)(23)H = (12)H = \{(1), (12)\}$$

Do you see the problem? Well $(13)H = (123)H$. Therefore, both of the above computations need to produce the same output. Since they do not, the operation is **not** well-defined.

Recall that a subgroup N of a group G is said to be a normal subgroup, written $N \triangleleft G$, if whenever $h \in N$ and $g \in G$, then $ghg^{-1} \in N$.

Theorem 10.12. Let G be a group and H a subgroup of G. Suppose that the operation on G/H defined by $(xH) \circ (yH) = xyH$ is well-defined. Then $H \triangleleft G$.

Proof: Suppose that the operation on G/H defined by $(xH) \circ (yH) = xyH$ is well-defined, let $h \in H$, and let $g \in G$. Then $(eH) \circ (g^{-1}H) = eg^{-1}H = g^{-1}H$ and $(hH) \circ (g^{-1}H) = hg^{-1}H$. Since $h = eh \in eH$, by Theorem 10.5 (3 → 4), $eH = hH$, and so, $g^{-1}H = hg^{-1}H$. Once again, by Theorem 10.5 (4 → 2), we have $ghg^{-1} = (g^{-1})^{-1}hg^{-1} \in H$. Since $h \in H$ and $g \in G$ were arbitrary, $H \triangleleft G$. □

So, in order for $(G/H, \circ)$ to be a group, it is necessary that H be normal in G. We will prove that the converse of this statement is true as well. We will first prove a preliminary technical lemma.

Lemma 10.13: Let $N \triangleleft G$. Then for all $x, y \in G$, $xN = yN$ if and only if $xy^{-1} \in N$.

Warning: Notice the subtle difference between Lemma 10.13 and Theorem 10.5 (and in particular, condition 2 in Theorem 10.5). We see from Theorem 10.5, that for **any** subgroup H of G, the condition $xH = yH$ is equivalent to the condition $x^{-1}y \in H$, but **not necessarily** to the condition $xy^{-1} \in H$. The equivalence to the latter condition requires that the subgroup be normal.

The key step in the proof of Lemma 10.13 is realizing that the normality of N can be used to show that we can interchange x^{-1} and y in condition 2 in Theorem 10.5. In other words, if N is normal in G, then $x^{-1}y \in N$ if and only if $yx^{-1} \in N$. We can interchange x^{-1} and y by "conjugating" appropriately. If we conjugate $x^{-1}y$ with x, we get $x(x^{-1}y)x^{-1}$, which simplifies to $(xx^{-1})yx^{-1} = eyx^{-1} = yx^{-1}$. Since N is a normal subgroup of G and $x^{-1}y \in N$, any conjugate of $x^{-1}y$ is also in N. Now yx^{-1} is the conjugate $x(x^{-1}y)x^{-1}$, and so, $yx^{-1} \in N$ as well.

Proof of Lemma 10.13: Let $N \triangleleft G$, and let $x, y \in G$. Suppose that $xN = yN$. By Theorem 10.5 (4 → 2), $y^{-1}x \in N$. Then since $N \triangleleft G$, $xy^{-1} = x(y^{-1}x)x^{-1} \in N$.

Conversely, assume that $xy^{-1} \in N$. Then since $N \triangleleft G$, we have $y^{-1}x = y^{-1}(xy^{-1})y \in N$. By Theorem 10.5 (2 → 4), $yN = xN$. □

Theorem 10.14. Let G be a group and let $N \triangleleft G$. Then $(G/N, \circ)$ is a group.

Proof: We first show that $\circ$ is well defined. Suppose $xN = yN$ and $zN = wN$. By Lemma 10.13, we have $xy^{-1} \in N$ and $zw^{-1} \in N$. Since $N \triangleleft G$ and $zw^{-1} \in N$, we have $y(zw^{-1})y^{-1} \in N$. So, $(xz)(yw)^{-1} = xzw^{-1}y^{-1} = (xy^{-1})(yzw^{-1}y^{-1}) \in N$. Again, by Lemma 10.13, $xzN = ywN$.

Closure and associativity are clear.

Note that $eN = N$ because for all $x \in G$, $x = ex$. It follows that $xN \circ N = xN \circ eN = xeN = xN$ and $N \circ xN = eN \circ xN = exN = xN$. Therefore, N is the identity element of G/N.

Finally, $(xN)^{-1} = x^{-1}N$ because $xN \circ x^{-1}N = xx^{-1}N = eN = N$. □

Theorems 10.12 and 10.14 together tell us that $(G/H, \circ)$ is a group if and only if $H \triangleleft G$.

If $H \triangleleft G$, then $(G/H, \circ)$ is called the **quotient group** (or **factor group**) of G by H.

Notes: (1) We will usually abbreviate $(xH) \circ (yH)$ as $(xH)(yH)$.

(2) If the group operation is addition, we will write $xH + yH = (x+y)H$.

(3) If $(G, +)$ is commutative, then by part 1 of Example 7.21, every subgroup H of G is normal, and so, for any subgroup H of G, $(G/H, +)$ is a group.

Since normal subgroups come up often in the study of group theory, it's nice to know that there are many equivalent definitions. We state and prove the equivalence of some of these definitions in the next theorem.

Theorem 10.15: Let G be a group and N a subgroup of G. The following are equivalent:

1. $N \triangleleft G$ (whenever $h \in N$ and $g \in G$, then $ghg^{-1} \in N$).
2. For all $g \in G$, $gN = Ng$.
3. Every left coset of N in G is a right coset of N in G.
4. For all $g \in G$, $gNg^{-1} = N$.
5. For all $g \in G$, $gNg^{-1} \subseteq N$.
6. For all $g \in G$, $N \subseteq gNg^{-1}$.

Proof: ($1 \to 2$) Suppose that $N \triangleleft G$, let $g \in G$, and let $x \in gN$. Then there is $h \in N$ with $x = gh$. Since $N \triangleleft G$, $ghg^{-1} \in N$. So, $x = gh = (gh)e = (gh)(g^{-1}g) = (ghg^{-1})g \in Ng$. Since $x \in gN$ was arbitrary, $gN \subseteq Ng$. Now, let $x \in Ng$. Then there is $h \in N$ with $x = hg$. Since $N \triangleleft G$, it follows that $g^{-1}hg = g^{-1}h(g^{-1})^{-1} \in N$. Therefore, $x = hg = e(hg) = (gg^{-1})(hg) = g(g^{-1}hg) \in gN$. Since $x \in Ng$ was arbitrary, $Ng \subseteq gN$. Since $gN \subseteq Ng$ and $Ng \subseteq gN$, it follows that $gN = Ng$.

($2 \to 3$) Assume that for all $g \in G$, $gN = Ng$, and let gN be a left coset of N in G. Then $gN = Ng$, and so, gN is also a right coset of N in G.

($3 \to 4$) Assume that every left coset of N in G is a right coset of N in G, and let $g \in G$. Then there is $k \in G$ such that $gN = Nk$. So, $g = ge \in gN = Nk$. By Note 2 following Theorem 10.5, $gk^{-1} \in N$. Since N is a group $(gk^{-1})^{-1} \in N$. By Problem 15 from Problem Set 2 (parts (i) and (ii)), we have $(gk^{-1})^{-1} = (k^{-1})^{-1}g^{-1} = kg^{-1}$. So, $kg^{-1} \in N$. Therefore,

$$gNg^{-1} = Nkg^{-1} = Nk(g^{-1}g)g^{-1} = N(kg^{-1})(gg^{-1}) = N(kg^{-1})e = N(kg^{-1}) = Ne = N.$$

($4 \to 5$) This is obvious.

($5 \to 6$) Assume that for all $g \in G$, $gNg^{-1} \subseteq N$ and let $g \in G$. Then $g^{-1} \in G$, and therefore, $g^{-1}Ng = g^{-1}N(g^{-1})^{-1} \subseteq N$. Let $x \in N$. Then $g^{-1}xg \in g^{-1}Ng$, and so, $g^{-1}xg \in N$. So, there is $h \in N$ such that $g^{-1}xg = h$. Thus, $x = exe = (gg^{-1})x(gg^{-1}) = g(g^{-1}xg)g^{-1} = ghg^{-1} \in gNg^{-1}$. Since $x \in N$ was arbitrary, we have $N \subseteq gNg^{-1}$.

($6 \to 1$) Assume that for all $g \in G$, $N \subseteq gNg^{-1}$, let $h \in N$, and let $g \in G$. Then $g^{-1} \in G$, and so, $h \in g^{-1}N(g^{-1})^{-1} = g^{-1}Ng$. Thus, there is $n \in N$ such that $h = g^{-1}ng$. It follows that we have $ghg^{-1} = g(g^{-1}ng)g^{-1} = (gg^{-1})n(gg^{-1}) = ene = n \in N$. □

Example 10.16:

1. Let $G = S_3 = \{(1), (12), (13), (23), (123), (132)\}$ and let $H = \{(1), (12)\}$ as in part 3 of Example 10.1. Since $(13)H = \{(13), (123)\}$ and $H(13) = \{(13), (132)\}$, by Theorem 10.15 (condition 2), H is not normal in G. It follows that $(G/H, \circ)$ is not a group ($\circ$ is not well-defined on G/H by Theorem 10.12).

   Let $K = \{(1), (123), (132)\}$. The two left cosets of K in G are

   $$K = \{(1), (123), (132)\} \qquad (12)K = \{(12), (23), (13)\}$$

   The two right cosets of H in G are

   $$K = \{(1), (123), (132)\} \qquad K(12) = \{(12), (23), (13)\}$$

   Since each left coset is the same as the corresponding right coset, by Theorem 10.15 (condition 2), $K \triangleleft G$. The multiplication table for $(G/K, \circ)$ is as follows:

$\circ$	K	$(12)K$
K	K	$(12)K$
$(12)K$	$(12)K$	K

   For example, we have $(12)K \circ K = (12)K \circ (1)K = (12)(1)K = (12)K$. As another example, we have $(12)K \circ (12)K = (12)(12)K = (1)K = K$.

2. Let $G = \mathbb{Z}$ and $H = 2\mathbb{Z} = \{\ldots, -6, -4, -2, 0, 2, 4, 6, \ldots\}$. Since G is commutative, every subgroup of G (and H in particular) is normal in G. By part 1 of Example 10.1, $G/H = \{2\mathbb{Z}, 2\mathbb{Z} + 1\}$. The multiplication table for $(\mathbb{Z}/2\mathbb{Z}, +)$ is as follows:

$+$	$2\mathbb{Z}$	$2\mathbb{Z} + 1$
$2\mathbb{Z}$	$2\mathbb{Z}$	$2\mathbb{Z} + 1$
$2\mathbb{Z} + 1$	$2\mathbb{Z} + 1$	$2\mathbb{Z}$

3. More generally, if $G = \mathbb{Z}$, $n \in \mathbb{Z}^+$, and $H = n\mathbb{Z} = \{\ldots, -3n, -2n, -n, 0, n, 2n, 3n, \ldots\}$, we see that $(\mathbb{Z}/n\mathbb{Z}, +)$ has essentially the same multiplication table as $(\mathbb{Z}_n, +)$ (the additive group of the integers modulo n). In fact, $(\mathbb{Z}/n\mathbb{Z}, +)$ is isomorphic to $(\mathbb{Z}_n, +)$ via the function $f: \mathbb{Z}/n\mathbb{Z} \to \mathbb{Z}_n$ defined by $f(n\mathbb{Z} + k) = [k]_n$. The verification that f is an isomorphism is straightforward, and so, I leave it to the reader.

4. Let $G = \mathbb{R}$ and $H = \mathbb{Z}$ (see part 5 of Example 10.1). Once again, every subgroup of $\mathbb{R}$ is normal (because addition is commutative in $\mathbb{R}$). In particular, $\mathbb{Z}$ is normal in $\mathbb{R}$, and so, $\mathbb{R}/\mathbb{Z}$ is a group. Let's do a few computations. Remember that we can always use a representative for a coset that is between 0 and 1.

   $$\mathbb{R} + \mathbb{R} = (0 + \mathbb{R}) + (0 + \mathbb{R}) = (0 + 0) + \mathbb{R} = 0 + \mathbb{R} = \mathbb{R}$$

   $$\left(\tfrac{1}{2} + \mathbb{R}\right) + \left(\tfrac{1}{2} + \mathbb{R}\right) = \left(\tfrac{1}{2} + \tfrac{1}{2}\right) + \mathbb{R} = 1 + \mathbb{R} = \mathbb{R}$$

   $$(0.7 + \mathbb{R}) + (0.9 + \mathbb{R}) = (0.7 + 0.9) + \mathbb{R} = 1.6 + \mathbb{R} = 0.6 + \mathbb{R}$$

   We see that $\mathbb{R}/\mathbb{Z}$ is isomorphic to the half open interval $[0, 1)$. An isomorphism $f: [0, 1) \to \mathbb{R}/\mathbb{Z}$ is given by $f(x) = x + \mathbb{Z}$.

Furthermore, $\mathbb{R}/\mathbb{Z}$ is isomorphic to the unit circle $\mathbb{S}$ in the Complex Plane (see Example 5.31). Note that $\mathbb{R}/\mathbb{Z}$ is a group under addition, whereas $\mathbb{S} = \{e^{i\theta} \in \mathbb{C} \mid 0 \leq \theta < 2\pi\}$ is a group under multiplication. An isomorphism $f: \mathbb{R}/\mathbb{Z} \to \mathbb{S}$ is given by $f(x + \mathbb{Z}) = e^{2\pi x i}$. To see that f is a homomorphism, observe that

$$f((x+y) + \mathbb{Z}) = e^{2\pi(x+y)i} = e^{2\pi xi + 2\pi yi} = e^{2\pi xi} \cdot e^{2\pi yi} = f(x + \mathbb{Z}) \cdot f(y + \mathbb{Z})$$

I leave it to the reader to verify that f is a bijection.

Quotient Rings

Let $(R, +, \cdot)$ be a ring and let $(H, +)$ be a subgroup of $(R, +)$. The definition of the set of cosets of H in R, $R/H = \{x + H \mid x \in R\}$, still makes perfect sense. Since $(R, +)$ is commutative (by the definition of a ring), $(H, +)$ is automatically normal in $(R, +)$, and so, the operation of addition of cosets, is well-defined. In other words, if we have $x + H = z + H$ and $y + H = w + H$, then it follows that $(x + y) + H = (z + w) + H$.

We would now like to define multiplication of cosets by $(xH)(yH) = xyH$. However, just like for the case of arbitrary subgroups of a group, it turns out this operation is not always well-defined.

Note: As mentioned at the end of Lesson 3, to avoid unnecessary clutter, we may abbreviate an element $[x]_n \in \mathbb{Z}_n$ as x. So, we may write $\mathbb{Z}_2 = \{0, 1\}$ instead of $\mathbb{Z}_2 = \{[0]_2, [1]_2\}$. Similarly, we may write $\mathbb{Z}_6 = \{0, 1, 2, 3, 4, 5\}$ and we may write $\mathbb{Z}_2 \times \mathbb{Z}_2 = \{(0,0), (0,1), (1,0), (1,1)\}$, etc.

Example 10.17. Let $R = \mathbb{Z}_2 \times \mathbb{Z}_2 = \{(0,0), (0,1), (1,0), (1,1)\}$ and $H = \{(0,0), (1,1)\}$. Observe that $(H, +)$ is a subgroup of $(\mathbb{Z}_2 \times \mathbb{Z}_2, +)$ (in fact, $(H, +, \cdot)$ is a subring of $(\mathbb{Z}_2 \times \mathbb{Z}_2, +, \cdot)$). Then $(\mathbb{Z}_2 \times \mathbb{Z}_2)/H = \{H, (0,1) + H\}$, where $(0,1) + H = \{(0,1), (1,0)\}$ (because $(0,1) + (0,0) = (0,1)$ and $(0,1) + (1,1) = (1,0)$). Now, addition of cosets is well-defined and the addition table is as follows:

+	H	$(0,1) + H$
H	H	$(0,1) + H$
$(0,1) + H$	$(0,1) + H$	H

However, multiplication is **not** well-defined. For example, we have the following:

$$((0,0) + H)((0,1) + H) = (0,0) + H = H.$$
$$((1,1) + H)((0,1) + H) = (0,1) + H \neq H.$$

Since $(0,0) + H = (1,1) + H = H$, we see that the operation is not well-defined.

Recall that a subset I of a ring R is is an ideal of R, written $I \triangleleft R$, if $(I, +)$ is a subgroup of $(R, +)$ and for all $a \in I$ and $x \in R$, $ax \in I$ and $xa \in I$ (we say that I **absorbs** R).

Theorem 10.18. Let $(R, +, \cdot)$ be a ring and $(H, +)$ a subgroup of $(R, +)$. Suppose that the operations on R/H defined by $(x + H) + (y + H) = (x + y) + H$, $(x + H)(y + H) = xy + H$ are well-defined. Then H is an ideal of R.

Note: The operation defined by $(x + H) + (y + H) = (x + y) + H$ is automatically well-defined because $(R, +)$ is commutative (and therefore, H is automatically a normal subgroup of R). So, it was not really necessary to mention this operation in the statement of the theorem.

Proof: Suppose that the operation on R/H defined by $(x + H)(y + H) = xy + H$ is well-defined, let $h \in H$, and let $x \in R$. Then $(0 + H)(x + H) = 0x + H = H$ and $(h + H)(x + H) = hx + H$. Since $h = 0 + h \in 0 + H$, by Theorem 10.5 (3 → 4), $0 + H = h + H$, and so, $H = hx + H$. Once again, by Theorem 10.5 (4 → 2), we have $hx = hx - 0 \in H$. Similarly, $xh \in H$. Since $h \in H$ and $x \in R$ were arbitrary, H is an ideal of R. □

So, in order for $(R/H, +, \cdot)$ to be a ring, it is necessary that H be an ideal of R. We will prove that the converse of this statement is true as well.

Theorem 10.19. Let R be a ring and let $I \triangleleft R$. Then $(R/I, +, \cdot)$ is a ring.

Proof: Since $(R, +)$ is a commutative group, by part 1 of Example 7.21, $(I, +)$ is a normal subgroup of $(R, +)$. So, $(R/I, +)$ is a group. Commutativity of addition is clear.

We next check that multiplication is well defined. Suppose $x + I = z + I$ and $y + I = w + I$. By Lemma 10.13, we have $x - z \in I$ and $y - w \in I$. Since $I \triangleleft R$, we have $(x - z)w \in I$ and $x(y - w) \in I$. Since $(I, +)$ is a group, we have

$$xy - zw = xy - xw + xw - zw = x(y - w) + (x - z)w \in I$$

By Lemma 10.13, $xy + I = zw + I$.

Associativity of multiplication is clear.

Finally, we show that $1 + I$ is the multiplicative identity of R/I. For all $x \in R$, $1x = x$ and $x \cdot 1 = x$. It follows that $(1 + I) \cdot (x + I) = 1x + I = x + I$ and $(x + I) \cdot (1 + I) = x \cdot 1 + I = x + I$. □

If I is an ideal of R, then $(R/I, +, \cdot)$ is called the **quotient ring** (or **factor ring**) of R by I.

Example 10.20:

1. Let $R = \mathbb{Z}$ and $H = 2\mathbb{Z} = \{\ldots, -6, -4, -2, 0, 2, 4, 6, \ldots\}$. By part 1 of Example 7.22, $2\mathbb{Z}$ is an ideal of $\mathbb{Z}$. By part 1 of Example 10.1, we have $\mathbb{Z}/2\mathbb{Z} = \{2\mathbb{Z}, 2\mathbb{Z} + 1\}$. The addition and multiplication tables for $(\mathbb{Z}/2\mathbb{Z}, +, \cdot)$ are as follows:

+	$2\mathbb{Z}$	$2\mathbb{Z} + 1$
$2\mathbb{Z}$	$2\mathbb{Z}$	$2\mathbb{Z} + 1$
$2\mathbb{Z} + 1$	$2\mathbb{Z} + 1$	$2\mathbb{Z}$

$\cdot$	$2\mathbb{Z}$	$2\mathbb{Z} + 1$
$2\mathbb{Z}$	$2\mathbb{Z}$	$2\mathbb{Z}$
$2\mathbb{Z} + 1$	$2\mathbb{Z}$	$2\mathbb{Z} + 1$

   Since $2\mathbb{Z}$ is the additive identity of $\mathbb{Z}/2\mathbb{Z}$, we will abbreviate $2\mathbb{Z}$ as $\bar{0}$. Similarly, since $2\mathbb{Z} + 1$ is the multiplicative identity of $\mathbb{Z}/2\mathbb{Z}$, we will abbreviate $2\mathbb{Z} + 1$ as $\bar{1}$. These abbreviations make the addition and multiplication tables much easier to read.

+	$\bar{0}$	$\bar{1}$
$\bar{0}$	$\bar{0}$	$\bar{1}$
$\bar{1}$	$\bar{1}$	$\bar{0}$

$\cdot$	$\bar{0}$	$\bar{1}$
$\bar{0}$	$\bar{0}$	$\bar{0}$
$\bar{1}$	$\bar{0}$	$\bar{1}$

Observe that these tables look just like the tables for $\mathbb{Z}_2$. In fact, the function $f: \mathbb{Z}/2\mathbb{Z} \to \mathbb{Z}_2$ defined by $f(\bar{k}) = [k]_2$ is an isomorphism. For all practical purposes, $\mathbb{Z}_2$ and $\mathbb{Z}/2\mathbb{Z}$ represent the same ring.

Since $\mathbb{Z}_2$ is a field and $\mathbb{Z}/2\mathbb{Z} \cong \mathbb{Z}_2$, it follows that $\mathbb{Z}/2\mathbb{Z}$ is a field.

2. More generally, if $R = \mathbb{Z}$, $n \in \mathbb{Z}^+$ and $H = n\mathbb{Z} = \{\ldots, -3n, -2n, -n, 0, n, 2n, 3n, \ldots\}$, we see that $(\mathbb{Z}/n\mathbb{Z}, +, \cdot)$ has essentially the same addition and multiplication tables as $(\mathbb{Z}_n, +, \cdot)$ (the ring of the integers modulo n). In fact, $(\mathbb{Z}/n\mathbb{Z}, +, \cdot)$ is isomorphic to $(\mathbb{Z}_n, +, \cdot)$ via the function $f: \mathbb{Z}/n\mathbb{Z} \to \mathbb{Z}_n$ defined by $f(\bar{k}) = [k]_n$. The verification that f is an isomorphism is straightforward, and so, I leave it to the reader.

Since $\mathbb{Z}_n$ is a field if and only if n is prime and $\mathbb{Z}/n\mathbb{Z} \cong \mathbb{Z}_n$, it follows that $\mathbb{Z}/n\mathbb{Z}$ is a field if and only if n is prime. Similarly, if n is not prime, then $\mathbb{Z}/n\mathbb{Z}$ is not an integral domain.

3. Let $R = \mathbb{Z}_2 \times \mathbb{Z}_6$ and $H = \langle(1,3)\rangle$ (since R is being considered as a ring, $\langle(1,3)\rangle$ means the principal ideal generated by $(1,3)$ and not the cyclic subgroup generated by $(1,3)$). To determine the elements of $\langle(1,3)\rangle$, we need to multiply $(1,3)$ by each element of $\mathbb{Z}_2 \times \mathbb{Z}_6$. If k is even, then $(1,3) \cdot (0,k) = (0,3k) = (0,0)$ and $(1,3) \cdot (1,k) = (1,3k) = (1,0)$. If k is odd, then $(1,3) \cdot (0,k) = (0,3k) = (0,3)$ and $(1,3) \cdot (1,k) = (1,3k) = (1,3)$. It follows that $\langle(1,3)\rangle = \{(0,0), (1,0), (0,3), (1,3)\}$.

Now, $|\mathbb{Z}_2 \times \mathbb{Z}_6| = |\mathbb{Z}_2| \cdot |\mathbb{Z}_6| = 2 \cdot 6 = 12$ and $|\langle(1,3)\rangle| = |\{(0,0), (1,0), (0,3), (1,3)\}| = 4$. Therefore, $|(\mathbb{Z}_2 \times \mathbb{Z}_6)/\langle(1,3)\rangle| = \frac{12}{4} = 3$. Let's list the three cosets:

$(0,0) + \langle(1,3)\rangle = \langle(1,3)\rangle = \{(0,0), (1,0), (0,3), (1,3)\}$

$(0,1) + \langle(1,3)\rangle = \{(0,1), (1,1), (0,4), (1,4)\}$

$(0,2) + \langle(1,3)\rangle = \{(0,2), (1,2), (0,5), (1,5)\}$

So, the quotient ring $(\mathbb{Z}_2 \times \mathbb{Z}_6)/\langle(1,3)\rangle = \{\langle(1,3)\rangle, (0,1) + \langle(1,3)\rangle, (0,2) + \langle(1,3)\rangle\}$. The multiplicative identity of this ring is $(0,1) + \langle(1,3)\rangle$. This ring is actually a field. Each of $(0,1) + \langle(1,3)\rangle$ and $(0,2) + \langle(1,3)\rangle$ is its own multiplicative inverse.

An ideal I of a commutative ring R is a **prime ideal** of R if $I \neq R$ and whenever $xy \in I$, either $x \in I$ or $y \in I$.

Example 10.21:

1. Let $R = \mathbb{Z}$ and $I = 2\mathbb{Z} = \{\ldots, -6, -4, -2, 0, 2, 4, 6, \ldots\}$. By part 1 of Example 7.22, $2\mathbb{Z}$ is an ideal of $\mathbb{Z}$. Let's show that $2\mathbb{Z}$ is a prime ideal of $\mathbb{Z}$. Suppose that $xy \in 2\mathbb{Z}$. If x and y were both odd, then by Problem 15 in Problem Set 8, xy would be odd. It follows that either x is even or y is even. So, $x \in 2\mathbb{Z}$ or $y \in 2\mathbb{Z}$.

$4\mathbb{Z}$ is an ideal of $\mathbb{Z}$ that is **not** a prime ideal of $\mathbb{Z}$. Indeed, $4 \in 4\mathbb{Z}$, and $4 = 2 \cdot 2$, but $2 \notin 4\mathbb{Z}$.

Similarly, $6\mathbb{Z}$ is an ideal of $\mathbb{Z}$ that is not a prime ideal of $\mathbb{Z}$ because $2 \cdot 3 = 6 \in 6\mathbb{Z}$, but neither 2 nor 3 is in $6\mathbb{Z}$.

For $n > 1$, $n\mathbb{Z}$ is a prime ideal of $\mathbb{Z}$ if and only if n is prime. This can be proved directly from the definition of a prime ideal together with the definition of a prime number. However, it turns out that an ideal I of a ring R is prime if and only if the quotient R/I is an integral domain (see Theorem 10.22 below). Since we know that $\mathbb{Z}_n$ is an integral domain if and only if n is prime and we know that $\mathbb{Z}/n\mathbb{Z}$ is isomorphic to $\mathbb{Z}_n$, the result follows.

By definition, $\mathbb{Z}$ is not a prime ideal of $\mathbb{Z}$ (no ring is a prime ideal of itself). $\{0\}$ is a prime ideal of $\mathbb{Z}$. Indeed, since $\mathbb{Z}$ is an integral domain (it has no zero divisors), whenever $xy = 0$, it must follow that $x = 0$ or $y = 0$.

2. Let $R = {}^{\mathbb{R}}\mathbb{R}$ (the ring of functions from $\mathbb{R}$ to itself, where addition and multiplication are defined pointwise), let $x \in \mathbb{R}$, and let $I_x = \{f \in {}^{\mathbb{R}}\mathbb{R} \mid f(x) = 0\}$. By Problem 19 in Problem Set 7, I_x is an ideal of ${}^{\mathbb{R}}\mathbb{R}$. Let's show that I_x is a prime ideal of ${}^{\mathbb{R}}\mathbb{R}$. Suppose that $fg \in I_x$. Then $(fg)(x) = 0$. So, $f(x)g(x) = (fg)(x) = 0$. Since $\mathbb{R}$ is an integral domain, $f(x) = 0$ or $g(x) = 0$. So, $f \in I_x$ or $g \in I_x$.

As promised, we will now show that an ideal I of a ring R is a prime ideal if and only if the quotient ring R/I is an integral domain.

Theorem 10.22: Let R be a commutative ring and let I be an ideal of R. Then I is a prime ideal of R if and only if R/I is an integral domain.

Proof: Let I be a prime ideal of R and let $x + I, y + I \in R/I$ such that $(x + I)(y + I) = I$ (remember that I is the additive identity of R/I). Since $(x + I)(y + I) = xy + I$, we have $xy + I = I$. Therefore, $xy \in I$. Since I is a prime ideal, $x \in I$ or $y \in I$. So, $x + I = I$ or $y + I = I$. Since $x + I, y + I \in R/I$ were arbitrary, R/I has no zero divisors. Therefore, R/I is an integral domain.

Conversely, assume that R/I is an integral domain and let $xy \in I$. Then $xy + I = I$. So, we have $(x + I)(y + I) = xy + I = I$. Since R/I is an integral domain, $x + I = I$ or $y + I = I$. Therefore, $x \in I$ or $y \in I$. So, I is a prime ideal of R. □

An ideal I of a commutative ring R is a **maximal ideal** of R if $I \neq R$ and whenever J is an ideal with $I \subseteq J \subseteq R$, either $J = I$ or $J = R$.

Example 10.23:

1. Let $R = \mathbb{Z}$ and $I = 2\mathbb{Z} = \{\ldots, -6, -4, -2, 0, 2, 4, 6, \ldots\}$. By part 1 of Example 10.21, $2\mathbb{Z}$ is a prime ideal of $\mathbb{Z}$. It is also a maximal ideal of $\mathbb{Z}$. To see this, let J be an ideal of $\mathbb{Z}$ with $2\mathbb{Z} \subseteq J \subseteq \mathbb{Z}$ and assume that $J \neq 2\mathbb{Z}$. Then there is an odd integer $x \in J$. Since x and 2 are relatively prime, by Theorem 8.18, there are integers $m, n \in \mathbb{Z}$ with $m(2) + nx = 1$. Since $2, x \in J$ and J absorbs $\mathbb{Z}$, we have $m(2), nx \in J$. Since J is a subgroup of $\mathbb{Z}$, we have $1 = m(2) + nx \in J$. Since 1 generates $\mathbb{Z}$, it follows that $\mathbb{Z} \subseteq J$. Therefore, $J = \mathbb{Z}$.

A similar argument can be used to show that if p is prime, then $p\mathbb{Z}$ is a maximal ideal of $\mathbb{Z}$. I leave the details to the reader.

$4\mathbb{Z}$ is an ideal of $\mathbb{Z}$ that is **not** a maximal ideal of $\mathbb{Z}$. Indeed, $4\mathbb{Z} \subset 2\mathbb{Z} \subset \mathbb{Z}$ (these are proper subsets).

Similarly, if $n > 1$ is **not** prime, then let p be a prime factor of n. We have $n\mathbb{Z} \subset p\mathbb{Z} \subset \mathbb{Z}$, showing that $n\mathbb{Z}$ is not a maximal ideal of $\mathbb{Z}$.

By definition, $\mathbb{Z}$ is not a maximal ideal of $\mathbb{Z}$ (no ring is a maximal ideal of itself). $\{0\}$ is also not a maximal ideal of $\mathbb{Z}$. For example, $\{0\} \subset 2\mathbb{Z} \subset \mathbb{Z}$. We see that $\{0\}$ is an example of a prime ideal that is **not** a maximal ideal.

2. Is there a maximal ideal of a commutative ring that is not a prime ideal? The answer is no. Every maximal ideal is a prime ideal. Let's prove this. Let R be a commutative ring and let I be a maximal ideal of R. Assume toward contradiction that I is **not** a prime ideal of R. Then there are $x, y \in R$ with $xy \in I$, $x \notin I$, and $y \notin I$. Let J be the ideal generated by x and I. In other words, J consists of all sums $ax + z$ with $a \in R$ and $z \in I$. Since $I \subset J$, by the maximality of I, we must have $J = R$. Since $1 \in R$, we have $1 \in J$. So, there is $a \in R$ and $z \in I$ with $1 = ax + z$. A similar argument shows that there is $b \in R$ and $w \in I$ with $1 = by + w$. So, we have

$$1 = 1 \cdot 1 = (ax + z)(by + w) = abxy + axw + zby + zw.$$

Since $xy, z,$ and $w \in I$, we have $1 \in I$. Therefore, $R \subseteq I$, and so, $I = R$, contradicting that I is maximal ideal of R. This contradiction proves that I is a prime ideal of R.

3. Let $R = {}^{\mathbb{R}}\mathbb{R}$, let $x \in \mathbb{R}$, and let $I_x = \{f \in {}^{\mathbb{R}}\mathbb{R} \mid f(x) = 0\}$. In part 2 of Example 10.21, we showed that I_x is a prime ideal of ${}^{\mathbb{R}}\mathbb{R}$. By Problem 19 in Problem Set 7, I_x is a maximal ideal of ${}^{\mathbb{R}}\mathbb{R}$.

We now show that an ideal I of a ring R is a maximal ideal if and only if the quotient ring R/I is a field.

Theorem 10.24: Let R be a commutative ring and let I be an ideal of R. Then I is a maximal ideal of R if and only if R/I is a field.

Proof: Let I be a maximal ideal of R and let $x + I \in R/I$ with $x \notin I$ (so that $x + I$ is not the additive identity of I). Let $J = \{ax + z \mid a \in R \text{ and } z \in I\}$ (so that J is the ideal generated by x and I). If $y \in I$, then $y = 0 \cdot x + y \in J$. Therefore, $I \subseteq J$. Since $x = 1x + 0$, $1 \in R$, and $0 \in I$, we have $x \in J$. It follows that $J \neq I$. Since I is a maximal ideal, $J = R$. Since $1 \in R$, we have $1 \in J$, and so, there is $a \in R$ and $z \in I$ such that $1 = ax + z$. So, $1 - ax = z \in I$. Thus, $1 + I = ax + I = (a + I)(x + I)$. It follows that $a + I$ is the inverse of $x + I$. Since $x + I \in R/I$ was an arbitrary nonzero (i.e., not I) element of R/I, we see that every nonzero element of R/I has a multiplicative inverse. Therefore, R/I is a field.

Conversely, assume that R/I is a field and let J be an ideal of R with $I \subseteq J \subseteq R$ and $I \neq J$. Let $x \in J \setminus I$. Since $x \notin I$, $x + I \neq I$. In other words, $x + I$ is not the zero element of R/I. Since R/I is a field, $x + I$ has a multiplicative inverse in R/I, say $y + I = (x + I)^{-1}$. So, $(x + I)(y + I) = 1 + I$. It follows that $xy + I = 1 + I$. So, $1 - xy \in I$. Since $I \subseteq J$, we have $1 - xy \in J$. Since $x \in J$ and J is an ideal of R, we have $xy \in J$. Since $(J, +)$ is a subgroup of $(R, +)$, $1 = (1 - xy) + xy \in J$. Therefore, $J = R$. Since J was an arbitrary ideal of R with $I \subseteq J \subseteq R$, I is a maximal ideal. □

Quotient Spaces

Let V be a vector space over a field F and let W be a subspace of V. As was the case for rings, the definition $V/W = \{v + W \mid v \in V\}$ of cosets of W in V still makes perfect sense. Since $(V, +)$ is commutative (by the definition of a vector space), $(W, +)$ is automatically normal in $(V, +)$, and so, the operation of addition of cosets is well-defined. In other words, if we have $u + W = v + W$ and $s + W = t + W$, then $(u + s) + W = (v + t) + W$.

We would now like to define a scalar multiplication on the set of cosets by $k(v + W) = kv + W$. Unlike the complexities that we underwent with quotient rings, this definition of scalar multiplication is always well-defined.

To see this, suppose that $u, v \in V$, $k \in F$, and $u + W = v + W$. Then $u - v \in W$. Since W is itself a vector space over the field F, $ku - kv = k(u - v) \in W$. Therefore, $ku + W = kv + W$.

If W is a subspace of V, then V/W is called the **quotient space** of V by W.

Example 10.25:

1. Let $V = \mathbb{R}^2$ and $W = \{(a, 0) \mid a \in \mathbb{R}\}$. By part 1 of Example 6.7, $W \leq V$. Geometrically, we can visualize $\mathbb{R}^2$ as the Cartesian plane and W as the x-axis in the Cartesian plane. Now, V/W consists of all cosets $(a, b) + W$, where (a, b) and (c, d) are in the same coset if and only if $(c - a, d - b) = (c, d) - (a, b) \in W$ if and only if $d - b = 0$ if and only if $d = b$ if and only if (a, b) and (c, d) lie on the same horizontal line. So, one way we can visualize V/W is as the set of all horizontal lines in the Cartesian plane.

   Now, if we define $f: V/W \to \mathbb{R}$ by $f\big((a, b) + W\big) = b$, then it is easy to verify that f is a well-defined isomorphism. I leave the details to the reader. So, the quotient space V/W is essentially just the vector space of real numbers.

2. If V is a vector space, then the quotient $V/\{0\}$ is isomorphic to V. Indeed, $v + \{0\} = w + \{0\}$ if and only if $v - w \in \{0\}$ if and only if $v - w = 0$ if and only if $v = w$. Therefore, the function $f: V/\{0\} \to V$ defined by $f(v + \{0\}) = v$ is an isomorphism.

   Also, the quotient V/V is isomorphic to $\{0\}$. We have $v + V = w + V$ if and only if $v - w \in V$, which is always true. Therefore, the function $f: V/V \to \{0\}$ defined by $f(v + V) = 0$ is an isomorphism.

Problem Set 10

Full solutions to these problems are available for free download here:
www.SATPrepGet800.com/AAFBTDW

LEVEL 1

1. Let $n \in \mathbb{Z}$. Compute $[\mathbb{Z}:n\mathbb{Z}]$.

2. Let G and H be groups such that $|G| = 14$ and $|H| = 45$. Compute $|G \cap H|$.

LEVEL 2

3. Let G be a group of finite order and let H and K be subgroups of G with H a subgroup of K. Prove that $[G:H] = [G:K] \cdot [K:H]$.

4. Consider $\mathbb{R}^* = \mathbb{R} \setminus \{0\}$ as a group under multiplication. Find a subgroup H of $\mathbb{R}^*$ such that $[\mathbb{R}^*:H] = 2$.

5. Let V be a vector space over a field F, let W be a subspace of V, and define $\pi: V \to V/W$ by $\pi(v) = v + W$. Prove that π is a linear transformation.

LEVEL 3

6. Consider the following subset of S_4:

$A_4 = \{(1),(123),(124),(134),(234),(132),(142),(143),(243),(12)(34),(13)(24),(14)(23)\}$.

Prove that A_4 is a subgroup of S_4 and that A_4 has no subgroup of order 6.

7. Let G be a group and let $H = \langle xyx^{-1}y^{-1} \mid x, y \in G \rangle$. Prove that H is a normal subgroup of G such that G/H is commutative. H is known as the **commutator subgroup** of G.

8. Let G be a group and let H be a subgroup of G such that $[G:H] = 2$. Prove that $H \triangleleft G$.

9. Prove that every element of $\mathbb{Q}/\mathbb{Z}$ has finite order.

10. Let R be a ring, let I be an ideal of R such that $I \neq R$ and R/I is commutative. Is R necessarily commutative? If so, prove it. If not, provide a counterexample.

11. Describe a ring with exactly 2 maximal ideals.

LEVEL 4

12. Let G be a group and let $H \triangleleft G$ such that $|H|$ and $|G/H|$ are finite. Prove that $|G|$ is finite.

13. Let G be a group, let H be a subgroup of G, and let Y be a selector for the collection of cosets of H in G. For each $x \in Y$, let $f_x: xH \to H$ be a bijection and define $g: G \to H \times \{zH \mid z \in G\}$ by $g(y) = (f_x(y), xH)$, where $x \in Y$ and $y \in xH$. Prove that g is a bijection.

14. Let $\mathbb{Z}_n = \{[k]_n \mid k \in \mathbb{Z}\}$, where $[k]_n$ is the equivalence class of k under the equivalence $\equiv_n$ and consider the commutative ring $(\mathbb{Z}_n, +, \cdot)$, where addition and multiplication are defined by $[x]_n + [y]_n = [x+y]_n$ and $[xy]_n = [x]_n \cdot [y]_n$ (see Problem 10 from Problem Set 3). Prove that $\mathbb{Z}/n\mathbb{Z} \cong \mathbb{Z}_n$. Find the ideals of $\mathbb{Z}/15\mathbb{Z}$ and $\mathbb{Z}_{15}$ and show that there is a natural one-to-one correspondence between them.

15. Let G be a group such that $G/Z(G)$ is cyclic. Prove that G is commutative.

16. Let H and K be finite subgroups of a group G and let $HK = \{hk \mid h \in H \wedge k \in K\}$. Prove that $|HK| \cdot |H \cap K| = |H| \cdot |K|$.

17. Prove that there are exactly two groups of order 6, up to isomorphism.

18. Let V be a vector space over a field F, let W be a subspace of V, and define $\pi: V \to V/W$ by $\pi(v) = v + W$. By Problem 5 above, π is a linear transformation. Suppose that U is a vector space over F and that $f: V \to U$ is a linear transformation such that $W \subseteq \ker(f)$. Prove that there is a unique linear transformation $g: V/W \to U$ such that $f = g \circ \pi$.

LEVEL 5

19. Let G be a group and let N be a normal subgroup of G such that N and G/N are finitely generated. Prove that G is finitely generated.

20. Let G be a group and let H be a subgroup of G such that H is the only subgroup of G with index $[G:H]$. Prove that H is normal in G.

21. Prove that there are exactly five groups of order 8, up to isomorphism.

22. Find all groups of order 10, up to isomorphism.

23. Let V be a finite-dimensional vector space over a field F and let W be a subspace of V. Prove that $\dim V/W = \dim V - \dim W$.

CHALLENGE PROBLEMS

24. Let G be a group of order $n \in \mathbb{Z}^+$ and let p be a prime number that divides n. Prove that G has a subgroup of order p (this result is known as **Cauchy's Theorem**).

25. Let M be a left module over a ring R and let N be a left submodule of M such that $M/N \cong \mathbb{Z}$. Prove that $M = \mathbb{Z}$ and $N = \{0\}$.

LESSON 11
STRUCTURE THEOREMS

Isomorphism Theorems

Let's quickly review a few things from previous lessons that will be useful here. Recall the following:

(1) If G and H are groups, then a function $f: G \to H$ is called a homomorphism if for all $x, y \in G$, $f(xy) = f(x)f(y)$. If the groups are commutative, we will use additive notation and write $f(x + y) = f(x) + f(y)$.

Similarly, if R and S are rings, then $f: R \to S$ is a homomorphism if for all $x, y \in R$, $f(x + y) = f(x) + f(y)$, $f(xy) = f(x)f(y)$, and $f(1_R) = 1_S$.

If V and W are vector spaces over a field F (or left modules over a ring R), then $f: V \to W$ is a homomorphism (or linear transformation) if for all $x, y \in V$ and all scalars k, $f(x + y) = f(x) + f(y)$ and $f(kv) = kf(v)$.

(2) An isomorphism is a bijective homomorphism.

(3) If $f: A \to B$ is a homomorphism, then the kernel of f is $\ker(f) = \{x \in G | f(x) = e_B\}$, where $e_B \in B$ is the identity of B. If f is a homomorphism of rings, then e_B is the additive identity.

(4) If $f: G \to H$ is a homomorphism of groups, then $\ker(f)$ is a normal subgroup of G. See Theorem 7.19 and part 2 of Example 7.21 for details

If $f: R \to S$ is a homomorphism of rings, then $\ker(f)$ is an ideal of R. See part 2 of Example 7.22 for details.

If $f: V \to W$ is a linear transformation of vector spaces over a field F (or left modules over a ring R), then $\ker(f)$ is a subspace of V (or a left submodule of V). See Note 5 following Theorem 7.19 for details.

(5) If $f: A \to B$ is a homomorphism, then the image of f is $f[A] = \{f(x) | x \in A\}$.

(6) If N is a normal subgroup of a group G, then the quotient group of G by N is the group $G/N = \{xN \mid x \in G\}$ consisting of the left cosets of N in G. By Theorem 10.15, we also have $G/N = \{Nx \mid x \in G\}$ (and so, it doesn't matter if we use left or right cosets). If G is commutative, we will use additive notation and write $G/N = \{x + N \mid x \in G\}$.

Similarly, if I is an ideal of a ring R, then the quotient ring of R by I is the ring $R/I = \{x + I \mid x \in R\}$ and if W is a subspace of a vector space V over a field F (or a left submodule of a module V over a ring R), then the quotient space (or quotient module) of V by W is the vector space (or left module) $V/W = \{x + W \mid x \in V\}$.

There are four important theorems relating isomorphisms and quotients that are very useful in the theories of groups, rings, vector spaces, and modules. These are called the **Isomorphism Theorems**. We will state and prove each theorem separately for groups, rings and vector spaces. In all cases, the proofs for left modules are identical to the proofs for vector spaces.

Given a group homomorphism $f: G \to H$, $\ker(f)$ is normal in G. Therefore, we can always form the quotient $G/\ker(f)$. The **First Isomorphism Theorem** (also known as the **Fundamental Homomorphism Theorem**) says that this quotient looks and acts just like the image of f.

Theorem 11.1(G) (First Isomorphism Theorem—Group Version): Let $f: G \to H$ be a homomorphism of groups. Then $G/\ker(f) \cong f[G]$.

Proof: Define $F: G/\ker(f) \to f[G]$ by $F(x \ker(f)) = f(x)$. Let's first show that F is well-defined. Suppose that $x \ker(f) = y \ker(f)$. By Theorem 10.5 (4 → 2), $x^{-1}y \in \ker(f)$. So, $f(x^{-1}y) = e_H$. Since f is a homomorphism, $f(x)^{-1}f(y) = f(x^{-1})f(y) = f(x^{-1}y) = e_H$. So,

$$f(y) = e_H f(y) = (f(x)f(x)^{-1})f(y) = f(x)(f(x)^{-1}f(y)) = f(x)e_H = f(x).$$

Therefore, F is well-defined.

To see that F is injective, suppose that $F(x \ker(f)) = F(y \ker(f))$, or equivalently, $f(x) = f(y)$. Then by direct substitution, we have $f(x)^{-1}f(y) = f(x)^{-1}f(x) = e_H$. Since f is a homomorphism, it follows that $f(x^{-1}y) = f(x^{-1})f(y) = f(x)^{-1}f(y) = e_H$. Therefore, $x^{-1}y \in \ker(f)$. By Theorem 10.5 (2 → 4), $x \ker(f) = y \ker(f)$. Since $x \ker(f), y \ker(f) \in G/\ker(f)$ were arbitrary, F is injective.

F is surjective because if $y \in f[G]$, then there is $x \in G$ with $y = f(x)$. So, $F(x \ker(f)) = f(x) = y$.

Finally, $F(x \ker(f) \cdot y \ker(f)) = F(xy \ker(f)) = f(xy) = f(x)f(y) = F(x \ker(f)) \cdot F(y \ker(f))$. So, F is a homomorphism. □

Note: In the case where the homomorphism f given in the statement of Theorem 11.1 is surjective, we see that H is isomorphic to $G/\ker(f)$.

For vector spaces, the first isomorphism theorem says the following:

Theorem 11.1(V) (First Isomorphism Theorem—Vector Space Version): Let $f: V \to W$ be a linear transformation of vector spaces over a field F. Then $V/\ker(f) \cong f[V]$.

The proof is quite similar to the one given for groups. There are a few modifications that need to be made. The first is that in the proof for vector spaces, additive notation should be used because a vector space is a commutative group. The second is that to check that F is a linear transformation, two computations need to be verified instead of just one (linearity and homogeneity). Also, we will use 0_V and 0_W for the identities of V and W, respectively. So, the proof looks as follows:

Proof of Theorem 11.1(V): Define $F: V/\ker(f) \to f[V]$ by $F(x + \ker(f)) = f(x)$. Let's first show that F is well-defined. Suppose that $x + \ker(f) = y + \ker(f)$. By Theorem 10.5 (4 → 2), $y - x \in \ker(f)$. So, $f(y - x) = 0_W$. Since f is a linear transformation, we see that

$$f(y) - f(x) = f(y - x) = 0_W.$$

So, $f(y) = 0_W + f(y) = (f(x) - f(x)) + f(y) = f(x) + (f(y) - f(x)) = f(x) + 0_W = f(x)$.

Therefore, F is well-defined.

To see that F is injective, suppose that $F(x + \ker(f)) = F(y + \ker(f))$, or equivalently, $f(x) = f(y)$. Then by direct substitution, we have $f(y) - f(x) = f(x) - f(x) = 0_W$. Since f is a linear transformation, it follows that $f(y - x) = f(y) - f(x) = 0_W$. Therefore, $y - x \in \ker(f)$. By Theorem 10.5 (2 → 4), $x + \ker(f) = y + \ker(f)$. Since $x + \ker(f), y + \ker(f) \in V/\ker(f)$ were arbitrary, F is injective.

F is surjective because if $y \in f[V]$, then there is $x \in V$ with $y = f(x)$. So, $F(x + \ker(f)) = f(x) = y$.

Finally,

$$F\big((x + \ker(f)) + (y + \ker(f))\big) = F\big((x + y) + \ker(f)\big)$$
$$= f(x + y) = f(x) + f(y) = F(x + \ker(f)) + F(y + \ker(f)).$$
$$F\big(k(x + \ker(f))\big) = F(kx + \ker(f)) = f(kx) = kf(x) = kF(x + \ker(f)).$$

So, F is a homomorphism. $\square$

For rings, the first isomorphism theorem says the following:

Theorem 11.1(R) (First Isomorphism Theorem—Ring Version): Let $f: R \to S$ be a homomorphism of rings. Then $R/\ker(f) \cong f[R]$.

The proof of this last version is most similar to the vector space version because the cosets of a subring always use additive notation. Of course, there is no scalar multiplication. Instead, we need to verify that the function F is well-behaved with respect to the ring multiplication. We also must check that the multiplicative identity of $R/\ker(f)$ is mapped to the multiplicative identity of S.

Proof of Theorem 11.1(R): Define $F: R/\ker(f) \to f[R]$ by $F(x + \ker(f)) = f(x)$. The proof that F is a well-defined bijection is identical to the vector space version of the theorem. Finally, we have

$$F\big((x + \ker(f)) + (y + \ker(f))\big) = F\big((x + y) + \ker(f)\big)$$
$$= f(x + y) = f(x) + f(y) = F(x + \ker(f)) + F(y + \ker(f)).$$
$$F\big((x + \ker(f)) \cdot (y + \ker(f))\big) = F(xy + \ker(f))$$
$$= f(xy) = f(x) \cdot f(y) = F(x + \ker(f)) \cdot F(y + \ker(f)).$$
$$F(1_R + \ker(f)) = f(1_R) = 1_S.$$

So, F is a homomorphism. $\square$

Example 11.2:

1. Define $f: \mathbb{Z} \to \mathbb{Z}_2$ by $f(k) = [k]_2$. Then f is a surjective ring homomorphism and $\ker(f) = 2\mathbb{Z}$ (see part 2 of Example 7.17). It follows from Theorem 11.1(R) that $\mathbb{Z}/2\mathbb{Z} \cong \mathbb{Z}_2$.

2. More generally, if $n \in \mathbb{Z}^+$, then $f: \mathbb{Z} \to \mathbb{Z}_n$ is a surjective ring homomorphism and $\ker(f) = n\mathbb{Z}$ (once again, see part 2 of Example 7.17). So, again by Theorem 11.1(R), we get $\mathbb{Z}/n\mathbb{Z} \cong \mathbb{Z}_n$.

3. Define $f: \mathbb{Z}_{36} \to \mathbb{Z}_9$ by $f([k]_{36}) = [k]_9$. Then f is a surjective ring homomorphism and $\ker(f) = \{0, 9, 18, 27\} = \langle 9 \rangle$. It follows from Theorem 11.1(R) that $\mathbb{Z}_{36}/\langle 9 \rangle \cong \mathbb{Z}_9$.

4. Let $G = S_3 = \{(1), (12), (13), (23), (123), (132)\}$ and define $f: G \to \mathbb{Z}_2$ as follows:
$$f = \{((1), 0), ((123), 0), ((132), 0), ((12), 1), ((13), 1), ((23), 1)\}$$
The reader can easily verify that f is a surjective homomorphism with kernel $N = \{(1), (123), (132)\}$. It follows from Theorem 11.1(G) that $S_3/N \cong \mathbb{Z}_2$. (We will see in Lesson 12 that $N = A_3$, the alternating group of degree 3.)

5. Let $\mathbb{S} = \{z \in \mathbb{C} \mid |z| = 1\}$ be the unit circle in the Complex Plane. Recall from Lesson 5 that the wrapping function $W: \mathbb{R} \to \mathbb{S}$ is defined by $W(x) = e^{ix}$. This function W is a surjective group homomorphism from $(\mathbb{R}, +)$ to $(\mathbb{S}, \cdot)$. To see that W is a homomorphism, observe that
$$W(x + y) = e^{i(x+y)} = e^{ix}e^{iy} = W(x) \cdot W(y).$$
The kernel of this homomorphism is $\ker(W) = \{2k\pi \mid k \in \mathbb{Z}\} = \langle 2\pi \rangle$. It follows from Theorem 11.1(G) that $\mathbb{R}/\langle 2\pi \rangle \cong \mathbb{S}$.

6. Let $T: \mathbb{R}^3 \to \mathbb{R}^2$ be the surjective linear transformation defined by $T(x, y, z) = (x, y)$. The kernel of T is $K = \{(0, 0, z) \mid z \in \mathbb{R}\}$. It follows from Theorem 11.1(V) that $\mathbb{R}^3/K \cong \mathbb{R}^2$. Since K is isomorphic to $\mathbb{R}$, we essentially showed that $\mathbb{R}^3/\mathbb{R} \cong \mathbb{R}^2$.

    Similarly, the surjective linear transformation $S: \mathbb{R}^3 \to \mathbb{R}$ defined by $S(x, y, z) = x$ has kernel $K = \{(0, y, z) \mid z \in \mathbb{R}\}$. Since $K \cong \mathbb{R}^2$, this essentially shows that $\mathbb{R}^3/\mathbb{R}^2 \cong \mathbb{R}$.

The converse of the first homomorphism theorem happens to be true as well.

Theorem 11.3(G): Let G be a group and let N be a normal subgroup of G. Then there is a surjective homomorphism $f: G \to G/N$.

Proof: Define $F: G \to G/N$ by $F(x) = xN$. Clearly F is surjective. If $x, y \in G$, then we have
$$F(xy) = xyN = (xN)(yN) = F(x)F(y).$$
Therefore, F is a homomorphism. □

Let's write the vector space and ring versions of Theorem 11.3. The proofs are just as straightforward as the group version, and so, I leave them to the reader.

Theorem 11.3(V): Let V be a vector space over a field F and let W be a subspace of V. Then there is a surjective linear transformation $f: V \to V/W$.

Theorem 11.3(R): Let R be a ring and let I be an ideal of R. Then there is a surjective homomorphism $f: R \to R/I$.

The **Second Isomorphism Theorem** (also known as the **Diamond Isomorphism Theorem**) will involve products (or sums in additive notation) and intersections of substructures. Before we introduce the theorem, we will want to make sure we understand all the definitions involved. Additionally, we will need some extra conditions on the substructures to make sure that everything "works."

Let's begin by defining sums, products, and intersections of structures.

If S and T are substructures of a structure A, then

$$ST = \{xy \mid x \in S \text{ and } y \in T\} \quad S + T = \{x + y \mid x \in S \text{ and } y \in T\} \quad S \cap T = \{x \mid x \in S \text{ and } y \in T\}$$

The structures we will be most concerned with here are groups, rings, and vector spaces.

For example, if S and T are subgroups of a group G, then we will want to consider ST (or $S + T$ if G is commutative) and $S \cap T$. These are clearly both subsets of G, but are they subgroups of G? Well, $S \cap T$ is a subgroup of G (you were asked to prove this in Problem 11 in Problem Set 6). However, ST need not be a subgroup of G.

Example 11.4: Let $G = S_3 = \{(1), (12), (13), (23), (123), (132)\}$ be the symmetric group on three elements, let $S = \{(1), (12)\}$, and let $T = \{(1), (13)\}$. We have $(1)(1) = 1$, $(1)(13) = (13)$, $(12)(1) = (12)$, and $(12)(13) = (132)$. So, $ST = \{(1), (12), (13), (132)\}$. By Lagrange's Theorem (or more specifically, by Theorem 10.9), ST is **not** a subgroup of S_3. Indeed, $|ST| = 4$, $|S_3| = 6$, and 4 does **not** divide 6.

We will see in Theorem 11.6 below that as long as one of the subgroups is a normal subgroup of G, then ST will be a subgroup of G.

We run into a similar issue with subrings S and T of a ring R. The intersection $S \cap T$ will always be a subring of R (again, by Problem 11 in Problem Set 6). However, $S + T$ need not be a subring of R.

Example 11.5: Let $R = \mathbb{R}$, let $S = \mathbb{Z}[\sqrt{2}]$, and let $T = \mathbb{Z}[\sqrt{3}]$, where $\mathbb{Z}[\sqrt{2}] = \{a + b\sqrt{2} \mid a, b \in \mathbb{Z}\}$ and $\mathbb{Z}[\sqrt{3}] = \{a + b\sqrt{3} \mid a, b \in \mathbb{Z}\}$. It is easy to verify that $\mathbb{Z}[\sqrt{2}]$ and $\mathbb{Z}[\sqrt{3}]$ are subrings of $\mathbb{R}$. However, $\mathbb{Z}[\sqrt{2}] + \mathbb{Z}[\sqrt{3}] = \{a + b\sqrt{2} + c\sqrt{3} \mid a, b, c \in \mathbb{Z}\}$ is **not** a subring of $\mathbb{R}$. For example, we have $\sqrt{2} \in \mathbb{Z}[\sqrt{2}] + \mathbb{Z}[\sqrt{3}]$ and $\sqrt{3} \in \mathbb{Z}[\sqrt{2}] + \mathbb{Z}[\sqrt{3}]$, but $\sqrt{2} \cdot \sqrt{3} = \sqrt{6} \notin \mathbb{Z}[\sqrt{2}] + \mathbb{Z}[\sqrt{3}]$, and therefore, $\mathbb{Z}[\sqrt{2}] + \mathbb{Z}[\sqrt{3}]$ is not closed under multiplication.

We will see in Theorem 11.8 below that as long as one of the subrings is an ideal of R, then $S + T$ will be a subring of R.

We don't run into any of these technical issues with vector spaces. If S and T are subspaces of a vector space V over a field F, then $S \cap T$ and $S + T$ are both subspaces of V. The same holds true if S and T are left submodules of a module over a ring R.

Let's now prove some preliminary theorems in preparation for the Second Isomorphism Theorem.

Theorem 11.6: Let G be a group, let S be a subgroup of G, and let N be a normal subgroup of G. Then $SN = \{xy \mid x \in S \text{ and } y \in N\}$ is a subgroup of G.

Proof: Since $e \in S$ and $e \in N$, we have $e = ee \in SN$. So, $SN \neq \emptyset$. Let $a, b \in SN$. Then there are $x, z \in S$ and $y, w \in N$ such that $a = xy$ and $b = zw$. So,

$$ab^{-1} = (xy)(zw)^{-1} = xyw^{-1}z^{-1} = x(z^{-1}z)yw^{-1}z^{-1} = (xz^{-1})(zyw^{-1}z^{-1}).$$

Since $x, z \in S$ and S is a subgroup of G, $xz^{-1} \in S$. Since $y, w \in N$ and N is a subgroup of G, $yw^{-1} \in N$. Since $yw^{-1} \in N$, $z \in G$, and N is a normal subgroup of G, $zyw^{-1}z^{-1} \in N$. It follows that $ab^{-1} \in SN$. By Problem 5 in Problem Set 6, SN is a subgroup of G. □

Notes: (1) If N is a normal subgroup of G, then N is also normal in any subgroup of G containing N. Let's check this. Suppose that $N \subseteq H \leq G$. Since N is a subset of H and N, H are both groups with the same operation, $N \leq H$. Now, let $y \in N$ and $x \in H$. Since $H \subseteq G$, $x \in G$. Since $N \triangleleft G$, $xyx^{-1} \in N$. So, $N \triangleleft H$.

(2) By Note 1 above and Theorem 11.6, if S is any subgroup of G and N is a normal subgroup of G, then $N \triangleleft SN$.

Theorem 11.7: Let G be a group, let S be a subgroup of G, and let N be a normal subgroup of G. Then $S \cap N$ is a normal subgroup of S.

Proof: By Problem 19 in Problem Set 1, $S \cap N \subseteq S$.

Since $e \in S$ and $e \in N$, we have $e \in S \cap N$. So, $S \cap N \neq \emptyset$. Let $a, b \in S \cap N$. Then $a, b \in S$ and $a, b \in N$. Since S and N are subgroups of G, $ab^{-1} \in S$ and $ab^{-1} \in N$. So, $ab^{-1} \in S \cap N$. By Problem 5 in Problem Set 6, $S \cap N$ is a subgroup of S.

Let $y \in S \cap N$ and $x \in S$. Then $y \in S$ and $y \in N$. Since S is a group, $xyx^{-1} \in S$. Since S is a subgroup of G, $x \in G$. Since N is a normal subgroup of G and $x \in G$, $xyx^{-1} \in N$. So, $xyx^{-1} \in S \cap N$. Therefore, $S \cap N$ is a normal subgroup of S. □

Theorem 11.8: Let R be a ring, let S be a subring of R, and let I be an ideal of R. Then $S + I = \{x + y \mid x \in S \text{ and } y \in I\}$ is a subring of R.

Proof: Since $0, 1 \in S$ and $0 \in I$, we have $0 = 0 + 0 \in S + I$ and $1 = 1 + 0 \in S + I$. Let $a, b \in S + I$. Then there are $x, z \in S$ and $y, w \in I$ such that $a = x + y$ and $b = z + w$. So,

$$a - b = (x + y) - (z + w) = (x - z) + (y - w).$$

Since $x, z \in S$ and $(S, +)$ is a subgroup of $(R, +)$, $x - z \in S$. Since $y, w \in I$ and $(I, +)$ is a subgroup of $(R, +)$, $y - w \in I$. It follows that $a - b \in S + I$. Also,

$$ab = (x + y)(z + w) = (x + y)z + (x + y)w = xz + yz + (x + y)w.$$

Since $x, z \in S$ and S is a subring of R, $xz \in S$. Since $y, w \in I$ and I absorbs R, $yz \in I$ and $(x + y)w \in I$. Since $(I, +)$ is a subgroup of $(R, +)$, $yz + (x + y)w \in I$. It follows that $ab \in S + I$. □

Notes: (1) If I is an ideal of R, then I is an ideal of any subring of R containing I. The verification of this is similar to what we did in Note 1 following Theorem 11.6. Therefore, I leave the details to the reader.

(2) By Note 1 above and Theorem 11.8, if S is any subring of R and I is an ideal of R, then I is an ideal of $S + I$.

Theorem 11.9: Let R be a ring, let S be a subring of R, and let I be an ideal of R. Then $S \cap I$ is an ideal of S.

Proof: By Problem 19 in Problem Set 1, $S \cap I \subseteq S$.

Since $0 \in S$ and $0 \in I$, we have $0 \in S \cap I$. Let $a, b \in S \cap I$. Then $a, b \in S$ and $a, b \in I$. Since S and I are subrings of R, $a - b \in S$ and $a - b \in I$. So, $a - b \in S \cap I$. By Problem 5 in Problem Set 6, $(S \cap I, +)$ is a subgroup of $(S, +)$.

Let $y \in S \cap I$ and $x \in S$. Then $y \in S$ and $y \in I$. Since S is a ring, $xy \in S$ and $yx \in S$. Since S is a subring of R, $x \in R$. Since I is an ideal of R, $y \in I$, and $x \in R$, we have $xy \in I$ and $yx \in I$. So, $xy \in S \cap I$ and $yx \in S \cap I$. Therefore, $S \cap I$ is an ideal of S. □

Theorem 11.10: Let V be a vector space over a field F and let U and W be subspaces of V. Then $U + W = \{x + y \mid x \in U \text{ and } y \in W\}$ and $U \cap V$ are subspaces of V.

The proof of Theorem 11.10 is essentially contained in the proof of Theorem 11.8 and Problem 11 in Problem Set 6. Therefore, I leave the details to the reader.

We are now ready to state and prove the Second Isomorphism Theorem. This theorem is also known as the Diamond Isomorphism Theorem because there is a nice visualization of this theorem as a diamond whose vertices are labelled with the four subgroups mentioned in the theorem. In the figure to the right, two pairs of vertices are circled because these pairs correspond to the two quotients that appear in the conclusion of the Second Isomorphism Theorem.

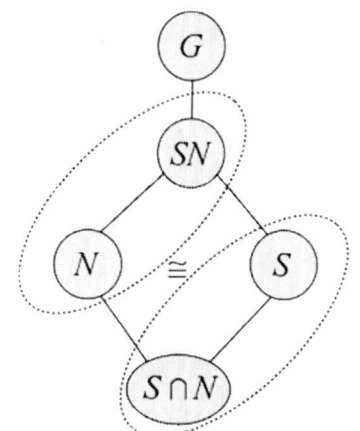

Theorem 11.11(G) (Second Isomorphism Theorem—Group Version): Let G be a group, let S be a subgroup of G and let N be a normal subgroup of G. Then $(SN)/N \cong S/(S \cap N)$.

Proof: By Note 2 following Theorem 11.6, $N \triangleleft SN$. Define $f : S \to (SN)/N$ by $f(x) = xN$. If $x, y \in S$, then $f(xy) = xyN = (xN)(yN) = f(x)f(y)$. So, f is a homomorphism. Also, we see that

$$\ker(f) = \{x \in S \mid f(x) = e_{(SN)/N}\} = \{x \in S \mid xN = N\} = \{x \in S \mid x \in N\} = S \cap N.$$

Let's show that f is surjective. To see this, let $xN \in (SN)/N$. Then $x \in SN$. So, there are $y \in S$ and $z \in N$ with $x = yz$. Since $z \in N$, $zN = N$. Therefore, $xN = yzN = yN$. So, $f(y) = yN = xN$.

By the First Isomorphism Theorem (Theorem 11.1(G)), $S/\ker(f) \cong f[S]$. Since $\ker(f) = S \cap N$ and f is surjective, $S/(S \cap N) \cong (SN)/N$. □

Theorem 11.11(V) (Second Isomorphism Theorem—Vector Space Version): Let V be a vector space over a field F and let U and W be subspaces of V. Then $(U + W)/W \cong U/(U \cap W)$.

Proof: Define $f : U \to (U + W)/W$ by $f(x) = x + W$. If $x, y \in U$, then we have
$$f(x + y) = (x + y) + W = (x + W) + (y + W) = f(x) + f(y)$$
$$f(kx) = kx + W = k(x + W) = kf(x).$$

So, f is a homomorphism. Also, we see that

$$\ker(f) = \{x \in U \mid f(x) = 0_{(U+W)/W}\} = \{x \in U \mid x + W = W\} = \{x \in U \mid x \in W\} = U \cap W.$$

Let's show that f is surjective. To see this, let $x + W \in (U + W)/W$. Then $x \in U + W$. So, there are $y \in U$ and $z \in W$ with $x = y + z$. Since $z \in W$, $z + W = W$. Therefore,

$$x + W = (y + z) + W = y + (z + W) = y + W.$$

So, $f(y) = y + W = x + W$.

By the First Isomorphism Theorem (Theorem 11.1(V)), $U/\ker(f) \cong f[U]$. Since $\ker(f) = U \cap W$ and f is surjective, $U/(U \cap W) \cong (U + W)/W$. □

Theorem 11.11(R) (Second Isomorphism Theorem—Ring Version): Let R be a ring, let S be a subring of R and let I be an ideal of R. Then $(S + I)/I \cong S/(S \cap I)$.

Proof: By Note 2 following Theorem 11.8, $I \triangleleft S + I$. Define $f: S \to (S + I)/I$ by $f(x) = x + I$. If $x, y \in S$, then we have $f(1) = 1 + I$, $f(x + y) = (x + y) + I = (x + I) + (y + I) = f(x) + f(y)$, and $f(xy) = xy + I = (x + I)(y + I) = f(x)f(y)$. So, f is a homomorphism. Also, we see that

$$\ker(f) = \{x \in S \mid f(x) = 0_{(S+I)/I}\} = \{x \in S \mid x + I = I\} = \{x \in S \mid x \in I\} = S \cap I.$$

Let's show that f is surjective. To see this, let $x + I \in (S + I)/I$. Then $x \in S + I$. So, there are $y \in S$ and $z \in I$ with $x = y + z$. Since $z \in I$, $z + I = I$. Therefore,

$$x + I = (y + z) + I = y + (z + I) = y + I.$$

So, $f(y) = y + I = x + I$.

By the First Isomorphism Theorem (Theorem 11.1(R)), $S/\ker(f) \cong f[S]$. Since $\ker(f) = S \cap I$ and f is surjective, $S/(S \cap I) \cong (S + I)/I$. □

Example 11.12:

1. Let $R = \mathbb{Z}$, $S = 4\mathbb{Z}$, and $I = 6\mathbb{Z}$. Then $S \cap I = \text{lcm}(4,6)\mathbb{Z} = 12\mathbb{Z}$ and $S + I = \gcd(4,6)\mathbb{Z} = 2\mathbb{Z}$ by Problem 13 below. It follows from Theorem 11.11(R) that $2\mathbb{Z}/6\mathbb{Z} \cong 4\mathbb{Z}/12\mathbb{Z}$.

2. More generally, if $m, n \in \mathbb{Z}^+$, $R = \mathbb{Z}$, $S = m\mathbb{Z}$, and $I = n\mathbb{Z}$, then $S \cap I = \text{lcm}(m,n)\mathbb{Z}$ and $S + I = \gcd(m,n)\mathbb{Z}$ by Problem 13 below. It follows from Theorem 11.11(R) that $\gcd(m,n)\mathbb{Z}/n\mathbb{Z} \cong m\mathbb{Z}/\text{lcm}(m,n)\mathbb{Z}$.

    For example, if m and n are relatively prime, then we get $\mathbb{Z}/n\mathbb{Z} \cong m\mathbb{Z}/mn\mathbb{Z}$.

3. In part 4 of Example 11.2, we showed that $S_3/A_3 \cong \mathbb{Z}_2$, where $A_3 = \{(1), (123), (132)\}$. We can verify this same result using Theorem 11.11(G) as well. If we let $S = \{(1), (12)\}$, then we see that $S \cap A_3 = \{(1)\}$ and $SA_3 = S_3$. By Theorem 11.11(G), $S_3/A_3 \cong S/\{(1)\}$. Now observe that $S/\{(1)\} \cong S$, which is a cyclic group of order 2. Therefore, $S \cong \mathbb{Z}_2$, and so, $S_3/A_3 \cong \mathbb{Z}_2$, as desired.

4. Let $V = \mathbb{R}^3$, $U = \{(a, b, 0) \mid a, b \in \mathbb{R}\}$, and $W = \{(0, b, c) \mid b, c \in \mathbb{R}\}$ (as vector spaces over $\mathbb{R}$). Then $U + W = \mathbb{R}^3$ and $U \cap W = \{(0, b, 0) \mid b \in \mathbb{R}\}$. By Theorem 11.11(V), we get $\mathbb{R}^3/W \cong U/U \cap W$. Notice that U and W are both isomorphic to $\mathbb{R}^2$ and $U \cap W$ is isomorphic to $\mathbb{R}$. So, we essentially showed that $\mathbb{R}^3/\mathbb{R}^2 \cong \mathbb{R}^2/\mathbb{R}$.

The **Third Isomorphism Theorem** (also known as the **Freshman Theorem**) will involve quotients of quotients. As we did for the Second Isomorphism Theorem, we will first prove some preliminary theorems in preparation for the Third Isomorphism Theorem.

Theorem 11.13: Let G be a group, let N be a normal subgroup of G, and let H be a subgroup of G with $N \subseteq H \subseteq G$. Then H/N is a subgroup of G/N. Furthermore, if H is a normal subgroup of G, then H/N is a normal subgroup of G/N.

Proof: Assume that N is a normal subgroup of G and H is a subgroup of G with $N \subseteq H \subseteq G$.

By Note 1 following Theorem 11.6, N is normal in H. Therefore, H/N is a group.

Now, if $xN \in H/N$, then $x \in H$. Since H is a subset of G, $x \in G$. Therefore, we have $xN \in G/N$. Since $xN \in H/N$ was arbitrary, $H/N \subseteq G/N$. Therefore, H/N is a subgroup of G/N.

Now, assume that H is a normal subgroup of G. Let $yN \in H/N$ and let $xN \in G/N$. Then $y \in H$ and $x \in G$. So, $xyx^{-1} \in H$. It follows that $(xN)(yN)(xN)^{-1} = xyx^{-1}N \in H/N$. Therefore, H/N is a normal subgroup of G/N. □

We now show that the converse of Theorem 11.13 is also true.

Theorem 11.14: Let G be a group and let N be a normal subgroup of G. Then every subgroup of G/N is of the form H/N for some subgroup H of G with $N \subseteq H \subseteq G$. Furthermore, every normal subgroup of G/N is of the form H/N for some normal subgroup H of G with $N \subseteq H \subseteq G$.

Proof: Assume that N is a normal subgroup of G and let K be a subgroup of G/N. We define H as follows: $H = \{x \in G \mid xN \in K\}$.

Let's show that H is a subgroup of G. Since K is a subgroup of G/N, $eN = N \in K$. Therefore, $e \in H$. So, $H \neq \emptyset$. Let $x, y \in H$. Then $xN, yN \in K$. Since K is a subgroup of G/N, by Problem 5 in Problem Set 6, $xy^{-1}N = (xN)(yN)^{-1} \in K$. So, $xy^{-1} \in H$. Again, By Problem 5 in Problem Set 6, H is a subgroup of G.

Next, let's show that $N \subseteq H$. To see this, let $x \in N$. Then $xN = N$. Since K is a subgroup of G/N and N is the identity of G/N, $xN = N \in K$. Therefore, $x \in H$.

We now show that $H/N = K$. Let $xN \in H/N$. Then $x \in H$. By the definition of H, $xN \in K$. Therefore, $H/N \subseteq K$. Conversely, let $xN \in K$. By the definition of H, $x \in H$. So, $xN \in H/N$. Thus, $K \subseteq H/N$. Since $H/N \subseteq K$ and $K \subseteq H/N$, we have $H/N = K$.

Finally, assume that K is a normal subgroup of G/N. Let $y \in H$ and $x \in G$. By the definition of H, we have $yN \in K$. Since K is a normal subgroup of G/N, $xyx^{-1}N = (xN)(yN)(xN)^{-1} \in K$. By the definition of H, $xyx^{-1} \in H$. Thus, H is a normal subgroup of G. □

The analogues of Theorems 11.13 and 11.14 for rings and vector spaces are also true and the proofs of these theorems are quite similar. Therefore, I will state these theorems and leave the proofs to the reader (see Problems 6 through 9 below).

Theorem 11.15: Let R be a ring, let I be an ideal of R, and let S be a subring of R with $I \subseteq S \subseteq R$. Then S/I is a subring of R/I. Furthermore, if S is an ideal of R, then S/I is an ideal of R/I.

Theorem 11.16: Let R be a ring and let I be an ideal of R. Then every subring of R/I is of the form S/I for some subring S of R with $I \subseteq S \subseteq R$. Furthermore, every ideal of R/I is of the form S/I for some ideal S of R with $I \subseteq S \subseteq R$.

Theorem 11.17: Let V be a vector space over a field F and let U and W be subspaces of V with $W \subseteq U \subseteq V$. Then U/W is a subspace of V/W. Conversely, if W is a subspace of the vector space V, then every subspace of V/W is of the form U/W for some subspace U of V with $W \subseteq U \subseteq V$.

We are now ready to state and prove the Third Isomorphism Theorem. This theorem is also known as the Freshman Theorem because of the way that the conclusion of the theorem shows that quotients can be simplified in the same way that a freshman would simplify a quotient of real numbers in a first year high school algebra course.

Theorem 11.18(G) (Third Isomorphism Theorem—Group Version): Let G be a group and let H and N be normal subgroups of G such that $N \subseteq H \subseteq G$. Then $(G/N)/(H/N) \cong G/H$.

Proof: By Theorem 11.13, H/N is a normal subgroup of G/N, and so, the quotient $(G/N)/(H/N)$ is a group. Define a function $F: G/N \to G/H$ by $F(xN) = xH$.

Let's first check that F is well-defined. If $xN = yN$, then by Theorem 10.5 (4 → 2), $x^{-1}y \in N$. Since $N \subseteq H$, we have $x^{-1}y \in H$. So, again by Theorem 10.5 (2 → 4), $xH = yH$.

Next, let's check that F is a homomorphism:
$$F\big((xN)(yN)\big) = F(xyN) = xyH = (xH)(yH) = F(xN)F(yN).$$

It's clear that F is surjective.

Let $K = \ker(F)$. By the First Isomorphism Theorem, $(G/N)/K \cong G/H$.

We will complete the proof by showing that $K = H/N$. Let $xN \in K$. Then $F(xN) = H$. Also, by definition, $F(xN) = xH$. So, we have $xH = H$, and therefore, $x \in H$. So, $xN \in H/N$. Since $xN \in K$ was arbitrary, we see that $K \subseteq H/N$. Conversely, let $xN \in H/N$. Then $x \in H$ and $F(xN) = xH$. Since $x \in H$, we have $xH = H$, and so, $F(xN) = H$. Therefore, $xN \in K$. Since $xN \in H/N$ was arbitrary, we have $H/N \subseteq K$. Since $K \subseteq H/N$ and $H/N \subseteq K$, it follows that $K = H/N$, as desired. □

Theorem 11.18(V) (Third Isomorphism Theorem—Vector Space Version): Let V be a vector space over a field F and let U and W be subspaces of V such that $W \subseteq U \subseteq V$. Then $(V/W)/(U/W) \cong V/U$.

Proof: By Theorem 11.17, U/W is a subspace of V/W, and so, the quotient $(V/W)/(U/W)$ is a vector space. Define a function $F: V/W \to V/U$ by $F(x + W) = x + U$.

Let's first check that F is well-defined. If $x + W = y + W$, then by Theorem 10.5 (4 → 2), $y - x \in W$. Since $W \subseteq U$, we have $y - x \in U$. So, again by Theorem 10.5 (2 → 4), $x + U = y + U$.

Next, let's check that F is a homomorphism:

$$F\big((x+W)+(y+W)\big) = F\big((x+y)+W\big) = (x+y)+U$$
$$= (x+U)+(y+U) = F(x+W)+F(y+W).$$
$$F\big(k(x+W)\big) = F(kx+W) = kx+U = k(x+U) = kF(x+W).$$

It's clear that F is surjective.

Let $K = \ker(F)$. By the First Isomorphism Theorem, $(V/W)/K \cong V/U$.

We will complete the proof by showing that $K = U/W$. Let $x+W \in K$. Then $f(x+W) = U$. Also, by definition, $F(x+W) = x+U$. So, we have $x+U = U$, and therefore, $x \in U$. So, $x+W \in U/W$. Since $x+W \in K$ was arbitrary, we see that $K \subseteq U/W$. Conversely, let $x+W \in U/W$. Then $x \in U$ and $F(x+W) = x+U$. Since $x \in U$, we have $x+U = U$, and so, $F(x+W) = U$. Therefore, $x+W \in K$. Since $x+W \in U/W$ was arbitrary, we have $U/W \subseteq K$. Since $K \subseteq U/W$ and $U/W \subseteq K$, it follows that $K = U/W$, as desired. □

Theorem 11.18(R) (Third Isomorphism Theorem—Ring Version): Let R be a ring and let S and I be ideals of R such that $I \subseteq S \subseteq R$. Then $(R/I)/(S/I) \cong R/S$.

Proof: By Theorem 11.15, S/I is an ideal of R/I, and so, the quotient $(R/I)/(S/I)$ is a ring. Define a function $F: R/I \to R/S$ by $F(x+I) = x+S$.

The rest of the proof is identical to the vector space version, except for the verification that F is a homomorphism. Let's check that now:

$$F\big((x+I)+(y+I)\big) = F\big((x+y)+I\big) = (x+y)+S$$
$$= (x+S)+(y+S) = F(x+I)+F(y+I).$$
$$F\big((x+I)(y+I)\big) = F(xy+I) = xy+S$$
$$= (x+S)(y+S) = F(x+I)F(y+I).$$
$$F(1+I) = 1+S$$

So, F is a homomorphism. □

Example 11.19:

1. Let $R = \mathbb{Z}$, $S = 2\mathbb{Z}$, and $I = 4\mathbb{Z}$. Then $4\mathbb{Z} \subseteq 2\mathbb{Z} \subseteq \mathbb{Z}$, and therefore, by Theorem 11.18(R), we have $(\mathbb{Z}/4\mathbb{Z})/(2\mathbb{Z}/4\mathbb{Z}) \cong \mathbb{Z}/2\mathbb{Z}$, or equivalently, $(\mathbb{Z}/4\mathbb{Z})/(2\mathbb{Z}/4\mathbb{Z}) \cong \mathbb{Z}_2$.

2. More generally, if $a, b, c \in \mathbb{Z}^+$, $a|b$, and $b|c$, then $c\mathbb{Z} \subseteq b\mathbb{Z} \subseteq a\mathbb{Z}$, and therefore, by Theorem 11.18(R), we have $(a\mathbb{Z}/c\mathbb{Z})/(b\mathbb{Z}/c\mathbb{Z}) \cong a\mathbb{Z}/b\mathbb{Z}$.

3. Let $V = \mathbb{R}^3$, $U = \{(a,b,0) \mid a,b \in \mathbb{R}\}$, and $W = \{(a,0,0) \mid a \in \mathbb{R}\}$ (as vector spaces over $\mathbb{R}$). Then $W \subseteq U \subseteq \mathbb{R}^3$. By Theorem 11.18(V), we get $(\mathbb{R}^3/W)/(U/W) \cong \mathbb{R}^3/U$. Notice that U is isomorphic to $\mathbb{R}^2$ and W is isomorphic to $\mathbb{R}$. So, we have essentially just showed that $(\mathbb{R}^3/\mathbb{R})/(\mathbb{R}^2/\mathbb{R}) \cong \mathbb{R}^3/\mathbb{R}^2$.

In part 6 of Example 11.2, we showed that $\mathbb{R}^3/\mathbb{R}^2 \cong \mathbb{R}$. By making the appropriate identifications, it follows that, $(\mathbb{R}^3/\mathbb{R})/(\mathbb{R}^2/\mathbb{R}) \cong \mathbb{R}$.

The **Fourth Isomorphism Theorem** (also known as the **Correspondence Theorem**) says that given a substructure A of a structure B there is a "nice" bijection between substructures of B containing A and all substructures of the quotient B/A.

Theorem 11.20(G) (Fourth Isomorphism Theorem—Group Version): Let G be a group, let N be a normal subgroup of G, let $\mathcal{X}$ be the set of all subgroups of G that contain N, let $\mathcal{Y}$ be the set of subgroups of G/N, and define $f: \mathcal{X} \to \mathcal{Y}$ by $f(H) = H/N$. Then f is a bijection.

Proof: Let's first check that f is a function from $\mathcal{X}$ to $\mathcal{Y}$. Let $H \in \mathcal{X}$. Then H is a subgroup of G containing N. By Theorem 11.13, H/N is a subgroup of G/N. So, $f(H) = H/N \in \mathcal{Y}$, as desired.

Next, let's check that f is injective. Suppose that $f(H) = f(K)$. Then $H/N = K/N$. Let $x \in H$. Then $x + N \in H/N$. Since $H/N = K/N$, $x + N \in K/N$. Therefore, $x \in K$. So, $H \subseteq K$. A symmetrical argument shows that $K \subseteq H$. Thus, $H = K$.

To see that f is surjective, let $K \in \mathcal{Y}$. Then K is a subgroup of G/N. By Theorem 11.14, $K = H/N$ for some subgroup H of G with $N \subseteq H \subseteq G$. So, $H \in \mathcal{X}$ and $f(H) = H/N = K$. □

The proofs for the vector space and ring versions of the fourth isomorphism theorem are just as easy. Therefore, I will simply state these versions of the theorem and leave the proofs for the reader (see Problem 10 below).

Theorem 11.20(V) (Fourth Isomorphism Theorem—Vector Space Version): Let V be a vector space over a field F, let W be a subspace of V, let $\mathcal{X}$ be the set of all subspaces of V that contain W, let $\mathcal{Y}$ be the set of subspaces of V/W, and define $f: \mathcal{X} \to \mathcal{Y}$ by $f(U) = U/W$. Then f is a bijection.

Theorem 11.20(R) (Fourth Isomorphism Theorem—Ring Version): Let R be a ring, let I be an ideal of R, let $\mathcal{X}$ be the set of all subrings of R that contain I, let $\mathcal{Y}$ be the set of subrings of R/I, and define $f: \mathcal{X} \to \mathcal{Y}$ by $f(S) = S/I$. Then f is a bijection.

Example 11.21:

1. Let $Q = \{1, i, j, k, -1, -i, -j, -k\}$. We make Q into a group by letting 1 be the identity, and defining multiplication so that $(-1)^2 = 1$, $i^2 = j^2 = k^2 = -1$, $(-1)i = -i$, $(-1)j = -j$, $(-1)k = -k$, $ij = k$, $jk = i$, $ki = j$, $ji = -k$, $kj = -i$, and $ik = -j$.

    Q is called the **Quaternion group**. Observe that this group is a noncommutative group of order 8. Q has 6 subgroups. They are as follows:

    the trivial subgroup: $\langle 1 \rangle = \{1\}$

    a subgroup of order 2: $\langle -1 \rangle = \{1, -1\}$

    three subgroups of order 4: $\langle i \rangle = \{1, -1, i, -i\}$, $\langle j \rangle = \{1, -1, j, -j\}$, and $\langle k \rangle = \{1, -1, k, -k\}$.

    The Quaternion group itself: $Q = \{1, i, j, k, -1, -i, -j, -k\}$

It is easily checked that all 6 subgroups of Q are normal in Q. In particular, $N = \langle -1 \rangle = \{1, -1\}$ is normal in Q. Theorem 11.20(G) gives us the following bijection between the subgroups of Q containing $\langle -1 \rangle$ and the subgroups of $Q/\langle -1 \rangle$.

$$\langle -1 \rangle = \{1, -1\} \longrightarrow \langle -1 \rangle / \langle -1 \rangle = \{\langle -1 \rangle\}$$
$$\langle i \rangle = \{1, -1, i, -i\} \longrightarrow \langle i \rangle / \langle -1 \rangle = \{\langle -1 \rangle, i\langle -1 \rangle\}$$
$$\langle j \rangle = \{1, -1, j, -j\} \longrightarrow \langle j \rangle / \langle -1 \rangle = \{\langle -1 \rangle, j\langle -1 \rangle\}$$
$$\langle k \rangle = \{1, -1, k, -k\} \longrightarrow \langle k \rangle / \langle -1 \rangle = \{\langle -1 \rangle, k\langle -1 \rangle\}$$
$$Q = \{1, i, j, k, -1, -i, -j, -k\} \longrightarrow Q/\langle -1 \rangle = \{\langle -1 \rangle, i\langle -1 \rangle, j\langle -1 \rangle, k\langle -1 \rangle\}$$

2. Letting $R = \mathbb{Z}$ and $I = 12\mathbb{Z}$, Theorem 11.20(R) gives us the following bijection:

$$12\mathbb{Z} \longrightarrow 12\mathbb{Z}/12\mathbb{Z}$$
$$6\mathbb{Z} \longrightarrow 6\mathbb{Z}/12\mathbb{Z}$$
$$4\mathbb{Z} \longrightarrow 4\mathbb{Z}/12\mathbb{Z}$$
$$3\mathbb{Z} \longrightarrow 3\mathbb{Z}/12\mathbb{Z}$$
$$2\mathbb{Z} \longrightarrow 2\mathbb{Z}/12\mathbb{Z}$$
$$\mathbb{Z} \longrightarrow \mathbb{Z}/12\mathbb{Z}$$

Notes: (1) Let $f: \mathcal{X} \to \mathcal{Y}$ be the bijection defined in the Fourth Isomorphism Theorem. Then $A \leq B$ if and only if $f(A) \leq f(B)$. You will be asked to prove this in part (1) of Problem 15 below.

The notation "$\leq$" here should be interpreted in the appropriate manner. For example, if we are thinking about the group version of the theorem, then $\leq$ is interpreted as "is a subgroup of."

(2) The bijection $f: \mathcal{X} \to \mathcal{Y}$ has many additional nice properties, some of which will be explored in Problems 15 and 16 below.

Fundamental Theorem of Finite Commutative Groups

We now introduce a theorem that allows us to classify all finite commutative groups, up to isomorphism. This result is known as the Fundamental Theorem of Finite Commutative Groups.

Note: Since we will be discussing only commutative groups in this section, we will assume that the group operation is addition.

Theorem 11.22 (The Fundamental Theorem of Finite Commutative Groups): Every finite commutative group is isomorphic to a direct product of cyclic groups of prime power order. Furthermore, the number of terms in the product and the orders (cardinalities) of the cyclic groups are uniquely determined by the group.

Observe that Theorem 11.22 has two parts:

(1) First, Theorem 11.22 says that every finite commutative group G can be expressed as a direct product of cyclic groups of prime power order. Since G is finite, each of the cyclic groups that make up the product must be finite as well.

As a simple example, $\mathbb{Z}_{12}$ is isomorphic to $\mathbb{Z}_4 \times \mathbb{Z}_3$ (Verify this!). Here $4 = 2^2$, $3 = 3^1$, and $2, 3 \in \mathbb{Z}^+$ are both prime.

(2) Second, Theorem 11.22 says that there is **essentially just one way** to write a finite commutative group G as a direct product of cyclic groups of prime power order: if G is isomorphic to both $H_1 \times H_2 \times \cdots \times H_{t_1}$ and $K_1 \times K_2 \times \cdots \times K_{t_2}$, where each H_i is cyclic, each K_i is cyclic, $|H_1| \geq |H_2| \geq \cdots \geq |H_{t_1}| \geq 2$, and $|K_1| \geq |K_2| \geq \cdots \geq |K_{t_2}| \geq 2$, then $t_1 = t_2$ and $|H_i| = |K_i|$ for each $i = 1, 2, \ldots, t_1$.

Continuing with the example of $\mathbb{Z}_{12}$ in (1) above, the second part of Theorem 11.22 says that $\mathbb{Z}_4 \times \mathbb{Z}_3$ is the **only** way to express $\mathbb{Z}_{12}$ as a product of cyclic groups of prime power order, where the first factor has order at least as big as the order of the second factor.

However, note that it is true that $\mathbb{Z}_{12}$ is also isomorphic to $\mathbb{Z}_3 \times \mathbb{Z}_4$. In other words, for the purposes of the uniqueness portion of Theorem 11.22, we consider different arrangements of the cyclic factors of prime power order to produce the same group. After all, when we rearrange the factors in a direct product of a group, we always wind up with a group that is isomorphic to the original. For example, as we already mentioned, $\mathbb{Z}_3 \times \mathbb{Z}_4$ is isomorphic to $\mathbb{Z}_4 \times \mathbb{Z}_3$.

Before discussing the proof of this theorem, let's look at some examples to help us understand what the theorem is saying and how it is used.

Example 11.23:

1. If p is prime, then there is exactly one commutative group of order p, up to isomorphism. It is the cyclic group $\mathbb{Z}_p$. In fact, by the discussion after Theorem 10.9, every group of prime order p is cyclic. It follows from Theorem 9.9 that every group of prime order p is isomorphic to the cyclic group $\mathbb{Z}_p$

    If G is a group of order p, then we can recognize G as a cyclic group "internally" by choosing any nonidentity element of $x \in G$. By Theorem 10.9, $|\langle x \rangle|$ must divide p. Since p is prime, it follows that $|\langle x \rangle| = p$. Therefore, $G = \langle x \rangle$.

2. If G is a commutative group of order pq, where p and q are prime, then Theorem 11.22 tells us that $G \cong \mathbb{Z}_p \times \mathbb{Z}_q$. For example, any commutative group of order $6 = 2 \cdot 3$ is isomorphic to $\mathbb{Z}_2 \times \mathbb{Z}_3$. Note that $\mathbb{Z}_3 \times \mathbb{Z}_2 \cong \mathbb{Z}_2 \times \mathbb{Z}_3$. Theorem 11.22 guarantees uniqueness only up to the order in which the factors appear. To make things as simple as possible we always agree to use the **canonical representation** (or **canonical form**), just as we do when writing the prime factorization of a natural number greater that 1 (see the discussion after Theorem 8.21). In other words, if p and q are distinct prime numbers with $p < q$, then $\mathbb{Z}_p$ will appear to the left of $\mathbb{Z}_q$ in the decomposition of a commutative group as a direct product. So, we will write a commutative group of order 6 as $\mathbb{Z}_2 \times \mathbb{Z}_3$.

    If $G \cong \mathbb{Z}_2 \times \mathbb{Z}_3$, then we call $\mathbb{Z}_2 \times \mathbb{Z}_3$ an **external direct product** of G.

    Since 2 and 3 are distinct prime numbers, it follows that $\gcd(2, 3) = 1$. So, by part (i) of Problem 19 from Problem Set 9, $\mathbb{Z}_2 \times \mathbb{Z}_3 \cong \mathbb{Z}_6$ (note that a ring homomorphism from $(R, +, \cdot)$ to $(S, +, \cdot)$ is automatically also a group homomorphism from $(R, +)$ to $(S, +)$). So, we can identify any commutative group of order 6 with either $\mathbb{Z}_2 \times \mathbb{Z}_3$ or $\mathbb{Z}_6$.

More generally, if p and q are prime, then we can identify any group of order pq with either $\mathbb{Z}_p \times \mathbb{Z}_q$ (and we will usually always write this so that $p < q$, as discussed above) or $\mathbb{Z}_{pq}$.

As in part 1 above, if G is a commutative group of order pq, where p and q are distinct primes, we can recognize G as a cyclic group "internally" as a direct product $\langle x_1 \rangle \times \langle x_2 \rangle$, where x_1 is any element of G of order p and x_2 is any element of G of order q (we will call this an **internal direct product** of G). For example, $\mathbb{Z}_6 \cong \langle 3 \rangle \times \langle 2 \rangle$ or $\mathbb{Z}_6 \cong \langle 3 \rangle \times \langle 4 \rangle$ (because 3 is the only element of order 2 in $\mathbb{Z}_6$ and $2, 4 \in \mathbb{Z}_6$ both have order 3.

Alternatively, we can also recognize G as a cyclic group "internally" by choosing any element of $x \in G$ of order pq and observing that $G = \langle x \rangle$. For example, in $\mathbb{Z}_2 \times \mathbb{Z}_3$, $(1,1)$ and $(1,2)$ both have order 6. So, we can write $\mathbb{Z}_2 \times \mathbb{Z}_3 = \langle (1,1) \rangle$ or $\mathbb{Z}_2 \times \mathbb{Z}_3 = \langle (1,2) \rangle$. Similarly, in $\mathbb{Z}_6$, 1 and 5 both have order 6. So, we can write $\mathbb{Z}_6 = \langle 1 \rangle$ or $\mathbb{Z}_6 = \langle 5 \rangle$.

3. If G is a commutative group of order p^2, where p is prime, then Theorem 11.22 tells us that $G \cong \mathbb{Z}_p \times \mathbb{Z}_p$ or $G \cong \mathbb{Z}_{p^2}$. For example, any commutative group of order $4 = 2^2$ is isomorphic to either $\mathbb{Z}_2 \times \mathbb{Z}_2$ or $\mathbb{Z}_4$. Similarly, any commutative group of order $9 = 3^2$ is isomorphic to either $\mathbb{Z}_3 \times \mathbb{Z}_3$ or $\mathbb{Z}_9$.

Notice that the groups $\mathbb{Z}_p \times \mathbb{Z}_p$ and $\mathbb{Z}_{p^2}$ are **not** isomorphic. This follows from Theorem 11.22, but it is also easy to see without using the theorem. $\mathbb{Z}_{p^2}$ is cyclic and therefore, it has an element of order p^2. On the other hand, every nonidentity element of $\mathbb{Z}_p \times \mathbb{Z}_p$ has order p. If two finite groups are isomorphic, then they must have the same **order diagram** (see Problem 18 below), where the order diagram of a finite group G is

$$d(G) = \{(n, x_n) \mid n \in \mathbb{Z}^+ \wedge x_n = |\{x \in G \mid |x| = n\}| > 0\}.$$

The idea is simple—the order diagram just specifies how many elements there are of order n for each positive integer n (and if there are no elements of order n, we don't mention it). For example, the order diagram of $\mathbb{Z}_2 \times \mathbb{Z}_2$ is $\{(1,1), (2,3)\}$ because $\mathbb{Z}_2 \times \mathbb{Z}_2$ has 1 element of order 1 (the identity element $(0,0)$) and 3 elements of order 2 (the elements $(0,1), (1,0)$, and $(1,1)$). The order diagram of $\mathbb{Z}_4$ is $\{(1,1), (2,1), (4,2)\}$ because $\mathbb{Z}_4$ has 1 element of order 1 (the identity element 0), 1 element of order 2 (the element 2), and 2 elements of order 4 (the elements 1 and 3).

In general, $\mathbb{Z}_{p^2}$ has an element of order p^2 and $\mathbb{Z}_p \times \mathbb{Z}_p$ does not. Therefore, these two groups have different order diagrams.

The order diagram also tells us how to recognize the isomorphism type of any group of order p^2 (p prime). If G is a commutative group of order p^2, write down the order diagram of G. If G has an element of order p^2, it is isomorphic to $\mathbb{Z}_{p^2}$. If not, it is isomorphic to $\mathbb{Z}_p \times \mathbb{Z}_p$.

Let's go over how to recognize a commutative group G of order p^2 as an internal direct product of cyclic groups. If G has an element x of order p^2, then $G = \langle x \rangle$. Otherwise, let x_1 be any nonidentity element of G. Then $|\langle x_1 \rangle| = p$. Choose any element $x_2 \in G \setminus \langle x_1 \rangle$. Then $|\langle x_2 \rangle| = p$ and we have $G \cong \langle x_1 \rangle \times \langle x_2 \rangle$. For example, $\mathbb{Z}_2 \times \mathbb{Z}_2 \cong \langle (0,1) \rangle \times \langle (1,0) \rangle$.

4. If G is a commutative group of order p^3, where p is prime, then Theorem 11.22 tells us that $G \cong \mathbb{Z}_p \times \mathbb{Z}_p \times \mathbb{Z}_p$, $G \cong \mathbb{Z}_{p^2} \times \mathbb{Z}_p$, or $G \cong \mathbb{Z}_{p^3}$. For example, any commutative group of order $8 = 2^3$ is isomorphic to one of the groups $\mathbb{Z}_2 \times \mathbb{Z}_2 \times \mathbb{Z}_2$, $\mathbb{Z}_4 \times \mathbb{Z}_2$, or $\mathbb{Z}_8$.

Note that $\mathbb{Z}_4 \times \mathbb{Z}_2 \cong \mathbb{Z}_2 \times \mathbb{Z}_4$. When we have a direct product of groups, each of whose order is a power of the same prime, we will place the groups in size order from largest to smallest as we go from the leftmost factor to the rightmost factor, as was suggested in observation (2) above, right after the statement of Theorem 11.22. Just like in part 2 above, we will call this the **canonical representation** (or **canonical form**) of the group. So, $\mathbb{Z}_4 \times \mathbb{Z}_2$ is in canonical form, whereas $\mathbb{Z}_2 \times \mathbb{Z}_4$ is not. Similarly, $\mathbb{Z}_{p^2} \times \mathbb{Z}_p$ is in canonical form.

As in part 3 above, we can easily see that the groups $\mathbb{Z}_p \times \mathbb{Z}_p \times \mathbb{Z}_p$, $\mathbb{Z}_{p^2} \times \mathbb{Z}_p$, and $\mathbb{Z}_{p^3}$ are **not** isomorphic by comparing their order diagrams: $\mathbb{Z}_{p^3}$ has elements of order p^3, while the other two groups do not. Also, $\mathbb{Z}_{p^2} \times \mathbb{Z}_p$ has elements of order p^2, while every nonidentity element of $\mathbb{Z}_p \times \mathbb{Z}_p \times \mathbb{Z}_p$ has order p.

Observe that the number of commutative groups of order p^3 (p prime), up to isomorphism, is in one-to-one correspondence with the number of **partitions** of 3. By a partition of 3, we mean a way of writing 3 as a sum of positive integers, where two sums are considered the same if the only difference between them is the order in which the integers are written.

There are 3 partitions of 3, namely $1 + 1 + 1$, $2 + 1$ and 3. Notice that the partition $1 + 2$ is the same as the partition $2 + 1$, just like $\mathbb{Z}_p \times \mathbb{Z}_{p^2} \cong \mathbb{Z}_{p^2} \times \mathbb{Z}_p$.

Important note: Do not confuse the definition of partition being used here with the definition given in Lesson 3. Although the two ideas are similar, they are different definitions.

As we did in part 3 above, let's go over how to recognize a commutative group G of order p^3 as an internal direct product of cyclic groups. If G has an element x of order p^3, then $G = \langle x \rangle$. Otherwise, check if G has an element x_1 of order p^2. If so, choose any element $x_2 \in G \setminus \langle x_1 \rangle$ of order p. Then $G \cong \langle x_1 \rangle \times \langle x_2 \rangle$. If G does not have an element of order p^2 or p^3, then let x_1 be any nonidentity element of G. We must have $|\langle x_1 \rangle| = p$. Choose any element $x_2 \in G \setminus \langle x_1 \rangle$. Then $|\langle x_2 \rangle| = p$ as well. Finally, choose $x_3 \in G \setminus \langle x_1, x_2 \rangle$. Then we have $|\langle x_3 \rangle| = p$ and $G \cong \langle x_1 \rangle \times \langle x_2 \rangle \times \langle x_3 \rangle$.

5. If G is a commutative group of order $p^2 q$, where p and q are prime, then Theorem 11.22 tells us that $G \cong \mathbb{Z}_p \times \mathbb{Z}_p \times \mathbb{Z}_q$ or $G \cong \mathbb{Z}_{p^2} \times \mathbb{Z}_q$. For example, any commutative group of order $12 = 2^2 \cdot 3$ is isomorphic to $\mathbb{Z}_2 \times \mathbb{Z}_2 \times \mathbb{Z}_3$ or $\mathbb{Z}_4 \times \mathbb{Z}_3$ (note that we wrote these in canonical form—see part 2 above).

   Once again, we can see that the groups $\mathbb{Z}_p \times \mathbb{Z}_p \times \mathbb{Z}_q$ and $\mathbb{Z}_{p^2} \times \mathbb{Z}_q$ are **not** isomorphic by comparing their order diagrams: $\mathbb{Z}_{p^2} \times \mathbb{Z}_q$ has elements of order p^2, while the other group does not. Notice how we need only analyze the orders of elements in $\mathbb{Z}_p \times \mathbb{Z}_p$ and $\mathbb{Z}_{p^2}$ to determine that $\mathbb{Z}_p \times \mathbb{Z}_p \times \mathbb{Z}_q$ and $\mathbb{Z}_{p^2} \times \mathbb{Z}_q$ are not isomorphic.

   Since p and q are distinct prime numbers, it follows that $\gcd(p, q) = 1$. So, by part (i) of Problem 19 from Problem Set 9, $\mathbb{Z}_p \times \mathbb{Z}_q \cong \mathbb{Z}_{pq}$ Therefore, we can identify $\mathbb{Z}_p \times \mathbb{Z}_p \times \mathbb{Z}_q$ with $\mathbb{Z}_p \times \mathbb{Z}_{pq}$. Similarly, we can identify $\mathbb{Z}_{p^2} \times \mathbb{Z}_q$ with $\mathbb{Z}_{p^2 q}$. For example, $\mathbb{Z}_2 \times \mathbb{Z}_2 \times \mathbb{Z}_3 \cong \mathbb{Z}_2 \times \mathbb{Z}_6$ and $\mathbb{Z}_4 \times \mathbb{Z}_3 \cong \mathbb{Z}_{12}$.

   If G is a commutative group of order $p^2 q$, where p and q are distinct primes, we can recognize G as an internal direct product of cyclic groups $H_1 \times H_2$, where H_1 is an internal direct product of order p^2 and H_2 is a cyclic group of order q.

For example, $\mathbb{Z}_{12} \cong \langle 3 \rangle \times \langle 4 \rangle$. Notice that 3 has order $4 = 2^2$ and 4 has order 3 in $\mathbb{Z}_{12}$. As another example, $\mathbb{Z}_2 \times \mathbb{Z}_6 \cong \langle (0,3) \rangle \times \langle (1,3) \rangle \times \langle (0,2) \rangle$. Notice that $\langle (0,3) \rangle \times \langle (1,3) \rangle$ has order $4 = 2^2$ and $\langle (0,2) \rangle$ has order 3 in $\mathbb{Z}_2 \times \mathbb{Z}_6$.

Alternatively, we can recognize G as an internal direct product of cyclic groups as follows: If G has an element x of order $p^2 q$, then $G = \langle x \rangle$. Otherwise, G has an element x_1 of order pq. Choose an element $x_2 \in G \setminus \langle x_1 \rangle$ such that x_2 has order p. Then we have $G \cong \langle x_1 \rangle \times \langle x_2 \rangle$. Using this alternate procedure, we see that $\mathbb{Z}_{12} \cong \langle 1 \rangle$ and $\mathbb{Z}_2 \times \mathbb{Z}_6 \cong \langle (0,1) \rangle \times \langle (1,0) \rangle$.

6. We now know all commutative groups of orders less than 16, up to isomorphism. Here they are, written in canonical form.

$$\{e\}, \mathbb{Z}_2, \mathbb{Z}_3, \mathbb{Z}_2 \times \mathbb{Z}_2, \mathbb{Z}_4, \mathbb{Z}_5, \mathbb{Z}_2 \times \mathbb{Z}_3 \cong \mathbb{Z}_6, \mathbb{Z}_7,$$
$$\mathbb{Z}_2 \times \mathbb{Z}_2 \times \mathbb{Z}_2, \mathbb{Z}_4 \times \mathbb{Z}_2, \mathbb{Z}_8, \mathbb{Z}_3 \times \mathbb{Z}_3, \mathbb{Z}_9,$$
$$\mathbb{Z}_2 \times \mathbb{Z}_5 \cong \mathbb{Z}_{10}, \mathbb{Z}_{11}, \mathbb{Z}_2 \times \mathbb{Z}_2 \times \mathbb{Z}_3 \cong \mathbb{Z}_2 \times \mathbb{Z}_6, \mathbb{Z}_4 \times \mathbb{Z}_3 \cong \mathbb{Z}_{12},$$
$$\mathbb{Z}_{13}, \mathbb{Z}_2 \times \mathbb{Z}_7 \cong \mathbb{Z}_{14}, \mathbb{Z}_3 \times \mathbb{Z}_5 \cong \mathbb{Z}_{15}$$

7. If G is a commutative group of order $p^3 q$, where p and q are prime, then Theorem 11.22 tells us that $G \cong \mathbb{Z}_p \times \mathbb{Z}_p \times \mathbb{Z}_p \times \mathbb{Z}_q$, $G \cong \mathbb{Z}_{p^2} \times \mathbb{Z}_p \times \mathbb{Z}_q$, or $G \cong \mathbb{Z}_{p^3} \times \mathbb{Z}_q$. Once again, by comparing the order diagrams of these groups, we see that they are pairwise nonisomorphic. For example, any commutative group of order $24 = 2^3 \cdot 3$ is isomorphic to $\mathbb{Z}_2 \times \mathbb{Z}_2 \times \mathbb{Z}_2 \times \mathbb{Z}_3$, $\mathbb{Z}_4 \times \mathbb{Z}_2 \times \mathbb{Z}_3$, or $\mathbb{Z}_8 \times \mathbb{Z}_3$. Note that we wrote these in canonical form—see parts 2 and 5 above. Part 2 told us that we should write $\mathbb{Z}_4 \times \mathbb{Z}_2$ instead of $\mathbb{Z}_2 \times \mathbb{Z}_4$, whereas part 5 told us that we should write $\mathbb{Z}_4 \times \mathbb{Z}_2 \times \mathbb{Z}_3$ instead of $\mathbb{Z}_3 \times \mathbb{Z}_4 \times \mathbb{Z}_2$.

If we wish, we can identify $\mathbb{Z}_p \times \mathbb{Z}_p \times \mathbb{Z}_p \times \mathbb{Z}_q$ with $\mathbb{Z}_p \times \mathbb{Z}_p \times \mathbb{Z}_{pq}$, we can identify $\mathbb{Z}_{p^2} \times \mathbb{Z}_p \times \mathbb{Z}_q$ with $\mathbb{Z}_p \times \mathbb{Z}_{p^2 q}$, and we can identify $\mathbb{Z}_{p^3} \times \mathbb{Z}_q$ with $\mathbb{Z}_{p^3 q}$. For example, we can identify $\mathbb{Z}_2 \times \mathbb{Z}_2 \times \mathbb{Z}_2 \times \mathbb{Z}_3$ with $\mathbb{Z}_2 \times \mathbb{Z}_2 \times \mathbb{Z}_6$, we can identify $\mathbb{Z}_4 \times \mathbb{Z}_2 \times \mathbb{Z}_3$ with $\mathbb{Z}_2 \times \mathbb{Z}_{12}$, and we can identify $\mathbb{Z}_8 \times \mathbb{Z}_3$ with $\mathbb{Z}_{24}$.

As in part 5 above, if G is a commutative group of order $p^3 q$, where p and q are distinct primes, we can recognize G as an internal direct product of cyclic groups $H_1 \times H_2$, where H_1 is an internal direct product of order p^3 and H_2 is a cyclic group of order q.

Let's also go over the alternate method for recognizing a commutative group G of order $p^3 q$ as an internal direct product of cyclic groups. If G has an element x of order $p^3 q$, then $G = \langle x \rangle$. Otherwise, check if G has an element x_1 of order $p^2 q$. If so, choose an element $x_2 \in G \setminus \langle x_1 \rangle$ such that x_2 has order p. Then we have $G \cong \langle x_1 \rangle \times \langle x_2 \rangle$. If G does not have an element of order $p^2 q$ or $p^3 q$, then let x_1 be any element of G of order pq. Choose an element $x_2 \in G \setminus \langle x_1 \rangle$ such that x_2 has order p. Finally, choose one more element $x_3 \in G \setminus \langle x_1, x_2 \rangle$ such that x_3 has order p. Then we have $G \cong \langle x_1 \rangle \times \langle x_2 \rangle \times \langle x_3 \rangle$.

8. Suppose that G is a commutative group of order p^4, where p is prime. The partitions of 4 are $1+1+1+1$, $2+1+1$, $2+2$, $3+1$, and 4 (see part 4 above for an explanation of partitions). So, $G \cong \mathbb{Z}_p \times \mathbb{Z}_p \times \mathbb{Z}_p \times \mathbb{Z}_p$, $G \cong \mathbb{Z}_{p^2} \times \mathbb{Z}_p \times \mathbb{Z}_p$, $G \cong \mathbb{Z}_{p^2} \times \mathbb{Z}_{p^2}$, $G \cong \mathbb{Z}_{p^3} \times \mathbb{Z}_p$, or $G \cong \mathbb{Z}_{p^4}$. For example, any commutative group of order $81 = 3^4$ is isomorphic to one of the following: $\mathbb{Z}_3 \times \mathbb{Z}_3 \times \mathbb{Z}_3 \times \mathbb{Z}_3$, $\mathbb{Z}_9 \times \mathbb{Z}_3 \times \mathbb{Z}_3$, $\mathbb{Z}_9 \times \mathbb{Z}_9$, $\mathbb{Z}_{27} \times \mathbb{Z}_3$, or $\mathbb{Z}_{81}$ (note that we wrote these in canonical form—see part 2 above).

I leave it to the reader to use order diagrams to show that these five groups are pairwise nonisomorphic.

We can recognize a commutative group G of order p^4 as an internal direct product of cyclic groups as follows: If G has an element x of order p^4, then $G = \langle x \rangle$. Otherwise, check if G has an element x_1 of order p^3. If so, choose an element $x_2 \in G \setminus \langle x_1 \rangle$ of order p. Then we have $G \cong \langle x_1 \rangle \times \langle x_2 \rangle$. If G does not have an element of order p^3 or p^4, check if G has an element x_1 of order p^2. If so, choose an element $x_2 \in G \setminus \langle x_1 \rangle$ of maximal order. If $|x_2| = p^2$, then we have $G \cong \langle x_1 \rangle \times \langle x_2 \rangle$. Otherwise, $|x_2| = p$ and we can choose an element $x_3 \in G \setminus \langle x_1, x_2 \rangle$ of order p and we have $G \cong \langle x_1 \rangle \times \langle x_2 \rangle \times \langle x_3 \rangle$. Finally, if G does not have an element of order p^2, p^3 or p^4, then let x_1 be any nonidentity element of G. We must have $|\langle x_1 \rangle| = p$. Choose any element $x_2 \in G \setminus \langle x_1 \rangle$. Then $|\langle x_2 \rangle| = p$ as well. Choose an element $x_3 \in G \setminus \langle x_1, x_2 \rangle$. Then $|\langle x_3 \rangle| = p$ too. Finally, choose $x_4 \in G \setminus \langle x_1, x_2, x_3 \rangle$. Again, $|\langle x_4 \rangle| = p$ and $G \cong \langle x_1 \rangle \times \langle x_2 \rangle \times \langle x_3 \rangle \times \langle x_4 \rangle$.

9. If p is prime and k is a positive integer, then by Theorem 11.22, up to isomorphism there is one group of order p^k for each partition of k (see part 4 above for an explanation of partitions). The partition $k = k_1 + k_2 + \cdots + k_n$ corresponds to the group $\mathbb{Z}_{p^{k_1}} \times \mathbb{Z}_{p^{k_2}} \times \cdots \times \mathbb{Z}_{p^{k_n}}$.

   We can recognize a commutative group G of order p^k as an internal direct product of cyclic groups by following these steps:

   (i) Choose $x_1 \in G$ of maximum order.

   (ii) If $G = \langle x_1 \rangle$, we are done. If not, choose $x_2 \in G \setminus \langle x_1 \rangle$ of maximum order p^{j_1} such that $p^{j_1} \leq \frac{|G|}{|\langle x_1 \rangle|}$.

   (iii) If $G \cong \langle x_1 \rangle \times \langle x_2 \rangle$, we are done. If not, choose $x_3 \in G \setminus \langle x_1, x_2 \rangle$ of maximum order p^{j_2} such that $p^{j_2} \leq \frac{|G|}{|\langle x_1, x_2 \rangle|}$.

   (iv) If $G \cong \langle x_1 \rangle \times \langle x_2 \rangle \times \langle x_3 \rangle$, we are done. If not, continue in this manner. At stage $i + 1$, choose $x_{i+1} \in G \setminus \langle x_1, x_2, \ldots, x_i \rangle$ of maximum order p^{j_i} such that $p^{j_i} \leq \frac{|G|}{|\langle x_1, x_2, \ldots, x_i \rangle|}$.

   (v) Since G is finite, this procedure must terminate in finitely many steps.

10. For the most general case, let $p_1, p_2, \ldots, p_t$ be distinct primes and let $k_1, k_2, \ldots, k_t$ be positive integers for each $i = 1, 2, \ldots, t$. To find all the groups of order $p_1^{k_1} p_2^{k_2} \cdots p_t^{k_t}$, we first find all groups of each of the orders $p_1^{k_1}, p_2^{k_2}, \ldots, p_t^{k_t}$ (see part 9 above). We then form all possible direct products of these groups. By the Fundamental Theorem of Arithmetic, every positive integer has a unique prime factorization, and so, we have found a method for determining all finite commutative groups.

    As a specific example, let's find all commutative groups of order $604{,}175 = 5^2 \cdot 11 \cdot 13^3$, up to isomorphism. The commutative groups of order 5^2 are $\mathbb{Z}_5 \times \mathbb{Z}_5$ and $\mathbb{Z}_{25}$. The only commutative group of order 11 is $\mathbb{Z}_{11}$. The commutative groups of order 13^3 are $\mathbb{Z}_{13} \times \mathbb{Z}_{13} \times \mathbb{Z}_{13}$, $\mathbb{Z}_{169} \times \mathbb{Z}_{13}$, and $\mathbb{Z}_{2197}$. So, a list of all the groups we are looking for is as follows:

$$\mathbb{Z}_5 \times \mathbb{Z}_5 \times \mathbb{Z}_{11} \times \mathbb{Z}_{13} \times \mathbb{Z}_{13} \times \mathbb{Z}_{13} \qquad \mathbb{Z}_{25} \times \mathbb{Z}_{11} \times \mathbb{Z}_{13} \times \mathbb{Z}_{13} \times \mathbb{Z}_{13}$$
$$\mathbb{Z}_5 \times \mathbb{Z}_5 \times \mathbb{Z}_{11} \times \mathbb{Z}_{169} \times \mathbb{Z}_{13} \qquad \mathbb{Z}_{25} \times \mathbb{Z}_{11} \times \mathbb{Z}_{169} \times \mathbb{Z}_{13}$$
$$\mathbb{Z}_5 \times \mathbb{Z}_5 \times \mathbb{Z}_{11} \times \mathbb{Z}_{2197} \qquad \mathbb{Z}_{25} \times \mathbb{Z}_{11} \times \mathbb{Z}_{2197}$$

If G is a commutative group of order $p_1^{k_1}, p_2^{k_2}, \ldots, p_t^{k_t}$, where $p_1, p_2, \ldots, p_t$ are distinct primes and $k_1, k_2, \ldots, k_t$ are positive integers for each $i = 1, 2, \ldots, t$, we can recognize G as an internal direct product of cyclic groups $H_1 \times H_2 \times \cdots \times H_t$, where H_i is an internal direct product of order p^{k_i} for each $i = 1, 2, \ldots, t$. We find each H_i by using the algorithm given in part 9 above.

We now finish this lesson by providing an outline to prove Theorem 11.22. The dedicated reader may want to attempt to fill in the details (see Problem 23 below).

Theorem 11.22 (The Fundamental Theorem of Commutative Groups): Every finite commutative group is isomorphic to a direct product of cyclic groups of prime power order. Furthermore, the number of terms in the product and the orders (cardinalities) of the cyclic groups are uniquely determined by the group.

Proof outline: The proof can be completed in the following 6 steps.

Step 1: Suppose that G is a finite commutative group of order $p^k n$, where p is prime, $k, n \in \mathbb{Z}^+$, and p does not divide n. Let $H = \{x \in G \mid x^{p^k} = e\}$ and let $K = \{x \in G \mid x^n = e\}$. Prove each of the following:

(i) H and K are subgroups of G.

(ii) $G = HK$.

(iii) $H \cap K = \{e\}$.

(iv) $G \cong H \times K$ (use (i), (ii), and (iii) above).

(v) $|H| = p^k$.

Step 2: Let G be a finite commutative group of order $p_1^{k_1} \cdot p_2^{k_2} \cdots p_t^{k_t}$, where $p_1, p_2, \ldots, p_t$ are distinct primes and $k_1, k_2, \ldots, k_t$ are positive integers for each $i = 1, 2, \ldots, t$. Use induction and the results from Step 1 to prove that $G \cong H_1 \times H_2 \times \cdots \times H_t$, where $H_i = \{x \in G \mid x^{p_i^{k_i}} = e\}$.

Step 3: Suppose that G is a finite commutative group of order p^k, where p is prime and $k \in \mathbb{Z}^+$. Let $x \in G$ be an element of maximum order. Prove that $G \cong \langle x \rangle \times K$ for some subgroup K of G.

Step 4: Suppose that G is a finite commutative group of order p^k, where p is prime and $k \in \mathbb{Z}^+$. Use induction and the result from Step 3 to prove that there is $s \in \mathbb{Z}^+$ and $x_1, x_2, \ldots, x_s \in G$ such that $G \cong \langle x_1 \rangle \times \langle x_2 \rangle \times \cdots \times \langle x_s \rangle$.

Step 5: Let G be a finite commutative group of order $p_1^{k_1} \cdot p_2^{k_2} \cdots p_t^{k_t}$, where $p_1, p_2, \ldots, p_t$ are distinct primes and $k_1, k_2, \ldots, k_t$ are positive integers for each $i = 1, 2, \ldots, t$. Use steps 2 and 4 together to prove that G is isomorphic to a direct product of cyclic groups of prime power order.

Step 6: Suppose that G is a finite commutative group of prime power order. Furthermore, suppose that $G \cong H_1 \times H_2 \times \cdots \times H_{t_1}$ and $G \cong K_1 \times K_2 \times \cdots \times K_{t_2}$, where each H_i and K_i is a cyclic subgroup of G such that $|H_1| \geq |H_2| \geq \cdots \geq |H_{t_1}| \geq 2$ and $|K_1| \geq |K_2| \geq \cdots \geq |K_{t_2}| \geq 2$. Use induction on $|G|$ to prove that $t_1 = t_2$ and $|H_i| = |K_i|$ for each $i = 1, 2, \ldots, t_1$.

Use the results from Steps 5 and 6 to complete the proof. □

Problem Set 11

Full solutions to these problems are available for free download here:
www.SATPrepGet800.com/AAFBTDW

LEVEL 1

1. Use the First Isomorphism Theorem for rings to prove each of the following:
   (i) $\mathbb{Z}_{15}/\{0, 3, 6, 9, 12\} \cong \mathbb{Z}_3$
   (ii) $(\mathbb{Z}_2 \times \mathbb{Z}_2)/\{(0,0), (0,1)\} \cong \mathbb{Z}_2$

2. Let G and H be groups. Prove that $(G \times H)/(G \times \{e_H\}) \cong H$.

3. List all the commutative groups of the given order, up to isomorphism:
   (i) 31
   (ii) 91

LEVEL 2

4. Let G be a commutative group. Prove that $G/\{x \in G \mid x^2 = e\} \cong \{x^2 \mid x \in G\}$.

5. List all the commutative groups of the given order, up to isomorphism:
   (i) 210
   (ii) 625
   (iii) 8575

LEVEL 3

6. Let R be a ring, let I be an ideal of R, and let S be a subring of R with $I \subseteq S \subseteq R$. Prove that S/I is a subring of R/I. Furthermore, prove that if S is an ideal of R, then S/I is an ideal of R/I.

7. Let R be a ring and let I be an ideal of R. Prove that every subring of R/I is of the form S/I for some subring S of R with $I \subseteq S \subseteq R$. Furthermore, prove that every ideal of R/I is of the form S/I for some ideal S of R with $I \subseteq S \subseteq R$.

8. Let V be a vector space over a field F and let U and W be subspaces of V with $W \subseteq U \subseteq V$. Prove that U/W is a subspace of V/W.

9. Let V be a vector space over a field F and let W be a subspace of V. Prove that every subspace of V/W is of the form U/W for some subspace U of V with $W \subseteq U \subseteq V$.

10. Prove the vector space and ring versions of the Fourth Isomorphism Theorem.

LEVEL 4

11. Let G be a group and for each $x \in G$, define the function $\phi_x: G \to G$ by $\phi_x(y) = xyx^{-1}$. Let $\text{Inn}(G) = \{\phi_x \mid x \in G\}$. Prove that $G/Z(G) \cong \text{Inn}(G)$, where $Z(G)$ is the center of G.

12. Let G and H be groups, let $N \triangleleft G$, and let $K \triangleleft H$. Prove that $(G \times H)/(N \times K) \cong G/N \times H/K$.

13. Let $m, n \in \mathbb{Z}^+$. Prove that $m\mathbb{Z} \cap n\mathbb{Z} = \text{lcm}(m,n)\mathbb{Z}$ and $m\mathbb{Z} + n\mathbb{Z} = \gcd(m,n)\mathbb{Z}$.

14. Let $a, b \in \mathbb{Z}^+$ with $\gcd(a, b) = 1$. Prove that the groups $(\mathbb{Z} \times \mathbb{Z})/\langle(a,b)\rangle$ and $\mathbb{Z}$ are isomorphic.

15. Let $f: \mathcal{X} \to \mathcal{Y}$ be the bijection defined in the Fourth Isomorphism Theorem and let $A, B \in \mathcal{X}$. Prove each of the following:

    (i) $A \leq B$ if and only if $f(A) \leq f(B)$.

    (ii) If $A \subseteq B$, then $[B:A] = [f(B):f(A)]$.

16. Let $f: \mathcal{X} \to \mathcal{Y}$ be the bijection defined in the group version of the Fourth Isomorphism Theorem. Let $H \in \mathcal{X}$. Prove that $H \triangleleft G$ if and only if $f(H) \triangleleft G/N$. State and prove the analogous result for the ring version of the Fourth Isomorphism Theorem.

LEVEL 5

17. $G = \{1, 4, 11, 14, 16, 19, 26, 29, 31, 34, 41, 44\}$ is a group under multiplication modulo 45. Express G as both an external and internal direct product of cyclic groups.

18. Prove that isomorphic finite groups have the same order diagram, where the order diagram of G is $d(G) = \{(n, x_n) \mid n \in \mathbb{Z}^+ \wedge x_n = |\{x \in G \mid |x| = n\}| > 0\}$.

19. List all the commutative groups of order 183,495,637, up to isomorphism.

20. Let G be a finite commutative group and let $k \in \mathbb{Z}^+$ such that k divides $|G|$. Prove that G has a subgroup of order k.

CHALLENGE PROBLEMS

21. A finite group G is **solvable** if there exist subgroups $G_0, G_1, \ldots, G_n$ such that
$$\{e\} = G_0 \triangleleft G_1 \triangleleft \cdots \triangleleft G_n = G,$$
where for each $i = 1, 2, \ldots, n$, the quotient G_i/G_{i-1} is commutative. Prove that if G is a finite solvable group, then every subgroup H of G is solvable and every quotient group G/H is solvable.

22. Let G be a finite group and let N be a normal subgroup of G. Prove that if N and G/N are solvable, then so is G. (See Problem 21 above for the definition of solvable).

23. Prove the Fundamental Theorem of Commutative Groups.

LESSON 12
PERMUTATIONS AND DETERMINANTS

Permutations on Finite Sets

Recall that if A is a set, then a permutation of A is a bijection from A to itself. In Lesson 4, we saw that $S(A)$, the set of permutations of A, is a group, where the group operation is composition of permutations (see Theorem 4.11). For each $n \in \mathbb{Z}^+$, we define S_n to be the group of permutations on $\{1, 2, \ldots, n\}$. We have seen several ways to visualize an element of S_n (see Example 4.12 and Note 4 following Example 9.3). Let's look at some more examples.

Example 12.1:

1. S_1 is the group of permutations on $\{1\}$. There is just one such permutation. It is the permutation $f: \{1\} \to \{1\}$ defined by $f(1) = 1$. We can also write $f = \{(1, 1)\}$. In cycle notation, f is the permutation (1). Here are a few ways to visualize this permutation.

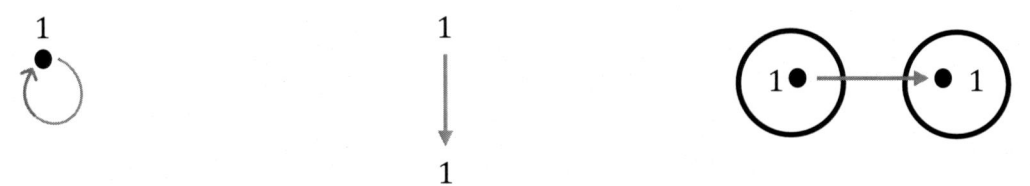

   $S_1 = \{(1)\}$ is isomorphic to the trivial group consisting of just an identity.

2. S_2 is the group of permutations on $\{1, 2\}$. There are two such permutations. The first permutation $f_1: \{1, 2\} \to \{1, 2\}$ is defined by $f_1(1) = 1$ and $f_1(2) = 2$. The second permutation $f_2: \{1, 2\} \to \{1, 2\}$ is defined by $f_2(1) = 2$ and $f_2(2) = 1$. We can also write $f_1 = \{(1, 1), (2, 2)\}$ and $f_2 = \{(1, 2), (2, 1)\}$. In cycle notation, f_1 is the permutation $(1)(2)$, which we abbreviate as (1) and f_2 is the permutation $(1\,2)$. Here are a few ways to visualize these permutations.

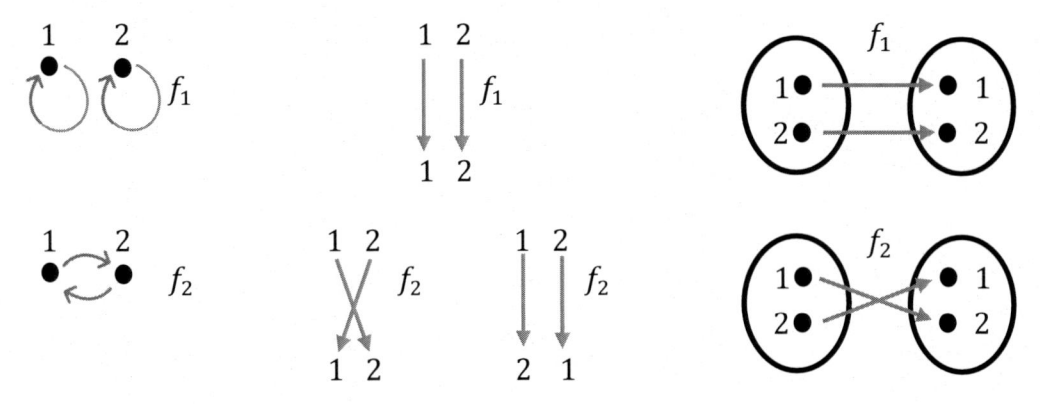

   Note that there are four visualizations of f_2, but only three of f_1. This is because for f_1 the two visualizations with the downward facing arrows are identical.

   One other thing worth noting is that in cycle notation, the identity is written as (1) in both S_1 and S_2. In fact, (1) is the cycle notation of the identity in S_n for all $n \in \mathbb{Z}^+$.

Similarly, the cycle (12) can be thought of as a permutation in S_2 (as shown in the second row of visualizations above) or as a permutation in S_n for any other $n \in \mathbb{Z}^+$ with $n > 2$. For example, in S_3, (12) is the permutation that sends 1 to 2, 2 to 1, and 3 to itself. See Example 4.12 for a visualization of (12) as a cycle in S_3.

When we multiply (or compose) the permutation (12) with itself, we get the identity permutation (1). See Example 4.12 to review how to multiply permutations in cycle notation. And of course, each permutation multiplied with the identity permutation is the identity permutation. Below is the full multiplication table for S_2.

∘	(1)	(12)
(1)	(1)	(12)
(12)	(12)	(1)

3. S_3 is the group of permutations on $\{1, 2, 3\}$. There are six such permutations. In Example 4.12, we saw the cycle notations of each of these permutations and we also saw how to visualize each element of S_3 as a cycle. Note that when drawing a cycle diagram, we usually omit the arrows for elements that are mapping to themselves to avoid clutter. Below is the full multiplication table for S_3.

∘	(1)	(12)	(13)	(23)	(123)	(132)
(1)	(1)	(12)	(13)	(23)	(123)	(132)
(12)	(12)	(1)	(132)	(123)	(23)	(13)
(13)	(13)	(123)	(1)	(132)	(12)	(23)
(23)	(23)	(132)	(123)	(1)	(13)	(12)
(123)	(123)	(13)	(23)	(12)	(132)	(1)
(132)	(132)	(23)	(12)	(13)	(1)	(123)

4. In general, $|S_n| = n! = 1 \cdot 2 \cdots n$ (see the Analysis after Theorem 8.9 for more information on $n!$). You will be asked to prove this in Problem 13 below. For example, we have $|S_1| = 1$, $|S_2| = 1 \cdot 2 = 2$, and $|S_3| = 1 \cdot 2 \cdot 3 = 6$, as we have seen above. Notice that S_4 is already quite large. We have $|S_4| = 1 \cdot 2 \cdot 3 \cdot 4 = 24$.

5. In S_4, (1234) is the permutation that sends 1 to 2, 2 to 3, 3 to 4, and 4 to 1. In other words, (1234) is the function $f: \{1, 2, 3, 4\} \to \{1, 2, 3, 4\}$ such that $f(1) = 2, f(2) = 3, f(3) = 4$, and $f(4) = 1$. We can also write $f = \{(1, 2), (2, 3), (3, 4), (4, 1)\}$. Below are a few ways to visualize this permutation.

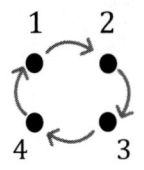

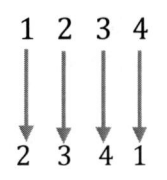

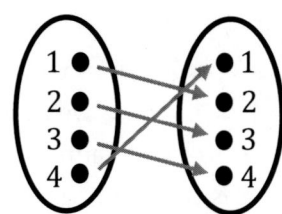

261

In Note 4 following Example 9.3, we saw several ways to visualize the permutation (1234) in S_5. Observe once again that the same notation is used for elements of different permutation groups.

From now on, we will usually use lowercase greek letters such as π (pi), σ (sigma), and τ (tau) to represent arbitrary permutations.

If π is a permutation, then π^2 is the permutation $\pi \circ \pi$. More generally, the powers of π are as follows:

$$\pi^0 = (1); \quad \pi^1 = \pi; \quad \pi^2 = \pi \circ \pi; \quad \pi^3 = (\pi \circ \pi) \circ \pi = \pi \circ (\pi \circ \pi); \quad \text{... and so on.}$$

In general, for $n > 0$, π^n is the composition of π with itself n times. We can write $\pi^{n+1} = \pi^n \circ \pi$. Furthermore, for $n > 0$, $\pi^{-n} = (\pi^n)^{-1} = (\pi^{-1})^n$ (by Corollary 5.15). So, for example, π^{-1} is the inverse of π and $\pi^{-2} = (\pi^2)^{-1} = (\pi \circ \pi)^{-1}$ is the inverse of $\pi^2 = \pi \circ \pi$.

Note: From now on, if π and σ are permutations, we may abbreviate the composition $\pi \circ \sigma$ by $\pi\sigma$. Therefore, we may also write $\pi\sigma(i)$ in place of $\pi(\sigma(i))$. We may still choose to use parentheses when composing a permutation with itself. So, we will usually write $\pi^2(i) = \pi(\pi(i))$, rather than $\pi\pi(i)$.

Let's now be a bit more rigorous in how we define a cycle, so that we can use the cycle notation of permutations to prove theorems. We'll start with the following observation:

If $\pi \in S_n$ and $i \in \{1, 2, \ldots, n\}$, then there can be at most n distinct elements of the set $\{\pi^k(i) \mid k \in \mathbb{Z}\}$ simply because this set is a subset of $\{1, 2, \ldots, n\}$, which has only n elements.

Example 12.2:

1. If $\pi \in S_3$ and $i = 1, 2$, or 3, then the set $\{i, \pi(i), \pi^2(i), \pi^3(i)\}$ can have at most 3 elements because each element is either 1, 2, or 3. Let's choose $\pi = (123)$ and $i = 1$. Then we have $\pi(1) = 2, \pi^2(1) = \pi(\pi(1)) = \pi(2) = 3$, and $\pi^3(1) = \pi(\pi^2(1)) = \pi(3) = 1$. So, we see that $\{1, \pi(1), \pi^2(1), \pi^3(1)\} = \{1, 2, 3, 1\} = \{1, 2, 3\}$. In this particular case, $\{i, \pi(i), \pi^2(i), \pi^3(i)\}$ has exactly 3 elements, the maximum possible amount. Another way to say this is that 3 is the least positive integer such that $\pi^3(1) \in \{1, \pi(1), \pi^2(1)\}$. Notice also that $\pi^3(1) = 1$.

    Let's look at the permutation $\sigma = (12)$. Then $\sigma(1) = 2$ and $\sigma^2(1) = \sigma(\sigma(1)) = \sigma(2) = 1$. Although it's not necessary, let's also compute $\sigma^3(1) = \sigma(\sigma^2(1)) = \sigma(1) = 2$. So, we see that $\{1, \sigma(1), \sigma^2(1), \sigma^3(1)\} = \{1, 2, 1, 2\} = \{1, 2\}$. So, in this case, $\{i, \sigma(i), \sigma^2(i), \sigma^3(i)\}$ has only 2 elements, and so, 2 is the least positive integer such that $\sigma^2(1) \in \{1, \sigma(1)\}$. Notice also that $\sigma^2(1) = 1$.

2. If $\pi \in S_5$ and $i = 1, 2, 3, 4$, or 5 then the set $\{i, \pi(i), \pi^2(i), \pi^3(i), \pi^4(i), \pi^5(i)\}$ can have at most 5 elements because each element is either 1, 2, 3, 4 or 5. Let's choose $\pi = (12)(345)$. If $i = 1$, then we have $\pi(1) = 2, \pi^2(1) = \pi(\pi(1)) = \pi(2) = 1$, and similarly, $\pi^3(1) = 2$, $\pi^4(1) = 1$, and $\pi^5(1) = 2$. So, we see that $\{1, \pi(1), \pi^2(1), \pi^3(1), \pi^4(1), \pi^5(1)\} = \{1, 2\}$. In this particular case, $\{i, \pi(i), \pi^2(i), \pi^3(i), \pi^4(i), \pi^5(i)\}$ has just 2 elements. Notice also that $\pi^2(1) = 1$. Similarly, if $i = 3$, we see that $\{3, \pi(3), \pi^2(3), \pi^3(3), \pi^4(3), \pi^5(3)\} = \{3, 4, 5\}$. So, in this case, we get 3 elements. Notice also that $\pi^3(3) = 3$.

Theorem 12.3: Let $\pi \in S_n$, let $i \in \{1, 2, \ldots, n\}$, and let m be the least positive integer such that $\pi^m(i) \in \{i, \pi(i), \pi^2(i), \ldots, \pi^{m-1}(i)\}$. Then $\pi^m(i) = i$.

Proof: Since $\{i, \pi(i), \pi^2(i), \ldots, \pi^n(i)\} \subseteq \{1, 2, \ldots, n\}$ and $|\{i, \pi(i), \pi^2(i), \ldots, \pi^n(i)\}| = n+1$, there are $j, k \in \mathbb{N}$ with $j < k$ such that $\pi^j(i) = \pi^k(i)$. So, $\pi^k(i) \in \{i, \pi(i), \pi^2(i), \ldots, \pi^{k-1}(i)\}$. By the Well Ordering Principle, there is a least positive integer m such that $\pi^m(i) \in \{i, \pi(i), \pi^2(i), \ldots, \pi^{m-1}(i)\}$.

Since $\pi^m(i) \in \{i, \pi(i), \pi^2(i), \ldots, \pi^{m-1}(i)\}$, there is $r \in \mathbb{Z}$ with $0 \le r < m$ such that $\pi^m(i) = \pi^r(i)$. So, we have

$$\pi^{m-r}(i) = \pi^m \pi^{-r}(i) = \pi^m (\pi^r)^{-1}(i) = \pi^m (\pi^m)^{-1}(i) = i.$$

Since $0 \le r < m$ and m is the least positive integer such that $\pi^m(i) \in \{i, \pi(i), \pi^2(i), \ldots, \pi^{m-1}(i)\}$, we must have $r = 0$ (if $r > 0$, then $0 < m - r < m$ and $\pi^{m-r}(i) = i \in \{i, \pi(i), \pi^2(i), \ldots, \pi^{m-r-1}(i)\}$, contradicting m being the least positive integer with this property). So, $\pi^m(i) = i$. □

We will call a permutation $\pi \in S_n$ a **cycle** if there are $m \in \mathbb{Z}^+$ and $i \in \{1, 2, \ldots, n\}$ such that

1. $\pi^m(i) = i$
2. for every $j \notin \{i, \pi(i), \pi^2(i), \ldots, \pi^{m-1}(i)\}$, $\pi(j) = j$.

If π is a permutation and m is the **least** positive integer satisfying 1 and 2 above, then we will call π an **m-cycle**. We will also say that m is the **length** of π. A 2-cycle is also called a **transposition**.

Example 12.4:

1. $\pi = (12) \in S_2$ is a transposition (or 2-cycle). Indeed, $\pi^2(1) = \pi(\pi(1)) = \pi(2) = 1$, and so, condition 1 is satisfied. Condition 2 is vacuously satisfied because $\{1, \pi(1)\} = \{1, 2\}$. Also, $\pi^1(1) = \pi(1) = 2$. Therefore, $\pi^1(1) \ne 1$, and so, 2 is the least positive value of m satisfying conditions 1 and 2.

   It's worth noting that 4 is also a positive integer satisfying conditions 1 and 2. Indeed, we have $\pi^4(1) = \pi^2(\pi^2(1)) = \pi^2(1) = 1$ and $\{1, \pi(1), \pi^2(1), \pi^3(1)\} = \{1, 2, 1, 2\} = \{1, 2\}$. But (12) is **not** a 4-cycle, because 4 is **not the least positive integer** satisfying conditions 1 and 2.

   Note that if we think of $\pi = (12)$ as a permutation in S_n for $n > 2$, then (12) is still a 2-cycle. In this case, we need to be a bit more careful checking condition 2. However, by definition of the cycle notation, (12) indicates that if $j > 2$, then $\pi(j) = j$, and so, condition 2 is satisfied.

2. The identity $(1) \in S_n$ is a 1-cycle. Indeed, $\pi^1(1) = 1$, and so, condition 1 is satisfied. If $j \notin \{1\}$, then $\pi(j) = j$ by definition of the cycle notation.

3. Every element of S_1, S_2, and S_3 is a cycle. S_1 consists of just the 1-cycle (1). S_2 consists of the 1-cycle (1) and the transposition (or 2-cycle) (12). S_3 consists of the 1-cycle (1), the transpositions $(12), (13)$, and (23), and the 3-cycles (123) and (132).

4. S_4 contains both cycles and permutations that are not cycles. For example, (1234) is a 4-cycle, whereas $\pi = (12)(34)$ is **not** a cycle. To see this, let's show that for each $m \in \mathbb{Z}^+$ and $i \in \{1,2,3,4\}$, condition 2 is not satisfied. Let $m \in \mathbb{Z}^+$. Then for $i = 1$ or 2, we see that $\{i, \pi(i), \pi^2(i), \ldots, \pi^{m-1}(i)\} \subseteq \{1,2\}$ and $\pi(3) = 4 \neq 3$. Similarly, for $i = 3$ or 4, we see that $\{i, \pi(i), \pi^2(i), \ldots, \pi^{m-1}(i)\} \subseteq \{3,4\}$ and $\pi(1) = 2 \neq 1$. So, condition 2 cannot be satisfied and $\pi = (12)(34)$ is **not** a cycle.

As we know, for $n \geq 3$, S_n is a noncommutative group. So, in general, permutations do **not** commute with each other. For example, let $\pi = (12)(34)$ and $\sigma = (123)$. Then $\pi\sigma = (12)(34)(123) = (243)$ and $\sigma\pi = (123)(12)(34) = (134)$. So, $\pi\sigma \neq \sigma\pi$.

However, as we will soon see, permutations that are "disjoint" from each other do commute. We say that permutations π and σ in S_n are **disjoint** if for all $i \in \{1, 2, \ldots, n\}$, $\pi(i) \neq i$ implies $\sigma(i) = i$.

In words, π and σ are disjoint if every positive integer moved by π is fixed by σ (note that this is equivalent to saying that every positive integer moved by σ is fixed by π).

Example 12.5:

1. (12) and (34) are disjoint transpositions.

2. $\pi = (124)$ and $\sigma = (345)$ are **not** disjoint 3-cycles. Indeed, they both move 4. We have $\pi(4) = 1$ and $\sigma(4) = 5$.

3. $(12)(567)$ and $(34)(89)$ are disjoint permutations that are not cycles.

4. The identity permutation $e = (1) \in S_n$ is disjoint from every other permutation because $e(i) = i$ for all $i \in \{1, 2, \ldots, n\}$.

Theorem 12.6: Disjoint permutations commute with each other.

Analysis: It's worth pointing out the following before writing the proof:

1. If π and σ are disjoint permutations in S_n and $i \in \{1, 2, \ldots, n\}$, then either $\pi(i) = i$ or $\sigma(i) = i$ (or both). In words, at least one of the permutations must fix i.

2. Permutations are injective. It follows that if σ is a permutation such that $\sigma(i) \neq i$, then we **must** have that $\sigma(\sigma(i)) \neq \sigma(i)$. In words, if σ does **not** fix i, then σ does **not** fix $\sigma(i)$ either.

Proof: Let $\pi, \sigma \in S_n$ be disjoint permutations and let $i \in \{1, 2, \ldots, n\}$. Since π and σ are disjoint, either $\pi(i) = i$ or $\sigma(i) = i$. Without loss of generality, assume that $\pi(i) = i$. Then $\sigma\pi(i) = \sigma(i)$. If σ also fixes i, then $\pi\sigma(i) = \pi(i) = i = \sigma(i) = \sigma\pi(i)$. If σ does not fix i, then $\sigma(i) \neq i$, and then since σ is injective, $\sigma(\sigma(i)) \neq \sigma(i)$. So, σ does not fix $\sigma(i)$. Since π and σ are disjoint, π must fix $\sigma(i)$. That is, $\pi(\sigma(i)) = \sigma(i) = \sigma(\pi(i))$. Since $i \in \{1, 2, \ldots n\}$ was arbitrary, $\pi\sigma = \sigma\pi$. □

Example 12.7:

1. Since (12) and (34) are disjoint transpositions, by Theorem 12.6, $(12)(34) = (34)(12)$.

2. The cycles (1356) and (27894) are disjoint. So, $(1356)(27894) = (27894)(1356)$ by Theorem 12.6.

3. $(124)(345) = (12453)$ and $(345)(124) = (12534)$. So, we see that (124) and (345) do not commute. Note that (124) and (345) are 3-cycles that are **not** disjoint.

4. (1234) and $(13)(24)$ are permutations that are **not** disjoint. Nonetheless, we have $(1234)(13)(24) = (1432)$ and $(13)(24)(1234) = (1432)$. So, the permutations (1234) and $(13)(24)$ commute even though they are not disjoint. This example shows that the converse of Theorem 12.6 is false. Observe that $(13)(24) = (1234)^2$.

   More generally, if π is a permutation and $m, n \in \mathbb{Z}$, then π^m and π^n are permutations that commute that are usually not disjoint.

5. The permutations $(12)(567)$ and $(34)(89)$ must commute by Theorem 12.6, and in fact, we can write the product of the four cycles in any order. For example, $(12)(567)(34)(89)$ is the same permutation as $(12)(34)(567)(89)$. There are $4! = 24$ different ways to write this permutation as a product of disjoint cycles, up to rearranging the elements in each individual cycle (for example, (12) is the same cycle as (21), and so, we consider $(21)(34)(567)(89)$ to be the same expression as $(12)(34)(567)(89)$).

In part 3 of Example 12.7, we were able to write the product $(124)(345)$ as the cycle (12453). In general, a product cannot necessarily be written as a single cycle. For example, $(12)(34)$ is completely simplified and is a product of two disjoint cycles. However, we will see shortly (in Theorem 12.9 below) that every permutation can be written as a product of disjoint cycles.

Note: When we say that $\pi = \pi_1 \pi_2 \cdots \pi_n$ is a product of disjoint cycles, we mean that if $i \neq j$, then π_i and π_j are disjoint. Perhaps it would be more accurate to say that π can be written as a product of *pairwise disjoint* cycles. However, we will stick with the "less accurate" terminology that is used by most authors.

Given a permutation $\pi \in S_n$, we define a relation $\sim$ on $\{1, 2, \ldots, n\}$ by $i \sim j$ if and only if there is $k \in \mathbb{Z}$ such that $\pi^k(i) = j$. Let's check that $\sim$ is an equivalence relation. Since $\pi^0(i) = i$, we see that $i \sim i$, and so, $\sim$ is reflexive. If $i \sim j$, then there is $k \in \mathbb{Z}$ such that $\pi^k(i) = j$. Then $\pi^{-k}(j) = (\pi^k)^{-1}(j) = i$, and so, $j \sim i$. Thus, $\sim$ is symmetric. Finally, if $i \sim j$ and $j \sim t$, then there are $k, m \in \mathbb{Z}$ such that $\pi^k(i) = j$ and $\pi^m(j) = t$. So, $\pi^{m+k}(i) = \pi^m \pi^k(i) = \pi^m(j) = t$. So, $i \sim t$. It follows that $\sim$ is transitive. Since $\sim$ is reflexive, symmetric, and transitive, $\sim$ is an equivalence relation.

The equivalence classes of $\sim$ are called the **orbits** of the equivalence relation. Since the orbits depend on the permutation π, we may call them orbits of the permutation π.

Example 12.8:

1. The identity permutation in S_n has n orbits. For example, in S_3, the orbits of (1) are $\{1\}$, $\{2\}$, and $\{3\}$.

2. The transposition $(12) \in S_2$ has 1 orbit, namely $\{1, 2\}$. However, if we think of $(12) \in S_3$, there are two orbits: $\{1, 2\}$ and $\{3\}$. In general, any number that does not appear in a cycle is in its own orbit.

3. An n-cycle in S_n has one orbit. For example, $(1352476) \in S_7$ has only one orbit. This orbit is $\{1, 2, 3, 4, 5, 6, 7\}$.

4. An n-cycle in S_m, where $m > n$ has $m - n + 1$ orbits. For example, $(135) \in S_5$ has 3 orbits ($5 - 3 + 1 = 3$). These orbits are $\{1, 3, 5\}$, $\{2\}$, and $\{4\}$.

5. In general, a product of disjoint cycles in S_n has $m + k$ orbits, where m is the number of cycles and k is the number of elements of $\{1, 2, \ldots, n\}$ that do not appear in a cycle. For example, in S_8, $(125)(38)$ has $2 + 3 = 5$ orbits. These orbits are $\{1, 2, 5\}$, $\{3, 8\}$, $\{4\}$, $\{6\}$, and $\{7\}$. Observe that there is one orbit for each cycle and one orbit for each element of $\{1, 2, \ldots, n\}$ that does not appear in either cycle.

6. A product of cycles that are not disjoint needs to be checked more carefully. For example, in S_4, the permutation $(12)(234)$ has just one orbit. Indeed, $(12)(234) = (1234)$, and so, the only orbit is $\{1, 2, 3, 4\}$.

In part 6 of Example 12.8 above, we saw two different ways to write the same permutation using cycle notation. The first way, $(12)(234)$ is a product of two cycles that are **not** disjoint. We transformed this product into the single cycle (1234) using a simple algorithm. This algorithm was described in Example 4.12. In general, this algorithm for multiplying cycles together will always result in a product of disjoint cycles.

Theorem 12.9: Every permutation can be written as a product of disjoint cycles.

The proof of Theorem 12.9 is a bit difficult to write down, but the idea is very simple. Given a permutation π, we look at the orbits of π, say $X_1, X_2, \ldots, X_m$. For each $i = 1, 2, \ldots, m$, we let π_i be the permutation that behaves the same as π on X_i and fixes every element not in X_i. It should be intuitively clear that $\pi = \pi_1 \pi_2 \cdots \pi_m$ and that for $i \neq j$, π_i and π_j are disjoint. Readers that do not wish to get caught up in the technical details can accept the informal argument just given and skip over the formal proof. However, for completeness, I will write out the details.

Proof: Let π be a permutation in S_n. If π is the identity permutation, then π can be written as the 1-cycle (1). So, assume that $\pi \neq (1)$. Then π has at least one orbit containing more than one element. Let $X_1, X_2, \ldots, X_m$ be the orbits of π that contain more than one element. For each $i = 1, 2, \ldots, m$, define $\sigma_i: X_i \to X_i$ by $\sigma_i(j) = \pi(j)$.

Let's show that for each $i = 1, 2, \ldots, k$, σ_i is a permutation of X_i. First, if $x \in X_i$, then $\pi(x) \in X_i$ by the definition of an orbit of π. Therefore, $\sigma_i(x) = \pi(x) \in X_i$. So, σ_i is indeed a function from X_i to X_i. Now, if $\sigma_i(x_1) = \sigma_i(x_2)$, then $\pi(x_1) = \pi(x_2)$. Since π is injective, $x_1 = x_2$. So, σ_i is injective. Finally, let $y \in X_i$ and let $x = \pi^{-1}(y)$. Then $x \sim y$, and so, $x \in X_i$. Also, $\sigma_i(x) = \pi(x) = \pi(\pi^{-1}(y)) = y$. So, σ_i is surjective.

For each $i = 1, 2, \ldots, m$, define $\pi_i \in S_n$ by $\pi_i(x) = \sigma_i(x)$ if $x \in X_i$ and $\pi_i(x) = x$ otherwise. If $i \neq j$, then π_i and π_j are disjoint because X_i and X_j are disjoint sets. For all $x \in X_i$, $\pi(x) = \pi_i(x)$ and $j \neq i$ implies $\pi_j(x) = x$ and $\pi_j(\pi_i(x)) = \pi_i(x)$. Therefore, we have $\pi = \pi_1 \pi_2 \cdots \pi_m$.

We now show that for each $i = 1, 2, \ldots, m$, $\pi_i \in S_n$ is a cycle. Fix i and let $j \in X_i$. Since X_i is finite, we can choose the least positive integer s such that $\pi^s(j) \in \{j, \pi(j), \pi^2(j), \ldots, \pi^{s-1}(j)\}$. By Theorem 12.3, $\pi^s(j) = j$. So, $Y_i = \{j, \pi(j), \pi^2(j), \ldots, \pi^{s-1}(j)\}$ is the orbit of the cycle $(j\ \pi(j)\ \pi^2(j) \cdots \pi^{s-1}(j))$. We will now show that $X_i = Y_i$.

To see this, let $k \in X_i$. Since $j \in X_i$, there is $t \in \mathbb{Z}$ such that $\pi^t(j) = k$. By the Division Algorithm, there are integers q and r such that $t = sq + r$ and $0 \leq r < s$. So, we have

$$k = \pi^t(j) = \pi^{sq+r}(j) = \pi^{r+sq}(j) = \pi^r \pi^{sq}(j) = \pi^r (\pi^s)^q(j) = \pi^r(j) \in Y_i.$$

So, $X_i \subseteq Y_i$. Since every element in Y_i is clearly in the orbit of j and $j \in X_i$, $Y_i \subseteq X_i$. It follows that $X_i = Y_i$.

Now, $\pi_i^s(j) = \pi^s(j) = j$ and if $k \notin \{j, \pi_i(j), \pi_i^2(j), \ldots, \pi_i^{s-1}(j)\} = X_i$, then $\pi_i(k) = k$. So, π_i is a cycle, as desired. $\square$

In addition to writing a permutation as a product of disjoint cycles, another particularly nice way to write a permutation is as a product of transpositions. In Example 12.10 below, we will see that this can always be done.

Example 12.10:

1. There are many ways to express the 3-cycle (123) as a product of transpositions. One way is (13)(12). Another way is (23)(13). Yet another (perhaps overly complicated) way is (13)(12)(13)(12)(13)(12)(13)(12). We see from this example that there are many ways to express a 3-cycle as a product of transpositions. Even the number of transpositions is not determined by the permutation. However, notice that each representation that we wrote down contains an even number of transpositions. We will see shortly that this not a coincidence.

    More generally, if $a, b, c \in \{1, 2, \ldots, n\}$, then one way to write the 3-cycle (abc) as a product of transpositions is $(ac)(ab)$. Another way is $(bc)(ac)$.

2. The 4-cycle (1234) can be written as (14)(13)(12). Another way to write (1234) as a product of transpositions is (23)(34)(14).

    More generally, if $a, b, c, d \in \{1, 2, \ldots, n\}$, then one way to write the 4-cycle $(abcd)$ as a product of transpositions is $(ad)(ac)(ab)$. Another way is $(bc)(cd)(ad)$.

3. If $a_1, a_2, \ldots, a_k \in \{1, 2, \ldots, n\}$, then one way to write the k-cycle $(a_1 a_2 \cdots a_k)$ as a product of transpositions is $(a_1 a_k)(a_1 a_{k-1}) \cdots (a_1 a_2)$. Another way is $(a_2 a_3)(a_3 a_4) \cdots (a_{k-1} a_k)(a_1 a_k)$. So, we see that for $k > 1$, every k-cycle can be written as a product of $k - 1$ transpositions.

4. Since $(123) = (13)(12)$ and $(4567) = (47)(46)(45)$, we can write $(123)(4567)$ as a product of transpositions by combining these results: $(123)(4567) = (13)(12)(47)(46)(45)$.

Part 3 of Example 12.10 shows that any cycle can be written as a product of transpositions. It follows that any permutation can be written as a product of transpositions (just use the method shown in part 4 of Example 12.10). We also saw in Example 12.10 that there are many ways to write a given permutation as a product of transpositions and in fact the number of transpositions that appear is not unique (see part 1 of Example 12.10). However, there is one thing that is unique. Given a permutation π, if we can write π as a product of an even number of transpositions, then whenever we write π as a product of transpositions, the number of transpositions will be even. Similarly, if we can write π as a product of an odd number of transpositions, then whenever we write π as a product of transpositions, the number of transpositions will be odd. We will prove this result shortly (see Theorem 12.12 below).

We will say that a permutation is **even** if it can be written as the product of an even number of transpositions. We will say that a permutation is **odd** if it can be written as the product of an odd number of transpositions.

Since every permutation can be written as a product of transpositions (by Example 12.10 and the following discussion), it follows that every permutation is either even or odd. We would now like to show that a permutation **cannot be both** even and odd.

Lemma 12.11: Let $(1) \in S_n$ be the identity and suppose that $(1) = \pi_1 \pi_2 \cdots \pi_k$, where for each $i = 1, 2, \ldots, k$, π_i is a transposition. Then k is even.

The proof of Lemma 12.11 is a bit tedious and so I leave it to the reader (see Problem 16 below).

Notes: (1) Lemma 12.11 tells us that $(1) \in S_n$ is **not** an odd permutation.

(2) If $n > 1$, then $(1) \in S_n$ can be written as $(1) = (12)(12)$, a product of two transpositions. Therefore, (1) is even.

(3) Since $(1) \in S_1$ cannot be written as a product of transpositions, technically speaking, it is neither even nor odd. To avoid this annoying exception, whenever we are speaking of odd and/or even permutations, we will assume that we are working in S_n for $n > 1$.

Theorem 12.12: If $\pi \in S_n$, then π cannot be both even and odd.

Proof: Suppose toward contradiction that $\pi = \sigma_1 \sigma_2 \cdots \sigma_{2j}$ and $\pi = \tau_1 \tau_2 \cdots \tau_{2k+1}$, where $j, k \in \mathbb{Z}^+$, σ_i is a transposition for each $i = 1, 2, \ldots, 2j$ and τ_i is a transposition for each $i = 1, 2, \ldots, 2k+1$. Since every transposition is its own inverse, we have

$$\pi^{-1} = (\tau_1 \tau_2 \cdots \tau_{2k+1})^{-1} = (\tau_{2k+1})^{-1} (\tau_{2k})^{-1} \cdots (\tau_2)^{-1} (\tau_1)^{-1} = \tau_{2k+1} \tau_{2k} \cdots \tau_2 \tau_1.$$

Therefore, $(1) = \pi \pi^{-1} = \sigma_1 \sigma_2 \cdots \sigma_{2j} \tau_{2k+1} \tau_{2k} \cdots \tau_2 \tau_1$ is a product of $2j + 2k + 1 = 2(j + k) + 1$ transpositions. Since $2(j + k) + 1$ is not even (it's odd), this contradicts Lemma 12.11. It follows that π cannot be both even and odd. □

It's worth noting that by part 3 of Example 12.10, we have the following:

(1) a k-cycle is even if and only if k is odd;

(2) a k-cycle is odd if and only if k is even.

The Alternating Group

Let $A_n = \{\pi \in S_n \mid \pi \text{ is even}\}$. Let's check that A_n is a subgroup of S_n. The identity permutation (1) is in A_n by Lemma 12.11. If $\pi, \sigma \in A_n$, then we can write $\pi = \pi_1 \pi_2 \cdots \pi_{2j}$ and $\sigma = \sigma_1 \sigma_2 \cdots \sigma_{2k}$, where $j, k \in \mathbb{Z}^+$, σ_i is a transposition for each $i = 1, 2, \ldots, 2j$ and τ_i is a transposition for each $i = 1, 2, \ldots, 2k$. Since every transposition is its own inverse, we have

$$\pi \sigma^{-1} = (\pi_1 \pi_2 \cdots \pi_{2j})(\sigma_1 \sigma_2 \cdots \sigma_{2k})^{-1} = \pi_1 \pi_2 \cdots \pi_{2j} (\sigma_{2k})^{-1} \cdots (\sigma_2)^{-1} (\sigma_1)^{-1}$$
$$= \pi_1 \pi_2 \cdots \pi_{2j} \sigma_{2k} \cdots \sigma_2 \sigma_1.$$

This is a product of $2j + 2k = 2(j + k)$ transpositions. Since $2(j + k)$ is even, $\pi\sigma^{-1} \in A_n$. By Problem 5 from Problem Set 6, A_n is a subgroup of S_n.

A_n is called the **alternating group on** $\{1, 2, \ldots, n\}$ (or the **alternating group of degree** n).

For each $n \in \mathbb{Z}^+$, define the function $\Delta_n: S_n \to \{-1, 1\}$ by $\Delta_n(\pi) = \begin{cases} -1 & \text{if } \pi \text{ is odd.} \\ 1 & \text{if } \pi \text{ is even.} \end{cases}$
In Problem 5 below, you will be asked to verify that for $n > 1$, Δ_n is a surjective homomorphism.

Another way to see that A_n is a subgroup of S_n (and in fact that A_n is a normal subgroup of S_n) is to observe that $A_n = \ker(\Delta_n)$.

Example 12.13:

1. $S_3 = \{(1), (12), (13), (23), (123), (132)\}$ and $A_n = \{(1), (123), (132)\}$. Let's explicitly write each element of A_3 as a product of an even number of transpositions. We have $(1) = (12)(12)$, $(123) = (13)(12)$, and $(132) = (12)(13)$.

2. S_4 contains 24 permutations. You were asked to list these elements in Problem 4 from Problem Set 4. Let's write down the subgroup A_4:

$A_4 = \{(1), (123), (124), (134), (234), (132), (142), (143), (243), (12)(34), (13)(24), (14)(23)\}$.

Theorem 12.14: If $n \in \mathbb{Z}$ with $n > 1$, then $|A_n| = \frac{n!}{2}$.

Proof: If $\pi \in S_n$ is an odd permutation, then $(12)\pi$ is an even permutation. Also, if $\pi \neq \sigma$, then by the left cancellation law for groups, $(12)\pi \neq (12)\sigma$. It follows that there are at least as many even permutations as there are odd permutations.

Similarly, if $\pi \in S_n$ is an even permutation, then $(12)\pi$ is an odd permutation. So, by the same reasoning as in the last paragraph, there are at least as many odd permutations as there are even permutations. So, $|A_n| = \frac{|S_n|}{2} = \frac{n!}{2}$. $\square$

By Lagrange's Theorem (Theorem 10.8), for $n \in \mathbb{Z}$ with $n > 1$, we have $|S_n| = |A_n| \cdot [S_n : A_n]$, and so, by Theorem 12.14, $n! = \frac{n!}{2} \cdot [S_n : A_n]$. Therefore, $[S_n : A_n] = 2$. Problem 8 from Problem Set 10 then gives us another way to see that A_n is a normal subgroup of S_n.

Permutation Matrices

Example 12.15: Consider the 2×2 matrix $P = \begin{bmatrix} 0 & 1 \\ 1 & 0 \end{bmatrix}$. Observe that the product of this matrix and a vector $\mathbf{x} = \begin{bmatrix} x_1 \\ x_2 \end{bmatrix}$ is

$$P\mathbf{x} = \begin{bmatrix} 0 & 1 \\ 1 & 0 \end{bmatrix}\begin{bmatrix} x_1 \\ x_2 \end{bmatrix} = \begin{bmatrix} 0 \cdot x_1 + 1 \cdot x_2 \\ 1 \cdot x_1 + 0 \cdot x_2 \end{bmatrix} = \begin{bmatrix} x_2 \\ x_1 \end{bmatrix}.$$

We see that multiplying a vector on the left by P interchanges the two entries of the vector. In other words, the matrix P behaves just like the permutation $(12) \in S_2$. We will call P a **permutation matrix**. Is there another permutation matrix that behaves just like $(1) \in S_2$. It should be clear that the identity matrix $I = \begin{bmatrix} 1 & 0 \\ 0 & 1 \end{bmatrix}$ has this behavior. Indeed,

$$I\mathbf{x} = \begin{bmatrix} 1 & 0 \\ 0 & 1 \end{bmatrix} \begin{bmatrix} x_1 \\ x_2 \end{bmatrix} = \begin{bmatrix} x_1 \\ x_2 \end{bmatrix}.$$

The above observations might lead us to expect that $M = \left\{ \begin{bmatrix} 1 & 0 \\ 0 & 1 \end{bmatrix}, \begin{bmatrix} 0 & 1 \\ 1 & 0 \end{bmatrix} \right\}$ is a group under matrix multiplication that is isomorphic to S_2. The identity of the group is $\begin{bmatrix} 1 & 0 \\ 0 & 1 \end{bmatrix}$. The following computation shows that M is closed under matrix multiplication and that each element is its own inverse:

$$\begin{bmatrix} 0 & 1 \\ 1 & 0 \end{bmatrix} \begin{bmatrix} 0 & 1 \\ 1 & 0 \end{bmatrix} = \begin{bmatrix} 1 & 0 \\ 0 & 1 \end{bmatrix}$$

The function $f: M \to S_2$ defined by $f\left(\begin{bmatrix} 1 & 0 \\ 0 & 1 \end{bmatrix}\right) = (1)$ and $f\left(\begin{bmatrix} 0 & 1 \\ 1 & 0 \end{bmatrix}\right) = (12)$ is an isomorphism.

In general, an $n \times n$ matrix is a **permutation matrix** if it can be attained by rearranging the rows of the $n \times n$ identity matrix (or equivalently, by rearranging the columns of the $n \times n$ identity matrix).

Example 12.16:

1. The 3×3 matrix $P = \begin{bmatrix} 0 & 1 & 0 \\ 0 & 0 & 1 \\ 1 & 0 & 0 \end{bmatrix}$ is a permutation matrix that behaves just like $(123) \in S_3$.

   Indeed, the product of this matrix and a vector $\mathbf{x} = \begin{bmatrix} x_1 \\ x_2 \\ x_3 \end{bmatrix}$ is

   $$P\mathbf{x} = \begin{bmatrix} 0 & 1 & 0 \\ 0 & 0 & 1 \\ 1 & 0 & 0 \end{bmatrix} \begin{bmatrix} x_1 \\ x_2 \\ x_3 \end{bmatrix} = \begin{bmatrix} x_2 \\ x_3 \\ x_1 \end{bmatrix}.$$

2. The 3×3 matrix $\begin{bmatrix} 1 & 0 & 0 \\ 0 & 0 & 1 \\ 1 & 0 & 0 \end{bmatrix}$ is **not** a permutation matrix. There can never be two identical rows in a permutation matrix. More specifically, in a permutation matrix, 1 appears exactly once in each row and column, and every other entry is 0.

3. The 4×4 matrix $Q = \begin{bmatrix} 0 & 0 & 1 & 0 \\ 0 & 0 & 0 & 1 \\ 1 & 0 & 0 & 0 \\ 0 & 1 & 0 & 0 \end{bmatrix}$ is a permutation matrix that behaves like $(13)(24) \in S_4$.

   Indeed, the product of this matrix and a vector $\mathbf{x} = \begin{bmatrix} x_1 \\ x_2 \\ x_3 \\ x_4 \end{bmatrix}$ is $P\mathbf{x} = \begin{bmatrix} 0 & 0 & 1 & 0 \\ 0 & 0 & 0 & 1 \\ 1 & 0 & 0 & 0 \\ 0 & 1 & 0 & 0 \end{bmatrix} \begin{bmatrix} x_1 \\ x_2 \\ x_3 \\ x_4 \end{bmatrix} = \begin{bmatrix} x_3 \\ x_4 \\ x_1 \\ x_2 \end{bmatrix}.$

Theorem 12.17: For each $n \in \mathbb{Z}^+$, the set of $n \times n$ permutation matrices forms a group under matrix multiplication that is isomorphic to S_n.

You will be asked to prove Theorem 12.17 in Problem 20 below.

Determinants

Let's begin by informally describing how to take the determinant of an $n \times n$ matrix. The following rules specify how to do this.

(1) We will be adding $n!$ terms. For example, for a 2×2 matrix, we will add $2! = 2$ terms, for a 3×3 matrix, we will add $3! = 1 \cdot 2 \cdot 3 = 6$ terms, and so on.

(2) There will be one term for each permutation of $\{1, 2, \ldots, n\}$. For example, for a 2×2 matrix, there will be one term for the identity permutation (1) and another term for the permutation (12).

(3) Each term will be equal to a product of entries from the matrix, one entry being chosen from each row. Each term may or may not be multiplied by -1.

(4) The permutation tells us which entry to choose in each row. Specifically, for the permutation π, in the ith row, we choose the entry in the $\pi(i)$th column. For example, for a 2×2 matrix, for the term corresponding to the identity permutation (1), we choose the 1st entry in the 1st row and the 2nd entry in the 2nd row. For the term corresponding to the permutation (12), we choose the 2nd entry in the 1st row and the 1st entry in the 2nd row.

(5) A term is multiplied by -1 if and only if it corresponds to an odd permutation.

If A is an $n \times n$ matrix, then the determinant of A will be abbreviated by $\det A$.

Example 12.18:

1. Consider the 2×2 matrix $A = \begin{bmatrix} 1 & 2 \\ 3 & 4 \end{bmatrix}$. Let's compute $\det A$ using the rules above. Since A is a 2×2 matrix, there will be 2 terms. For the term corresponding to the permutation (1), we choose the 1st entry in the 1st row and the 2nd entry in the 2nd row to form the product $1 \cdot 4$. Since (1) is an even permutation, we do **not** multiply this term by -1. For the term corresponding to the permutation (12), we choose the 2nd entry in the 1st row and the 1st entry in the 2nd row to form the product $2 \cdot 3$. Since (12) is an odd permutation, we **do** multiply this term by -1 to get $-1 \cdot 2 \cdot 3$. We get the final answer by adding these two terms to get

$$\det(A) = 1 \cdot 4 - 2 \cdot 3 = 4 - 6 = -2.$$

Following the procedure in the above paragraph, we see that in general, the determinant of the 2×2 matrix $A = \begin{bmatrix} a_{11} & a_{12} \\ a_{21} & a_{22} \end{bmatrix}$ is

$$\det(A) = \det\left(\begin{bmatrix} a_{11} & a_{12} \\ a_{21} & a_{22} \end{bmatrix}\right) = a_{11}a_{22} - a_{12}a_{21}.$$

2. Consider the 3×3 matrix $B = \begin{bmatrix} 1 & 2 & 3 \\ 4 & 3 & 2 \\ 1 & 5 & 1 \end{bmatrix}$. Let's compute det B using the rules above. Since B is a 3×3 matrix, there will be $3! = 6$ terms. Let's list each permutation and the corresponding product:

Permutation	Even or Odd	Term
(1)	Even	$1 \cdot 3 \cdot 1$
(123)	Even	$2 \cdot 2 \cdot 1$
(132)	Even	$3 \cdot 4 \cdot 5$
(13)	Odd	$-1 \cdot 3 \cdot 3 \cdot 1$
(23)	Odd	$-1 \cdot 1 \cdot 2 \cdot 5$
(12)	Odd	$-1 \cdot 2 \cdot 4 \cdot 1$

We get the final answer by adding these six terms to get

$$\det(B) = 1 \cdot 3 \cdot 1 + 2 \cdot 2 \cdot 1 + 3 \cdot 4 \cdot 5 - 3 \cdot 3 \cdot 1 - 1 \cdot 2 \cdot 5 - 2 \cdot 4 \cdot 1$$
$$= 3 + 4 + 60 - 9 - 10 - 8 = \mathbf{40}.$$

Following the procedure in the above paragraph, we see that in general, the determinant of the 3×3 matrix $B = \begin{bmatrix} a_{11} & a_{12} & a_{13} \\ a_{21} & a_{22} & a_{23} \\ a_{31} & a_{32} & a_{33} \end{bmatrix}$ is

$$\det(B) = \det\left(\begin{bmatrix} a_{11} & a_{12} & a_{13} \\ a_{21} & a_{22} & a_{23} \\ a_{31} & a_{32} & a_{33} \end{bmatrix}\right)$$

$$= a_{11}a_{22}a_{33} + a_{12}a_{23}a_{31} + a_{13}a_{21}a_{32} - a_{13}a_{22}a_{31} - a_{11}a_{23}a_{32} - a_{12}a_{21}a_{33}.$$

Let's now give a precise definition of the determinant of an $n \times n$ matrix.

Recall that for $n \in \mathbb{Z}^+$, $\Delta_n: S_n \to \{-1, 1\}$ is defined by $\Delta_n(\pi) = \begin{cases} -1 & \text{if } \pi \text{ is odd.} \\ 1 & \text{if } \pi \text{ is even.} \end{cases}$

Let $n \in \mathbb{Z}^+$ and let A be an $n \times n$ matrix over a field F. The **determinant** of A, written det A, is

$$\det A = \sum_{\pi \in S_n} \Delta_n(\pi) a_{1,\pi(1)} a_{2,\pi(2)} \cdots a_{n,\pi(n)}$$

Notes: (1) See Notes 1 and 2 before Example 7.13 for a detailed explanation about how this notation works.

(2) I added commas in the subscripts of the expression $a_{1,\pi(1)} a_{2,\pi(2)} \cdots a_{n,\pi(n)}$ for easier readability. For example, $a_{1,\pi(1)}$ is the entry in row 1 and column $\pi(1)$ of the matrix A.

For example, if π is the permutation (12), then $a_{1,\pi(1)} = a_{12}$, and so, $a_{1,\pi(1)}$ is the entry in row 1 and column 2 of the matrix A.

(3) Observe that the determinant is a function from $n \times n$ matrices over the given field F to the field itself. In other words, $\det: M_{n,n}^F \to F$.

Example 12.19: Let's analyze the formal definition of a determinant carefully for a 2×2 matrix $A = \begin{bmatrix} a_{11} & a_{12} \\ a_{21} & a_{22} \end{bmatrix}$. There are two permutations is S_2, namely $\pi_1 = (1)$ and $\pi_2 = (12)$. Since π_1 is an even permutation, $\Delta(\pi_1) = 1$. Since π_2 is an odd permutation, $\Delta(\pi_2) = -1$. So, we get

$$\det A = \sum_{\pi \in S_2} \Delta_2(\pi) a_{1,\pi(1)} a_{2,\pi(2)} = \Delta_2(\pi_1) a_{1,\pi_1(1)} a_{2,\pi_1(2)} + \Delta_2(\pi_2) a_{1,\pi_2(1)} a_{2,\pi_2(2)}$$

$$= 1 \cdot a_{11} a_{22} + (-1) a_{12} a_{21} = a_{11} a_{22} - a_{12} a_{21}.$$

The dedicated reader should go through a similar analysis for a 3×3 matrix to help solidify their understanding of the definition of the determinant.

Matrix Groups

In Lesson 7, we saw that $M_{n,n}^F$, the set of $n \times n$ matrices with entries in a field F is a ring under matrix addition and matrix multiplication (see the notes following Example 7.14). In particular, $(M_{n,n}^F, +)$ is a commutative group. On the other hand, $(M_{n,n}^F, \cdot)$ is only a monoid, which is **not** commutative. See part 3 of Example 7.13 for a pair of 2×2 matrices that do not commute. Let's give a simple example of a 2×2 matrix that is not invertible.

Example 12.20: The matrix $\begin{bmatrix} 1 & 0 \\ 0 & 0 \end{bmatrix} \in M_{2,2}^{\mathbb{R}}$ has no multiplicative inverse. To see this, suppose toward contradiction that there is a matrix $\begin{bmatrix} a_{11} & a_{12} \\ a_{21} & a_{22} \end{bmatrix} \in M_{2,2}^{\mathbb{R}}$ such that $\begin{bmatrix} 1 & 0 \\ 0 & 0 \end{bmatrix} \begin{bmatrix} a_{11} & a_{12} \\ a_{21} & a_{22} \end{bmatrix} = \begin{bmatrix} 1 & 0 \\ 0 & 1 \end{bmatrix}$. The entry in row 2 and column 2 of the product on the left is $0 a_{12} + 0 a_{22} = 0$. The entry in row 2 and column 2 of the matrix on the right is 1. So, $0 = 1$, which is a contradiction. Therefore, $\begin{bmatrix} 1 & 0 \\ 0 & 0 \end{bmatrix}$ has no inverse.

It is natural to ask if there are any subsets X of $M_{n,n}^F$ such that $(X, \cdot)$ is a group. Let's begin by analyzing $M_{2,2}^F$, the set of 2×2 matrices over a field F. We begin with the following useful result.

Theorem 12.21: Let $A, B \in M_{2,2}^F$. Then $\det AB = (\det A)(\det B)$.

Proof: We prove this by direct computation. We have

$$AB = \begin{bmatrix} a_{11} & a_{12} \\ a_{21} & a_{22} \end{bmatrix} \begin{bmatrix} b_{11} & b_{12} \\ b_{21} & b_{22} \end{bmatrix} = \begin{bmatrix} a_{11}b_{11} + a_{12}b_{21} & a_{11}b_{12} + a_{12}b_{22} \\ a_{21}b_{11} + a_{22}b_{21} & a_{21}b_{12} + a_{22}b_{22} \end{bmatrix}.$$

Therefore, we have the following:

$$\det AB = (a_{11}b_{11} + a_{12}b_{21})(a_{21}b_{12} + a_{22}b_{22}) - (a_{11}b_{12} + a_{12}b_{22})(a_{21}b_{11} + a_{22}b_{21})$$
$$= (a_{11}b_{11}a_{21}b_{12} + a_{11}b_{11}a_{22}b_{22} + a_{12}b_{21}a_{21}b_{12} + a_{12}b_{21}a_{22}b_{22})$$
$$- (a_{11}b_{12}a_{21}b_{11} + a_{11}b_{12}a_{22}b_{21} + a_{12}b_{22}a_{21}b_{11} + a_{12}b_{22}a_{22}b_{21})$$
$$= (a_{11}b_{11}a_{21}b_{12} - a_{11}b_{12}a_{21}b_{11}) + (a_{12}b_{21}a_{22}b_{22} - a_{12}b_{22}a_{22}b_{21})$$
$$+ (a_{11}b_{11}a_{22}b_{22} + a_{12}b_{21}a_{21}b_{12}) - (a_{11}b_{12}a_{22}b_{21} + a_{12}b_{22}a_{21}b_{11})$$
$$= 0 + 0 + a_{11}a_{22}b_{11}b_{22} - a_{11}a_{22}b_{12}b_{21} - a_{12}a_{21}b_{11}b_{22} + a_{12}a_{21}b_{12}b_{21}$$
$$= (a_{11}a_{22} - a_{12}a_{21})(b_{11}b_{22} - b_{12}b_{21}) = (\det A)(\det B). \qquad \square$$

It's easy to see that the determinant of the 2 × 2 identity matrix is 1. Indeed, we have

$$\det I = \det \begin{bmatrix} 1 & 0 \\ 0 & 1 \end{bmatrix} = 1 \cdot 1 - 0 \cdot 0 = 1.$$

This observation makes the next result quite easy to prove.

Corollary 12.22: Let $A \in M_{2,2}^F$ have a multiplicative inverse. Then $\det(A) \neq 0$ and $\det(A^{-1}) = \frac{1}{\det(A)}$.

Proof: Since $AA^{-1} = I$, by Theorem 12.21, we have $(\det A)(\det A^{-1}) = \det(AA^{-1}) = \det I = 1$. If $\det A = 0$, we get the contradiction $1 = (\det A)(\det A^{-1}) = 0 \cdot (\det A^{-1}) = 0$. So, $\det(A) \neq 0$ and $\det(A^{-1}) = \frac{1}{\det(A)}$. $\qquad \square$

Theorem 12.23: A matrix $A \in M_{2,2}^F$ has a multiplicative inverse if and only if $\det A \neq 0$.

Proof: Let $A = \begin{bmatrix} a_{11} & a_{12} \\ a_{21} & a_{22} \end{bmatrix}$. First assume that A has a multiplicative inverse. Then by Corollary 12.22, $\det A \neq 0$.

Conversely, assuming that $\det A \neq 0$, we can let $B = \frac{1}{\det A}\begin{bmatrix} a_{22} & -a_{12} \\ -a_{21} & a_{11} \end{bmatrix}$. Then we have

$$AB = \begin{bmatrix} a_{11} & a_{12} \\ a_{21} & a_{22} \end{bmatrix} \cdot \frac{1}{\det A}\begin{bmatrix} a_{22} & -a_{12} \\ -a_{21} & a_{11} \end{bmatrix} = \frac{1}{\det A}\begin{bmatrix} a_{11}a_{22} - a_{12}a_{21} & a_{11}(-a_{12}) + a_{12}a_{11} \\ a_{21}a_{22} - a_{22}a_{21} & a_{21}(-a_{12}) + a_{22}a_{11} \end{bmatrix}$$

$$= \frac{1}{\det A}\begin{bmatrix} \det A & 0 \\ 0 & \det A \end{bmatrix} = \begin{bmatrix} 1 & 0 \\ 0 & 1 \end{bmatrix} = I \text{ (where } I \text{ is the 2 × 2 identity matrix).}$$

Similarly, we have $BA = I$ (Check this!), verifying that B is the multiplicative inverse of A. $\qquad \square$

Theorem 12.24: Let F be a field and let $GL_2(F) = \{A \in M_{2,2}^F \mid \det A \neq 0\}$. Then $(GL_2(F), \cdot)$ is a group.

Proof: Let $A, B \in GL_2(F)$. By Theorem 12.21, $\det(AB) = (\det A)(\det B)$. Since $\det A \neq 0$, $\det B \neq 0$, and F is a field, $(\det A)(\det B) \neq 0$. Therefore, $\det(AB) \neq 0$, and so, $AB \in GL_2(F)$. It follows that $GL_2(F)$ is closed under matrix multiplication.

Since matrix multiplication is associative in $M_{2,2}^F$ and $GL_2(F) \subseteq M_{2,2}^F$, matrix multiplication is associative in $GL_2(F)$.

Since $\det I = 1 \neq 0$, I is an identity for $GL_2(F)$.

By Theorem 12.23, every element of $GL_2(F)$ has a multiplicative inverse.

It follows that $(GL_2(F), \cdot)$ is a group. $\square$

$(GL_2(F), \cdot)$ is known as the **general linear group of degree 2.**

More generally, given $n \in \mathbb{Z}^+$, we let $GL_n(F) = \{A \in M_{n,n}^F \mid \det A \neq 0\}$. The structure $(GL_n(F), \cdot)$ is also a group known as the **general linear group of degree n.** The proof that $(GL_n(F), \cdot)$ is a group is identical to the proof of Theorem 12.24. However, the proofs of the analogues of Theorems 12.21 and 12.23 are much more difficult and require theorems from linear algebra that are outside the scope of this book. Therefore, I leave these theorems as challenging problems for the interested reader (see Problems 21 and 22 below).

A **matrix group** is a subgroup of $GL_n(F)$ for some $n \in \mathbb{Z}^+$ and some field F. The group operation is matrix multiplication.

Example 12.25: Let F be a field.

1. For each $n \in \mathbb{Z}^+$, $GL_n(F)$ is clearly a matrix group. At the other extreme, $\{I\}$, where I is the $n \times n$ identity matrix is the **trivial matrix group**.

2. Let $SL_2(F) = \{A \in M_{2,2}^F \mid \det A = 1\}$. To see that $SL_2(F)$ is a matrix group, first observe that $SL_2(F) \subseteq GL_2(F)$ and $I \in SL_2(F)$. If $A, B \in SL_2(F)$, then by Theorem 12.21, Corollary 12.22, and Theorem 12.23, $\det(AB^{-1}) = (\det A)(\det B^{-1}) = 1 \cdot \frac{1}{1} = 1$. So, by Problem 5 in Problem Set 6, we have $SL_2(F) \leq GL_2(F)$.

3. More generally, $SL_n(F) = \{A \in M_{n,n}^F \mid \det A = 1\}$ is a matrix group. Once again, the proof of this is the same as the proof in part 2 above, except that we need to use Problems 21 and 22 below.

4. A 2×2 matrix A is **upper triangular** if the lower left entry of the matrix is zero ($a_{21} = 0$).

    A 2×2 matrix A is **lower triangular** if the upper right entry of the matrix is zero ($a_{12} = 0$).

    A 2×2 matrix is **diagonal** if each entry that is **not** on the main diagonal of A is zero ($a_{21} = 0$ and $a_{12} = 0$).

    Let $U_2 = \{A \in GL_2(F) \mid A \text{ is upper triangular}\}$, $L_2 = \{A \in GL_2(F) \mid A \text{ is lower triangular}\}$, and $D_2 = \{A \in GL_2(F) \mid A \text{ is diagonal}\}$. It is not too hard to verify that U_2, L_2, and D_2 are matrix groups. See Problem 7 below.

5. An $n \times n$ matrix A is **upper triangular** if $a_{ij} = 0$ whenever $i > j$. An $n \times n$ matrix A is **lower triangular** if $a_{ij} = 0$ whenever $i < j$. An $n \times n$ matrix A is **diagonal** if $a_{ij} = 0$ whenever $i \neq j$. Let $U_n = \{A \in GL_n(F) \mid A \text{ is upper triangular}\}$, $L_n = \{A \in GL_n(F) \mid A \text{ is lower triangular}\}$, and $D_n = \{A \in GL_n(F) \mid A \text{ is diagonal}\}$. Then U_n, L_n, and D_n are matrix groups. See Problem 23 below.

Problem Set 12

Full solutions to these problems are available for free download here:
www.SATPrepGet800.com/AAFBTDW

LEVEL 1

1. Compute the order of each permutation.

    (i) (12345)

    (ii) (12)(345)(6789)

2. Express each permutation as a product of transpositions.

    (i) (123456)

    (ii) (123)(4567)

    (iii) (123)(345)(25)(1576)

3. Determine if each permutation is even or odd.

    (i) (1837942)

    (ii) (123)(34)(4567)

LEVEL 2

4. Prove each of the following:

    (i) The product of two even permutations is even.

    (ii) The product of two odd permutations is even.

    (iii) The product of an even permutation and an odd permutation is odd.

5. Prove that for $n > 1$, the function $\Delta_n : S_n \to \{-1, 1\}$ is a surjective homomorphism, where Δ_n is defined by $\Delta_n(\pi) = \begin{cases} -1 & \text{if } \pi \text{ is odd.} \\ 1 & \text{if } \pi \text{ is even.} \end{cases}$

6. Prove that $GL_2(\mathbb{Z}_2)$ is isomorphic to S_3.

7. Let $U_2 = \{A \in GL_2(F) \mid A \text{ is upper triangular}\}$, $L_2 = \{A \in GL_2(F) \mid A \text{ is lower triangular}\}$, and $D_2 = \{A \in GL_2(F) \mid A \text{ is diagonal}\}$. Prove that U_2, L_2, and D_2 are matrix groups.

LEVEL 3

8. Let π and σ be disjoint permutations. Prove that for each $n \in \mathbb{Z}^+$, $(\pi\sigma)^n = \pi^n \sigma^n$.

9. Let $n \geq 3$ and let $s, t \in \{1, 2, \ldots, n\}$ with $s \neq t$. Prove that $A_n = \langle (stu) \mid 1 \leq u \leq n \wedge u \neq s, t \rangle$.

10. Write down a formula for the determinant of a 4×4 matrix in terms of the entries of the matrix.

LEVEL 4

11. Let π be a product of k disjoint cycles of lengths $n_1, n_2, \ldots, n_k$, respectively. Prove that the order of π is $\text{lcm}(n_1, n_2, \ldots, n_k)$.

12. Prove that two cycles in S_n have the same length if and only if they are conjugates of each other.

13. Prove that $|S_n| = n!$

14. Let $n \geq 3$ and let N be a normal subgroup of A_n containing a 3-cycle. Prove that $N = A_n$.

15. Prove that the determinant of an $n \times n$ upper triangular matrix A is the product of the entries of A that lie on the main diagonal of A. Is the same result true for a lower triangular matrix A?

LEVEL 5

16. Let $(1) \in S_n$ be the identity and suppose that $(1) = \pi_1 \pi_2 \cdots \pi_k$, where for each $i = 1, 2, \ldots, k$, π_i is a transposition. Prove that k is even. (This is Lemma 12.11.)

17. Prove that $S_n = \langle (12), (12 \cdots n) \rangle$.

18. Prove that if H is a subgroup of S_n such that $[S_n : H] = 2$, then $H = A_n$.

19. Prove that for $n \geq 5$, A_n has no proper normal subgroups.

20. Prove that for each $n \in \mathbb{Z}^+$, the set of $n \times n$ permutation matrices forms a group under matrix multiplication that is isomorphic to S_n.

CHALLENGE PROBLEMS

21. Let $n \in \mathbb{Z}^+$, let F be a field, and let $A, B \in M_{n,n}^F$. Prove that $\det AB = (\det A)(\det B)$.

22. Let $n \in \mathbb{Z}^+$ and let F be a field. Prove that a matrix $A \in M_{n,n}^F$ has a multiplicative inverse if and only if $\det A \neq 0$.

23. Let $U_n = \{A \in GL_n(F) \mid A \text{ is upper triangular}\}$, $L_n = \{A \in GL_n(F) \mid A \text{ is lower triangular}\}$, and $D_n = \{A \in GL_n(F) \mid A \text{ is diagonal}\}$. Prove that U_n, L_n, and D_n are matrix groups.

LESSON 13
SYLOW THEORY AND GROUP ACTIONS

Conjugacy Classes

Let G be a group and let $h, k \in G$. Recall that k is called a conjugate of h if there is a $g \in G$ such that $k = ghg^{-1}$. We define a relation $\sim$ on G by $h \sim k$ if and only if k is a conjugate of h. It is easy to verify that $\sim$ is an equivalence relation (see Problem 4 below). The equivalence classes for the relation $\sim$ will be called **conjugacy classes**. If $h \in G$, we will let $[h]_C$ be the conjugacy class of h. In other words,

$$[h]_C = \{k \in G \mid h \sim k\} = \{k \in G \mid k = ghg^{-1} \text{ for some } g \in G\} = \{ghg^{-1} \mid g \in G\}.$$

Example 13.1:

1. If G is a commutative group, then the conjugacy class of each element consists of only that element. Indeed, if $h \in G$, then for every $g \in G$, we have $ghg^{-1} = hgg^{-1} = he = h$. Therefore, $[h]_C = \{h\}$. So, in the case of a commutative group G, the number of conjugacy classes is equal to the order of G.

2. $S_3 = \{(1), (12), (13), (23), (123), (132)\}$ has 3 conjugacy classes.

$$[(1)]_C = \{1\}$$
$$[(12)]_C = \{(12), (13), (23)\}$$
$$[(123)]_C = \{(123), (132)\}$$

   For example, the computation $(123)(12)(123)^{-1} = (123)(12)(132) = (23)$ shows that $(12) \sim (23)$. Similarly, $(132)(12)(132)^{-1} = (132)(12)(123) = (13)$ shows $(12) \sim (13)$. I leave it to the reader to show that $(123) \sim (132)$ and that $(12) \not\sim (123)$.

   These computations aren't actually necessary to determine the conjugacy classes in this example. We already know from Problem 12 in Problem Set 12 that two cycles are conjugates of each other if and only if they have the same length.

3. In any group G with identity e, $[e]_C = \{e\}$. Indeed, if $g \in G$, then $geg^{-1} = gg^{-1} = e$.

4. More generally, for any group G, if $h \in Z(G)$ (the center of G—see part 3 of Example 6.5), then $[h]_C = \{h\}$. Indeed, if $h \in Z(G)$ and $g \in G$, then we have $ghg^{-1} = hgg^{-1} = he = h$.

   Also, if $h \notin Z(G)$, then there is $g \in G$ such that $gh \neq hg$, or equivalently, $ghg^{-1} \neq h$. So, ghg^{-1} is an element of $[h]_C$ distinct from h.

   The previous two paragraphs together show that $|[h]_C| = 1$ (the cardinality of the conjugacy class of h is 1) if and only if $h \in Z(G)$.

Let G be a group and let $h \in G$. The **centralizer** of h in G is $C_h(G) = \{x \in G \mid hx = xh\}$. In Problem 4 in Problem Set 6, you were asked to prove that $C_h(G)$ is a subgroup of G. We will now prove that $[G : C_h(G)] = |[h]_G|$ (the index of $C_h(G)$ in G is equal to the cardinality of the conjugacy class of h).

Theorem 13.2: Let G be a group and let $h \in G$. Then $|[h]_G| = [G : C_h(G)]$.

Proof: Let $h \in G$. We define a function $f: [h]_G \to G/C_h(G)$ by $f(ghg^{-1}) = gC_h(G)$. We have $g_1 h g_1^{-1} = g_2 h g_2^{-1}$ if and only if $(g_2^{-1} g_1) h = h(g_2^{-1} g_1)$ if and only if $g_2^{-1} g_1 \in C_h(G)$ if and only if $g_1 C_h(G) = g_2 C_h(G)$ if and only if $f(g_1 h g_1^{-1}) = f(g_2 h g_2^{-1})$. Therefore, f is both well-defined and injective. Surjectivity is obvious, and so, f is a bijection. Thus, $|[h]_G| = |G/C_h(G)| = [G:C_h(G)]|$. □

Corollary 13.3: Let G be a finite group and let $h \in G$. Then $|[h]_G|$ divides $|G|$.

Proof: By Lagrange's Theorem (Theorem 10.8), $|G| = |C_h(G)| \cdot [G:C_h(G)]$. So, by Theorem 13.2, $|G| = |C_h(G)| \cdot |[h]_G|$. Therefore, $|[h]_G|$ divides $|G|$. □

Corollary 13.4: Let G be a finite group. Then $|G| = |Z(G)| + \sum_{i=1}^{n}[G:C_{h_i}(G)]$, where $h_1, h_2, \ldots, h_n$ are representatives of the n distinct conjugacy classes of cardinality greater than 1.

Note that $\sum_{i=1}^{n}[G:C_{h_i}(G)] = [G:C_{h_1}(G)] + [G:C_{h_2}(G)] + \cdots + [G:C_{h_n}(G)]$ (see Notes 1 and 2 before Example 7.13 for a detailed explanation about how "sigma notation" works).

The equation $|G| = |Z(G)| + \sum_{i=1}^{n}[G:C_{h_i}(G)]$ is called the **class equation**.

Proof of Corollary 13.4: Let $h_1, h_2, \ldots, h_n$ be representatives of the n distinct conjugacy classes of cardinality greater than 1. Since the conjugacy classes form a partition of G (by Theorem 3.18), we have $|G| = x + |[h_1]_C| + \cdots + |[h_n]_C|$, where x is the number of elements of G whose conjugacy class has cardinality 1. By part 4 of Example 13.1, $|[h]_C| = 1$ if and only if $h \in Z(G)$, and if $h \in Z(G)$, then $[h]_C = \{h\}$. So, it follows that $x = |Z(G)|$. By Theorem 13.2, for each $i = 1, 2, \ldots, n$, we have $|[h_i]_C| = [G:C_{h_i}(G)]$ and so, $|[h_1]_C| + \cdots + |[h_n]_C| = [G:C_{h_1}(G)] + \cdots + [G:C_{h_n}(G)]$. It follows that

$$|G| = |Z(G)| + [G:C_{h_1}(G)] + \cdots + [G:C_{h_n}(G)] = |Z(G)| + \sum_{i=1}^{n}[G:C_{h_i}(G)].$$

This proves the result. □

Example 13.5:

1. If G is a commutative group, then for each $h \in G$, $C_h(G) = G$ because every element of G commutes with h. Also, for each $h \in G$, by part 1 of Example 13.1, $[h]_C = \{h\}$. So, the class equation tells us that $|G| = |Z(G)|$. This is obvious of course, as $G = Z(G)$ in this case.

2. If G is a group and $h \in G$, then clearly h commutes with itself. So, $h \in C_h(G)$.

   More generally, for any $n \in \mathbb{Z}$, h^n commutes with h. Indeed, $h^n h = h^{n+1} = h^{1+n} = h h^n$. It follows that $\langle h \rangle \subseteq C_h(G)$.

3. Let's find $C_\pi(S_3)$ for each permutation $\pi \in S_3 = \{(1), (12), (13), (23), (123), (132)\}$.

   Since (1) commutes with every element of S_3, $\boldsymbol{C_{(1)}(S_3) = S_3}$.

   Now, we have $(12)(13) = (132)$ and $(13)(12) = (123)$. So, $(12)(13) \neq (13)(12)$. Therefore, $(13) \notin C_{(12)}(S_3)$. Similar computations show that $(23), (123), (132) \notin C_{(12)}(S_3)$. So, $\boldsymbol{C_{(12)}(S_3) = \langle(12)\rangle = \{(1), (12)\}}$.

   Similarly, it is easy to show the following:

$$C_{(13)}(S_3) = \langle(13)\rangle = \{(1),(13)\}$$
$$C_{(23)}(S_3) = \langle(23)\rangle = \{(1),(23)\}$$
$$C_{(123)}(S_3) = \langle(123)\rangle = \{(1),(123),(132)\}$$
$$C_{(132)}(S_3) = \langle(132)\rangle = \{(1),(123),(132)\}$$

Now, $Z(S_3) = \{(1)\}$. Recall from part 2 of Example 13.1 that $[(12)]_C = \{(12),(13),(23)\}$ and $[(123)]_C = \{(123),(132)\}$. Let's choose one representative from each conjugacy class, say (12) and (123). We have $[S_3:C_{(12)}(S_3)] = \frac{6}{2} = 3$ and $[S_3:C_{(123)}(S_3)] = \frac{6}{3} = 2$. Therefore, we have $|Z(S_3)| + [S_3:C_{(12)}(S_3)] + [S_3:C_{(123)}(S_3)] = 1 + 3 + 2 = 6 = |S_3|$, verifying Corollary 13.4 for the group S_3.

Groups of Prime Power Order

If p is a prime number, then a **p-group** is a group in which the order of every element is a power of p.

Example 13.6:

1. For each prime number p, $(\mathbb{Z}_p, +)$ is a p-group. To see this, note that by Theorem 10.9, the order of any element must divide $|\mathbb{Z}_p| = p$, and therefore, the order of any nonidentity element is p (and the order of the identity element is $p^0 = 1$).

   Furthermore, if p is prime, then $(\mathbb{Z}_p, +)$ is the only p-group, up to isomorphism. This follows from Theorem 9.9 together with the discussion following Theorem 10.9 (where we showed that if p is prime, then every group of order p is cyclic).

2. There are exactly two groups of order $4 = 2^2$, up to isomorphism. You were asked to prove this in Problem 18 in Problem Set 2. By the Fundamental Theorem of Commutative Groups (Theorem 11.22), these two groups are $\mathbb{Z}_2 \times \mathbb{Z}_2$ and $\mathbb{Z}_4$.

   In Theorem 13.8 below, we will see that every group of order p^2 (p prime) is commutative. Once again, it follows from the Fundamental Theorem of Commutative Groups (Theorem 11.22) that $\mathbb{Z}_p \times \mathbb{Z}_p$ and $\mathbb{Z}_{p^2}$ are the only two groups of order p^2, up to isomorphism.

3. There are exactly five groups of order $8 = 2^3$, up to isomorphism. You were asked to prove this in Problem 21 in Problem Set 10. There are three commutative groups of order 8 and two noncommutative groups of order 8. By the Fundamental Theorem of Commutative Groups, the three commutative groups of order 8 are $\mathbb{Z}_2 \times \mathbb{Z}_2 \times \mathbb{Z}_2$, $\mathbb{Z}_4 \times \mathbb{Z}_2$, and $\mathbb{Z}_8$. The two noncommutative groups of order 8 are the dihedral group D_4 (see part 4 of Example 6.8) and the Quaternion group (see part 1 of Example 11.21).

   More generally, there are five groups of order p^3 (p prime), up to isomorphism. By the Fundamental Theorem of Commutative Groups, there are three commutative groups of order p^3, namely $\mathbb{Z}_p \times \mathbb{Z}_p \times \mathbb{Z}_p$, $\mathbb{Z}_{p^2} \times \mathbb{Z}_p$, and $\mathbb{Z}_{p^3}$. You will be asked to describe the remaining two groups and prove that there are no others in Problem 19 below.

Theorem 13.7: Let p be a prime number and let G be a p-group with identity e. Then $Z(G) \neq \{e\}$.

Proof: By Corollary 13.4, $|G| = |Z(G)| + \sum_{i=1}^{n}[G:C_{h_i}(G)]$, where $h_1, h_2, \ldots, h_n$ are representatives of the n distinct conjugacy classes of cardinality greater than 1. By Theorem 13.2, for each $i = 1, 2, \ldots, n$, we have $[G:C_{h_i}(G)] = |[h_i]_G|$. So, $|G| = |Z(G)| + \sum_{i=1}^{n}[h_i]_G$. By Corollary 13.3, for each $i = 1, 2, \ldots, n$, $|[h_i]_G|$ divides $|G|$. Since for each i, $|[h_i]_G| > 1$ and $|G| = p^j$ for some $j \in \mathbb{Z}^+$, we must have $|[h_i]_G| = p^{k_i}$ for some integer k_i with $0 < k_i < j$. So, substituting into $|G| = |Z(G)| + \sum_{i=1}^{n}[h_i]_G$ gives us $p^j = |Z(G)| + \sum_{i=1}^{n} p^{k_i}$, or equivalently, $|Z(G)| = p^j - \sum_{i=1}^{n} p^{k_i} = p(p^{j-1} - \sum_{i=1}^{n} p^{k_i - 1})$. This shows that $|Z(G)| \geq p > 1$. Therefore, $Z(G) \neq \{e\}$. □

Theorem 13.8 Let p be a prime number and let G be a group of order p^2. Then G is commutative.

Proof: For every $a \in G$ with $a \neq e$, either $|a| = p$ or $|a| = p^2$. If there is an $a \in G$ such that $|a| = p^2$, then G is cyclic and therefore, G is commutative. So, we may assume that every nonidentity element of G has order p. Now, $|Z(G)|$ divides the order of G. By Theorem 13.7, $|Z(G)| \neq 1$. Assume toward contradiction that $|Z(G)| = p$. Then there is $x \in G$ with $x \notin Z(G)$. By our assumption, $|x| = p$, and so, $|\langle x \rangle| = p$. Since $Z(G) \subseteq C_x(G)$, $\langle x \rangle \subseteq C_x(G)$, and $x \notin Z(G)$, $|C_x(G)| > p$. Since $C_x(G)$ is a subgroup of G, $|C_x(G)| = p^2 = |G|$, and so, $C_x(G) = G$. Therefore, $x \in Z(G)$, contrary to our assumption. This contradiction proves that $|Z(G)| = p^2 = |G|$, and so, $Z(G) = G$. Therefore, G is commutative. □

Sylow Subgroups

Let G be a group and let p be a prime number such that p divides $|G|$. A **p-subgroup** of G is a subgroup of G that is a p-group. A **Sylow p-subgroup** of G is a p-subgroup of G that is not a proper subgroup of any other p-subgroup of G.

Note: If p is a prime number such that p divides $|G|$, then we can write $|G| = kp^n$, where $k, n \in \mathbb{Z}^+$ and k is **not** divisible by p. If we express $|G|$ this way, then a Sylow p-subgroup is a subgroup of G of order p^n.

Example 13.9:

1. Let $G = \mathbb{Z}_{12}$. Then $|G| = 2^2 \cdot 3$. The 2-subgroups of G are $H_1 = \{0, 6\}$ and $H_2 = \{0, 3, 6, 9\}$. Since $|H_2| = 2^2$, H_2 is a Sylow 2-subgroup. The only 3-subgroup of G is $H_3 = \{0, 4, 8\}$. Since $|H_3| = 3^1$, H_3 is a Sylow 3-subgroup.

2. Let $G = D_6 = \langle r, s \mid r^6 = e, s^2 = e, sr = r^5 s \rangle$, the dihedral group of degree 6. Recall from part 4 of Example 6.8 that geometrically, we can think of D_6 as all the symmetries of a regular 6-sided polygon (a 6-sided polygon is called a **hexagon**). The complete set of elements of D_6 is

$$D_6 = \{e, r, r^2, r^3, r^4, r^5, s, rs, r^2s, r^3s, r^4s, r^5s\}.$$

$|D_6| = 12 = 2^2 \cdot 3$. Therefore, a Sylow 2-subgroup of D_6 has order 4 and a Sylow 3-subgroup of D_6 has order 3. The Sylow 2-subgroups are $\{e, r^3, s, r^3s\}, \{e, r^3, rs, r^4s\}$, and $\{e, r^3, r^2s, r^5s\}$. The only Sylow 3-subgroup is $\{e, r^2, r^4\}$.

We now introduce the **Sylow Theorems**. These theorems provide information about the subgroups of a given group. This information can often be used to find all groups of certain orders, up to isomorphism. The proofs of the Sylow Theorems are challenging, and so, I leave these proofs as optional exercises for the reader (see Problem 20 below).

281

Theorem 13.10 (First Sylow Theorem): Let G be a finite group, let p be a prime number, and let k be a positive integer such that p^k divides the order of G. Then G has at least one subgroup of order p^k.

In particular, Theorem 13.10 implies that if p is a prime dividing the order of G, then G has at least one Sylow p-subgroup.

Example 13.11:

1. Let G be a group of order 15. Since $15 = 3 \cdot 5$, a Sylow 3-subgroup has 3 elements and a Sylow 5-subgroup has 5 elements. By the First Sylow Theorem, G has at least one subgroup of order 3 and at least one subgroup of order 5. Since 3 and 5 are prime, both of these subgroups are cyclic.

   More generally, a group of order pq, where p and q are distinct primes has at least one subgroup of order p and at least one subgroup of order q.

2. Let G be a group of order 45. Since $45 = 3^2 \cdot 5$, a Sylow 3-subgroup has $3^2 = 9$ elements and a Sylow 5-subgroup has 5 elements. Therefore, by the First Sylow Theorem, G has at least one subgroup of order 3, at least one subgroup of order 9 and at least one subgroup of order 5. Note that the First Sylow Theorem does **not** tell us if G has a subgroup of order $3 \cdot 5 = 15$.

   More generally, a group of order $p^2 q$, where p and q are distinct primes has at least one subgroup of order p, at least one subgroup of order p^2 and at least one subgroup of order q. Note that the First Sylow Theorem does **not** tell us if G has a subgroup of order pq.

3. Let G be a group of order $2^4 \cdot 3^3 \cdot 5^2 \cdot 11$. Then G has at least one subgroup of each of the following orders: $2, 4, 8, 16, 3, 9, 27, 5, 25, 11$. The Sylow subgroups have orders $16, 27, 25,$ and 11. Note that the First Sylow Theorem does **not** tell us if G has subgroups of orders $6, 12, 24,$ and so on.

4. Let G be a group of order p^k, where p is prime and $k \in \mathbb{Z}^+$. Then G has a subgroup of order p^t for each $t = 1, 2, \ldots, k$. In other words, the First Sylow Theorem tells us that the converse of Theorem 10.9 is true for groups whose order is a power of a prime.

Theorem 13.12 (Second Sylow Theorem): Let G be a finite group of order kp^n, where p is a prime number, n is a positive integer, and k is an integer not divisible by p. If m is the number of Sylow p-subgroups of G. Then $m \equiv_p 1$.

Example 13.13:

1. Let G be a group of order 15. By the Second Sylow Theorem, the number of subgroups of G of order 3 must be one of the following: $1, 4, 7, 10, 13, \ldots$ The number of subgroups of order 5 must be one of the following: $1, 6, 11, 16, 21, \ldots$

   More generally, let G be a group of order pq, where p and q are distinct primes. The number of subgroups of G of order p must be one of the following: $1, p+1, 2p+1, 3p+1, \ldots$ The number of subgroups of order q must be one of the following: $1, q+1, 2q+1, 3q+1, \ldots$

2. Let G be a group of order 45. Since $45 = 3^2 \cdot 5$. By the Second Sylow Theorem, the number of subgroups of G of order 9 must be one of the following: $1, 4, 7, 10, 13, \ldots$ The number of subgroups of order 5 must be one of the following: $1, 6, 11, 16, 21, \ldots$

More generally, let G be a group of order p^2q, where p and q are distinct primes. The number of subgroups of G of order p^2 must be one of the following: $1, p+1, 2p+1, 3p+1, \ldots$ The number of subgroups of order q must be one of the following: $1, q+1, 2q+1, 3q+1, \ldots$

3. Let G be a group of order $2^4 \cdot 3^3 \cdot 5^2 \cdot 11$. By the Second Sylow Theorem, the number of subgroups of G of order 16 must be one of the following: $1, 3, 5, 7, 9, \ldots$ The number of subgroups of G of order 27 must be one of the following: $1, 4, 7, 10, 13, \ldots$ The number of subgroups of order 25 must be one of the following: $1, 6, 11, 16, 21, \ldots$ Finally, the number of subgroups of order 11 must be one of the following: $1, 12, 23, 34, 45, \ldots$

Theorem 13.14 (Third Sylow Theorem): Let G be a finite group and let H be any p-subgroup of G. Then H is contained in some Sylow p-subgroup of G.

Example 13.15:

1. Let $G = \mathbb{Z}_{12}$. Then $|G| = 2^2 \cdot 3$. We saw in part 1 of Example 13.9 that the 2-subgroups of G are $H_1 = \{0, 6\}$ and $H_2 = \{0, 3, 6, 9\}$. Notice that 2-subgroup H_1 is contained in the Sylow 2-subgroup H_2.

2. Let $G = D_6 = \langle r, s \mid r^6 = e, s^2 = e, sr = r^5 s \rangle$, the dihedral group of degree 6. Recall from part 2 of Example 13.9 that the Sylow 2-subgroups of D_6 are $\{e, r^3, s, r^3 s\}, \{e, r^3, rs, r^4 s\}$, and $\{e, r^3, r^2 s, r^5 s\}$. The subgroups of D_6 of order 2 are $\{e, s\}, \{e, rs\}, \{e, r^2 s\}, \{e, r^3 s\}, \{e, r^4 s\}$, and $\{e, r^5 s\}$ (Check this!). Notice how each of these 2-subgroups is contained in a Sylow 2-subgroup. For example, $\{e, s\} \subseteq \{e, r^3, s, r^3 s\}$, $\{e, rs\} \subseteq \{e, r^3, rs, r^4 s\}$, and so on.

If H is a subgroup of a group G, then $gHg^{-1} = \{ghg^{-1} \mid h \in H\}$ is called **a conjugate subgroup** of H or simply a **conjugate** of H. I leave it to the reader to show that the conjugate of a subgroup of a group G is also a subgroup of G.

Theorem 13.16 (Fourth Sylow Theorem): Let G be a finite group and let S and T be Sylow p-subgroups of G. Then S and T are conjugate. Furthermore, the number of Sylow p-subgroups of G divides $|G|$.

Example 13.17:

1. Let G be a group of order 15. Recall from part 1 of Example 13.13 that the number of subgroups of G of order 3 must be one of the following: $1, 4, 7, 10, 13, \ldots$ (by the Second Sylow Theorem). Now, by the Fourth Sylow Theorem, the number of subgroups of G of order 3 must be 1, 3, 5, or 15. Putting these two results together, we see that G has **exactly** 1 subgroup of order 3. Let P be this unique Sylow 3-subgroup.

   Similarly, the number of subgroups of G of order 5 must be one of $1, 6, 11, 16, 21, \ldots$ (by the Second Slow Theorem) and also must be one of $1, 3, 5, 15$ (by the Fourth Sylow Theorem). So, once again, there is a unique Sylow 5-subgroup, which we will name Q.

   Now, if we let $x \in G$ be any element of order 3, then $\langle x \rangle$ is a Sylow 3-subgroup, and therefore, $\langle x \rangle = P$. Similarly, if $y \in G$ is any element of order 5, then $\langle y \rangle = Q$. Thus, G has one element of order 1, two elements of order 3, and four elements of order 5. It follows that the other eight elements must have order 15. So, G is cyclic. It follows that the only group of order 15, up to isomorphism, is the cyclic group $\mathbb{Z}_{15}$.

2. Let G be a group and let H be a finite subgroup of G. If $g \in G$, then the conjugate gHg^{-1} is also a subgroup of G and $|H| = |gHg^1|$. It follows that if H is a Sylow p-subgroup, then so is any conjugate gHg^{-1}. So, if H is the only Sylow p-subgroup of G, then for any $g \in G$, we have $gHg^{-1} = H$, and so, by Theorem 10.15 (4 → 1), H is a normal subgroup of G.

3. If H is a normal Sylow p-subgroup of a group G, then every conjugate of H is equal to H. Therefore, by the Fourth Sylow Theorem, H is the only Sylow p-subgroup of G.

4. Putting parts 2 and 3 above together gives us the following result: A Sylow p-subgroup of a group G is normal in G if and only if it is the only Sylow p-subgroup of G.

5. Let G be a group of order 6. By the Second Sylow Theorem, the number of Sylow 3-subgroups (of order 3) is $1, 4, 7, \ldots$. By the Fourth Sylow Theorem, the number of Sylow 3-subgroups is $1, 2, 3$, or 6. So, there is exactly 1 Sylow 3-subgroup P. By part 2 (or part 4) above, P is a normal subgroup of G. Since a group of order 3 is cyclic, there is $x \in G$ with $P = \langle x \rangle$.

    Also, by the Second Sylow Theorem, the number of Sylow 2-subgroups (of order 2) is $1, 3, 5, \ldots$. By the Fourth Sylow Theorem, the number of Sylow 2-subgroups is $1, 2, 3$, or 6. So, the number of Sylow 2-subgroups is 1 or 3. Unfortunately, we do not necessarily have a unique Sylow 2-subgroup in this case. In the case where there is only 1 Sylow 2-subgroup, we can use the same reasoning as in part 1 above to prove that P is isomorphic to $\mathbb{Z}_6$.

    However, if there are 3 Sylow 2-subgroups, then we have 1 element of order 1, 3 elements of order 2, and 2 elements of order 3, and so, G is **not** cyclic (as there are no elements of order 6). Let $y \in G$ be an element of order 2. Since $P = \langle x \rangle$ is a normal subgroup of G, $yxy^{-1} \in \langle x \rangle$. Therefore, $yxy^{-1} = e$, $yxy^{-1} = x$, or $yxy^{-1} = x^2$.

    If $yxy^{-1} = e$, then $x = y^{-1}ey = y^{-1}y = e$, which is impossible because x has order 3. So, this case cannot happen.

    If $yxy^{-1} = x$, then $yx = xy$. So, we get the following:
    $$(xy)^2 = (xy)(xy) = x(yx)y = x(xy)y = (xx)(yy) = x^2y^2 = x^2e = x^2$$
    $$(xy)^3 = (xy)^2(xy) = x^2(xy) = x^3y = y$$
    $$(xy)^4 = (xy)^3(xy) = y(xy) = y(yx) = (yy)x = y^2x = ex = x$$
    $$(xy)^5 = (xy)^4(xy) = x(xy) = (xx)y = x^2y$$
    $$(xy)^6 = (xy)^5(xy) = (x^2y)(xy) = x^2(yx)y = x^2(xy)y = x^3y^2 = ee = e$$
    So, $\langle xy \rangle = \{e, xy, x^2, y, x, x^2y\} = G$. Therefore, G is cyclic and so, $G \cong \mathbb{Z}_6$.

    Finally, if $yxy^{-1} = x^2$, then $G \cong D_3$, the Dihedral group of degree 3 (or equivalently, $G \cong S_3$, the symmetric group on $\{1, 2, 3\}$). I leave the details to the reader.

The Fundamental Theorem of Finite Commutative Groups (Theorem 11.22) provides us with a complete classification of all finite commutative groups, up to isomorphism. Classifying all finite noncommutative groups is much more difficult. Parts 1 and 5 of Example 13.17 above illustrate how the Sylow Theorems can be used to provide a complete classification of the groups of a certain order, up to isomorphism. You will be asked to prove similar results in the problems below (see for example Problems 14, 16, 17, 18, and 19). **Hint:** Begin by using the Sylow Theorems.

Group Actions

In the beginning of this lesson, we discussed conjugation and conjugacy classes. Conjugation is one example of the more general notion of a "group action." In the case of conjugation, a group G is "acting" on itself by taking the pair $(g, h) \in G \times G$ to the element $ghg^{-1} \in G$. Informally, we can say that for $g, h \in G$, we have $g \cdot h = ghg^{-1}$. More formally, we can define a function $f: G \times G \to G$ by $f(g, h) = ghg^{-1}$. The "action" of conjugation has the following two properties:

1. $e \cdot h = ehe^{-1} = h$.
2. For all $g, k, h \in G$, $g \cdot (k \cdot h) = g \cdot (khk^{-1}) = g(khk^{-1})g^{-1} = (gk)h(gk)^{-1} = (gk) \cdot h$.

Let G be a group and let A be a set. Formally, a **group action** of G on A is a function $f: G \times A \to A$ satisfying the following two conditions.

1. **(Identity)** For all $a \in A$, $f(e, a) = a$.
2. **(Compatibility)** For all $x, y \in G$ and $a \in A$, $f\bigl(x, f(y, a)\bigr) = f(xy, a)$.

In practice, for easier readability, we will abbreviate $f(x, a)$ as $x \cdot a$ or xa. Using this friendlier notation, the two conditions become the following:

1. For all $a \in A$, $e \cdot a = a$.
2. For all $x, y \in G$ and $a \in A$, $x \cdot (y \cdot a) = (xy) \cdot a$.

Be careful here—when we use this abbreviated notation, there is no mention of the function f. However, there is a function there. We are simply not mentioning its name.

Notes: (1) Instead of saying that f is a group action of G on A, we may also say that **G acts on A** or that **A is a G-set**.

(2) Technically speaking, we defined a **left group action**. A **right group action** is a function $f: A \times G \to A$ satisfying $a \cdot e = a$ and $(a \cdot y) \cdot x = a \cdot (yx)$. We will not be concerned with right group actions in this book.

(3) If $f: G \times A \to A$ is a group action, then define $h: G \to Sym(A)$ by $h(x) = \pi_x$, where $\pi_x(a) = x \cdot a$ (remember that $x \cdot a$ is an abbreviation for $f(x, a)$). To see that π_x is injective, observe that $x \cdot a = x \cdot b$ implies $a = e \cdot a = (x^{-1}x) \cdot a = x^{-1} \cdot (x \cdot a) = x^{-1} \cdot (x \cdot b) = (x^{-1}x) \cdot b = e \cdot b = b$. To see that π_x is surjective, let $b \in A$ Then $\pi_x(x^{-1} \cdot b) = x \cdot (x^{-1} \cdot b) = (xx^{-1}) \cdot b = e \cdot b = b$. So, π_x is a bijection, and therefore, $\pi_x \in Sym(A)$.

It turns out that h is a homomorphism. To see this, note that for any $a \in A$, we have

$$\pi_{xy}(a) = (xy) \cdot a = x \cdot (y \cdot a) = \pi_x(y \cdot a) = \pi_x \pi_y(a).$$

Since $a \in A$ was arbitrary, $\pi_{xy} = \pi_x \pi_y$. So, $h(xy) = h(x)h(y)$, and thus, h is a homomorphism.

Conversely, suppose that $h: G \to Sym(A)$ is a homomorphism. We can then define a group action $f: G \times A \to A$ by $f(x, y) = h(x)(y)$ (or equivalently, $x \cdot y = \pi_x(y)$, where $h(x) = \pi_x$).

So, an alternate definition of a group action of G on A is a homomorphism $h: G \to Sym(A)$.

Example 13.18:

1. Let $G = S_n$ (the symmetric group on $\{1, 2, \ldots, n\}$) and let $A = \{1, 2, \ldots, n\}$. Then G acts on A by the formula $\pi \cdot k = \pi(k)$. Let's check that the two properties for a group action hold. For each $k \in A$, we have $e \cdot k = e(k) = k$. So, the first condition holds. For all $\pi, \sigma \in S_n$ and all $k \in A$, we have $\pi \cdot (\sigma \cdot k) = \pi(\sigma(k)) = (\pi\sigma)(k) = (\pi\sigma) \cdot k$. So, the second condition holds.

2. Let G be a group. Then G acts on itself by left multiplication. In other words, if $x, y \in G$, we define $x \cdot y = xy$. We have $e \cdot y = ey = y$. So, the first condition holds. For all $x, z, y \in G$, we have $x \cdot (z \cdot y) = x \cdot (zy) = x(zy) = (xz)y = (xz) \cdot y$. So, the second condition holds.

3. Let G be a group and let H be a subgroup of G. Then G acts on $G/H = \{xH \mid x \in G\}$ by left multiplication. In other words, if $x \in G$ and $yH \in G/H$, then $x \cdot yH = (xy)H$. We have $e \cdot yH = (ey)H = yH$. So, the first condition holds. For all $x, z \in G$ and $yH \in G/H$, we have $x \cdot (z \cdot yH) = x \cdot (zy)H = (x(zy))H = ((xz)y)H = (xz) \cdot yH$. So, the second condition holds. Note that in this example, H didn't need to be a normal subgroup of G. G/H still makes sense in this case. Since we did not attempt to define a binary operation on G/H, we don't run into any issues similar to what we experienced in Lesson 10.

4. Let G be a group. Then G acts on itself by conjugation, as we discussed in the beginning of this lesson. In other words, if $x, y \in G$, we define $x \cdot y = xyx^{-1}$. We have $e \cdot y = eye^{-1} = y$ and for all $x, y, z \in G$, $x \cdot (z \cdot y) = x \cdot (zyz^{-1}) = x(zyz^{-1})x^{-1} = (xz)y(xz)^{-1} = (xz) \cdot y$. So, both conditions hold.

5. Let G be a group and let $S = \{H \mid H \leq G\}$ be the set of subgroups of G. Then G acts on S by conjugation. That is, if $x \in G$ and $H \in S$, then $x \cdot H = xHx^{-1}$. We have $e \cdot H = eHe^{-1} = H$. So, the first condition holds. For all $x, y \in G$ and $H \in S$, we have
$$x \cdot (y \cdot H) = x \cdot (yHy^{-1}) = x(yHy^{-1})x^{-1} = (xy)H(xy)^{-1} = (xy) \cdot H.$$
So, the second condition holds.

6. Let G be a group, let H be a subgroup of G, and let $A = \{gHg^{-1} \mid g \in G\}$ be the set of conjugates of H in G. Then G acts on A by conjugation. That is, if $x \in G$ and $gHg^{-1} \in A$, then $x \cdot gHg^{-1} = (xg)H(xg)^{-1}$. We have $e \cdot gHg^{-1} = (eg)H(eg)^{-1} = gHg^{-1}$. So, the first condition holds. For all $x, y \in G$ and $gHg^{-1} \in A$, we have
$$x \cdot (y \cdot gHg^{-1}) = x \cdot ((yg)H(yg)^{-1}) = (x(yg))H(x(yg))^{-1}$$
$$= ((xy)g)H((xy)g)^{-1} = (xy) \cdot gHg^{-1}.$$
So, the second condition holds.

Suppose that a group G acts on a set A and let $a \in A$. The **stabilizer** of a is the set
$$G_a = \{x \in G \mid x \cdot a = a\}.$$

If $x \cdot a = a$, we will say that **x fixes a**. So, G_a is the set of group elements that fix a.

We will look at some examples in just a moment, but let's first show that stabilizers are groups.

Theorem 13.19: Suppose that a group G acts on a set A, let $a \in A$, and let G_a be the stabilizer of a. Then $G_a \leq G$.

Proof: Since $e \cdot a = a$, $e \in G_a$. Let $x, y \in G_a$. Then $(xy) \cdot a = x \cdot (y \cdot a) = x \cdot a = a$. Thus, $xy \in G_a$. If $x \in G_a$, then $x^{-1} \cdot a = x^{-1} \cdot (x \cdot a) = (x^{-1}x) \cdot a = e \cdot a = a$. So, $x^{-1} \in A$. It follows that $G_a \leq G$. $\square$

Example 13.20:

1. Let $G = S_n$ and let $A = \{1, 2, \ldots, n\}$. By part 1 of Example 13.18, G acts on A by the formula $\pi \cdot k = \pi(k)$. G_n consists of the permutations that fix n. Notationally, in cycle notation, a permutation in S_n that fixes n is indistinguishable from a permutation in S_{n-1}. For example, the permutation (12) represents a permutation in both S_2 and $G_3 \leq S_3$. It follows that $G_n \cong S_{n-1}$. By renaming the elements of $\{1, 2, \ldots n\}$ as needed, we also see that for each $i = 1, 2, \ldots, n$, $G_i \cong S_{n-1}$.

   As a specific example, let $G = S_4$. Then $G_4 = \{(1), (12), (13), (23), (123), (132)\} \cong S_3$. You may argue that it looks like $G_4 = S_3$, and you would "essentially" be correct. However, even though the notation being used is the same in G_4 and S_3, the interpretation is slightly different. For example, in G_4 the permutation (123) sends 1 to 2, 2 to 3, 3 to 1, and 4 to 4. The number 4 isn't included in the notation because it is fixed by the permutation. On the other hand, in S_3 the permutation (123) doesn't do anything to 4 because $4 \notin \{1, 2, 3\}$.

2. Let G be a group. By part 2 of Example 13.18, G acts on itself by left multiplication $(x \cdot y = xy)$. For any $h \in G$, $G_h = \{e\}$. Indeed, if $g \in G_h$, then $g \cdot h = h$, and therefore, $gh = h$. So,
$$g = ge = g(hh^{-1}) = (gh)h^{-1} = hh^{-1} = e.$$
Thus, $g \in \{e\}$. So, $G_h \subseteq \{e\}$. Since $e \cdot h = h$, $e \in G_h$, and so, $\{e\} \subseteq G_h$.

3. Let G be a group and let H be a subgroup of G. By part 3 of Example 13.18, G acts on $G/H = \{xH \mid x \in G\}$ by left multiplication $(x \cdot yH = (xy)H)$. For any $aH \in G/H$, $G_{aH} = aHa^{-1}$. Indeed, $g \in G_{aH}$ if and only if $g \cdot aH = aH$ if and only if $(ga)H = aH$ if and only if $a^{-1}gaH = H$ if and only if $a^{-1}ga \in H$ if and only if there is $h \in H$ with $a^{-1}ga = h$ if and only if there is $h \in H$ with $g = aha^{-1}$ if and only if $g \in aHa^{-1}$.

4. Let G be a group. By part 4 of Example 13.18, G acts on itself by conjugation $(x \cdot y = xyx^{-1})$. For any $h \in G$, $G_h = C_h(G) = \{x \in G \mid hx = xh\}$ (The centralizer of y in G). Indeed, we have $x \in G_h$ if and only if $x \cdot h = h$ if and only if $xhx^{-1} = h$ if and only if $xh = hx$ if and only if $x \in C_h(G)$.

5. Let G be a group and let $S = \{H \mid H \leq G\}$ be the set of subgroups of G. By part 5 of Example 13.18, G acts on S by conjugation $(x \cdot H = xHx^{-1})$. If $H \in S$, then $G_H = \{x \in G \mid xHx^{-1} = H\}$. Indeed, $x \in G_H$ if and only if $x \cdot H = H$ if and only if $xHx^{-1} = H$. We call G_H the **normalizer** of H in G and we write $N_H(G) = G_H = \{x \in G \mid xHx^{-1} = H\}$. Observe that $H \leq N_H(G)$. After all, if $x \in H$, then $xHx^{-1} = H$, and so, $x \in N_H(G)$. In fact, H is a normal subgroup of $N_H(G)$. To see this, let $x \in H$ and $y \in N_H(G)$. Then $yHy^{-1} = H$, and so,
$$(xyx^{-1})H(xyx^{-1})^{-1} = xy(x^{-1}Hx)y^{-1}x^{-1} = x(yHy^{-1})x^{-1} = xHx^{-1} = H.$$
Furthermore, $N_H(G)$ is the largest subgroup of G containing H in which H is normal (see Problem 3 below). In particular, H is normal in G if and only if $N_H(G) = G$.

6. Let G be a group, let H be a subgroup of G, and let $A = \{gHg^{-1} \mid g \in G\}$ be the set of conjugates of H in G. By part 5 of Example 13.18, G acts on A by conjugation $(x \cdot gHg^{-1} = (xg)H(xg)^{-1})$. If $gHg^{-1} \in A$, then $G_{gHg^{-1}} = gN_H(G)g^{-1}$, where $N_H(G)$ is the normalizer of H in G (as defined in part 5 above). To see this, observe that $x \in G_{gHg^{-1}}$ if and only if $x \cdot gHg^{-1} = gHg^{-1}$ if and only if $(xg)H(xg)^{-1} = gHg^{-1}$ if and only if $g^{-1}xgH(g^{-1}xg)^{-1} = H$ if and only if $g^{-1}xg \in N_H(G)$ if and only if $x \in gN_H(G)g^{-1}$.

Suppose that the group G acts on a set A. Define a relation $\sim$ on A by $a \sim b$ if and only if there is a $g \in G$ with $b = g \cdot a$. It is easy to verify that $\sim$ is an equivalence relation (see Problem 4 below). The equivalence classes for the relation $\sim$ will be called **orbits**. If $a \in A$, we will let $\text{orb}(a)$ be the orbit of a. In other words,

$$\text{orb}(a) = \{b \in A \mid a \sim b\} = \{b \in A \mid b = g \cdot a \text{ for some } g \in G\} = \{g \cdot a \mid g \in G\}.$$

Example 13.21:

1. Let $G = S_n$ and let $A = \{1, 2, \ldots, n\}$. By part 1 of Example 13.18, G acts on A by the formula $\pi \cdot k = \pi(k)$. There is just one orbit for this group action. Indeed, if $i, j \in \{1, 2, \ldots, n\}$, then the transposition $\pi = (ij)$ satisfies $\pi(i) = j$. So, $i \sim j$. A group action with just one orbit (such as in this example) is called a **transitive action**.

2. Let G be a group. By part 2 of Example 13.18, G acts on itself by left multiplication ($x \cdot y = xy$). This is another example of a transitive action. Given any $g, h \in G$, we have $hg^{-1} \in G$ and $(hg^{-1}) \cdot g = hg^{-1}g = he = h$. This shows that for all $g, h \in G$, $g \sim h$, and so, this action has just one orbit.

3. Let G be a group and let H be a subgroup of G. By part 3 of Example 13.18, G acts on $G/H = \{xH \mid x \in G\}$ by left multiplication ($x \cdot yH = (xy)H$). This is yet another transitive action. Given $xH, yH \in G/H$, we have $yx^{-1} \in G$ and $(yx^{-1}) \cdot xH = yx^{-1}xH = yeH = yH$. This shows that for all $xH, yH \in G/H$, $xH \sim yH$, and so, this action has just one orbit.

4. Let G be a group. By part 4 of Example 13.18, G acts on itself by conjugation ($x \cdot y = xyx^{-1}$). The orbits in this case are called conjugacy classes (as defined in the beginning of this lesson). In other words, if $h \in G$, then $\text{orb}(h) = [h]_C = \{ghg^{-1} \mid g \in G\}$.

5. Let G be a group and let $S = \{H \mid H \leq G\}$ be the set of subgroups of G. By part 5 of Example 13.18, G acts on S by conjugation ($x \cdot H = xHx^{-1}$). The orbits of a subgroup H of G is the set of subgroups conjugate to H. In other words, $\text{orb}(H) = \{gHg^{-1} \mid g \in G\}$.

Theorem 13.22 (Orbit-Stabilizer Theorem): Suppose that a group G acts on a set A and let $a \in A$. Then

$$|\text{orb}(a)| = [G : G_a].$$

Proof: Let $a \in A$. We define a function $f : \text{orb}(a) \to G/G_a$ by $f(g \cdot a) = gG_a$. We have $g \cdot a = h \cdot a$ if and only if $(h^{-1}g) \cdot a = h^{-1} \cdot (g \cdot a) = h^{-1} \cdot (h \cdot a) = (h^{-1}h) \cdot a = e \cdot a = a$ if and only if $h^{-1}g \in G_a$ if and only if $gG_a = hG_a$ if and only if $f(g \cdot a) = f(h \cdot a)$. Therefore, f is both well-defined and injective. Surjectivity is obvious, and so, f is a bijection. Thus, $|\text{orb}(a)| = |G/G_a| = [G : G_a]$. □

Example 13.23:

1. Let $G = S_n$ and let $A = \{1, 2, \ldots, n\}$. By part 1 of Example 13.18, G acts on A by the formula $\pi \cdot k = \pi(k)$. By part 1 of Example 13.21, this action is transitive, and so, there is just one orbit. So, if $i \in A$, $\text{orb}(i) = \{1, 2, \ldots, n\}$. So, $|\text{orb}(i)| = n$. By part 1 of Example 13.20, $G_i \cong S_{n-1}$. It follows that $[S_n : G_i] = \frac{|S_n|}{|S_{n-1}|} = \frac{n!}{(n-1)!} = n$. This verifies the Orbit-Stabilizer Theorem for this particular group action.

2. Let G be a group. By part 2 of Example 13.18, G acts on itself by left multiplication ($x \cdot y = xy$). By part 2 of Example 13.21, this action is transitive, and so, there is just one orbit. So, if $x \in A$, $\text{orb}(x) = G$. So, $|\text{orb}(x)| = |G|$. By part 2 of Example 13.20, $G_x = \{e\}$. It follows that $[G : G_x] = \frac{|G|}{|\{e\}|} = |G|$. This verifies the Orbit-Stabilizer Theorem for this particular group action.

3. Let G be a group and let H be a subgroup of G. By part 3 of Example 13.18, G acts on $G/H = \{xH \mid x \in G\}$ by left multiplication ($x \cdot yH = (xy)H$). By part 3 of Example 13.21, this action is transitive, and so, there is just one orbit. So, if $x \in G$, $\text{orb}(g) = G/H$. Therefore, $|\text{orb}(x)| = |G/H| = [G : H]$ (the index of H in G). By part 3 of Example 13.20, $G_{xH} = xHx^{-1}$. It follows that $[G : G_{xH}] = [G : xHx^{-1}]$. By the Orbit-Stabilizer Theorem, $|G : H| = [G : xHx^{-1}]$. This proves that given any subgroup H of a group G, any conjugate of H has the same index in G as H does.

4. Let G be a group. By part 4 of Example 13.18, G acts on itself by conjugation ($x \cdot y = xyx^{-1}$). By part 4 of Example 13.21, if $h \in G$, then $\text{orb}(h) = [h]_C = \{ghg^{-1} \mid g \in G\}$. By part 4 of Example 13.20, $G_h = C_h(G) = \{x \in G \mid hx = xh\}$. In this case, the Orbit-Stabilizer Theorem tells us that $|[h]_C| = [G : C_h(G)]$. This is precisely Theorem 13.2.

5. Let G be a group and let $S = \{H \mid H \leq G\}$ be the set of subgroups of G. By part 5 of Example 13.18, given $H \in S$, $\text{orb}(H) = \{gHg^{-1} \mid g \in G\}$. By part 5 of Example 13.20, $G_H = N_H(G)$ (the normalizer of H in G). In this case, the Orbit-Stabilizer Theorem tells us that $|\{gHg^{-1} \mid g \in G\}| = [G : N_H(G)]$.

6. Each of the Sylow Theorems can be proved using various group actions. Just to give one example, let's use a group action together with the Orbit-Stabilizer Theorem to prove the last part of the Fourth Sylow Theorem. Let G be a finite group. We will assume that we already know that any two Sylow p-subgroups are conjugate, and we will prove that the number of Sylow p-subgroups of G divides $|G|$. To this end, let S be the set of Sylow p-subgroups of G, assume that $|S| = n$, and let G act on S by conjugation, as in part 5 above. Since any two Sylow p-subgroups are conjugate, there is only one orbit. So, $|\text{orb}(H)| = n$ for any $H \in S$. By the Orbit-Stabilizer Theorem, $[G : G_H] = |\text{orb}(H)| = n$. By Lagrange's Theorem, $|G| = |G_H|[G : G_H]$, and therefore, $|G| = |G_H| \cdot n$, proving that n divides the order of G.

Problem Set 13

Full solutions to these problems are available for free download here:
www.SATPrepGet800.com/AAFBTDW

LEVEL 1

1. Describe the 5 conjugacy classes of S_4.

2. Let G be a group of order 40. Prove that G has a normal subgroup of order 5.

3. Let G be a group and let H and K be subgroups of G. Prove that if H is a normal subgroup of K, then K is a subgroup of $N_H(G)$.

LEVEL 2

4. Let G be a group and define a relation $\sim$ on G by $h \sim k$ if and only if k is a conjugate of h. Prove that $\sim$ is an equivalence relation on G. Then generalize this result to an arbitrary group action.

5. Let G be a group and let $h, k \in G$ be conjugates of each other. Prove that for each $j \in \mathbb{Z}$, $|h^j| = |k^j|$.

6. Determine if the following statement is true or false. If it is true, prove it. If it is false, provide a counterexample. "If G is a group and $h, k \in G$ have the same order, then h and k are in the same conjugacy class."

7. Let G be a group, let p be prime, and let H be a p-subgroup of G such that p does **not** divide $[G:H]$. Prove that H is a Sylow p-subgroup of G.

LEVEL 3

8. Let G be a finite group, let p be prime, let H be a Sylow p-subgroup of G, and let N be a normal subgroup of G. Prove that $H \cap N$ is a Sylow p-subgroup of N.

9. Let G be a finite group, let p be prime, let H be a Sylow p-subgroup of G, and let N be a normal subgroup of G. Prove that HN/N is a Sylow p-subgroup of G/N.

10. Let p, q be prime with $q > p$. Prove that a group of order pq has a normal Sylow q-subgroup.

11. Let G be a group acting on a set A. The group action is said to be **faithful** if for all $x \in G$ with $x \neq e$, there is $a \in A$ such that $x \cdot a \neq a$. Provide an example of a group action that is **not** faithful.

LEVEL 4

12. Let G be a group of order 12. Prove that G has either a normal Sylow 2-subgroup or a normal Sylow 3-subgroup.

13. Prove that a group of order 105 has a subgroup of order 35.

14. Prove that there are exactly two groups of order 99, up to isomorphism.

15. Prove that Lagrange's Theorem follows from the Orbit-Stabilizer Theorem by defining an appropriate group action.

LEVEL 5

16. Let p be an odd prime. Prove that there are exactly 2 groups of order $2p$, up to isomorphism.

17. Prove that there are exactly 2 groups of order 21, up to isomorphism.

CHALLENGE PROBLEMS

18. Find all groups of order 30, up to isomorphism.

19. Let p be prime. Prove that there are exactly five groups of order p^3, up to isomorphism. Describe the two noncommutative groups.

20. Prove the four Sylow Theorems.

LESSON 14
POLYNOMIALS

Polynomials Over a Ring

A polynomial in the variable x over $\mathbb{Z}$ is an expression of the form $a_k x^k + a_{k-1} x^{k-1} + \cdots + a_1 x + a_0$, where $a_0, a_1, \ldots, a_{k-1}, a_k \in \mathbb{Z}$ and x is just a new symbol that is not in $\mathbb{Z}$. If $a_k \neq 0$, then the natural number k is called the **degree** of the polynomial. The **zero** polynomial, 0, has no degree. The integers $a_0, a_1, \ldots, a_{k-1}, a_k$ are called the **coefficients** of the polynomial and if $a_k \neq 0$, then a_k is the **leading coefficient**. The expressions $a_k x^k, a_{k-1} x^{k-1}, \ldots, a_1 x, a_0$ are called the **terms** of the polynomial and if $a_k \neq 0$, then $a_k x^k$ is the **leading term**. a_0 is called the **constant term**. In general, for each $i = 0, 1, \ldots, k$, a_i is called the i**th** term. We will usually represent polynomials with expressions such as $a(x), f(x)$, or $p(x)$. For example, we might write $a(x) = a_k x^k + a_{k-1} x^{k-1} + \cdots + a_1 x + a_0$.

Rewriting the terms of a polynomial in a different order does not change the polynomial. For example, we can write the polynomial $a_k x^k + a_{k-1} x^{k-1} + \cdots + a_1 x + a_0$ as $a_0 + a_1 x + \cdots + a_{k-1} x^{k-1} + a_k x^k$ instead. Two polynomials are considered to be equal if the coefficients of each power of x are equal. For example, $3x^2 + 2x + 5$, $2x + 5 + 3x^2$, and $5 + 2x + 3x^2$ all represent the same polynomial. Are there any more ways to represent this polynomial? If so, what are they?

Example 14.1:

1. Each integer is a polynomial in x over $\mathbb{Z}$. For example, $a(x) = 3$ is a polynomial. The degree of this polynomial is 0. A degree 0 polynomial is also called a **constant polynomial**. The leading coefficient and constant term of this polynomial are the same in this case. They are both 3. The leading term is also 3. Some other degree 0 polynomials are $b(x) = 57$, $c(x) = -1$, and $d(x) = -2536$. Note that the polynomial $e(x) = 0$ is **not** a degree 0 polynomial. It is the only polynomial with **no degree**.

2. $f(x) = 1x + 1$ (which we will abbreviate as $f(x) = x + 1$) is a polynomial in x over $\mathbb{Z}$ of degree 1. A degree 1 polynomial is also called a **linear polynomial**. The two coefficients of this polynomial are both 1, the leading term is x and the constant term is 1. Note that $f(x)$ can also be written as $f(x) = 1 + x$. Some other linear polynomials are $g(x) = 2x + 5$, $h(x) = x$ (which is the same as $h(x) = 1x + 0$), $j(x) = -37 + 16x$, and $k(x) = 4x + (-12)$ (which we will abbreviate as $4x - 12$).

3. $p(x) = x^2 + 2x + 3$ is a polynomial in x over $\mathbb{Z}$ of degree 2. A degree 2 polynomial is also called a **quadratic polynomial**. The coefficients of this polynomial are 1, 2, and 3. 1 is the leading coefficient. The leading term of this polynomial is x^2, the first term is $2x$, and the constant term is 3. Note that $p(x)$ can be written in six different equivalent ways. For example, we can write $p(x) = 3 + 2x + x^2$. Some other quadratic polynomials are $q(x) = -2x^2 + 3x - 5$, $r(x) = 3 + 0x - 2x^2$ (which we abbreviate as $r(x) = 3 - 2x^2$), and $s(x) = 17x^2$.

4. $t(x) = 5x^3 + 2x^2 - x + 5$ is a degree 3 polynomial (or **cubic polynomial**) in x over $\mathbb{Z}$. $u(x) = x^4 - 2x + 7$ is a degree 4 polynomial (or **quartic polynomial**) in x over $\mathbb{Z}$, and $v(x) = -6x^5 + x^4 - x^3 + x$ is a degree 5 polynomial (or **quintic polynomial**) in x over $\mathbb{Z}$. $w(x) = x^{1256} - x^{73} + 2x + 1$ is a degree 1256 polynomial in x over $\mathbb{Z}$.

The polynomial $a(x) = a_k x^k + a_{k-1} x^{k-1} + \cdots + a_1 x + a_0$ is said to be written in **descending order of exponents**, whereas $a(x) = a_0 + a_1 x + \cdots + a_{k-1} x^{k-1} + a_k x^k$ is the same polynomial written in **ascending order of exponents**.

It is often most convenient to write polynomials in ascending order of exponents. For example, using ascending order of exponents, we can abbreviate the polynomial $a(x)$ using sigma notation as follows:

$$a(x) = a_0 + a_1 x + \cdots + a_{k-1} x^{k-1} + a_k x^k = \sum_{i=0}^{k} a_i x^i$$

Notes: (1) See Notes 1 and 2 before Example 7.13 for a detailed explanation about how this notation works.

(2) Notice that we have implicitly made the following identifications: $a_0 = a_0 x^0$ and $a_1 x = a_1 x^1$.

Observe that if $j < k$, then the polynomial $a(x) = a_0 + a_1 x + \cdots + a_{j-1} x^{j-1} + a_j x^j$ can be written as $a(x) = a_0 + a_1 x + \cdots + a_{j-1} x^{j-1} + a_j x^j + a_{j+1} x^{j+1} + \cdots + a_{k-1} x^{k-1} + a_k x^k$, where $a_{j+1}, \ldots, a_k$ are all 0. In general, the degree of the polynomial $a(x) = a_0 + a_1 x + \cdots + a_{k-1} x^{k-1} + a_k x^k$ is **less than or equal to** k. In other words, we allow for the possibility that $a_k = 0$ (and we will allow any of the other coefficients to be 0 as well). Just remember that the degree of the polynomial is the greatest value of k for which $a_k \neq 0$. For example, the polynomial $3 - 2x + 5x^2$ can also be written as $3 - 2x + 5x^2 + 0x^3$ or as $3 - 2x + 5x^2 + 0x^3 + 0x^4 + 0x^5 + 0x^6 + 0x^7$. Whichever way we write this polynomial, its degree is 2.

We would now like to define addition and multiplication of polynomials in x over $\mathbb{Z}$.

Let $a(x)$ and $b(x)$ be the following polynomials:

$$a(x) = a_0 + a_1 x + \cdots + a_{k-1} x^{k-1} + a_k x^k = \sum_{i=0}^{k} a_i x^i$$

$$b(x) = b_0 + b_1 x + \cdots + b_{k-1} x^{k-1} + b_k x^k = \sum_{i=0}^{k} b_i x^i$$

We first define **polynomial addition**. The **sum** of the polynomials $a(x)$ and $b(x)$ is

$$(a+b)(x) = (a_0 + b_0) + (a_1 + b_1)x + \cdots + (a_{k-1} + b_{k-1})x^{k-1} + (a_k + b_k)x^k = \sum_{i=0}^{k} (a_i + b_i) x^i$$

Example 14.2: Let $a(x) = 1 - 2x + 3x^2$ and $b(x) = 3 + 4x - x^2$. Then $(a+b)(x)$ is the following polynomial:

$$(a+b)(x) = (1+3) + (-2+4)x + (3-1)x^2 = 4 + 2x + 2x^2.$$

Theorem 14.3: With the definition of polynomial addition just given, $(\mathbb{Z}[x], +)$ is a commutative group.

Proof: To see that $\mathbb{Z}[x]$ is closed under addition, let $a(x) = \sum_{i=0}^{k} a_i x^i$ and $b(x) = \sum_{i=0}^{k} b_i x^i$ be in $\mathbb{Z}[x]$. Then $a_i, b_i \in \mathbb{Z}$ for each $i = 0, 1, \ldots, k$. Since $\mathbb{Z}$ is closed under addition, $a_i + b_i \in \mathbb{Z}$ for each $i = 0, 1, \ldots, k$. Therefore, $(a+b)(x) = \sum_{i=0}^{k}(a_i + b_i)x^i \in \mathbb{Z}[x]$, as desired.

Let's also show in detail that addition is commutative in $\mathbb{Z}[x]$. Let $a(x) = \sum_{i=0}^{k} a_i x^i$ and $b(x) = \sum_{i=0}^{k} b_i x^i$ be in $\mathbb{Z}[x]$. Since addition is commutative in $\mathbb{Z}$, we have $a_i + b_i = b_i + a_i$ for each $i = 0, 1, \ldots, k$. So, $(a+b)(x) = \sum_{i=0}^{k}(a_i + b_i)x^i = \sum_{i=0}^{k}(b_i + a_i)x^i = (b+a)(x)$, as desired.

Similar computations show that addition is associative in $\mathbb{Z}[x]$, $0 = \sum_{i=0}^{k} 0 x^i = 0 + 0x \cdots + 0x^k$ is an additive identity in $\mathbb{Z}[x]$, and the additive inverse of $\sum_{i=0}^{k} a_i x^i = a_0 + a_1 x + \cdots + a_{k-1} x^{k-1} + a_k x^k$ is $\sum_{i=0}^{k}(-a_i)x^i = -a_0 - a_1 x - \cdots - a_{k-1} x^{k-1} - a_k x^k$. I leave it to the reader to write out the details. □

The definition of **polynomial multiplication** is much more complicated to write down. The idea, however, is quite simple. Let's start with a specific example to help motivate the definition:

Example 14.4: As in Example 14.2, let $a(x) = 1 - 2x + 3x^2$ and $b(x) = 3 + 4x - x^2$. The degree of the product $c(x)$ will be equal to the sum of the degrees of $a(x)$ and $b(x)$. So, in this case, $c(x)$ will have degree $2 + 2 = 4$. Let's find the coefficients of $c(x) = c_0 + c_1 x + c_2 x^2 + c_3 x^3 + c_4 x^4$.

To get the constant term of the product, we multiply the constant terms of $a(x)$ and $b(x)$ to get $1 \cdot 3 = 3$. So, $c_0 = 3$.

$$a(x)b(x) = (1 - 2x + 3x^2)(3 + 4x - x^2) = \boxed{3} + c_1 x + c_2 x^2 + c_3 x^3 + c_4 x^4$$

Next, to get the first term of the product, we look at all possible ith terms of $a(x)$ and jth terms of $b(x)$ so that $i + j = 1$. There are two ways that this can happen: $0 + 1 = 1$ and $1 + 0 = 1$. Let's do each separately:

$0 + 1 = 1$: In this case, we have $i = 0$ and $j = 1$. So, we multiply the 0th coefficient of $a(x)$ with the 1st coefficient of $b(x)$ to get $1 \cdot 4 = 4$.

$1 + 0 = 1$: In this case, we have $i = 1$ and $j = 0$. So, we multiply the 1st coefficient of $a(x)$ with the 0th coefficient of $b(x)$ to get $-2 \cdot 3 = -6$.

We now add these two results to get $c_1 = 4 + (-6) = -2$.

$$a(x)b(x) = (1 - 2x + 3x^2)(3 + 4x - x^2) = 3 \boxed{-2x} + c_2 x^2 + c_3 x^3 + c_4 x^4$$

Next, to get the second term of the product, we look at all possible ith terms of $a(x)$ and jth terms of $b(x)$ so that $i + j = 2$. There are three ways that this can happen: $0 + 2 = 2$, $1 + 1 = 2$, and $2 + 0 = 2$. Once again, let's do each one separately:

$0 + 2 = 2$: In this case, we have $i = 0$ and $j = 2$. So, we multiply the 0th coefficient of $a(x)$ with the 2nd coefficient of $b(x)$ to get $1 \cdot (-1) = -1$.

$1 + 1 = 2$: In this case, we have $i = 1$ and $j = 1$. So, we multiply the 1st coefficient of $a(x)$ with the 1st coefficient of $b(x)$ to get $-2 \cdot 4 = -8$.

$2 + 0 = 2$: In this case, we have $i = 2$ and $j = 0$. So, we multiply the 2nd coefficient of $a(x)$ with the 0th coefficient of $b(x)$ to get $3 \cdot 3 = 9$.

We now add these three results to get $c_2 = -1 + (-8) + 9 = 0$.

$$a(x)b(x) = (1 - 2x + 3x^2)(3 + 4x - x^2) = 3 - 2x + \boxed{0}x^2 + c_3x^3 + c_4x^4$$

For the third term of the product, we look at all possible ith terms of $a(x)$ and jth terms of $b(x)$ so that $i + j = 3$. There are two ways that this can happen: $1 + 2 = 3$ and $2 + 1 = 3$. It then follows that $c_3 = (-2)(-1) + 3 \cdot 4 = 2 + 12 = 14$.

$$a(x)b(x) = (1 - 2x + 3x^2)(3 + 4x - x^2) = 3 - 2x + 0x^2 + \boxed{14}x^3 + c_4x^4$$

Finally, for the fourth term of the product, we look at all possible ith terms of $a(x)$ and jth terms of $b(x)$ so that $i + j = 4$. There is just one way this can happen: $2 + 2 = 4$. So, $c_4 = 3(-1) = -3$.

$$a(x)b(x) = (1 - 2x + 3x^2)(3 + 4x - x^2) = 3 - 2x + 0x^2 + 14x^3 \boxed{-3}x^4$$

So, we found that $c_0 = 3, c_1 = -2, c_2 = 0, c_3 = 14$, and $c_4 = -3$. Therefore, the product of the polynomials $a(x)$ and $b(x)$ is

$$a(x)b(x) = c(x) = c_0 + c_1x + c_2x^2 + c_3x^3 + c_4x^4 = 3 - 2x + 0x^2 + 14x^3 - 3x^4.$$

We will abbreviate this by deleting the term with coefficient 0 to get

$$\mathbf{a(x)b(x) = 3 - 2x + 14x^3 - 3x^4}.$$

Let's now describe the general procedure for multiplying the two polynomials $a(x)$ and $b(x)$ over $\mathbb{Z}$.

Suppose that $a(x)$ and $b(x)$ are the polynomials given by

$$a(x) = \sum_{i=0}^{k} a_i x^i \qquad b(x) = \sum_{i=0}^{k} b_i x^i$$

The **product** of the polynomials $a(x)$ and $b(x)$ is

$$(ab)(x) = c_0 + c_1x + \cdots + c_{2k-1}x^{2k-1} + c_{2k}x^{2k} = \sum_{i=0}^{2k} c_i x^i,$$

where for each $i = 0, 1, \ldots, 2k$,

$$c_i = a_0 b_i + a_1 b_{i-1} + \cdots + a_{i-1} b_1 + a_i b_0 = \sum_{j=0}^{i} a_j b_{i-j}.$$

Notes: (1) Here is what c_i looks like for $i = 0, 1, \ldots, k$:

$$c_0 = a_0 b_0$$
$$c_1 = a_0 b_1 + a_1 b_0$$
$$c_2 = a_0 b_2 + a_1 b_1 + a_2 b_0$$
$$c_3 = a_0 b_3 + a_1 b_2 + a_2 b_1 + a_3 b_0$$
$$\vdots \qquad \vdots \qquad \vdots$$
$$c_k = a_0 b_k + a_1 b_{k-1} + \cdots + a_{k-1} b_1 + a_k b_0$$

(2) Now, $c_{k+1} = a_0 b_{k+1} + a_1 b_k + \cdots a_k b_1 + a_{k+1} b_0$. However, since $a_{k+1} = 0$ and $b_{k+1} = 0$, c_{k+1} can be simplified to $c_{k+1} = a_1 b_k + a_2 b_{k-1} + \cdots + a_{k-1} b_2 + a_k b_1$ (notice that we deleted the first and last term).

Similarly, we have the following:

$$c_{k+2} = a_2 b_k + a_3 b_{k-1} + \cdots + a_{k-1} b_3 + a_k b_2$$
$$\vdots \qquad \vdots \qquad \vdots$$
$$c_{2k-1} = a_{k-1} b_k + a_k b_{k-1}$$
$$c_{2k} = a_k b_k$$

(3) We can also write the definition of c_i as follows:

$$c_i = \sum_{s+t=i} a_s b_t.$$

For example, for $i = 0$, there is just one way to choose s and t. Since $0 + 0 = 0$ is the only sum that works, we must have $s = 0$ and $t = 0$. So, $c_0 = a_0 b_0$.

For $i = 1$, there are two ways to choose s and t. We have $0 + 1 = 1$ and $1 + 0 = 1$. So, we get $c_0 = a_0 b_1 + a_1 b_0$.

At the other extreme, for $i = 2k$, there is just one way to choose s and t. Since $k + k = 2k$, we can choose $s = k$ and $t = k$. It is pointless to increase one of these values because a_{k+1} and b_{k+1} are both zero. So, for example, if we choose $s = k + 1$ and $t = k - 1$, we have $a_{k+1} b_{k-1} = 0 b_{k-1} = 0$. So, we get $c_{2k} = a_k b_k$.

With the definition of polynomial multiplication just given, $(\mathbb{Z}[x], \cdot)$ is a commutative monoid and in fact, together with polynomial addition, $(\mathbb{Z}[x], +, \cdot)$ is a commutative ring. This follows from Problem 16 below.

Actually, there is nothing so special about the ring $\mathbb{Z}$. Given any ring R, we can define polynomial addition and multiplication the exact same way (except that the coefficients of the polynomials belong to R) and we have the following more general result:

Theorem 14.5: Let R be a ring. Then $(R[x], +, \cdot)$ is a ring, where $+$ and $\cdot$ are polynomial addition and polynomial multiplication, respectively. If R is commutative, then so is $R[x]$.

This follows from Problem 16 below. $(R[x], +, \cdot)$ is called the **ring of polynomials over R**.

Example 14.6: Let $a(x) = 1 + 3x + 2x^2$ and $b(x) = 2 + x + 3x^2$.

1. If we consider $a(x), b(x) \in \mathbb{Z}_4[x]$, then $(a+b)(x)$ and $(ab)(x)$ are the following polynomials:
$$(a+b)(x) = (1+2) + (3+1)x + (2+3)x^2 = 3 + 0x + x^2 = 3 + x^2.$$
$$(ab)(x) = (1 \cdot 2) + (1 \cdot 1 + 3 \cdot 2)x + (1 \cdot 3 + 3 \cdot 1 + 2 \cdot 2)x^2 + (3 \cdot 3 + 2 \cdot 1)x^3 + (2 \cdot 3)x^4$$
$$= 2 + (1+2)x + (3+3+0)x^2 + (1+2)x^3 + 2x^4 = 2 + 3x + 2x^2 + 3x^3 + 2x^4.$$

2. If we consider $a(x), b(x) \in \mathbb{Z}_5[x]$, then $(a+b)(x)$ and $(ab)(x)$ are the following polynomials:
$$(a+b)(x) = (1+2) + (3+1)x + (2+3)x^2 = 3 + 4x + 0x^2 = 3 + 4x.$$
$$(ab)(x) = (1 \cdot 2) + (1 \cdot 1 + 3 \cdot 2)x + (1 \cdot 3 + 3 \cdot 1 + 2 \cdot 2)x^2 + (3 \cdot 3 + 2 \cdot 1)x^3 + (2 \cdot 3)x^4$$
$$= 2 + (1+1)x + (3+3+4)x^2 + (4+2)x^3 + 1x^4 = 2 + 2x + 0x^2 + 1x^3 + 1x^4$$
$$= 2 + 2x + x^3 + x^4.$$

3. If we consider $a(x), b(x) \in \mathbb{Z}_6[x]$, then $(a+b)(x)$ and $(ab)(x)$ are the following polynomials:
$$(a+b)(x) = (1+2) + (3+1)x + (2+3)x^2 = 3 + 4x + 5x^2.$$
$$(ab)(x) = (1 \cdot 2) + (1 \cdot 1 + 3 \cdot 2)x + (1 \cdot 3 + 3 \cdot 1 + 2 \cdot 2)x^2 + (3 \cdot 3 + 2 \cdot 1)x^3 + (2 \cdot 3)x^4$$
$$= 2 + (1+0)x + (3+3+4)x^2 + (3+2)x^3 + 0x^4 = 2 + x + 4x^2 + 5x^3.$$

Notes: (1) In part 2 of Example 14.6, we added two polynomials of degree 2 and wound up with a polynomial of degree 1. In general, whenever we add two polynomials over any ring R, the degree of the sum will be less than or equal to the maximum of the degrees of the two polynomials we are adding. Symbolically, we have

$$\text{degree}\,(a(x) + b(x)) \leq \max\{\text{degree } a(x), \text{degree } b(x)\}.$$

(2) In part 3 of Example 14.6, we multiplied two polynomials of degree 2 and wound up with a polynomial of degree 3. One might initially guess that the degree should have been $2 + 2 = 4$, as it turned out to be in parts 1 and 2. However, due to the fact that 2 and 3 are zero divisors in $\mathbb{Z}_6$ ($2 \cdot 3 = 0$), the x^4 term vanishes when taking the product. In general, whenever we multiply two polynomials over a ring R, the degree of the product will be less than or equal to the sum of the degrees of the two polynomials we are multiplying. Symbolically, we have

$$\text{degree}\,(a(x)b(x)) \leq \text{degree } a(x) + \text{degree } b(x).$$

We will see shortly that if R is a domain, then $R[x]$ is also a domain, and we can replace $\leq$ with $=$ in the last formula.

In Problem 22 in Problem Set 2, you were asked to prove that $R[x]$ is a left R-module (as well as a right R-module). In fact, combining this result with Theorem 14.5, we have that $(R[x], +, \cdot)$ is an R-algebra, which we now define.

A **left R-algebra** over a ring R is a triple $(A, +, \cdot)$, where $(A, +)$ is a left R-module, $(A, +, \cdot)$ is a ring, and for all $u, v \in A$ and $r \in R$, $r(uv) = (ru)v = u(rv)$.

We will call the last property "**compatibility of scalar and vector multiplication**."

Notes: (1) The definition of a left R-algebra is identical to the definition of a linear algebra given in Lesson 7, except that the scalars come from a ring instead of a field.

(2) A right R-algebra can be defined similarly. In this case, the compatibility condition would be written "for all $u, v \in A$ and $r \in R$, $(uv)r = u(vr) = (ur)v$.

Theorem 14.7: Let R be a ring. Then $(R[x], +, \cdot)$ is a left R-algebra, where $+$ and $\cdot$ are polynomial addition and polynomial multiplication, respectively. If R is commutative, then so is $(R[x], +, \cdot)$.

You will be asked to prove Theorem 14.7 in Problem 16 below. The hardest part is proving that polynomial multiplication is associative in $R[x]$.

Recall: If a and b are nonzero elements in a ring R such that $ab = 0$, then we say that a and b are **zero divisors**. A ring that does **not** contain any zero divisors is called a **domain**. A commutative domain is called an **integral domain**.

The following theorem will prove to be very useful.

Theorem 14.8: If $(R, +, \cdot)$ is a domain, then $(R[x], +, \cdot)$ is a domain.

Proof: Let $a(x), b(x) \in R[x]$ with $a(x) \neq 0$ and $b(x) \neq 0$. Let a_j be the leading coefficient of $a(x)$ and let b_k be the leading coefficient of $b(x)$. By the definition of a leading coefficient, $a_j \neq 0$ and $b_k \neq 0$. Since R is a domain, $a_j b_k \neq 0$. So, $a_j b_k$ is a nonzero coefficient of the polynomial $(ab)(x) = a(x)b(x)$, and so, $a(x)b(x) \neq 0$. Therefore, $(R[x], +, \cdot)$ is a domain. □

If $(R, +, \cdot)$ is a domain, then $(R[x], +, \cdot)$ is called the **domain of polynomials over R**. Note that if $(R, +, \cdot)$ is an integral domain, then by Theorems 14.5 and 14.8, $(R[x], +, \cdot)$ is also an integral domain.

As we saw in Note 2 following Example 14.6, if R is a ring and $a(x), b(x) \in R$, then the polynomial product $(ab)(x)$ has degree less than or equal to the sum of the degrees of $a(x)$ and $b(x)$. We can write

$$\text{degree}\big(a(x)b(x)\big) \leq \text{degree } a(x) + \text{degree } b(x).$$

The reason that we cannot replace $\leq$ by $=$ is because the leading coefficients of $a(x)$ and $b(x)$ could be zero divisors (see part 3 of Example 14.6).

However, if R is a domain, then we do get the nicer formula

$$\text{degree}\big(a(x)b(x)\big) = \text{degree } a(x) + \text{degree } b(x).$$

Polynomials Over a Field

Let F be a field. It would be nice if the polynomial ring $F[x]$ were also a field. Unfortunately, this is **never** the case. Indeed, if we multiply x by any polynomial $p(x) \in F[x]$, then $xp(x)$ is a polynomial of degree at least 1, and therefore, $xp(x) \neq 1$. So, x has no multiplicative inverse in $F[x]$.

However, since every field is an integral domain (by part (iv) of Problem 7 in Problem Set 2), it follows from Theorems 14.5 and 14.8 that $F[x]$ is an integral domain.

Since $F[x]$ is never a field, you may wonder why we are dedicating a whole section to this special case, rather than simply working with polynomials over an integral domain. One reason is because in $F[x]$, there are nice analogues of the Division Algorithm, Euclid's Principle, and the Fundamental Theorem of Arithmetic that may not hold in the domain of polynomials over an arbitrary integral domain. We will explore these ideas now.

A polynomial $a(x) \in F[x]$ is **divisible** by a polynomial $b(x) \in F[x]$, written $b(x)|a(x)$, if there is another polynomial $c(x) \in F[x]$ such that $a(x) = b(x)c(x)$. We also say that $b(x)$ is a **factor** of $a(x)$, $b(x)$ is a **divisor** of $a(x)$, $b(x)$ **divides** $a(x)$, or $a(x)$ is a **multiple** of $b(x)$.

Example 14.9:

1. In $\mathbb{Q}[x]$, $x^4 - 9$ is divisible by $x^2 - 3$ because $x^4 - 9 = (x^2 - 3)(x^2 + 3)$. Note that this also shows that $x^4 - 9$ is divisible by $x^2 + 3$ in $\mathbb{Q}[x]$.

2. In $\mathbb{R}[x]$, $x^2 - 2$ is divisible by $x - \sqrt{2}$ because $x^2 - 2 = (x - \sqrt{2})(x + \sqrt{2})$. Note that this also shows that $x^2 - 2$ is divisible by $x + \sqrt{2}$ in $\mathbb{R}[x]$.

3. In $\mathbb{Q}[x]$, $x^2 - 2$ does **not** have any linear divisors. If $x^2 - 2 = (ax + b)(cx + d)$, then we have $x^2 - 2 = acx^2 + (ad + bc)x + bd$. So, $ac = 1$, $ad + bc = 0$, and $bd = -2$. We then have that $0 = cd(ad + bc) = (ac)d^2 + (bd)c^2 = 1d^2 - 2c^2$. So, $d^2 = 2c^2$, and therefore, $\left(\frac{d}{c}\right)^2 = \frac{d^2}{c^2} = 2$. Since $\mathbb{Q}$ is a field, $\left(\frac{d}{c}\right)^2 = \frac{d^2}{c^2} = d^2(c^2)^{-1} \in \mathbb{Q}$. Since there is no rational number whose square is 2 (see Problem 17 below), there are no rational values for $a, b, c,$ and d such that $x^2 - 2 = (ax + b)(cx + d)$.

4. In $\mathbb{C}[x]$, $x^2 + 1$ is divisible by $x - i$ because $x^2 + 1 = (x - i)(x + i)$. Note that this also shows that $x^2 + 1$ is divisible by $x + i$ in $\mathbb{C}[x]$.

5. In $\mathbb{Z}_2[x]$, $x^2 + 1$ is divisible by $x + 1$ because $x^2 + 1 = (x + 1)(x + 1) = (x + 1)^2$ (note that $(x + 1)(x + 1) = x^2 + (1 + 1)x + 1^2 = x^2 + 0x + 1 = x^2 + 1$). Similarly, $x^4 + 1$ is divisible by $x + 1$ because $x^4 + 1 = (x^2 + 1)(x^2 + 1) = (x^2 + 1)^2 = ((x + 1)^2)^2 = (x + 1)^4$. Can you generalize this further?

6. If F is any field, $a(x) \in F[x]$, and $k \in F[x]$ is any nonzero constant polynomial, then $a(x)$ is divisible by k. Indeed, if $a(x) = a_0 + a_1 x + \cdots + a_n x^n$, let $c(x) = \frac{a_0}{k} + \frac{a_1}{k} x + \cdots + \frac{a_n}{k} x^n$. Then
$$a(x) = a_0 + a_1 x + \cdots + a_n x^n = k\left(\frac{a_0}{k} + \frac{a_1}{k} x + \cdots + \frac{a_n}{k} x^n\right) = k \cdot c(x).$$

Theorem 14.10: Let F be a field. The invertible elements of $F[x]$ are precisely the nonzero constant polynomials.

Proof: If $k \in F[x]$ is a nonzero constant polynomial, then so is $\frac{1}{k}$ and $k \cdot \frac{1}{k} = \frac{1}{k} \cdot k = 1$. Conversely, suppose that $a(x) \in F[x]$ is invertible. Then there is a polynomial $b(x) \in F[x]$ so that $a(x)b(x) = 1$. So, degree $a(x)$ + degree $b(x)$ = degree $1 = 0$. Therefore, $a(x)$ and $b(x)$ must both have degree 0, and so, they are both constant polynomials. □

We would now like to give the definition of a "prime polynomial," as we did for positive integers in Lesson 8. It's worth exploring for a moment why we restricted our attention to positive integers when we gave this definition. Let's take the prime number 5 as an example. 5 is prime because its only **positive** factors are 1 and itself. However, 5 does have two additional factors, namely -1 and -5. So, in $\mathbb{Z}$, 5 can be factored as $1 \cdot 5$ and as $(-1)(-5)$.

Note that in $\mathbb{Z}$, there are exactly two invertible elements, namely 1 and -1. We can define an integer a to be **irreducible** if whenever we write $a = bc$ for some integers b and c, then either b or c is invertible. So, for example, 5 is irreducible because both factorizations of 5 involve an invertible element (either 1 or -1). 6 on the other hand is **reducible**, because $6 = 2 \cdot 3$ and neither 2 nor 3 is invertible.

Note: For our purposes here, the words "prime" and "irreducible" are interchangeable (when working in $\mathbb{Z}$ or $F[x]$, they are equivalent). However, in arbitrary integral domains they have slightly different meanings and they may not be equivalent. So, I will choose to use the technically correct terminology.

In the case of $F[x]$, there are many more invertible elements than there are in $\mathbb{Z}$. Indeed, by Theorem 14.10, every nonzero constant polynomial is invertible. Due to this, any polynomial can be factored in infinitely many trivial ways. For example, the polynomial $x + 1$ can be factored as $\frac{1}{2}(2x + 2)$ or as $2\left(\frac{1}{2}x + \frac{1}{2}\right)$. In fact, if c is any nonzero constant polynomial, then $x + 1 = c\left(\frac{1}{c}x + \frac{1}{c}\right)$. However, since c is invertible, these factorizations do **not** imply that $x + 1$ is reducible. In fact, $x + 1$ is irreducible in $F[x]$. Indeed, if $x + 1 = b(x)c(x)$, where $b(x)$ and $c(x)$ are not invertible (so that $x + 1$ is reducible), then $b(x)$ and $c(x)$ both have degree at least 1. It follows that the degree of $x + 1$ is at least $1 + 1 = 2$, which of course is not true ($x + 1$ has degree 1). This contradiction proves that $x + 1$ is irreducible over F.

Let's formally define reducible and irreducible in $F[x]$.

A nonconstant polynomial $a(x) \in F[x]$ is **reducible** over F if there are nonconstant polynomials $b(x), c(x) \in F[x]$ such that $a(x) = b(x)c(x)$. Otherwise, $a(x)$ is **irreducible** over F.

Note: Just as we do for the integers with 0 and 1 (these integers are neither prime nor composite), we exclude constant polynomials from consideration when discussing reducible and irreducible.

Example 14.11:

1. The polynomial $x^4 - 9$ is reducible over $\mathbb{Q}$ because $x^4 - 9 = (x^2 - 3)(x^2 + 3)$, as we saw in part 1 of Example 14.9.

2. The polynomial $x^2 - 2$ is reducible over $\mathbb{R}$ because $x^2 - 2 = (x - \sqrt{2})(x + \sqrt{2})$, as we saw in part 2 of Example 14.9. However, $x^2 - 2$ is irreducible over $\mathbb{Q}$, as we saw in part 3 of Example 14.9.

3. The polynomial $x^2 + 1$ is reducible over both $\mathbb{C}$ and $\mathbb{Z}_2$, as we saw in parts 4 and 5 of Example 14.9. I leave it to the reader to prove that $x^2 + 1$ is irreducible over $\mathbb{R}$ (see Problem 13 below).

We now prove the analogue of Theorem 8.7 for the polynomial domain $F[x]$.

Theorem 14.12: Every nonconstant polynomial over a field F can be written as a product of irreducible factors.

Proof: Suppose toward contradiction that there exists a nonconstant polynomial $a(x)$ that cannot be written as a product of irreducible factors. Furthermore, we may assume that $a(x)$ is a polynomial of the least degree that does not satisfy the conclusion. Since $a(x)$ cannot be written as a product of irreducible factors, then in particular, $a(x)$ is not irreducible. So, we can write $a(x) = b(x)c(x)$ with $b(x), c(x) \in F[x]$ and $0 <$ degree $b(x) <$ degree $a(x)$ and $0 <$ degree $c(x) <$ degree $a(x)$. Since $a(x)$ is a polynomial of least degree that cannot be written as a product of irreducible factors, $b(x)$ and $c(x)$ can both be written as products of irreducible factors. But then $a(x) = b(x)c(x)$ is also a product of irreducible factors, contradicting our choice of $a(x)$. This contradiction shows that every nonconstant polynomial can be written as a product of irreducible factors. $\square$

The factorization given by Theorem 14.12 is **not** unique. For example, $x^2 - 1$ can be factored as $(x-1)(x+1)$, but it can also be factored as $(2x - 2)\left(\frac{1}{2}x + \frac{1}{2}\right)$. More generally, given any nonzero constant k, $x^2 - 1$ can be factored as $(kx - k)\left(\frac{1}{k}x + \frac{1}{k}\right)$. However, the first factorization seems to be the most "natural" way to factor $x^2 - 1$. The factorization $(x-1)(x+1)$ is special in the sense that the leading coefficients of the factors are both 1.

A polynomial $p(x) \in F[x]$ is called a **monic polynomial** if its leading coefficient is 1.

Note: Every nonconstant polynomial $a(x)$ over a field F can be factored **uniquely** as a product $kp_1(x)p_2(x) \cdots p_n(x)$, where k is a constant and for each $i = 1, 2, \ldots, n$, $p_i(x)$ is a monic polynomial, up to the order in which the factors are written. See Theorem 14.23 below for details.

The Division Algorithm for Polynomials

Theorem 14.13 (The Division Algorithm for Polynomials): Let F be a field and let $a(x), b(x) \in F[x]$ with $b(x) \neq 0$. Then there are polynomials $k(x), r(x) \in F[x]$ such that $a(x) = b(x)k(x) + r(x)$ and $r = 0$ or degree $r(x) <$ degree $b(x)$.

The proof of this theorem is a bit tricky and requires some cleverness. If you find the proof below difficult at first, it's perfectly okay to skip over it for now and to simply accept that the theorem is true.

We will prove this theorem by contradiction. The main idea is to assume that for a fixed $b(x) \in F[x]$, there is a polynomial $a(x) \in F[x]$ such that the conclusion of the theorem fails (and we will assume that $a(x)$ is of smallest possible degree). We will then define a new polynomial $c(x) \in F[x]$ in terms of $a(x)$ and $b(x)$ that has a smaller degree than $a(x)$. Thus, the conclusion of the theorem holds for $c(x)$. We get our contradiction by using $c(x)$ to show that the conclusion of the theorem holds for $a(x)$.

Proof of Theorem 14.13: Let $b(x) \in F[x]$ with $b(x) \neq 0$ and assume toward contradiction that there is a polynomial $a(x)$ that does not satisfy the conclusion of the theorem. Furthermore, we may assume that $a(x)$ is a polynomial of the least degree that does not satisfy the conclusion of the theorem (note that $a(x)$ cannot be the 0 polynomial because $0 = b(x) \cdot 0 + 0$).

We cannot have degree $a(x)$ < degree $b(x)$, because then we would have $a(x) = b(x) \cdot 0 + a(x)$. Therefore, degree $a(x) \geq$ degree $b(x)$. Let degree $a(x) = n$ and degree $b(x) = m$. Then we can write

$$a(x) = \sum_{i=0}^{n} a_i x^i \qquad b(x) = \sum_{i=0}^{m} b_i x^i$$

Here $a_n \neq 0$ and $b_m \neq 0$. Letting $c(x) = a(x) - \frac{a_n}{b_m} x^{n-m} b(x)$, we get the following:

$$c(x) = a(x) - \frac{a_n}{b_m} x^{n-m} b(x) = a(x) - \left(\frac{a_n}{b_m} b_0 x^{n-m} + \frac{a_n}{b_m} b_1 x^{n-m+1} + \cdots + \frac{a_n}{b_m} b_m x^{n-m+m} \right)$$

$$= (a_0 + a_1 x + \cdots + a_n x^n) - \left(\frac{a_n}{b_m} b_0 x^{n-m} + \frac{a_n}{b_m} b_1 x^{n-m+1} + \cdots + \frac{a_n}{b_m} b_{m-1} x^{n-1} + a_n x^n \right)$$

$$= (a_0 + a_1 x + \cdots + a_{n-1} x^{n-1}) - \left(\frac{a_n}{b_m} b_0 x^{n-m} + \frac{a_n}{b_m} b_1 x^{n-m+1} + \cdots + \frac{a_n}{b_m} b_{m-1} x^{n-1} \right)$$

Notice how the terms $a_n x^n$ and $-a_n x^n$ cancelled out. It follows that $c(x)$ is a polynomial of degree less than n. Since $a(x)$ is a polynomial of the least degree that does not satisfy the conclusion of the theorem, $c(x)$ **does** satisfy the conclusion of the theorem. So, there are polynomials $k(x), r(x) \in F[x]$ such that $c(x) = b(x)k(x) + r(x)$ and $r = 0$ or degree $r(x) <$ degree $b(x)$.

Now, adding $\frac{a_n}{b_m} x^{n-m} b(x)$ to each side of the first equality above and letting $j(x) = \frac{a_n}{b_m} x^{n-m}$ gives us

$$a(x) = c(x) + \frac{a_n}{b_m} x^{n-m} b(x) = \big(b(x)k(x) + r(x)\big) + j(x)b(x) = b(x)\big(k(x) + j(x)\big) + r(x).$$

Since the sum of two polynomials over F is a polynomial over F, $k(x) + j(x) \in F[x]$. It follows that $a(x)$ **does** satisfy the conclusion of the theorem, contradicting our choice of $a(x)$. □

Notes: (1) In order to define $c(x)$, we needed to divide a_n by b_m, or equivalently, multiply a_n by b_m^{-1}. In order to do this, we need to know that b_m has a multiplicative inverse. This is the step where it is important that F is a field (and of course that $b_m \neq 0$, which is true). If F were simply an integral domain, then the proof would fail here.

(2) The leading term of $a(x)$ is $a_n x^n$ and the leading term of $b(x)$ is $b_m x^m$. When we multiply $b(x)$ by $\frac{a_n}{b_m} x^{n-m}$, the new leading term becomes $\left(\frac{a_n}{b_m} x^{n-m} \right)(b_m x^m) = a_n x^n$, the same leading term as $a(x)$! Thus, when we subtract the polynomial $\frac{a_n}{b_m} x^{n-m} b(x)$ from the polynomial $a(x)$, the nth term vanishes.

(3) Given $a(x), b(x) \in F[x]$, the polynomials $k(x), r(x) \in F[x]$ are unique (just like k and r are unique in the Division Algorithm from Lesson 8). To see this, suppose that $a(x) = b(x)k_1(x) + r_1(x)$ and $r_1 = 0$ or degree $r_1(x) <$ degree $b(x)$. Also, suppose that $a(x) = b(x)k_2(x) + r_2(x)$ and $r_2 = 0$ or degree $r_2(x) <$ degree $b(x)$. By direct substitution, $b(x)k_1(x) + r_1(x) = b(x)k_2(x) + r_2(x)$.

Subtracting $b(x)k_2(x)$ from each side of the equation and simultaneously subtracting $r_1(x)$ from each side of the equation, we get $b(x)k_1(x) - b(x)k_2(x) = r_2(x) - r_1(x)$.

Factoring $b(x)$ on the left gives $b(x)(k_1(x) - k_2(x)) = r_2(x) - r_1(x)$.

Note that degree $(r_2(x) - r_1(x)) <$ degree $b(x)$.

If $k_1(x) - k_2(x) \neq 0$, then since $F[x]$ is an integral domain, degree $b(x)(k_1(x) - k_2(x))$ is greater than or equal to degree $b(x)$. This implies that degree $(r_1(x) - r_2(x)) \geq$ degree $b(x)$, which is impossible. Therefore, $k_1(x) - k_2(x) = 0$, and so, $k_1(x) = k_2(x)$.

It follows that $r_2(x) - r_1(x) = b(x)(k_1(x) - k_2(x)) = 0$, or equivalently, $r_2(x) = r_1(x)$.

(4) An integral domain R is a **Euclidean domain** if there exists a function $f: R \setminus \{0\} \to \mathbb{N}$ such that if $a, b \in R$ and $b \neq 0$, then there exist $k, r \in R$ such that $a = bk + r$ and either $r = 0$ or $f(r) < f(b)$. The function f is sometimes called a **Euclidean function** (or **degree function**). Theorem 14.13 says that if F is a field, then $F[x]$ is a Euclidean domain with Euclidean function $f(a(x)) =$ degree $a(x)$.

Principal Ideal Domains

Recall that if $(R, +, \cdot)$ is a commutative ring, a subset I of R is an **ideal** of R if $(I, +)$ is a subgroup of $(R, +)$ and I absorbs R (for every $a \in I$ and $r \in R$, $ar \in I$). An ideal of R is a **principal ideal** if it has the form $\langle a \rangle = \{ar \mid r \in R\}$ (see part 3 of Example 7.22).

A commutative ring R is called a **PIR (Principal Ideal Ring)** if every ideal of R is principal. A PIR that is a domain is called a **PID (Principal Ideal Domain)**. We will mostly be interested in PIDs.

Example 14.14:

1. $\mathbb{Z}$ is a PID. Let's verify this in detail. We already know that $\mathbb{Z}$ is an integral domain. Let I be an ideal of $\mathbb{Z}$. If $I = \{0\}$, then $I = \langle 0 \rangle$, and so, I is principal. If $I \neq 0$, then since I is closed under additive inverses (because $(I, +)$ is a subgroup of $(\mathbb{Z}, +)$), I must contain a positive integer. Let a be the least positive integer in I. Then since I absorbs $\mathbb{Z}$, $\langle a \rangle = \{an \mid n \in \mathbb{Z}\} \subseteq I$. Conversely, let $b \in I$. By the Division Algorithm, there are unique integers k and r such that $b = ak + r$ and $0 \leq r < a$. Then $r = b - ak$. Since $a \in I$ and I absorbs $\mathbb{Z}$, $ak \in I$. Since $b, ak \in I$ and $(I, +)$ is a subgroup of $(\mathbb{Z}, +)$, $r = b - ak \in I$. Since $r < a$ and a is the least positive integer in I, $r = 0$. Therefore, $b = ak \in \langle a \rangle$. So, $I \subseteq \langle a \rangle$. Since $\langle a \rangle \subseteq I$ and $I \subseteq \langle a \rangle$, we have $I = \langle a \rangle$. Thus, I is principal. Since I was an arbitrary ideal of $\mathbb{Z}$, we see that $\mathbb{Z}$ is a PID.

2. Every field is a PID. This follows from the fact that a field F has only two ideals: $\{0\}$ and F itself (see part 7 of Example 7.22). Note that F itself is a principal ideal because $F = \langle 1 \rangle$.

3. If F is a field, then $F[x]$ is a PID. We already know that $F[x]$ is an integral domain by Theorems 14.5 and 14.8. Let I be an ideal of $F[x]$. If $I = \{0\}$, then $I = \langle 0 \rangle$, and so, I is principal. If $I \neq 0$, let $p(x)$ be any polynomial of lowest degree in I. Then since I absorbs $F[x]$, we have $\langle p(x) \rangle = \{p(x)a(x) \mid a(x) \in F[x]\} \subseteq I$.

   Conversely, let $a(x) \in I$. By the Division Algorithm for polynomials, there are polynomials $k(x), r(x) \in F[x]$ such that $a(x) = p(x)k(x) + r(x)$ and either $r = 0$ or degree $r(x) <$ degree $p(x)$.

Since $p(x) \in I$ and I absorbs $F[x]$, $p(x)k(x) \in I$. Since $a(x), p(x)k(x) \in I$, and $(I, +)$ is a subgroup of $(F[x], +)$, $r(x) = a(x) - p(x)k(x) \in I$. Since $p(x)$ is a polynomial in I of lowest degree, we must have $r = 0$. So, $a(x) = p(x)k(x) \in \langle p(x) \rangle$. Thus, $I \subseteq \langle p(x) \rangle$.

Since $\langle p(x) \rangle \subseteq I$ and $I \subseteq \langle p(x) \rangle$, we have $I = \langle p(x) \rangle$. Thus, I is principal. Since I was an arbitrary ideal of $F[x]$, we see that $F[x]$ is a PID.

4. Every Euclidean domain (see Note 4 following Theorem 14.13) is a PID. You will be asked to prove this in Problem 6 below. Parts 1, 2, and 3 above are special cases of this result.

Unique Factorization Domains

Let R be an integral domain and let $a \in R$ be nonzero and noninvertible. a is said to be **reducible** if there are noninvertible elements $b, c \in R$ such that $a = bc$. Otherwise, a is **irreducible**.

Example 14.15:

1. Let $R = \mathbb{Z}$. The only invertible elements of $\mathbb{Z}$ are 1 and −1. The integer 6 is reducible because $6 = 2 \cdot 3$ and both 2 and 3 are nonzero and noninvertible. Similarly, −6 is reducible because $(-6) = (-2)(3)$. Observe that the reducible elements of $\mathbb{Z}$ are precisely the composite numbers and their negatives.

    The integer 5 is irreducible because the only two ways to factor 5 in $\mathbb{Z}$ (up to the order in which the factors are written) is $5 = 1 \cdot 5$ and $5 = (-1)(-5)$. Observe that the irreducible elements of $\mathbb{Z}$ are precisely the prime numbers and their negatives.

    By definition, 0, 1, and −1 are neither reducible nor irreducible.

2. Let $R = \mathbb{Z}[x]$. As was the case for $\mathbb{Z}$, the only invertible elements of $\mathbb{Z}[x]$ are 1 and −1 The polynomial $2x + 2$ is reducible because $2x + 2 = 2(x + 2)$ and both 2 and $x + 2$ are nonzero and noninvertible. In this case we say that $2x + 2$ is reducible over $\mathbb{Z}$.

3. Let $R = \mathbb{Q}[x]$. By Theorem 14.10, the invertible elements of $\mathbb{Q}[x]$ are precisely the nonzero constant polynomials. In particular, 2 is invertible in $\mathbb{Q}[x]$. So, although $2x + 2 = 2(x + 2)$, this equation does **not** show that $2x + 2$ is reducible over $\mathbb{Q}$. In fact, all linear polynomials are irreducible over $\mathbb{Q}$. To see this, let $ax + b \in \mathbb{Q}[x]$ with $a \neq 0$ and suppose that there are polynomials $p(x), q(x) \in \mathbb{Q}[x]$ with $ax + b = p(x)q(x)$. Since the degree of $p(x)q(x)$ is equal to the sum of the degrees of $p(x)$ and $q(x)$, either $p(x)$ or $q(x)$ must have degree 0. So, either $p(x)$ or $q(x)$ is a nonzero constant polynomial, which is invertible.

Let R be an integral domain and let $a, b \in R$. a and b are **associates** if there is an invertible element $u \in R$ such that $b = au$.

Example 14.16:

1. Let $R = \mathbb{Z}$. The integers 6 and −6 are associates because $-6 = -1 \cdot 6$ and −1 is invertible. More generally, for any nonzero integer n, n and $-n$ are associates because $-n = -1 \cdot n$. In fact, any two nonzero integers n and m are associates if and only if $m = -n$.

2. Let $R = \mathbb{Z}[x]$. Although $2x + 2 = 2(x + 2)$, the polynomials $x + 2$ and $2x + 2$ are **not** associates. Neither 2 nor $x + 2$ are invertible elements of $\mathbb{Z}[x]$. Two polynomials in $\mathbb{Z}[x]$ are associates if and only if they are negatives of each other. For example, $x + 2$ and $-x - 2$ are associates because $-x - 2 = -(x + 2)$.

3. Let $R = \mathbb{Q}[x]$. The polynomials $x + 2$ and $2x + 2$ are associates because $2x + 2 = 2(x + 2)$ and 2 is invertible in $\mathbb{Q}[x]$ (remember that the invertible elements in $F[x]$ (F a field) are precisely the nonzero constant polynomials).

   In fact, if F is a field, then polynomials $a(x), b(x) \in F[x]$ are associates if and only if they are nonzero constant multiples of each other.

Let R be an integral domain and define a relation $\sim$ on R by $a \sim b$ if and only if a and b are associates. In Problem 2 below, you will be asked to show that $\sim$ is an equivalence relation on R.

Theorem 14.17: Let F be a field. Two polynomials in $F[x]$ are associates if and only if they divide each other.

Proof: Let $a(x), b(x) \in F[x]$ be associates. Then there is $k \in F$ such that $k \neq 0$ and $a(x) = kb(x)$. So, $b(x) | a(x)$. The equation $a(x) = kb(x)$ is equivalent to $b(x) = \frac{1}{k}a(x)$, and so, $a(x) | b(x)$.

Conversely, suppose that $a(x), b(x) \in F[x]$ divide each other. Then there are polynomials $c(x), d(x) \in F[x]$ such that $a(x) = b(x)c(x)$ and $b(x) = a(x)d(x)$. So, we have

$$a(x) \cdot 1 = a(x) = b(x)c(x) = a(x)d(x)c(x).$$

Since $F[x]$ is an integral domain, it satisfies the left cancellation law. So, $1 = d(x)c(x)$.

Since degree $d(x)c(x)$ = degree $d(x)$ + degree $c(x)$ and degree $1 = 0$, $d(x)$ and $c(x)$ must both have degree 0. So, $d(x)$ and $c(x)$ are nonzero constant polynomials. In particular, $a(x)$ is a nonzero constant multiple of $b(x)$. So, $a(x)$ and $b(x)$ are associates. □

An integral domain R is called a **UFD (Unique Factorization Domain)** if every nonzero element of R can be written uniquely as a product of irreducible elements of R and an invertible element of R, up to the order in which the factors are written and the factors being associates of each other.

When we write an element r of an integral domain as a product of other elements of R, we call that product a **factorization** of r. If all the factors in the product are irreducible or invertible, we call the product an **irreducible factorization** of r.

An irreducible factorization of r can be written in the form $r = up_1p_2 \cdots p_n$, where u is invertible and $p_1, p_2, \ldots, p_n \in R$ are irreducible. In a UFD, if $r = vq_1q_2 \cdots q_m$ is another irreducible factorization, then $m = n$ and we can reorder $q_1, q_2, \ldots, q_n$ so that p_i and q_i are associates for each $i = 1, 2, \ldots, n$.

Example 14.18:

1. $\mathbb{Z}$ is a UFD. This follows from Theorem 8.21 (The Fundamental Theorem of Arithmetic).

2. Every field is a UFD. The proof is trivial. Since every nonzero element of a field is invertible, each nonzero element itself is its own irreducible factorization.

3. If F is a field, then $F[x]$ is a UFD. We will state this formally in Theorem 14.23 below. We already proved that every nonconstant polynomial has an irreducible factorization (Theorem 14.12). We still need to show that the factorization is unique. The argument is analogous to what we did in Lesson 8 and so, the details are left as an exercise (see Problem 11 below).

Recall that a polynomial $p(x) \in F[x]$ is called a **monic polynomial** if its leading coefficient is 1.

Every nonzero polynomial $a(x)$ has a unique monic associate. To see this, let $a(x)$ be a polynomial of degree $n \in \mathbb{N}$, say $a(x) = a_0 + a_1 x + \cdots + a_{n-1} x^{n-1} + a_n x^n$ with $a_n \neq 0$. Then

$$a(x) = a_0 + a_1 x + \cdots + a_{n-1} x^{n-1} + a_n x^n = a_n \left(\frac{a_0}{a_n} + \frac{a_1}{a_n} x + \cdots + \frac{a_{n-1}}{a_n} x^{n-1} + x^n \right).$$

If we let $p(x) = \frac{a_0}{a_n} + \frac{a_1}{a_n} x + \cdots + \frac{a_{n-1}}{a_n} x^{n-1} + x^n$, then we see that $p(x)$ is a monic polynomial that is an associate of $a(x)$.

On the other hand, if $p(x)$ and $q(x)$ are associates, then there is a nonzero constant k such that $p(x) = k q(x)$. If the leading coefficient of $p(x)$ and $q(x)$ are both 1, then $1 = k \cdot 1 = k$, and so, $p(x) = q(x)$.

Let $a(x), b(x) \in F[x]$. We say that $c(x) \in F[x]$ is a **common divisor** (or **common factor**) of $a(x)$ and $b(x)$ if $c(x)$ is a factor of both $a(x)$ and $b(x)$. We say that $d(x) \in F[x]$ is a **greatest common divisor** (or **greatest common factor**) of $a(x)$ and $b(x)$ if $d(x)$ is a common divisor of $a(x)$ and $b(x)$ such that every common divisor of $a(x)$ and $b(x)$ divides $d(x)$.

Observe that a greatest common divisor of $a(x)$ and $b(x)$ is not unique. However, by Theorem 14.17, any two greatest common divisors are associates and of all these associates there is a unique monic polynomial. We define this monic polynomial to be **the greatest common divisor** of $a(x)$ and $b(x)$, and we denote it by $\gcd(a(x), b(x))$.

Example 14.19:

1. In $\mathbb{Q}[x]$, let $a(x) = x^2 - 1$ and $b(x) = x^2 - 2x + 1$. Then $d(x) = x - 1$ is a common divisor of $a(x)$ and $b(x)$ because $a(x) = (x-1)(x+1)$ and $b(x) = (x-1)(x-1)$. Since $a(x)$ does not divide $b(x)$ and $b(x)$ does not divide $a(x)$, $d(x)$ is a greatest common divisor of $a(x)$ and $b(x)$. Since $d(x)$ is monic, it is the greatest common divisor of $a(x)$ and $b(x)$, and we can write $\gcd(a(x), b(x)) = d(x) = x - 1$.

   Note that every nonzero constant polynomial is a common divisor of $a(x)$ and $b(x)$. However, none of them are greatest common divisors because $x - 1$ is a common divisor of greater degree ($x - 1$ has degree 1, whereas each nonzero constant polynomial has degree 0). Also, every nonzero constant multiple of $x - 1$ is a greatest common divisor of $a(x)$ and $b(x)$. Only the monic polynomial of degree 1, $x - 1$, is **the** greatest common divisor of $a(x)$ and $b(x)$.

2. In $\mathbb{Q}[x]$, let $a(x) = x^2 - 2$ and $b(x) = x^3$. Since $a(x)$ does not divide $b(x)$ and $b(x)$ does not divide $a(x)$, the greatest common divisor of $a(x)$ and $b(x)$ has degree at most 1. By part 3 of Example 14.9, $a(x)$ does not have any divisors of degree 1. Therefore, the greatest common divisors of $a(x)$ and $b(x)$ are the nonzero constant polynomials. So, $\gcd(a(x), b(x)) = 1$. In this case, we say that $a(x)$ and $b(x)$ are relatively prime.

Specifically, we say that two polynomials $a(x), b(x) \in F[x]$ are **relatively prime** if $\gcd(a(x), b(x)) = 1$, or equivalently, if the only common divisors of $a(x)$ and $b(x)$ are the nonzero constant polynomials.

Let $a(x), b(x) \in F[x]$. A **linear combination** of $a(x)$ and $b(x)$ is an expression of the form $c(x)a(x) + d(x)b(x)$ with $c(x), d(x) \in F[x]$.

Theorem 14.20: Let $a(x), b(x) \in F[x]$ be nonzero polynomials. Then $\gcd(a(x), b(x))$ can be expressed as a linear combination of $a(x)$ and $b(x)$.

Proof: Let $a(x), b(x) \in F[x]$ be nonzero polynomials and let

$$I = \{c(x)a(x) + d(x)b(x) \mid c(x), d(x) \in F[x]\}.$$

By Problem 8 below, I is an ideal of $F[x]$. By part 3 of Example 14.14, I is a principal ideal. So, there is $p(x) \in F[x]$ such that $I = \langle p(x) \rangle$. Since polynomials that are associates of each other generate the same ideal, we may assume that $p(x)$ is monic. Since $a(x) = 1a(x) + 0b(x)$, $a(x) \in I = \langle p(x) \rangle$. So, there is $q(x) \in F[x]$ such that $a(x) = p(x)q(x)$. Therefore, $p(x) | a(x)$. A similar argument shows that $p(x) | b(x)$.

Now let $q(x) \in F[x]$ be any common divisor of $a(x)$ and $b(x)$. Since $q(x)$ is a divisor of $a(x)$, there is $e(x) \in F[x]$ such that $a(x) = q(x)e(x)$. Since $q(x)$ is a divisor of $b(x)$, there is $f(x) \in F[x]$ such that $b(x) = q(x)f(x)$. Since $p(x) \in I$, there are $g(x), h(x) \in F[x]$ such that

$$p(x) = g(x)a(x) + h(x)b(x) = g(x)q(x)e(x) + h(x)q(x)f(x) = q(x)[g(x)e(x) + h(x)f(x)].$$

Therefore, $q(x) | p(x)$. It follows that $\gcd(a(x), b(x)) = p(x)$. □

Theorem 14.21 (Euclid's Principle for Polynomials): Let $a(x), b(x), c(x) \in F[x]$ with $a(x)$ and $b(x)$ relatively prime and suppose that $a(x) | b(x)c(x)$. Then $a(x) | c(x)$.

Theorem 14.22: Let $p(x) \in F[x]$ be an irreducible polynomial and let $a_1(x), a_2(x), \ldots, a_n(x) \in F[x]$. Suppose that $p(x) | a_1(x)a_2(x) \cdots a_n(x)$. Then there is an integer j with $1 \leq j \leq n$ such that $p(x) | a_j(x)$.

Theorem 14.23: $F[x]$ is a UFD. More specifically, every nonconstant polynomial $a(x)$ over a field F can be factored **uniquely** as a product $kp_1(x)p_2(x) \cdots p_n(x)$, where k is a constant and for each $i = 1, 2, \ldots, n$, $p_i(x)$ is a monic irreducible polynomial, up to the order in which the factors are written.

The proofs of Theorems 14.21, 14.22, and 14.23 are similar to the proofs of Theorems 8.19, 8.20, and 8.21, and so we leave them to the reader (see Problems 9, 10, and 11 below).

Polynomial Functions

Let F be a field and let $a(x) \in F[x]$, say $a(x) = a_0 + a_1 x + \cdots + a_n x^n$. If $c \in F$, we define $a(c)$ to be $a_0 + a_1 c + \cdots + a_n c^n$. Notice that $a(c) \in F$ (because F is closed under addition and multiplication). So, we get a function $a: F \to F$. In other words, for each $c \in F$, $a(c) = a_0 + a_1 c + \cdots a_n c^n$. This function a is called a **polynomial function**.

Every polynomial $a(x) \in F[x]$ gives us a polynomial function $a: F \to F$. Note that under this correspondence, two different polynomials could lead to the same polynomial function. For example, let $F = \mathbb{Z}_3$ and consider the polynomials $a(x), b(x) \in \mathbb{Z}_3[x]$ defined by $a(x) = x - 2$ and $b(x) = x^3 + 1$. Clearly $a(x)$ and $b(x)$ are different polynomials. However, the corresponding polynomial functions a and b are equal. Indeed, in $\mathbb{Z}_3$, we have $a(0) = 0 - 2 = 1 = 0^3 + 1 = b(0)$, $a(1) = 1 - 2 = 2 = 1 + 1 = 1^3 + 1 = b(1)$, and $a(2) = 2 - 2 = 0 = 2 + 1 = 2^3 + 1 = b(2)$.

This anomaly of different polynomials representing the same polynomial function is unique to polynomials over finite fields. We will see in Theorem 14.29 below that if F is an infinite field and $a(x), b(x) \in F[x]$ are distinct, then the corresponding polynomial functions a and b are also distinct.

If F is a field and $a(x) \in F[x]$, then an element $c \in F$ is called a **root** (or **solution** or **zero**) of $a(x)$ over F if $a(c) = 0$.

Example 14.24:

1. Let $a(x) \in \mathbb{Q}[x]$ be defined by $a(x) = x^3 - 2x^2 - 3x$. Then
$$a(3) = 3^3 - 2(3)^2 - 3(3) = 27 - 2 \cdot 9 - 9 = 27 - 18 - 9 = 0.$$
   It follows that 3 is a root of $a(x)$ over $\mathbb{Q}$.

2. Let $b(x) \in \mathbb{R}[x]$ be defined by $b(x) = x^2 - 2$. Then $b(\sqrt{2}) = (\sqrt{2})^2 - 2 = 2 - 2 = 0$.

   So, $\sqrt{2}$ is a root of $b(x)$ over $\mathbb{R}$. Note that although $b(x) \in \mathbb{Q}[x]$, $\sqrt{2}$ is **not** a root of $b(x)$ over $\mathbb{Q}$ because $\sqrt{2} \notin \mathbb{Q}$.

3. Let $c(x) \in \mathbb{Z}_3[x]$ be defined by $c(x) = 2x^2 + x + 1$. Then $c(0) = 1$, $c(1) = 2 + 1 + 1 = 1$, and $c(2) = 2 \cdot 2^2 + 2 + 1 = 2 \cdot 1 + 0 = 2$. So, $c(x)$ has no roots over $\mathbb{Z}_3$.

Theorem 14.25 (The Factor Theorem): Let F be a field, $a(x) \in F[x]$, and $c \in F$. Then c is a root of $a(x)$ over F if and only if $x - c$ is a factor of $a(x)$.

Proof: First suppose that c is a root of $a(x)$ over F. By the Division Algorithm for Polynomials, there are polynomials $k(x), r(x) \in F[x]$ such that $a(x) = (x - c)k(x) + r(x)$ and either $r = 0$ or degree $r(x) <$ degree $(x - c)$. Since the degree of $x - c$ is 1, $r(x)$ must be a constant, say $r(x) = r$. So, $0 = a(c) = (c - c)k(c) + r = 0 \cdot k(c) + r = 0 + r = r$. Thus, $a(x) = (x - c)k(x)$. Therefore, $x - c$ is a factor of $a(x)$.

Conversely, assume that $x - c$ is a factor of $a(x)$. Then there is a polynomial $k(x) \in F[x]$ such that $a(x) = (x - c)k(x)$. So, $a(c) = (c - c)k(c) = 0 \cdot k(c) = 0$, as desired. □

Theorem 14.25 tells us that we can find all the roots of a polynomial by finding all the linear factors of the polynomial and vice versa.

Example 14.26:

1. Let $a(x) \in \mathbb{Q}[x]$ be defined by $a(x) = x^3 - 2x^2 - 3x$, as in part 1 of Example 14.24. We can factor $a(x)$ as follows:

   $$a(x) = x^3 - 2x^2 - 3x = x(x^2 - 2x - 3) = x(x-3)(x+1) = (x-0)(x-3)(x+1)$$

   So, we see that $a(x)$ has three linear factors: $x - 0$, $x - 3$, and $x + 1$. By Theorem 14.25, $a(x)$ has three roots over $\mathbb{Q}$, namely 0, 3, and -1 (note that $x + 1 = x - (-1)$).

2. Let $b(x) \in \mathbb{R}[x]$ be defined by $b(x) = x^2 - 2$, as in part 2 of Example 14.24. We can factor $b(x)$ as follows: $b(x) = (x - \sqrt{2})(x + \sqrt{2})$

   So, we see that $b(x)$ has two linear factors: $x - \sqrt{2}$ and $x + \sqrt{2}$. By Theorem 14.25, $b(x)$ has two roots over $\mathbb{R}$, namely $\sqrt{2}$ and $-\sqrt{2}$ (note that $x + \sqrt{2} = x - (-\sqrt{2})$).

3. Let $c(x) \in \mathbb{Z}_3[x]$ be defined by $c(x) = 2x^2 + x + 1$, as in part 3 of Example 14.24. We saw in that same example that $c(x)$ has no roots over $\mathbb{Z}_3$. It follows from Theorem 14.25 that $c(x)$ has no linear factors.

Theorem 14.27: Let F be a field, let $a(x) \in F[x]$, and let $c_1, c_2, \ldots, c_n \in F$ be distinct roots of $a(x)$ over F. Then $(x - c_1)(x - c_2) \cdots (x - c_n)$ is a factor of $a(x)$.

Proof: We will prove this by induction on n, where n is the number of distinct roots of $a(x)$ over F. The base case $k = 1$ follows immediately from Theorem 14.25. Assume that the conclusion of the theorem holds for k distinct roots. Let $a(x) \in F[x]$ and let $c_1, c_2, \ldots, c_{k+1} \in F$ be distinct roots of $a(x)$ over F. By the inductive hypothesis, $(x - c_1)(x - c_2) \cdots (x - c_k)$ is a factor of $a(x)$. It follows that $a(x) = (x - c_1)(x - c_2) \cdots (x - c_k)k(x)$ for some polynomial $k(x) \in F(x)$. Since c_{k+1} is distinct from $c_1, c_2, \ldots, c_k$, we have $(c_{k+1} - c_1)(c_{k+1} - c_2) \cdots (c_{k+1} - c_k) \neq 0$. Since $a(c_{k+1}) = 0$, we must have $k(c_{k+1}) = 0$. By Theorem 14.25 again, $k(x) = (x - c_{k+1})j(x)$ for some polynomial $j(x) \in F(x)$. It follows that $a(x) = (x - c_1)(x - c_2) \cdots (x - c_k)(x - c_{k+1})j(x)$, as desired. □

Theorem 14.28: Let F be a field, let $a(x) \in F[x]$, and assume that degree $a(x) = n$. Then $a(x)$ has at most n roots over F.

We will prove this theorem by contrapositive.

Proof: Assume that $a(x)$ has $n + 1$ roots over F, say $c_1, c_2, \ldots, c_{n+1}$. By Theorem 14.27, $(x - c_1)(x - c_2) \cdots (x - c_n)(x - c_{n+1})$ is a factor of $a(x)$. The leading term of this expression is x^{n+1}. Therefore, degree $a(x) \geq n + 1$. In particular, degree $a(x) \neq n$. □

Theorem 14.29: Let F be an infinite field and let $a(x), b(x)$ be distinct polynomials in $F[x]$. Then the polynomial functions $a, b: F \to F$ are not equal.

Proof: Let $a(x), b(x) \in F[x]$ and assume that $a = b$. Then for all $c \in F$, $a(c) = b(c)$. Let $p(x) \in F[x]$ be the polynomial defined by $p(x) = a(x) - b(x)$. Then for all $c \in F$, $p(c) = 0$. Since F is infinite, $p(x)$ has infinitely many roots. By Theorem 14.28, the degree of $p(x)$ is not equal to n for any $n \in \mathbb{N}$. So, $p(x)$ must be the zero polynomial. It follows that $a(x) = b(x)$. □

Problem Set 14

Full solutions to these problems are available for free download here:
www.SATPrepGet800.com/AAFBTDW

LEVEL 1

1. Prove that if $(R, +, \cdot)$ is a ring, but **not** an integral domain, then $(R[x], +, \cdot)$ is **not** an integral domain.

2. Let R be an integral domain and define a relation $\sim$ on R by $a \sim b$ if and only if a and b are associates. Prove that $\sim$ is an equivalence relation on R.

3. Prove that every field is a Euclidean domain.

LEVEL 2

4. Let F be a field and let $p(x) \in F[x]$ be an irreducible polynomial. Prove that $\langle p(x) \rangle$ is a maximal ideal of $F[x]$.

5. Let $p(x) \in \mathbb{Q}[x]$. Prove that there is $q(x) \in \mathbb{Z}[x]$ such that $q(x)$ has the same roots as $p(x)$.

6. Prove that every Euclidean domain is a PID.

LEVEL 3

7. Prove that $\mathbb{Z}[x]$ is **not** a PID.

8. Let F be a field, let $a(x), b(x) \in F[x]$, and let $I = \{c(x)a(x) + d(x)b(x) \mid c(x), d(x) \in F[x]\}$. Prove that I is an ideal of $F[x]$.

9. Let F be a field and let $a(x), b(x), c(x) \in F[x]$ with $a(x)$ and $b(x)$ relatively prime, and suppose that $a(x) | b(x)c(x)$. Prove that $a(x) | c(x)$.

10. Let F be a field, let $p(x) \in F[x]$ be irreducible, and let $a_1(x), a_2(x), \ldots, a_n(x) \in F[x]$. Suppose that $p(x) | a_1(x)a_2(x) \cdots a_n(x)$. Prove that there is an integer j with $1 \leq j \leq n$ such that $p(x) | a_j(x)$.

11. Prove that every nonconstant polynomial $a(x)$ over a field F can be factored **uniquely** as a product $kp_1(x)p_2(x) \cdots p_n(x)$, where k is a constant and for each $i = 1, 2, \ldots, n$, $p_i(x)$ is a monic irreducible polynomial, up to the order in which the factors are written.

12. Let $a(x) \in \mathbb{R}[x]$. Prove that if a complex number z is a root of $a(x)$, then the conjugate of z is also a root of $a(x)$.

LEVEL 4

13. Prove that $x^2 + 1$ is irreducible over $\mathbb{R}$.

14. Prove that the ring of Gaussian integers, $\mathbb{Z}[i] = \{a + bi \mid a, b \in \mathbb{Z}\}$, is a PID.

15. Let $a(x) = a_0 + a_1 x + \cdots + a_n x^n \in \mathbb{Z}[x]$ and let $\frac{m}{k} \in \mathbb{Q}$ with $\gcd(m, k) = 1$. Suppose that $\frac{m}{k}$ is a root of $a(x)$. Prove that $m | a_0$ and $k | a_n$.

LEVEL 5

16. Let $(R, +, \cdot)$ be a ring. Prove that $(R[x], +, \cdot)$ is a left R-algebra. Then prove that $(R[x], +, \cdot)$ is commutative if and only if R is commutative.

17. Prove that there is no rational number x such that $x^2 = 2$.

18. A commutative ring R is called a **Noetherian ring** if whenever $I_1 \subseteq I_2 \subseteq \cdots$ is a chain of ideals, then there is $n \in \mathbb{Z}^+$ such that for all $k > n$, $I_k = I_n$. Prove that a PID is a Noetherian ring.

19. Prove that every PID is a UFD.

CHALLENGE PROBLEMS

20. Let R be a UFD. Prove that $R[x]$ is a UFD. Is $R[x]$ necessarily a PID?

21. Let $a(x) = a_0 + a_1 x + \cdots + a_n x^n \in \mathbb{Z}[x]$. Suppose that there is a prime number p such that p divides each of $a_0, a_1, \ldots, a_{n-1}$, p does **not** divide a_n, and p^2 does **not** divide a_0. Prove that $a(x)$ is irreducible over $\mathbb{Q}$. (This is known as **Eisenstein's Irreducibility Theorem.**)

LESSON 15
FIELD THEORY

Field Extensions

Recall that a set F is a **subfield** of a field E, written $F \leq E$, if F is itself a field with respect to the same operations of E. Equivalently, $F \leq E$ if

1. $F^* \neq \emptyset$.
2. $F \subseteq E$.
3. F is closed under subtraction.
4. F is closed under multiplication.
5. F^* is closed under taking multiplicative inverses.

Notes: (1) $F^* = F \setminus \{0\}$ consists of all the nonzero elements of F.

(2) It is straightforward to check that the two definitions of a subfield given above are equivalent. You will be asked to do so in Problem 1 below.

If F is a subfield of E, we will also say that E is an **extension field** of F, or more simply that E is an **extension** of F (as long as it is clear that F and E are both fields).

Example 15.1:

1. $\mathbb{Q}$ is a subfield of $\mathbb{R}$. Therefore, $\mathbb{R}$ is an extension field of $\mathbb{Q}$.
2. Similarly, $\mathbb{C}$ is an extension field of $\mathbb{R}$.
3. Let $\mathbb{Q}(\sqrt{2}) = \{a + b\sqrt{2} \mid a, b \in \mathbb{Q}\}$, where addition and multiplication are defined as follows:
$$(a + b\sqrt{2}) + (c + d\sqrt{2}) = (a + c) + (b + d)\sqrt{2}$$
$$(a + b\sqrt{2})(c + d\sqrt{2}) = (ac + 2bd) + (ad + bc)\sqrt{2}$$
First, let's check that $\mathbb{Q}(\sqrt{2})$ is a subfield of $\mathbb{R}$. Since $1 = 1 + 0\sqrt{2}$, we have $1 \in \mathbb{Q}(\sqrt{2})$. Therefore, $\mathbb{Q}(\sqrt{2})^* \neq \emptyset$. Since $\mathbb{R}$ is a field containing $\mathbb{Q}$ and $\sqrt{2}$, we have $\mathbb{Q}(\sqrt{2}) \subseteq \mathbb{R}$. If $a + b\sqrt{2}, c + d\sqrt{2} \in \mathbb{Q}(\sqrt{2})$, then we have
$$(a + b\sqrt{2}) - (c + d\sqrt{2}) = (a - c) + (b - d)\sqrt{2} \in \mathbb{Q}(\sqrt{2});$$
$$(a + b\sqrt{2})(c + d\sqrt{2}) = (ac + 2bd) + (ad + bc)\sqrt{2} \in \mathbb{Q}(\sqrt{2}).$$
If $a + b\sqrt{2} \in \mathbb{Q}(\sqrt{2})^*$, then $\frac{a - b\sqrt{2}}{a^2 - 2b^2} = \frac{a}{a^2 - 2b^2} - \frac{b}{a^2 - 2b^2}\sqrt{2} \in \mathbb{Q}(\sqrt{2})^*$ (why is $a^2 - 2b^2 \neq 0$?) and we have
$$(a + b\sqrt{2})\left(\frac{a - b\sqrt{2}}{a^2 - 2b^2}\right) = \frac{a^2 - 2b^2}{a^2 - 2b^2} = 1.$$

So, $(a+b\sqrt{2})^{-1}$ exists and is equal to $\frac{a}{a^2-2b^2} - \frac{b}{a^2-2b^2}\sqrt{2}$.

It follows that $\mathbb{Q}(\sqrt{2}) \leq \mathbb{R}$, or equivalently, $\mathbb{R}$ is an extension field of $\mathbb{Q}(\sqrt{2})$.

Note that if $a = a + 0\sqrt{2} \in \mathbb{Q}$ and $b = b + 0\sqrt{2} \in \mathbb{Q}$, then

$$(a + 0\sqrt{2}) - (b + 0\sqrt{2}) = a - b \in \mathbb{Q};$$
$$(a + 0\sqrt{2})(b + 0\sqrt{2}) = (ab + 2 \cdot 0) + (a \cdot 0 + 0 \cdot b)\sqrt{2} = ab \in \mathbb{Q}.$$

If $a = a + 0\sqrt{2} \in \mathbb{Q}^*$, then $a^{-1} = \frac{a - 0\sqrt{2}}{a^2 - 2 \cdot 0^2} = \frac{a}{a^2} = \frac{1}{a} \in \mathbb{Q}$.

So, $\mathbb{Q} \leq \mathbb{Q}(\sqrt{2})$, or equivalently, $\mathbb{Q}(\sqrt{2})$ is an extension field of $\mathbb{Q}$.

Let F and E be fields with $F \leq E$ and let $a \in E$. We define a function $f_a: F[x] \to E$ by $f_a(p(x)) = p(a)$. We will call f_a the **evaluation function at** a. The idea is very simple. We are simply plugging the element $a \in E$ in for x in each polynomial in $F[x]$. Notice that a is an element of the "bigger" field E, but it is only being substituted into polynomials with coefficients in the "smaller" field F.

Example 15.2: Consider the field extension $\mathbb{Q} \leq \mathbb{Q}(\sqrt{2})$. Let $p(x) = x^2 - 2$. Then we have

$$f_{\sqrt{2}}(p(x)) = (\sqrt{2})^2 - 2 = 2 - 2 = 0.$$

Each evaluation function $f_a: F[x] \to E$ is a ring homomorphism. Indeed, if $p(x), q(x) \in F[x]$, then

$$f_a(p(x) + q(x)) = f_a((p+q)(x)) = (p+q)(a) = p(a) + q(a) = f_a(p(x)) + f_a(q(x)),$$
$$f_a(p(x)q(x)) = f_a((pq)(x)) = (pq)(a) = p(a)q(a) = f_a(p(x))f_a(q(x)).$$

Also, $f_a(1) = 1$.

Let $K_a = \ker(f_a)$. Then $K_a = \{p(x) \in F[x] \mid p(a) = 0\}$. In words, the kernel of f_a is the set of all polynomials over F that have a as a root. Note that by part 2 of Example 7.22, K_a is an ideal in $F[x]$.

The computation in Example 15.2 shows us that $x^2 - 2 \in K_{\sqrt{2}}$.

Let F and E be fields with $F \leq E$. An element $a \in E$ is said to be **algebraic over** F if a is a root of some nonzero polynomial in $F[x]$. Otherwise, a is said to be **transcendental over** F. If every $a \in E$ is algebraic over F, then E is said to be an **algebraic extension** of F. If there is an $a \in E$ that is transcendental over F, then E is said to be a **transcendental extension** of F.

Note: $a \in E$ is algebraic over F if and only if $K_a \neq \{0\}$, whereas $a \in E$ is transcendental over F if and only if $K_a = \{0\}$.

Example 15.3:

1. Consider the field extension $\mathbb{Q} \leq \mathbb{Q}(\sqrt{2})$. Then $\sqrt{2} \in \mathbb{Q}(\sqrt{2})$ is algebraic over $\mathbb{Q}$. After all, in Example 15.2, we saw that $\sqrt{2}$ is a root of $x^2 - 2 \in \mathbb{Q}[x]$. The same polynomial shows that $-\sqrt{2}$ is also algebraic over $\mathbb{Q}$. Indeed, $(-\sqrt{2})^2 - 2 = 2 - 2 = 0$.

In fact, $\mathbb{Q}(\sqrt{2})$ is an algebraic extension of $\mathbb{Q}$. In other words, **every** element of $\mathbb{Q}(\sqrt{2})$ is a root of some polynomial in $\mathbb{Q}[x]$. To see this, let $a + b\sqrt{2} \in \mathbb{Q}(\sqrt{2})$. We see that

$$((a+b\sqrt{2})^2 - a^2 - 2b^2)^2 - 8a^2b^2 = (a^2 + 2ab\sqrt{2} + 2b^2 - a^2 - 2b^2)^2 - 8a^2b^2$$
$$= (2ab\sqrt{2})^2 - 8a^2b^2 = 8a^2b^2 - 8a^2b^2 = 0.$$

It follows that $a + b\sqrt{2}$ is a root of the polynomial $(x^2 - a^2 - 2b^2)^2 - 8a^2b^2 \in \mathbb{Q}[x]$.

2. Consider the field extension $\mathbb{Q} \leq \mathbb{R}$. As in part 1 above, $\sqrt{2} \in \mathbb{R}$ is algebraic over $\mathbb{Q}$ (since $\mathbb{Q}(\sqrt{2}) \leq \mathbb{R}$, the computation is the same, as in part 1).

   However, $\mathbb{R}$ is a transcendental extension of $\mathbb{Q}$. For example, $\pi \in \mathbb{R}$ is transcendental over $\mathbb{Q}$. The proof of this is quite difficult, and so, I leave it as a challenging problem for the reader to explore (see Problem 18 below).

3. Consider the field extension $\mathbb{R} \leq \mathbb{C}$. Then $i \in \mathbb{C}$ is algebraic over $\mathbb{R}$. Indeed, i is a root of the polynomial $x^2 + 1 \in \mathbb{R}[x]$. The same polynomial shows that $-i$ is also algebraic over $\mathbb{Q}$.

   $\mathbb{C}$ is an algebraic extension of $\mathbb{R}$. In other words, **every** element of $\mathbb{C}$ is a root of some polynomial in $\mathbb{R}[x]$. To see this, let $a + bi \in \mathbb{C}$. We see that

$$((a+bi)^2 - a^2 + b^2)^2 + 4a^2b^2 = (a^2 + 2abi - b^2 - a^2 + b^2)^2 + 4a^2b^2$$
$$= (2abi)^2 + 4a^2b^2 = -4a^2b^2 + 4a^2b^2 = 0.$$

It follows that $a + bi$ is a root of the polynomial $(x^2 - a^2 + b^2)^2 + 4a^2b^2 \in \mathbb{R}[x]$.

Note that $\mathbb{C}$ is a transcendental extension of $\mathbb{Q}$, simply because $\pi \in \mathbb{C}$.

Algebraic Elements

Let F and E be fields with $F \leq E$. Suppose that $a \in E$ is algebraic over F. Then there is a nonzero polynomial $p(x) \in F[x]$ such that $p(x)$ is in the ideal K_a. In this case, there will be infinitely many nonzero polynomials in K_a. Indeed, if $q(x)$ is any polynomial, then the product $q(x)p(x)$ is a polynomial and $q(a)p(a) = q(a) \cdot 0 = 0$, and so, $q(x)p(x) \in K_a$. For example, in Example 15.2, we saw that $\sqrt{2} \in \mathbb{Q}(\sqrt{2})$ is a root of the polynomial $x^2 - 2 \in \mathbb{Q}[x]$, showing that $\sqrt{2}$ is algebraic over $\mathbb{Q}$. We also have $x^3 + x^2 - 2x - 2 \in \mathbb{Q}[x]$ and $(\sqrt{2})^3 + (\sqrt{2})^2 - 2\sqrt{2} - 2 = 2\sqrt{2} + 2 - 2\sqrt{2} - 2 = 0$. So, $x^3 + x^2 - 2x - 2 \in K_{\sqrt{2}}$ and the polynomial $x^3 + x^2 - 2x - 2 \in \mathbb{Q}[x]$ can also be used to show that $\sqrt{2}$ is algebraic over $\mathbb{Q}$. However, there are two things that are "special" about the polynomial $x^2 - 2$. The first is that it is a polynomial of the lowest possible degree in $\mathbb{Q}[x]$ that has $\sqrt{2}$ as a root. The second is that it is a monic polynomial (the leading coefficient is 1). It turns out that every other polynomial in $K_{\sqrt{2}}$ is a multiple of this "special" one. For example, we have

$$x^3 + x^2 - 2x - 2 = x^2(x+1) - 2(x+1) = (x+1)(x^2 - 2).$$

Let $a \in E$ be algebraic over F. Since the kernel K_a is an ideal of $F[x]$ (by part 2 of Example 7.22) and $F[x]$ is a principal ideal domain (by part 3 of Example 14.14), it follows that there is a polynomial $p(x) \in F[x]$ such that K_a is generated by $p(x)$. Since $\langle p(x) \rangle$ contains all multiples of $p(x)$, by multiplying by the reciprocal of the leading coefficient of $p(x)$ if necessary, we can assume that $p(x)$ is a monic polynomial. This polynomial $p(x)$ is the unique irreducible monic polynomial of lowest degree such that a is a root of $p(x)$, as we now show.

Theorem 15.4: Let F and E be fields with $F \leq E$, let $a \in E$ be algebraic over F, let $f_a: F[x] \to E$ be the evaluation function at a, and let $K_a = \ker(f_a)$. Then there is a unique irreducible monic polynomial $p(x) \in F[x]$ such that $K_a = \langle p(x) \rangle$. Furthermore, if $q(x) \in K_a$, then degree $q(x) \geq$ degree $p(x)$.

Proof: Since $a \in E$ is algebraic over F and $F[x]$ is a principal ideal domain, there is a nonzero polynomial $r(x) \in F[x]$ such that $K_a = \langle r(x) \rangle$. If $q(x) \in K_a$, then $r(x) | q(x)$, and therefore, degree $q(x) \geq$ degree $r(x)$. If $r(x) = s(x)t(x)$, then since $r(x) \in K_a$, $s(a)t(a) = r(a) = 0$. Since E is a field (and therefore an integral domain), $s(a) = 0$ or $t(a) = 0$. So, $s(x) \in K_a$ or $t(x) \in K_a$. Without loss of generality, assume that $t(x) \in K_a$. Then $r(x) | t(x)$. Since $t(x) | s(x)t(x)$, we have $t(x) | r(x)$. By Theorem 14.17, $r(x)$ and $t(x)$ are associates. So, there is a nonzero constant $c \in F$ such that $r(x) = ct(x)$. Therefore, $s(x)t(x) = ct(x)$. Since $F[x]$ is an integral domain, $s(x) = c$. It follows that $r(x)$ is irreducible. Let b be the leading coefficient of $r(x)$. Then $b \neq 0$, and so, we can let $p(x) = \frac{1}{b} r(x)$. Then $p(x)$ is monic, has the same degree as $r(x)$, and is irreducible. Furthermore, $\langle p(x) \rangle = \langle r(x) \rangle$, proving the existence of an irreducible monic polynomial $p(x)$ such that $K_a = \langle p(x) \rangle$.

To see that $p(x)$ is unique, suppose that $q(x)$ is another irreducible monic polynomial such that $K_a = \langle q(x) \rangle$. Then $\langle p(x) \rangle = \langle q(x) \rangle$. So, $p(x) | q(x)$ and $q(x) | p(x)$. By Theorem 14.17, $p(x)$ and $q(x)$ are associates. So, there is a nonzero constant $c \in F$ such that $p(x) = cq(x)$. Since $p(x)$ and $q(x)$ both have leading coefficient 1, $c = 1$, and so, $p(x) = q(x)$. □

The **minimal polynomial of a over F** is the unique irreducible monic polynomial $p(x) \in F[x]$ such that $K_a = \langle p(x) \rangle$. The existence and uniqueness of this polynomial is guaranteed by Theorem 15.4.

Example 15.5:

1. Consider the field extension $\mathbb{Q} \leq \mathbb{Q}(\sqrt{2})$. The minimal polynomial of $\sqrt{2}$ over $\mathbb{Q}$ is $x^2 - 2$. To see that $x^2 - 2 \in K_{\sqrt{2}}$, simply observe that $(\sqrt{2})^2 - 2 = 2 - 2 = 0$. By part 3 of Example 14.9, $x^2 - 2$ does not have any divisors of degree 1 in $\mathbb{Q}[x]$. It follows that $K_{\sqrt{2}} = \langle x^2 - 2 \rangle$. Since $x^2 - 2$ is a monic polynomial, it is indeed the minimal polynomial of $\sqrt{2}$ over $\mathbb{Q}$.

2. Consider the field extension $\mathbb{R} \leq \mathbb{C}$. The minimal polynomial of i over $\mathbb{R}$ is $x^2 + 1$. To see that $x^2 + 1 \in K_i$, simply observe that $i^2 + 1 = -1 + 1 = 0$. It is straightforward to check that $x^2 + 1$ does not have any divisors of degree 1 in $\mathbb{R}[x]$. It follows that $K_i = \langle x^2 + 1 \rangle$. Since $x^2 + 1$ is a monic polynomial, it is indeed the minimal polynomial of i over $\mathbb{R}$.

In Lesson 6, we defined what it means for a substructure to be generated by a set. Let's now repeat that definition, specifically for fields.

Let E be a field and let $S \subseteq E$. The subfield of E **generated** by S, written $\langle S \rangle$, is the intersection of all subfields of E containing S. In Problem 11 in Problem Set 6, you were asked to prove that an arbitrary intersection of subfields of a field E is also a subfield of E. It follows that $\langle S \rangle$ is the "smallest" subfield of E containing S. Stated more formally, $\langle S \rangle$ is a subfield of E such that if K is any subfield of E containing S, then $\langle S \rangle \subseteq K$ (Check this!).

The dedicated reader should also try to give an "explicit definition" of $\langle S \rangle$, as we did for other types of structures in Lesson 6 (this is part (i) of Problem 22 in Problem Set 6).

If $F \leq E$ and $a \in E$, we will write $F(a)$ in place of $\langle F \cup \{a\}\rangle$. We will sometimes say that we are **adjoining** a to F, and we will call $F(a)$ a **simple extension** of F.

Theorem 15.6: Let F and E be fields with $F \leq E$ and let $a \in E$. Then $F(a) = f_a[F[x]]$.

Since the equation $F(a) = f_a[F[x]]$ might seem confusing, let's briefly review all the notation that appears there.

- $F[x]$ is the ring of polynomials with coefficients in F. Since F is a field, by Theorems 14.5 and 14.8, $F[x]$ is an integral domain.

- $f_a: F[x] \to E$ is the evaluation function at a. We input a polynomial $p(x)$ into this function and get the output $p(a)$. For example, $f_{\sqrt{2}}(x^4 + 1) = (\sqrt{2})^4 + 1 = 4 + 1 = 5$.

- $f_a[F[x]]$ is the range of the function f_a. It consists of all expressions of the form $p(a)$, where $p(x)$ is an arbitrary polynomial in $F[x]$. If $p(x) = b_0 + b_1 x + \cdots + b_{k-1} x^{k-1} + b_k x^k$, then $p(a) = b_0 + b_1 a + \cdots + b_{k-1} a^{k-1} + b_k a^k$. So, $f_a[F[x]]$ consists of all expressions of the form $b_0 + b_1 a + \cdots + b_{k-1} a^{k-1} + b_k a^k$, where $b_0, b_1, \ldots, b_k \in F$.

- $F(a)$ is the subfield of E generated by F and a. This means that if K is any subfield of E that contains $F \cup \{a\}$, then $F(a) \subseteq K$.

We will prove Theorem 15.6 in three steps as follows:

(1) We will first prove that $f_a[F[x]]$ is a subfield of E. It's almost immediate that $f_a[F[x]]$ is a ring (because f_a is a ring homomorphism). The hard part is showing that every nonzero element of $f_a[F[x]]$ has a multiplicative inverse.

(2) Next, we will show that $f_a[F[x]] \subseteq F(a)$. This is pretty easy to see, as any element $p(a)$ in $f_a[F[x]]$ involves only sums and products of a with itself and elements of F.

(3) Finally, we show that $F(a) \subseteq f_a[F[x]]$. Since we already know that $f_a[F[x]]$ is a field (by (1) above), it suffices to show that $f_a[F[x]]$ contains F and contains a (by the last bullet point above).

Proof of Theorem 15.6: Let $p(x)$ be the minimal polynomial of a over F. So, $K_a = \ker(f_a) = \langle p(x)\rangle$.

We first show that $f_a[F[x]]$ is a subfield of E. Since $f_a: F[x] \to E$ is a ring homomorphism, by Theorem 7.18, $f_a[F[x]]$ is a subring of E. So, we need only show that $(f_a[F[x]])^*$ is closed under multiplicative inverses. To see this, let $q(a) \in (f_a[F[x]])^*$. Since $q(a) \neq 0$, $q(x) \notin K_a = \langle p(x)\rangle$. So, $q(x)$ is not a constant multiple of $p(x)$. Since $p(x)$ is irreducible, $\gcd(p(x), q(x)) = 1$. By Theorem 14.20, there are polynomials $s(x)$ and $t(x)$ over F such that $s(x)p(x) + t(x)q(x) = 1$. Since $p(x) \in K_a$, $p(a) = 0$. Therefore, we have $1 = s(a)p(a) + t(a)q(a) = s(a) \cdot 0 + t(a)q(a) = t(a)q(a)$. This shows that $q(a)$ is invertible with inverse $t(a)$. Therefore, $f_a[F[x]]$ is a subfield of E.

Now, let $p(a) \in f_a[F[x]]$, say $p(a) = b_0 + b_1 a + \cdots + b_{k-1} a^{k-1} + b_k a^k$, where $b_0, b_1, \ldots, b_k \in F$. Since $F(a)$ is a field containing F and a, it must contain $b_0 + b_1 a + \cdots + b_{k-1} a^{k-1} + b_k a^k = p(a)$. Therefore, $f_a[F[x]] \subseteq F(a)$.

To prove the reverse inclusion, first let $b \in F$. Then $b \in F[x]$ and $f_a(b) = b$. So, $b \in f_a[F[x]]$. Also, $x \in F[x]$ and $f_a(x) = a$. So, $a \in f_a[F[x]]$. Therefore, $f_a[F[x]]$ is a field containing F and a. Since $F(a)$ is the subfield of E generated by $F \cup \{a\}$, $F(a) \subseteq f_a[F[x]]$.

Since $f_a[F[x]] \subseteq F(a)$ and $F(a) \subseteq f_a[F[x]]$, we have $F(a) = f_a[F[x]]$. □

Let F be a field and let $p(x) \in F[x]$ be an irreducible polynomial. By Problem 4 in Problem Set 14, $\langle p(x) \rangle$ is a maximal ideal of $F[x]$. So, by Theorem 10.24, $F[x]/\langle p(x) \rangle$ is a field. Define the function $f: F \to F[x]/\langle p(x) \rangle$ by $F(a) = a + \langle p(x) \rangle$. The following computations show that f is a field homomorphism:

$$f(a+b) = (a+b) + \langle p(x) \rangle = (a + \langle p(x) \rangle) + (b + \langle p(x) \rangle) = f(a) + f(b)$$
$$f(a \cdot b) = ab + \langle p(x) \rangle = (a + \langle p(x) \rangle)(b + \langle p(x) \rangle) = f(a) \cdot f(b)$$
$$f(1) = 1 + \langle p(x) \rangle$$

Furthermore, by Theorem 7.23, f is injective.

Now, $f[F] = \{a + \langle p(x) \rangle \mid a \in F\}$ and $F \cong f[F]$. So, we can think of the field F as a subset of $F[x]/\langle p(x) \rangle$ by identifying each $a \in F$ with the coset $a + \langle p(x) \rangle \in F[x]/\langle p(x) \rangle$. Furthermore, by the Note following Theorem 7.18, $f[F]$ is a subfield of $F[x]/\langle p(x) \rangle$. In this way, we can think of F as a subfield of $F[x]/\langle p(x) \rangle$. Equivalently, we see that given any irreducible polynomial $p(x)$ in $F[x]$, we can think of $F[x]/\langle p(x) \rangle$ as an extension field of F (once again, by identifying F with its image $f[F]$).

Note: If $a \in F$, then $a \in F[x]$. However, we think of a slightly differently in each of these sets. In F, we think of a as an element of the field F, whereas in $F[x]$, we think of a as a constant polynomial. Finally, in $F[x]/\langle p(x) \rangle$, $a + \langle p(x) \rangle$ is the coset of the constant polynomial a.

Let $p(x) \in F[x]$, say $p(x) = a_0 + a_1 x + \cdots + a_{n-1} x^{n-1} + a_n x^n$, where $a_0, a_1, \ldots, a_{n-1}, a_n \in F$. Recall that if we are thinking of F as a subfield of $F[x]/\langle p(x) \rangle$, then the coefficients $a_0, a_1, \ldots, a_n$ are actually the cosets $a_0 + \langle p(x) \rangle$, $a_1 + \langle p(x) \rangle, \ldots, a_n + \langle p(x) \rangle$ (specifically, these cosets are the images of $a_0, a_1, \ldots, a_n$ under the homomorphism f). So, as a polynomial over $F[x]/\langle p(x) \rangle$ (or equivalently, as an element of $(F[x]/\langle p(x) \rangle)[x]$, the polynomial $p(x)$ looks as follows:

$$(a_0 + \langle p(x) \rangle) + (a_1 + \langle p(x) \rangle)x + \cdots + (a_{n-1} + \langle p(x) \rangle)x^{n-1} + (a_n + \langle p(x) \rangle)x^n$$

Next, observe that since $x \in F[x]$, we have $x + \langle p(x) \rangle \in F[x]/\langle p(x) \rangle$.

In fact, we now show that $x + \langle p(x) \rangle$ is a root of the polynomial $p(x)$. Indeed, we have

$$p(x + \langle p(x) \rangle) = (a_0 + \langle p(x) \rangle) + (a_1 + \langle p(x) \rangle)(x + \langle p(x) \rangle)$$
$$+ \cdots + (a_{n-1} + \langle p(x) \rangle)(x + \langle p(x) \rangle)^{n-1} + (a_n + \langle p(x) \rangle)(x + \langle p(x) \rangle)^n$$
$$= (a_0 + \langle p(x) \rangle) + (a_1 x + \langle p(x) \rangle) + \cdots + (a_{n-1} x^{n-1} + \langle p(x) \rangle) + (a_n x^n + \langle p(x) \rangle)$$
$$= (a_0 + a_1 x + \cdots a_{n-1} x^{n-1} + a_n x^n) + \langle p(x) \rangle$$
$$= p(x) + \langle p(x) \rangle = \langle p(x) \rangle.$$

The first equality above is just a direct substitution of $x + \langle p(x) \rangle$ for x into the polynomial. The second equality follows from the definition of multiplication of cosets. The third equality follows from the definition of addition of cosets. The fourth equality is true because $p(x) \in \langle p(x) \rangle$.

We have just proved the following:

Theorem 15.7: Let F be a field, let $p(x) \in F[x]$ be an irreducible polynomial, let $E = F[x]/\langle p(x) \rangle$, and define $f: F \to E$ by $F(a) = a + \langle p(x) \rangle$. Then F is isomorphic to $f[F]$. By identifying F with its image in E, we see that E is a field extension of F such that $p(x)$ has a root in E.

Let F be a field and let $a(x) \in F[x]$ be an arbitrary nonconstant polynomial. By Theorem 14.23, $a(x)$ can be factored as a product $kp_1(x)p_2(x) \cdots p_n(x)$, where k is a constant and for each $i = 1, 2, \ldots, n$, $p_i(x)$ is an irreducible polynomial. Applying Theorem 15.7 to the irreducible polynomial $p_1(x)$, we get a field extension E of F such that $p_1(x)$ has a root c in E. Then $a(c) = kp_1(c)p_2(c) \cdots p_n(c) = 0$ (because $p_1(c) = 0$). So, c is also a root of $a(x)$. In this way, we see that we can eliminate the condition that $p(x)$ is irreducible in Theorem 15.7 to get the following Fundamental Theorem of Field Extensions.

Theorem 15.8 (Fundamental Theorem of Field Extensions): Let F be a field and let $a(x) \in F[x]$ be a nonconstant polynomial. Then there is a field extension E of F such that $a(x)$ has a root in E.

Example 15.9: Let $F = \mathbb{R}$, let $a(x) = x^2 + 1$, and let $E = F[x]/\langle a(x) \rangle = \mathbb{R}[x]/\langle x^2 + 1 \rangle$. We can think of $\mathbb{R}$ as "living inside" E by identifying each real number r with the coset $r + \langle x^2 + 1 \rangle$. A general element of E has the form $(a + bx) + \langle x^2 + 1 \rangle$. To see this, observe that

$$(a + bx + cx^2) + \langle x^2 + 1 \rangle = \big((a - c) + bx\big) + c(x^2 + 1) + \langle x^2 + 1 \rangle = \big((a - c) + bx\big) + \langle x^2 + 1 \rangle.$$

Note how $c(x^2 + 1)$ gets "absorbed" by the ideal $\langle x^2 + 1 \rangle$. Powers of x higher than 2 get absorbed in a similar manner (Check this!).

Finally, let's check that $\mathbb{R}[x]/\langle x^2 + 1 \rangle$ is isomorphic to $\mathbb{C}$. Define the function $f: \mathbb{C} \to \mathbb{R}[x]/\langle x^2 + 1 \rangle$ by $f(a + bi) = (a + bx) + \langle x^2 + 1 \rangle$. The following computations show that f is a homomorphism.

$$f\big((a + bi) + (c + di)\big) = f\big((a + c) + (b + d)i\big) = \big((a + c) + (b + d)x\big) + \langle x^2 + 1 \rangle$$
$$= \big((a + bx) + (c + dx)\big) + \langle x^2 + 1 \rangle = \big((a + bx) + \langle x^2 + 1 \rangle\big) + \big((c + dx) + \langle x^2 + 1 \rangle\big)$$
$$= f(a + bi) + f(c + di)$$
$$f\big((a + bi)(c + di)\big) = f\big((ac - bd) + (ad + bc)i\big) = \big((ac - bd) + (ad + bc)x\big) + \langle x^2 + 1 \rangle$$
$$= \big((ac - bd) + (ad + bc)x\big) + bd(x^2 + 1) + \langle x^2 + 1 \rangle$$
$$= (ac + (ad + bc)x + bdx^2) + \langle x^2 + 1 \rangle = (a + bx)(c + dx) + \langle x^2 + 1 \rangle$$
$$= \big((a + bx) + \langle x^2 + 1 \rangle\big)\big((c + dx) + \langle x^2 + 1 \rangle\big) = f(a + bi)f(c + di)$$
$$f(1) = f(1 + 0i) = (1 + 0x) + \langle x^2 + 1 \rangle = 1 + \langle x^2 + 1 \rangle.$$

Now, suppose that $f(a + bi) = f(c + di)$. Then $(a + bx) + \langle x^2 + 1 \rangle = (c + dx) + \langle x^2 + 1 \rangle$, and so, $(a - c) + (b - d)x = (a + bx) - (c + dx) \in \langle x^2 + 1 \rangle$. Since the degree of $(a - c) + (b - d)x$ is less than 2, we must have $(a - c) + (b - d)x = 0$. So, $a - c = 0$ and $b - d = 0$. Therefore, $a = c$ and $b = d$, and so, it follows that $a + bi = c + di$. This shows that f is injective.

Since surjectivity is clear, we see that f is an isomorphism.

Field Extensions as Vector Spaces

Let F and E be fields with $F \leq E$. Then E is a vector space over F, where vector addition is simply the ordinary field addition in E and scalar multiplication is ordinary field multiplication in E with the exception that the scalars are restricted to elements of F. For example, $\mathbb{C}$ is a field extension of $\mathbb{R}$ and so, $\mathbb{C}$ is a vector space over $\mathbb{R}$, where we define addition and scalar multiplication as follows: if $a + bi, c + di \in \mathbb{C}$ and $k \in \mathbb{R}$, then

$$(a + bi) + (c + di) = (a + c) + (b + d)i \quad \text{and} \quad k(a + bi) = ka + kbi.$$

See Example 2.17 for the complete verification that with these definitions, $\mathbb{C}$ is a vector space over $\mathbb{R}$.

The general proof that E is a vector space over F for arbitrary fields F and E with $F \leq E$ is nearly identical to the verification that $\mathbb{C}$ is a vector space over $\mathbb{R}$. The dedicated reader should write out the details.

If the dimension of E over F is a positive integer n (in other words, there exists a basis consisting of n elements in F), then we write $[E:F] = n$, and we say that the **degree of E over F is n** or that **E is an extension of degree n over F**. We can also say that **E is a finite extension of F**. If the dimension of E over F is infinite, then we say that **E is an infinite extension of F** or the **degree of E over F is infinite**.

Note: By Problem 20 in Problem Set 6, every vector has a basis. It follows that for fields F and E with $F \leq E$, the degree of E over F is always defined. However, this degree need not be finite.

Example 15.10:

1. $[\mathbb{C}:\mathbb{R}] = 2$. To see this, just observe that $\mathcal{B} = \{1, i\}$ is a basis of $\mathbb{C}$ (as a vector space over $\mathbb{R}$). To see that 1 and i are linearly independent just observe that i is **not** a scalar multiple of 1 (there is no **real number** k such that $i = k \cdot 1$). It is clear from the definition of $\mathbb{C}$ that $\text{span}\{1, i\} = \mathbb{C}$ (after all, $\mathbb{C} = \{a + bi \mid a, b \in \mathbb{R}\}$). So, $\mathbb{C}$ is a finite extension of $\mathbb{R}$. Specifically, the degree of $\mathbb{C}$ over $\mathbb{R}$ is 2.

2. $\mathbb{R}$ is an infinite extension of $\mathbb{Q}$. To see this, assume toward contradiction that $[\mathbb{R}:\mathbb{Q}] = n$ and let $\{a_1, a_2, \ldots, a_n\}$ be a basis of $\mathbb{R}$ (as a vector space over $\mathbb{Q}$). Then $\mathbb{R} = \text{span}\{a_1, a_2, \ldots, a_n\}$. Define a function $f: \mathbb{R} \to {}^n\mathbb{Q}$ by $f(k_1 a_1 + k_2 a_2 + \cdots k_n a_n) = (k_1, k_2, \ldots, k_n)$. By Problem 17 in Problem Set 6, f is injective. Since ${}^n\mathbb{Q}$ is a countable set (Check this!), it follows that $\mathbb{R}$ is countable. This contradiction tells us that $[\mathbb{R}:\mathbb{Q}] \neq n$ for any $n \in \mathbb{N}$, and so, $\mathbb{R}$ is an infinite extension of $\mathbb{Q}$.

3. Let F and E be fields with $F \leq E$, let $a \in E$ be algebraic over F, and let $p(x)$ be the minimal polynomial of a over F. Then $[F(a):F] = \text{degree } p(x)$. To see this, suppose that degree $p(x) = n$. We will show that $\{1, a, a^2, \ldots, a^{n-1}\}$ is a basis of $F(a)$ over F.

   To see that $F(a) = \text{span}\{1, a, a^2, \ldots, a^{n-1}\}$, let $b \in F(a)$. By Theorem 15.6, $b \in f_a[F[x]]$, and so, $b = q(a)$ for some polynomial $q(x) \in F[x]$. By Theorem 14.13, there are polynomials $k(x), r(x) \in F[x]$ such that $q(x) = p(x)k(x) + r(x)$ and $r = 0$ or degree $r(x) < n$. Then we have $q(a) = p(a)k(a) + r(a) = 0 \cdot k(a) + r(a) = r(a) \in \text{span}\{1, a, a^2, \ldots, a^{n-1}\}$.

To see that $\{1, a, a^2, \ldots, a^{n-1}\}$ is linearly independent, suppose there are $c_0, c_1, \ldots, c_{n-1} \in F$ such that $c_0 + c_1 a + c_2 a^2 + \cdots + c_{n-1} a^{n-1} = 0$. If $c_0, c_1, \ldots, c_{n-1}$ were not all 0, then a would be the root of a polynomial over F of degree less than n, contradicting that the minimal polynomial of a over F has degree n. So, $c_0 = c_1 = \cdots = c_{n-1} = 0$.

Note that part 1 above is a specific example of this more general situation. Indeed, we have $\mathbb{C} = \mathbb{R}(i)$ and the minimal polynomial of i over $\mathbb{R}$ is $x^2 + 1$, which has degree 2. It follows that $[\mathbb{C}:\mathbb{R}] = [\mathbb{R}(i):\mathbb{R}] = 2$.

As another example, the minimal polynomial of $\sqrt{2}$ over $\mathbb{Q}$ is $x^2 - 2$, and so, $[\mathbb{Q}(\sqrt{2}):\mathbb{Q}] = 2$.

Theorem 15.11: Let F, E, and K be fields with $F \leq E \leq K$. Suppose that E is a finite extension of F and K is a finite extension of E. Then $[K:F] = [K:E][E:F]$.

Proof: Let $\{a_1, a_2, \ldots, a_n\}$ be a basis of E over F and let $\{b_1, b_2, \ldots, b_m\}$ be a basis of K over E. We will show that $\{a_i b_j \mid 1 \leq i \leq n, 1 \leq j \leq m\}$ is a basis of K over F.

To see that $\text{span}\{a_i b_j \mid 1 \leq i \leq n, 1 \leq j \leq m\} = K$, let $x \in K$. Since $\{b_1, b_2, \ldots, b_m\}$ is a basis of K over E, there are $c_1, c_2, \ldots, c_m \in E$ such that $x = c_1 b_1 + c_2 b_2 + \cdots + c_m b_m$. Since $\{a_1, a_2, \ldots, a_n\}$ is a basis of E over F, for each $i = 1, 2, \ldots, m$, there are $d_{i1}, d_{i2}, \ldots, d_{in} \in F$ such that $c_i = d_{i1} a_1 + \cdots + d_{in} a_n$. So,

$$x = c_1 b_1 + c_2 b_2 + \cdots + c_m b_m$$
$$= (d_{11} a_1 + \cdots + d_{1n} a_n) b_1 + (d_{21} a_1 + \cdots + d_{2n} a_n) b_2 + \cdots + (d_{m1} a_1 + \cdots + d_{mn} a_n) b_m$$

$$= d_{11}(a_1 b_1) + d_{12}(a_2 b_1) + \cdots + d_{1n}(a_n b_1)$$
$$+ d_{21}(a_1 b_2) + d_{22}(a_2 b_2) + \cdots + d_{2n}(a_n b_2)$$
$$\vdots \qquad \vdots \qquad \vdots$$
$$+ d_{m1}(a_1 b_m) + d_{m2}(a_2 b_m) + \cdots + d_{mn}(a_n b_m)$$

Therefore, $x \in \text{span}\{a_i b_j \mid 1 \leq i \leq n, 1 \leq j \leq m\}$, and so, $\text{span}\{a_i b_j \mid 1 \leq i \leq n, 1 \leq j \leq m\} = K$.

To see that the elements of $\{a_i b_j \mid 1 \leq i \leq n, 1 \leq j \leq m\}$ are linearly independent, suppose that

$$d_{11}(a_1 b_1) + \cdots d_{1n}(a_n b_1) + d_{21}(a_1 b_2) + \cdots d_{2n}(a_n b_2) + \cdots + d_{m1}(a_1 b_m) + \cdots d_{mn}(a_n b_m) = 0.$$

Then

$$(d_{11} a_1 + \cdots + d_{1n} a_n) b_1 + (d_{21} a_1 + \cdots + d_{2n} a_n) b_2 + \cdots + (d_{m1} a_1 + \cdots + d_{mn} a_n) b_m = 0.$$

Since $b_1, b_2, \ldots, b_m$ is linearly independent, we have

$$d_{11} a_1 + \cdots + d_{1n} a_n = 0, \; d_{21} a_1 + \cdots + d_{2n} a_n = 0, \ldots, d_{m1} a_1 + \cdots + d_{mn} a_n = 0.$$

Since $a_1, a_2, \ldots, a_n$ is linearly independent, we have

$$d_{11} = \cdots = d_{1n} = d_{21} = \cdots = d_{2n} = d_{m1} = \cdots = d_{mn} = 0.$$

So, the elements of $\{a_i b_j \mid 1 \leq i \leq n, 1 \leq j \leq m\}$ are linearly independent.

Since $|\{a_i b_j \mid 1 \le i \le n, 1 \le j \le m\}| = mn$, we have $[K:F] = mn = [K:E][E:F]$. $\square$

Example 15.12:

1. We saw in part 3 of Example 15.10 that $[\mathbb{Q}(\sqrt{2}):\mathbb{Q}] = 2$. Once again, this follows from the fact that the minimal polynomial of $\sqrt{2}$ over $\mathbb{Q}$ is the degree 2 polynomial $x^2 - 2$.

   Recall also that $\mathbb{Q}(\sqrt{2}) = \{a + b\sqrt{2} \mid a, b \in \mathbb{Q}\}$. It follows that $\sqrt{3} \notin \mathbb{Q}(\sqrt{2})$. To see this, assume toward contradiction that there are $a, b \in \mathbb{Q}$ with $\sqrt{3} = a + b\sqrt{2}$. Clearly a and b must be nonzero. Then $3 = (a + b\sqrt{2})^2 = a^2 + 2b^2 + 2ab\sqrt{2}$. So, $2ab\sqrt{2} = 3 - a^2 - 2b^2$. Multiplying by the inverse of $2ab$ yields $\sqrt{2} = \frac{1}{2ab}(3 - a^2 - 2b^2)$. But the right-hand side of this equation is a rational number, whereas $\sqrt{2}$ is not. This is a contradiction.

   Since $\sqrt{3} \notin \mathbb{Q}(\sqrt{2})$, the degree of the minimal polynomial of $\sqrt{3}$ over $\mathbb{Q}(\sqrt{2})$ must be greater than 1. Since $\sqrt{3}$ is a root of the polynomial $x^2 - 3$, it follows that $x^2 - 3$ is the minimal polynomial of $\sqrt{3}$ over $\mathbb{Q}(\sqrt{2})$. Therefore, $[\mathbb{Q}(\sqrt{2})(\sqrt{3}):\mathbb{Q}(\sqrt{2})] = 2$. We will abbreviate $\mathbb{Q}(\sqrt{2})(\sqrt{3})$ with the easier to read notation $\mathbb{Q}(\sqrt{2}, \sqrt{3})$.

   By Theorem 15.11, we have $[\mathbb{Q}(\sqrt{2}, \sqrt{3}):\mathbb{Q}] = [\mathbb{Q}(\sqrt{2}, \sqrt{3}):\mathbb{Q}(\sqrt{2})][\mathbb{Q}(\sqrt{2}):\mathbb{Q}] = 2 \cdot 2 = 4$.

   Note that $\{1, \sqrt{2}\}$ is a basis of $\mathbb{Q}(\sqrt{2})$ over $\mathbb{Q}$ and $\{1, \sqrt{3}\}$ is a basis of $\mathbb{Q}(\sqrt{2}, \sqrt{3})$ over $\mathbb{Q}(\sqrt{2})$. The proof of Theorem 15.11 tells us that $\{1, \sqrt{2}, \sqrt{3}, \sqrt{6}\}$ is a basis of $\mathbb{Q}(\sqrt{2}, \sqrt{3})$ over $\mathbb{Q}$. So, $\mathbb{Q}(\sqrt{2}, \sqrt{3}) = \{a + b\sqrt{2} + c\sqrt{3} + d\sqrt{6} \mid a, b, c, d \in \mathbb{Q}\}$.

   It's worth mentioning that we can obtain $\mathbb{Q}(\sqrt{2}, \sqrt{3})$ from $\mathbb{Q}$ by adjoining the single real number $\sqrt{2} + \sqrt{3}$. In other words, $\mathbb{Q}(\sqrt{2}, \sqrt{3}) = \mathbb{Q}(\sqrt{2} + \sqrt{3})$. I leave the verification of this as an exercise for the reader. See Problem 3 below.

2. In general, if a is algebraic over F and b is algebraic over $F(a)$, we define $F(a, b)$ to be $F(a)(b)$. It's easy to check that $F(a)(b) = F(b)(a)$, as they are both equal to $\langle F \cup \{a, b\}\rangle$. Now, $[F(a):F]$ is equal to the degree of the minimal polynomial of a over F, $[F(a, b):F(a)]$ is equal to the degree of the minimal polynomial of b over $F(a)$, and by Theorem 15.11, we have

   $$[F(a, b):F] = [F(a, b):F(a)][F(a):F].$$

   More generally, we can define $F(a_1, a_2, \ldots, a_n)$ to be $\langle F \cup \{a_1, a_2, \ldots, a_n\}\rangle$, or equivalently, we can define $F(a_1, a_2, \ldots, a_n)$ inductively as $F(a_1, a_2, \ldots, a_{k+1}) = F(a_1, a_2, \ldots, a_k)(a_{k+1})$ for $k = 1, 2, \ldots, n-1$. In this case, $F(a_1, a_2, \ldots, a_n)$ is called an **iterated extension of length n**. We can visualize this iterated extension as follows:

   $$F \subseteq F(a_1) \subseteq F(a_1, a_2) \subseteq \cdots \subseteq F(a_1, a_2, \ldots, a_n)$$

   If at each stage of this construction, we insist that a_{k+1} is algebraic over $F(a_1, a_2, \ldots, a_k)$, then by Theorem 15.11 and induction, this iterated extension is a finite extension.

   Conversely, every finite extension of F is an iterated extension of F. To see this, suppose that $[E:F] = n$. Then there is a basis of E over F consisting of n elements of E, say $\{a_1, a_2, \ldots, a_n\}$. It is then easy to check that $E = F(a_1, a_2, \ldots, a_n)$.

Furthermore, every finite extension of F is an algebraic extension of F. To see this, suppose that $[E:F] = n$ and let $a \in E$. Then $\{1, a, a^2, \ldots, a^n\}$ has $n + 1$ elements, and therefore, is linearly dependent. It follows that there are weights $b_0, b_1, \ldots, b_n \in F$ such that $b_0 + b_1 a + \cdots + b_n a^n = 0$. Therefore, a is a root of the polynomial $b_0 + b_1 x + \cdots + b_n x^n$, proving that a is algebraic over F.

Splitting Fields

Let F and E be fields with $F \leq E$ and let $a(x) \in F[x]$ be a nonconstant polynomial. We say that E is a **splitting field** (or **root field**) for $a(x)$ over F if there are $c_1, c_2, \ldots, c_n \in E$ and $c \in F$ such that $a(x) = c(x - c_1)(x - c_2) \cdots (x - c_n)$ and $E = F(c_1, c_2, \ldots, c_n)$.

In general, if there are $c_1, c_2, \ldots, c_n \in E$ and $c \in F$ such that $a(x) = c(x - c_1)(x - c_2) \cdots (x - c_n)$, then we say that $a(x)$ **splits** in E.

Note that if $a(x) \in F[x]$ splits in E, it does **not** necessarily mean that E is a splitting field for $a(x)$ over F. In fact, if E is a splitting field for F, $E \leq K$, and $E \neq K$, then $a(x)$ splits in K, but K is **not** a splitting field for $a(x)$ over F.

Example 15.13:

1. Let $a(x) \in \mathbb{Q}[x]$ be the polynomial $a(x) = x^2 - 2$. Then $\mathbb{Q}(\sqrt{2})$ is a splitting field for $a(x)$ over $\mathbb{Q}$. To see this, observe $a(x) = (x - \sqrt{2})(x + \sqrt{2})$ (so that the roots of $a(x)$ are $\sqrt{2}$ and $-\sqrt{2}$) and $\mathbb{Q}(\sqrt{2}) = \mathbb{Q}(\sqrt{2}, -\sqrt{2})$. To verify the last claim, simply note that $-\sqrt{2} \in \mathbb{Q}(\sqrt{2})$ because $-\sqrt{2}$ is the additive inverse of $\sqrt{2}$.

   We also have $\sqrt{2} \in \mathbb{R}$, $-\sqrt{2} \in \mathbb{R}$, and $a(x) = (x - \sqrt{2})(x + \sqrt{2})$, so that $a(x)$ splits in $\mathbb{R}$. However, $\mathbb{R}$ is **not** a splitting field for $a(x)$ over $\mathbb{Q}$ because $\mathbb{R} \neq \mathbb{Q}(\sqrt{2}, -\sqrt{2})$. Indeed, we have for example, $\sqrt{3} \in \mathbb{R} \setminus \mathbb{Q}(\sqrt{2}, -\sqrt{2})$ (see part 1 of Example 15.12).

2. Let $b(x) \in \mathbb{Q}[x]$ be the polynomial $b(x) = x^4 - 8x^2 + 15 = (x^2 - 3)(x^2 - 5)$. Then $\mathbb{Q}(\sqrt{3}, \sqrt{5})$ is a splitting field for $b(x)$ over $\mathbb{Q}$.

   Once again, although $b(x)$ splits in $\mathbb{R}$, $\mathbb{R}$ is **not** a splitting field for $b(x)$ over $\mathbb{Q}$

3. Let $p(x) \in \mathbb{Q}[x]$ be the polynomial $a(x) = x^4 - 6x^2 - 7 = (x^2 + 1)(x^2 - 7)$. Then $\mathbb{Q}(\sqrt{7}, i)$ is a splitting field for $p(x)$ over $\mathbb{Q}$. Notice that $\mathbb{Q}(\sqrt{7}, i) \neq \mathbb{C}$, and therefore, although $p(x)$ splits in $\mathbb{C}$, $\mathbb{C}$ is **not** a splitting field for $p(x)$ over $\mathbb{Q}$.

4. Let $q(x) \in \mathbb{Q}[x]$ be the polynomial $q(x) = x^2 - 2x - 3 = (x + 1)(x - 3)$. Then $\mathbb{Q}$ is a splitting field for $q(x)$ over $\mathbb{Q}$. After all, the roots of $q(x)$ are -1 and 3, which are both in $\mathbb{Q}$.

5. Let $r(x) \in \mathbb{Q}[x]$ be the polynomial $r(x) = x^6 + 1$. The roots of this polynomial are the six 6th roots of unity (see Lesson 5). They are $\sqrt[6]{1} e^{\frac{k\pi}{6}i} = e^{\frac{k\pi}{6}i}$ for $k = 0, 1, 2, 3, 4, 5$. When $k = 1$, we have $e^{\frac{k\pi}{6}i} = e^{\frac{\pi}{6}i} = \cos\frac{\pi}{6} + i\sin\frac{\pi}{6} = \frac{\sqrt{3}}{2} + \frac{1}{2}i$. It follows that the splitting field for $r(x)$ over $\mathbb{Q}$ is $\mathbb{Q}(\sqrt{3}, i)$.

Theorem 15.14: Let F be a field and let $a(x) \in F[x]$ be a nonconstant polynomial. Then there exists a splitting field for $a(x)$ over F.

The proof of Theorem 15.14 follows from the Fundamental Theorem of Field Extensions (Theorem 15.8) and induction. I leave the details to the reader (see Problem 10 below).

It turns out that given a field F and a polynomial $a(x) \in F[x]$, there is essentially only one splitting field for $a(x)$ over F.

Theorem 15.15: Let F be a field and let $a(x) \in F[x]$ be a nonconstant polynomial. Then any two splitting fields for $a(x)$ over F are isomorphic.

The proof of Theorem 15.15 is much more difficult than the proof that splitting fields exist (Theorem 15.14). Therefore, I leave it as a challenge problem for the reader (see Problem 19 below).

Note: If F and K are fields with $E \leq K$, and $a(x)$ is a polynomial that splits in K, then there is a **unique** splitting field E for $a(x)$ over F such that $F \leq E \leq K$. However, if we take two arbitrary splitting fields E_1 and E_2 for $a(x)$ over F, then they need not be the same if they lie inside different extension fields. However, Theorem 15.15 guarantees that they will be isomorphic.

Problem Set 15

Full solutions to these problems are available for free download here:

www.SATPrepGet800.com/AAFBTDW

LEVEL 1

1. Let E be a field. Prove that F is a subfield of E if and only if $F^* \neq \emptyset$, $F \subseteq E$, F is closed under subtraction and multiplication, and F^* is closed under taking multiplicative inverses.

2. Find the minimal polynomial of $\sqrt{11}$ over $\mathbb{Q}$.

3. Prove that $\mathbb{Q}(\sqrt{2}, \sqrt{3}) = \mathbb{Q}(\sqrt{2} + \sqrt{3})$.

4. Prove that each of the following numbers is algebraic over the given field:

    (i) $2i$ over $\mathbb{Q}$

    (ii) $\sqrt{3} + \sqrt{5}$ over $\mathbb{Q}$

    (iii) $\sqrt{\pi}$ over $\mathbb{Q}(\pi)$

LEVEL 2

5. Find the minimal polynomial of $\sqrt{2} + i$ over each of the following fields:

    (i) $\mathbb{Q}$

    (ii) $\mathbb{Q}(i)$

    (iii) $\mathbb{R}$

6. Find the splitting field for $x^4 + 1$ over both $\mathbb{Q}$ and $\mathbb{R}$.

7. Let F be a field and let $a(x) = a_0 + a_1 x + a_2 x^2 + \cdots + a_n x^n \in F[x]$. The **derivative** of $a(x)$ is the polynomial in $F[x]$ defined by $a'(x) = a_1 + 2a_2 x + \cdots + na_n x^{n-1}$. Prove each of the following:

    (i)   If $a(x), b(x) \in F[x]$, then $(a+b)'(x) = a'(x) + b'(x)$.

    (ii)   If $a(x) \in F[x]$ and $k \in F$, then $(ka)'(x) = k(a'(x))$.

    (iii) If $a(x), b(x) \in F[x]$, then $(ab)'(x) = a(x)b'(x) + b(x)a'(x)$.

LEVEL 3

8. Compute $[\mathbb{Q}(\sqrt[3]{2}, \sqrt[4]{5}) : \mathbb{Q}]$.

9. Let F and E be fields with $F \leq E$ and let $A = \{x \in E \mid x \text{ is algebraic over } F\}$. Prove that $A \leq E$. A is called the **algebraic closure** of F in E.

10. Let F be a field and let $a(x) \in F[x]$ be a nonconstant polynomial. Prove that there exists a splitting field for $a(x)$ over F.

11. Prove that the splitting fields for $x^2 - 2x - 2$ and $x^2 - 3$ over $\mathbb{Q}$ are the same.

LEVEL 4

12. Let F be a field. Prove that every element of $F(c_1, c_2, \ldots, c_n)$ is a finite sum of terms of the form $kc_1^{j_1} c_2^{j_2} \cdots c_n^{j_n}$, where $k \in F$ and $j_1, j_2, \ldots, j_n \in \mathbb{N}$.

13. Let F be a field of characteristic 0. We say that c is a multiple root of a polynomial $a(x) \in F[x]$ if $(x - c)^k$ is a factor of $a(x)$ for some $k \in \mathbb{Z}^+$ with $k > 1$. Let $p(x) \in F[x]$ be an irreducible polynomial. Use the derivative of $p(x)$ to prove that $p(x)$ does **not** have any multiple roots (see Problem 7 above for the definition of the derivative of $p(x)$).

14. Prove that the algebraic closure of $\mathbb{Q}$ in $\mathbb{C}$ is **not** equal to $\mathbb{C}$ (see Problem 9 above for the definition of algebraic closure).

15. A field that has no proper algebraic extension is said to be **algebraically closed**. Let F and E be fields with $F \leq E$ and let A be the algebraic closure of F in E (see Problem 9 above). Prove that A is algebraically closed.

LEVEL 5

16. Let F be a field of characteristic 0 and let E be a finite extension of F. Prove that there is $c \in E$ such that $E = F(c)$.

17. Let $E = \mathbb{Q}(\sqrt[3]{2}, \sqrt[5]{2}, \ldots, \sqrt[2n+1]{2}, \ldots)$. Prove that E is an infinite algebraic extension of $\mathbb{Q}$ that is not equal to the algebraic closure of $\mathbb{Q}$ in $\mathbb{C}$ (see Problem 9 above for the definition of algebraic closure).

CHALLENGE PROBLEMS

18. Prove that π is transcendental over $\mathbb{Q}$.

19. Let F be a field and let $a(x) \in F[x]$ be a nonconstant polynomial. Prove that any two splitting fields for $a(x)$ over F are isomorphic.

20. Prove that $\mathbb{C}$ is algebraically closed (see Problem 15 above for the definition of algebraically closed).

LESSON 16
GALOIS THEORY

Field Automorphisms

Recall that the **characteristic** of a ring is the least positive integer n such that $\underbrace{1 + 1 + \cdots + 1}_{n \text{ times}} = 0$. If $1 + 1 + \cdots + 1$ is never 0, then the characteristic of the ring is 0.

Since every field is a ring, it also makes sense to talk about the characteristic of a field. By part 3 of Example 3.19, the characteristic of a field is either 0 or a prime number. **In this final lesson, our emphasis will be on fields that have characteristic 0.**

Let E be a field. Recall that an automorphism of E is an isomorphism from E to itself. In other words, $f: E \to E$ is an automorphism if f is a bijection from E to itself such that $f(1) = 1$ and for all $x, y \in E$, we have $f(x + y) = f(x) + f(y)$ and $f(xy) = f(x)f(y)$.

If we let $\text{Aut}(E)$ be the set of automorphisms of a field E, then by part (ii) of Problem 8 in Problem Set 7, $(\text{Aut}(E), \circ)$ is a group, where $\circ$ is composition of automorphisms.

If F and E are fields and $F \leq E$, then we say that $f \in \text{Aut}(E)$ **fixes** F if for all $x \in F$, $f(x) = x$. We define $G(E/F) = \{f \in \text{Aut}(E) \mid f \text{ fixes } F\}$. In Problem 1 below, you will be asked to prove that $G(E/F)$ is a subgroup of $\text{Aut}(E)$. $G(E/F)$ is called the **Galois group of E over F**.

Now, let F be a field and let E be the splitting field for an irreducible polynomial $p(x)$ over F with roots $c_1, c_2, \ldots, c_n$ (so, $E = F(c_1, c_2, \ldots, c_n)$). In this section, we will show the following:

(1) Given $1 \leq i, j \leq n$, there is $f \in G(E/F)$ such that $f(c_i) = c_j$.

(2) If $f \in G(E/F)$, then the restriction of f to $\{c_1, c_2, \ldots, c_n\}$ is a permutation of $\{c_1, c_2, \ldots, c_n\}$.

Notes: (1) The restriction of f to $\{c_1, c_2, \ldots, c_n\}$ is the function $\pi: \{c_1, c_2, \ldots, c_n\} \to \{c_1, c_2, \ldots, c_n\}$ such that for each $i = 1, 2, \ldots, n$, $\pi(c_i) = f(c_i)$.

(2) The second result above implies that $G(E/F)$ is isomorphic to a subgroup of S_n.

(3) A common misconception among students is that **every** permutation of $\{c_1, c_2, \ldots, c_n\}$ can be extended to an element $f \in G(E/F)$. Although this **could** be true for some field extensions $F \leq E$, it **does not need to be true** (even though E is the splitting field for an irreducible polynomial). If it is true, then $G(E/F) \cong S_n$. If not, then $G(E/F)$ is isomorphic to a proper subgroup of S_n.

Example 16.1:

1. Let $a(x) \in \mathbb{Q}[x]$ be the polynomial $a(x) = x^2 - 2$. By part 1 of Example 15.13, $\mathbb{Q}(\sqrt{2})$ is the splitting field for $a(x)$ over $\mathbb{Q}$. The Galois group of $\mathbb{Q}(\sqrt{2})$ over $\mathbb{Q}$ is $G(\mathbb{Q}(\sqrt{2})/\mathbb{Q}) = \{f_1, f_2\}$, where $f_1, f_2: \mathbb{Q}(\sqrt{2}) \to \mathbb{Q}(\sqrt{2})$ are defined by $f_1(a + b\sqrt{2}) = a + b\sqrt{2}$ (in other words, f_1 is the identity automorphism) and $f_2(a + b\sqrt{2}) = a - b\sqrt{2}$.

Clearly f_1 is an automorphism of $\mathbb{Q}(\sqrt{2})$ that fixes $\mathbb{Q}$. So, $f_1 \in G(\mathbb{Q}(\sqrt{2})/\mathbb{Q})$. Also, it is easy to verify that $f_2 \in G(\mathbb{Q}(\sqrt{2})/\mathbb{Q})$ (Check this!) To see that f_2 fixes $\mathbb{Q}$, observe that if $a \in \mathbb{Q}$, then $f_2(a) = f_2(a + 0\sqrt{2}) = a - 0\sqrt{2} = a$.

Observe that $f_1(\sqrt{2}) = \sqrt{2}$, $f_1(-\sqrt{2}) = -\sqrt{2}$ and $f_2(\sqrt{2}) = -\sqrt{2}$, $f_2(-\sqrt{2}) = \sqrt{2}$. This shows that (1) above holds. It also shows that when we restrict f_1 to $\{\sqrt{2}, -\sqrt{2}\}$, we get the identity permutation on $\{\sqrt{2}, -\sqrt{2}\}$ and when we restrict f_2 to $\{\sqrt{2}, -\sqrt{2}\}$, we get the permutation that interchanges $\sqrt{2}$ and $-\sqrt{2}$. So, (2) above holds as well. In this case, $G(\mathbb{Q}(\sqrt{2})/\mathbb{Q}) \cong S_2$.

2. Similar to part 1 above, the Galois group of $\mathbb{C} = \mathbb{R}(i)$ over $\mathbb{R}$ is $G(\mathbb{C}/\mathbb{R}) = \{f_1, f_2\}$, where f_1 is the identity automorphism and f_2 is complex conjugation (see part 2 of Example 7.4). In other words, $f_2: \mathbb{C} \to \mathbb{C}$ is defined by $f_2(a + bi) = a - bi$.

   Observe that the roots of the minimal polynomial of $\mathbb{C}$ over $\mathbb{R}$, namely $x^2 + 1$, are i and $-i$ and there are two permutations of these roots. They are the identity permutation π_1 and the permutation π_2 that interchanges i and $-i$. π_1 can be extended to $f_1 \in G(\mathbb{C}/\mathbb{R})$ and π_2 can be extended to $f_2 \in G(\mathbb{C}/\mathbb{R})$. Equivalently, when we restrict f_1 to $\{i, -i\}$, we get π_1 and when we restrict f_2 to $\{i, -i\}$, we get π_2. In this case, $G(\mathbb{C}/\mathbb{R}) \cong S_2$.

3. Let $b(x) \in \mathbb{Q}[x]$ be the polynomial $b(x) = x^4 - 8x^2 + 15 = (x^2 - 3)(x^2 - 5)$. By part 2 of Example 15.13, $\mathbb{Q}(\sqrt{3}, \sqrt{5})$ is the splitting field for $b(x)$ over $\mathbb{Q}$. The Galois group of $\mathbb{Q}(\sqrt{3}, \sqrt{5})$ over $\mathbb{Q}$ is $G(\mathbb{Q}(\sqrt{3}, \sqrt{5})/\mathbb{Q}) = \{f_1, f_2, f_3, f_4\}$, where $f_1, f_2, f_3, f_4: \mathbb{Q}(\sqrt{3}, \sqrt{5}) \to \mathbb{Q}(\sqrt{3}, \sqrt{5})$ are defined as follows:

$$f_1(a + b\sqrt{3} + c\sqrt{5} + d\sqrt{15}) = a + b\sqrt{3} + c\sqrt{5} + d\sqrt{15}$$
$$f_2(a + b\sqrt{3} + c\sqrt{5} + d\sqrt{15}) = a - b\sqrt{3} + c\sqrt{5} - d\sqrt{15}$$
$$f_3(a + b\sqrt{3} + c\sqrt{5} + d\sqrt{15}) = a + b\sqrt{3} - c\sqrt{5} - d\sqrt{15}$$
$$f_4(a + b\sqrt{3} + c\sqrt{5} + d\sqrt{15}) = a - b\sqrt{3} - c\sqrt{5} + d\sqrt{15}$$

I leave it to the reader to verify that $f_1, f_2, f_3, f_4 \in G(\mathbb{Q}(\sqrt{3}, \sqrt{5})/\mathbb{Q})$.

Observe that the automorphisms f_1, f_2, f_3, f_4 correspond to permutations $\pi_1, \pi_2, \pi_3, \pi_4$ of the set $\{\sqrt{3}, -\sqrt{3}, \sqrt{5}, -\sqrt{5}\}$, which are defined as follows:

$$\pi_1(\sqrt{3}) = \sqrt{3} \quad \pi_1(-\sqrt{3}) = -\sqrt{3} \quad \pi_1(\sqrt{5}) = \sqrt{5} \quad \pi_1(-\sqrt{5}) = -\sqrt{5}$$
$$\pi_2(\sqrt{3}) = -\sqrt{3} \quad \pi_2(-\sqrt{3}) = \sqrt{3} \quad \pi_2(\sqrt{5}) = \sqrt{5} \quad \pi_2(-\sqrt{5}) = -\sqrt{5}$$
$$\pi_3(\sqrt{3}) = \sqrt{3} \quad \pi_3(-\sqrt{3}) = -\sqrt{3} \quad \pi_3(\sqrt{5}) = -\sqrt{5} \quad \pi_3(-\sqrt{5}) = \sqrt{5}$$
$$\pi_4(\sqrt{3}) = -\sqrt{3} \quad \pi_4(-\sqrt{3}) = \sqrt{3} \quad \pi_4(\sqrt{5}) = -\sqrt{5} \quad \pi_4(-\sqrt{5}) = \sqrt{5}$$

Notice that only 4 of the 24 permutations of $\{\sqrt{3}, -\sqrt{3}, \sqrt{5}, -\sqrt{5}\}$ can be extended to elements of $G(\mathbb{Q}(\sqrt{3}, \sqrt{5})/\mathbb{Q})$. There can be no automorphism in $G(\mathbb{Q}(\sqrt{3}, \sqrt{5})/\mathbb{Q})$ sending $\sqrt{3}$ to $\sqrt{5}$ because the polynomial $x^4 - 30x^2 + 15$ is **not** irreducible and $\sqrt{3}$ and $\sqrt{5}$ are roots of different irreducible factors of $x^4 - 30x^2 + 15$. So, $G(\mathbb{Q}(\sqrt{3}, \sqrt{5})/\mathbb{Q}) \not\cong S_4$. An element of $G(\mathbb{Q}(\sqrt{3}, \sqrt{5})/\mathbb{Q})$ will only permute roots of the same **irreducible** polynomial.

In this case, we may want to think of the set $\{\sqrt{3}, -\sqrt{3}, \sqrt{5}, -\sqrt{5}\}$ as being split into the two sets $\{\sqrt{3}, -\sqrt{3}\}$ and $\{\sqrt{5}, -\sqrt{5}\}$ for the purposes of trying to determine the permutations that can be extended to $(\mathbb{Q}(\sqrt{3}, \sqrt{5})/\mathbb{Q})$.

We would now like to prove the two claims made before Example 16.1. We first set out to prove that in a splitting field E for an irreducible polynomial $p(x)$ over F, given any two roots c and d of $p(x)$, there is an automorphism of E that fixes F and sends c to d. We will begin by looking at some more general results.

The first result we will prove is that any isomorphism between two fields can be extended to an isomorphism between the two corresponding polynomial domains.

Theorem 16.2: Let F_1 and F_2 be fields and let $f: F_1 \to F_2$ be a field isomorphism. The function $g: F_1[x] \to F_2[x]$ defined by $g(a_0 + a_1 x + \cdots + a_n x^n) = f(a_0) + f(a_1)x + \cdots + f(a_n)x^n$ is a ring isomorphism.

Proof: Let $a(x) = a_0 + a_1 x + \cdots + a_n x^n$ and $b(x) = b_0 + b_1 x + \cdots + b_n x^n$ be in $F_1[x]$. Then we have

$$g(a(x) + b(x)) = g((a_0 + b_0) + (a_1 + b_1)x + \cdots + (a_n + b_n)x^n)$$
$$= f(a_0 + b_0) + f(a_1 + b_1)x + \cdots + f(a_n + b_n)x^n$$
$$= (f(a_0) + f(b_0)) + (f(a_1) + f(b_1))x + \cdots + (f(a_n) + f(b_n))x^n$$
$$= (f(a_0) + f(a_1)x + \cdots + f(a_n)x^n) + (f(b_0) + f(b_1)x + \cdots + f(b_n)x^n)$$
$$= g(a(x)) + g(b(x)).$$
$$g(a(x)b(x)) = g(a_0 b_0 + (a_0 b_1 + a_1 b_0)x + \cdots + a_n b_n x^{2n})$$
$$= f(a_0 b_0) + f(a_0 b_1 + a_1 b_0)x + \cdots + f(a_n b_n)x^{2n}$$
$$= f(a_0)f(b_0) + (f(a_0)f(b_1) + f(a_1)f(b_0))x + \cdots + f(a_n)f(b_n)x^{2n}$$
$$(f(a_0) + f(a_1)x + \cdots + f(a_n)x^n)(f(b_0) + f(b_1)x + \cdots + f(b_n)x^n)$$
$$= g(a(x))g(b(x)).$$
$$g(1_{F_1}) = f(1_{F_1}) = 1_{F_2}.$$

Therefore, g is a homomorphism.

To see that g is injective, simply observe that if $a(x) \neq b(x)$, then there is $i \in \{0, 1, \ldots, n\}$ such that $a_i \neq b_i$. Since f is injective, $f(a_i) \neq f(b_i)$. So, $g(a(x)) \neq g(b(x))$ (because their ith terms disagree).

Finally, to see that g is surjective, let $b(x) = b_0 + b_1 x + \cdots + b_n x^n \in F_2[x]$. Since f is surjective, there are $a_0, a_1, \ldots, a_n \in F_1$ such that $f(a_0) = b_0, f(a_1) = b_1, \ldots, f(a_n) = b_n$. Therefore, we have

$$g(a_0 + a_1 x + \cdots + a_n x^n) = f(a_0) + f(a_1)x + \cdots + f(a_n)x^n$$
$$= b_0 + b_1 x + \cdots + b_n x^n = b(x).$$

Since f is a homomorphism that is injective and surjective, f is an isomorphism. □

The next result is similar, but the proof is a little bit trickier. This time we want to extend an isomorphism between two fields F_1 and F_2 to an isomorphism between the two field extensions $F_1(c)$ and $F_2(d)$, where c and d are roots of the "corresponding" irreducible polynomials over F_1 and F_2.

Theorem 16.3: Let F_1 and F_2 be fields, let $f: F_1 \to F_2$ be an isomorphism, and let $p(x) \in F_1[x]$ be irreducible. Furthermore, let c and d be roots of $p(x)$ and $g(p(x))$, respectively (where $g: F_1[x] \to F_2[x]$ is defined by $g(a_0 + a_1 x + \cdots + a_n x^n) = f(a_0) + f(a_1)x + \cdots + f(a_n)x^n$, c lies in an extension field of F_1, and d lies in an extension field of F_2). Then f can be extended to an isomorphism $h: F_1(c) \to F_2(d)$ such that $h(c) = d$.

To prove Theorem 16.3, we define $h: F_1(c) \to F_2(d)$ by
$$h(a_0 + a_1 c + \cdots + a_n c^n) = f(a_0) + f(a_1)d + \cdots + f(a_n)d^n.$$

The reader will be asked to prove that h is a well-defined isomorphism in Problem 11 below.

Corollary 16.4: Let F be a field, let $p(x) \in F[x]$ be irreducible, and let c and d be roots of $p(x)$ in an extension field of F. Then there is an isomorphism $h: F(c) \to F(d)$ such that h fixes F and $h(c) = d$.

Proof: Let $F_1 = F$, let $F_2 = F$, and let $f: F \to F$ be the identity function. By Theorem 16.3, f can be extended to an isomorphism $h: F(c) \to F(d)$ such that $h(c) = d$. Since f is the identity function on F, h fixes F. □

The next theorem will make it easy to prove the first main result of this lesson. It says that any isomorphism between two extension fields of a field F that are contained in the same splitting field E for some polynomial over F can be extended to an automorphism of E that fixes F.

Theorem 16.5: Let F, K, L, E be fields of characteristic 0 such that K and L are extensions of F and E is a common extension of K and L with E the splitting field for some polynomial over F. Then any isomorphism $h: K \to L$ that fixes F can be extended to an element of $G(E/F)$.

The reader will be asked to prove Theorem 16.5 in Problem 8 below.

And now, we are ready to prove the first main result mentioned in the beginning of this lesson.

Theorem 16.6: Let F be a field of characteristic 0, let E be the splitting field for an irreducible polynomial $p(x)$ over F with roots $c_1, c_2, \ldots, c_n$, and let $1 \leq i, j \leq n$. Then there is $f \in G(E/F)$ such that $f(c_i) = c_j$.

Proof: Since $c_i, c_j \in E$, $F(c_i) \leq E$ and $F(c_j) \leq E$. By Corollary 16.4, there is an isomorphism $h: F(c_i) \to F(c_j)$ such that h fixes F and $h(c_i) = c_j$. By Theorem 16.5, h can be extended to $f \in G(E/F)$. □

We would now like to prove that if $f \in G(E/F)$, where E is the splitting field for an irreducible polynomial $p(x)$ over F, then f permutes the roots of $p(x)$. This is the second main result mentioned at the beginning of this lesson. It is a special case of Theorem 16.8 below. We first prove a preliminary result that will make the proof of Theorem 16.8 less cumbersome.

Theorem 16.7: Let F, K, and L be fields with K and L extending F, let $a(x) \in F[x]$, and let $f: K \to L$ be an isomorphism that fixes F. If $c \in K$ is a root of $a(x)$, then $f(c)$ is also a root of $a(x)$.

Proof: Suppose that $a(x) = a_0 + a_1 x + \cdots + a_n x^n$. Since f is an isomorphism that fixes F, we have
$$a(f(c)) = a_0 + a_1 f(c) + \cdots + a_n f(c)^n = f(a_0) + f(a_1) f(c) + \cdots + f(a_n) f(c)^n$$
$$= f(a_0 + a_1 c + \cdots + a_n c^n) = f(a(c)) = f(0) = 0.$$

The first and fourth equalities follow by substitution. The second equality is true because f fixes F. The third equality is true because f is a homomorphism. The fifth equality is true because c is a root of $a(x)$. The sixth equality follows from Theorem 7.2.

Since $a(f(c)) = 0$, $f(c)$ is a root of $a(x)$. □

Theorem 16.8: Let F, E, and L be fields with E and L extending F and with E the splitting field for a polynomial $a(x)$ over F. Suppose that $c_1, c_2, \ldots, c_n$ are the roots of $a(x)$ and let $f: E \to L$ be an isomorphism that fixes F. Then the restriction π of f to $\{c_1, c_2, \ldots, c_n\}$ is a permutation of $\{c_1, c_2, \ldots, c_n\}$. Furthermore, $f \in G(E/F)$ and f is the only automorphism of E fixing F whose restriction to $\{c_1, c_2, \ldots, c_n\}$ is π.

Proof: Since E is the splitting field for $a(x)$ over F, all the roots of $a(x)$ are in E. By Theorem 16.7, each of the roots $c_1, c_2, \ldots, c_n$ must be sent to another root by f. Since f is a bijection, f must permute the roots. So, the restriction π of f to $\{c_1, c_2, \ldots, c_n\}$ is a permutation of $\{c_1, c_2, \ldots, c_n\}$.

Now, by Problem 12 in Problem Set 15, every element of $E = F(c_1, c_2, \ldots, c_n)$ is a finite sum of terms of the form $k c_1^{j_1} c_2^{j_2} \cdots c_n^{j_n}$, where $k \in F$ and $j_1, j_2, \ldots, j_n \in \mathbb{N}$. Since f fixes F, $f(k) = k$. Since f permutes the roots $c_1, c_2, \ldots, c_n$, $f(k c_1^{j_1} c_2^{j_2} \cdots c_n^{j_n}) = k f(c_1)^{j_1} f(c_2)^{j_2} \cdots f(c_n)^{j_n} = k c_1^{t_1} c_2^{t_2} \cdots c_n^{t_n}$, where for each $i = 1, 2, \ldots, n$, $t_i = j_s$ for some $s = 1, 2, \ldots, n$. So, $f(k c_1^{j_1} c_2^{j_2} \cdots c_n^{j_n}) \in E$. Since E is closed under addition, $f(x) \in E$ for every $x \in E$. Therefore, $L = E$, and so, $f \in G(E/F)$.

If $g \in G(E/F)$ satisfies $g(c_i) = f(c_i)$ for each $i = 1, 2, \ldots, n$, then we have
$$g(k c_1^{j_1} c_2^{j_2} \cdots c_n^{j_n}) = k g(c_1)^{j_1} g(c_2)^{j_2} \cdots g(c_n)^{j_n} = k f(c_1)^{j_1} f(c_2)^{j_2} \cdots f(c_n)^{j_n} = f(k c_1^{j_1} c_2^{j_2} \cdots c_n^{j_n}).$$

A similar computation shows that g and f produce the same result when applied to a finite sum of terms of the form $k c_1^{j_1} c_2^{j_2} \cdots c_n^{j_n}$. So, $g = f$. Therefore, f is the only automorphism of E fixing F whose restriction to $\{c_1, c_2, \ldots, c_n\}$ is π. □

The Galois Correspondence

We previously saw that if E is the splitting field for a polynomial $a(x)$ over a field F, then every automorphism of E that fixes F permutes the roots of $a(x)$ (this I sthe special case of Theorem 16.8 where we start with $L = E$). However, it is is **very important** to understand that it is **not** always true that every permutation of the roots of $a(x)$ can be extended to an automorphism of E that fixes F (**even if $a(x)$ is irreducible**). As a starting point, the next theorem will tell us exactly how many automorphisms there actually are in $G(E/F)$.

Theorem 16.9: Let E be the splitting field for a polynomial $a(x)$ over a field F of characteristic 0. Then $|G(E/F)| = [E:F]$.

The reader will be asked to prove Theorem 16.9 in Problem 6 below.

Let F be a field and let E be the splitting field for a polynomial $a(x)$ over F. If K is a field such that $F \subseteq K \subseteq E$, then we call K an **intermediate field of $a(x)$ between F and E**. Observe that if K is an intermediate field of $a(x)$ between F and E, then E is also the splitting field for $a(x)$ over K. If it is clear what $F, E,$ and $a(x)$ are, we may simply refer to K as an **intermediate field**.

Notes: When we say that K is an intermediate field of $a(x)$ between F and E, we are automatically implying that F is a field and that E is the splitting field for $a(x)$ over E.

Given an intermediate field K of $a(x)$ between F and E, we define the **fixer of K** to be the Galois group

$$K^* = G(E/K).$$

Notice that K^* is a subgroup of $F^* = G(E/F)$. After all, since $F \subseteq K$, any automorphism of E that fixes K must also fix F.

More generally, if K and L are intermediate fields and $K \subseteq L$, then L^* is a subgroup of K^*.

Conversely, given a subgroup H of $G(E/F)$, we define the **fixed field** of H to be

$$H_* = \{x \in E \mid f(x) = x \text{ for all } f \in H\}.$$

In Problem 3 below, you will be asked to show that H_* is an intermediate field of $a(x)$ between F and E (where E is the splitting field for the polynomial $a(x)$ over F).

We would now like to show that there is a one-to-one correspondence between the subgroups of $G(E/F)$ and the intermediate fields between F and E. This will follow immediately from the next theorem.

Theorem 16.10: Let F be a field of characteristic 0, let E be the splitting field for some polynomial $p(x)$ over F, let H be a subgroup of $G(E/F)$, and let K be an intermediate field of $p(x)$ between F and E. Then $H = K^*$ if and only if $K = H_*$.

Notes: (1) In words, the conclusion of Theorem 16.10 says "H is the fixer of K if and only if K is the fixed field of H."

(2) In order to show that $K = H_*$ implies $H = K^*$, we will need the following technical result:

If H is a subgroup of $G(E/F)$ and $K = H_*$, then $|H| = [E:K]$.

You will be asked to prove this technical result in Problem 12 below.

Proof: Let's first assume that $H = K^*$. We will show that $K = H_*$.

Let $x \in K$. Since $H = K^*$, $f(x) = x$ for all $f \in H$. So, $x \in H_*$. Therefore, $K \subseteq H_*$.

To prove that $H_* \subseteq K$, we will prove that $c \notin K$ implies $c \notin H_*$. So, let $c \in E$ with $c \notin K$. Since $c \in E \setminus K$, the minimal polynomial $q(x)$ of c over K must have degree at least 2. So, there is a root d of $q(x)$ with $d \neq c$. By Corollary 16.4, there is an isomorphism $h: K(c) \to K(d)$ such that h fixes K and $h(c) = d$. By Theorem 16.5, h can be extended to $f \in G(E/K) = K^* = H$. It follows that $f(c) = h(c) = d \neq c$. So, $c \notin H_*$. Therefore, $H_* \subseteq K$.

Since $K \subseteq H_*$ and $H_* \subseteq K$, we have $K = H_*$, as desired.

Next, assume that $K = H_*$. We will show that $H = K^*$.

Let $f \in H$. Since $K = H_*$, $f(x) = x$ for all $x \in K$. So, $f \in K^*$. Therefore, $H \subseteq K^*$.

By Problem 12 below, $|H| = [E:K]$ and $|K^*| = [E:(K^*)_*]$. We already proved above that $(K^*)_* = K$. Therefore, $|K^*| = [E:K]$. So, $|H| = |K^*|$.

Since H, K^* are finite, $H \subseteq K^*$, and $|H| = |K^*|$, it follows that $H = K^*$, as desired. □

Note: The conclusion of Theorem 16.10 can be stated more briefly as $(K^*)_* = K$ and $(H_*)^* = H$. The dedicated reader should verify this.

Theorem 16.11 (The Fundamental Theorem of Galois Theory): Let F be a field, let E be the splitting field for some polynomial $p(x)$ over F, let $\mathcal{X}$ be the set of all subgroups of $G(E/F)$, let $\mathcal{Y}$ be the set of intermediate fields of $p(x)$ between F and E, and define $\Phi: \mathcal{X} \to \mathcal{Y}$ by $\Phi(H) = H_*$ Then Φ is a bijection.

Proof: To see that Φ is injective, let $\Phi(H_1) = \Phi(H_2)$. Then $(H_1)_* = (H_2)_*$. By Theorem 16.10, we have
$$H_1 = ((H_1)_*)^* = ((H_2)_*)^* = H_2.$$

To see that Φ is surjective, let $K \in \mathcal{Y}$. Then $K^* \in \mathcal{X}$ and $\Phi(K^*) = (K^*)_* = K$. □

The function Φ defined in Theorem 16.11 is called a **Galois correspondence**.

Note: A Galois correspondence maps smaller subgroups of $G(E/F)$ to larger subfields between F and E and larger subgroups of $G(E/F)$ to smaller subfields between F and E. For example, $\Phi(\{e\}) = E$, $\Phi(G(E/F)) = F$, and if $\{e\} < H < K < G(E/F)$, then $F < \Phi(K) < \Phi(H) < E$.

Example 16.12:

1. In part 1 of Example 16.1, we saw that $G(\mathbb{Q}(\sqrt{2})/\mathbb{Q}) = \{f_1, f_2\}$, where $f_1, f_2: \mathbb{Q}(\sqrt{2}) \to \mathbb{Q}(\sqrt{2})$ are defined by $f_1(a + b\sqrt{2}) = a + b\sqrt{2}$ and $f_2(a + b\sqrt{2}) = a - b\sqrt{2}$.

    The subgroups of $G(\mathbb{Q}(\sqrt{2})/\mathbb{Q})$ are $\{f_1\}$ and $G(\mathbb{Q}(\sqrt{2})/\mathbb{Q}) = \{f_1, f_2\}$.

    The intermediate fields between $\mathbb{Q}$ and $\mathbb{Q}(\sqrt{2})$ are just $\mathbb{Q}$ and $\mathbb{Q}(\sqrt{2})$.

    We have the following Galois correspondence: $\Phi(\{f_1\}) = \mathbb{Q}(\sqrt{2})$ and $\Phi(\{f_1, f_2\}) = \mathbb{Q}$.

    We can visualize this Galois correspondence as follows:

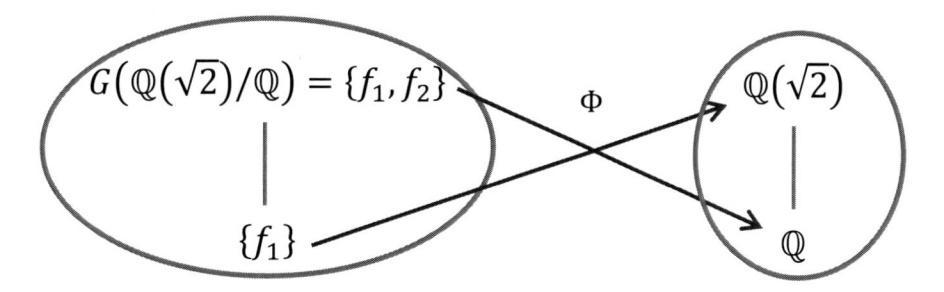

Here we have an example of a Galois group that is isomorphic to S_2.

2. In part 3 of Example 16.1, we saw that $G(\mathbb{Q}(\sqrt{3},\sqrt{5})/\mathbb{Q}) = \{f_1, f_2, f_3, f_4\}$, where $f_1, f_2, f_3, f_4 \colon \mathbb{Q}(\sqrt{3},\sqrt{5}) \to \mathbb{Q}(\sqrt{3},\sqrt{5})$ are defined as follows:

$$f_1(a + b\sqrt{3} + c\sqrt{5} + d\sqrt{15}) = a + b\sqrt{3} + c\sqrt{5} + d\sqrt{15}$$
$$f_2(a + b\sqrt{3} + c\sqrt{5} + d\sqrt{15}) = a - b\sqrt{3} + c\sqrt{5} - d\sqrt{15}$$
$$f_3(a + b\sqrt{3} + c\sqrt{5} + d\sqrt{15}) = a + b\sqrt{3} - c\sqrt{5} - d\sqrt{15}$$
$$f_4(a + b\sqrt{3} + c\sqrt{5} + d\sqrt{15}) = a - b\sqrt{3} - c\sqrt{5} + d\sqrt{15}$$

The subgroups of $G(\mathbb{Q}(\sqrt{3},\sqrt{5})/\mathbb{Q})$ are $\{f_1\}$, $\{f_1, f_2\}$, $\{f_1, f_3\}$, $\{f_1, f_4\}$, and $\{f_1, f_2, f_2, f_3\}$ (this last subgroup is equal to $G(\mathbb{Q}(\sqrt{3},\sqrt{5})/\mathbb{Q})$).

The intermediate fields between $\mathbb{Q}$ and $\mathbb{Q}(\sqrt{3},\sqrt{5})$ are $\mathbb{Q}$, $\mathbb{Q}(\sqrt{3})$, $\mathbb{Q}(\sqrt{5})$, $\mathbb{Q}(\sqrt{15})$, and $\mathbb{Q}(\sqrt{3},\sqrt{5})$.

We have the following Galois correspondence: $\Phi(\{f_1\}) = \mathbb{Q}(\sqrt{3},\sqrt{5})$, $\Phi(\{f_1, f_2\}) = \mathbb{Q}(\sqrt{5})$, $\Phi(\{f_1, f_3\}) = \mathbb{Q}(\sqrt{3})$, $\Phi(\{f_1, f_4\}) = \mathbb{Q}(\sqrt{15})$, and $\Phi(\{f_1, f_2, f_2, f_3\}) = \mathbb{Q}$.

We can visualize the Galois correspondence as follows:

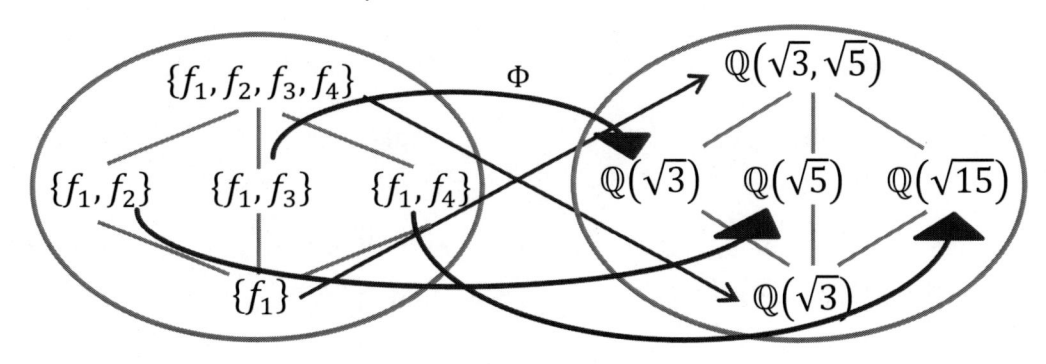

Solvability

A finite group G is **solvable** if there exist subgroups $G_0, G_1, \ldots, G_n$ such that

$$\{e\} = G_0 \triangleleft G_1 \triangleleft \cdots \triangleleft G_n = G,$$

where for each $i = 1, 2, \ldots, n$, the quotient G_i/G_{i-1} is commutative.

Example 16.13:

1. All commutative groups are solvable. Indeed, if G is commutative, then $\{e\} \triangleleft G$ and $G/\{e\}$ is commutative because it is isomorphic to G.

2. Let p be prime and let G be a group of order p^3. Then G is solvable. To see this, let $G_0 = \{e\}$, $G_1 = Z(G)$ (the center of G), and $G_2 = G$. If $Z(G) = G$, then G is commutative, and so, by part 1 above, G is solvable. So, assume that $Z(G) \neq G$. It is easy to see that $\{e\} \triangleleft Z(G) \triangleleft G$ (Check this!).

    $Z(G)/\{e\}$ is commutative because it is isomorphic to $Z(G)$. By Theorem 13.7, $Z(G) \neq \{e\}$. So, $|Z(G)| = p$ or p^2. If $|Z(G)| = p$, then $|G/Z(G)| = \frac{p^3}{p} = p^2$, and so, by Theorem 13.8, $G/Z(G)$ is commutative. If $|Z(G)| = p^2$, then $|G/Z(G)| = \frac{p^3}{p^2} = p$, and so, $G/Z(G)$ is cyclic, and therefore, commutative.

3. S_3 is solvable. Indeed, we have $\{e\} \triangleleft A_3 \triangleleft S_3$ and since $|S_3/A_3| = \frac{6}{2} = 3$, S_3/A_3 is cyclic, and therefore, commutative (and it should be clear that $|A_3/\{e\}| = |A_3| = 3$ as well).

4. S_5 is **not** solvable. To see this, assume that we have
$$\{(1)\} = G_0 \triangleleft G_1 \triangleleft \cdots \triangleleft G_n = S_5.$$
Let k be the least integer such that G_k contains all 3-cycles in S_5. Then $k > 0$ because certainly $\{(1)\}$ does not contain any 3-cycles. We will show that G_k/G_{k-1} is **not** commutative. Suppose toward contradiction that G_k/G_{k-1} is commutative. By Problem 7 below, G_{k-1} contains the commutator subgroup of G_k (the commutator subgroup of G_k is $\langle xyx^{-1}y^{-1} \mid x, y \in G_k \rangle$). Let (abc) be a 3-cycle and let $d, e \in \{1, 2, 3, 4, 5\}$ with $d, e \neq a, b, c$. Then, we have
$$(abc) = (adc)(cbe)(acd)(ceb) = (adc)(cbe)(adc)^{-1}(cbe)^{-1} \in G_{k-1}.$$
So, G_{k-1} contains all 3-cycles in S_5, contradicting that k was the least such integer.

    It follows that G_k/G_{k-1} is **not** commutative, and so, S_5 is **not** solvable.

Let F and E be fields with $F \leq E$ and let $c \in E$. The simple extension $F(c)$ is called a **simple radical extension** of F if there is $k \in \mathbb{Z}^+$ such that $c^k \in F$.

More generally, if $c_1, c_2, \ldots, c_n \in E$, then the iterated extension $F(c_1, c_2, \ldots, c_n)$ is a **radical extension** of F if for each $i = 1, 2, \ldots, n$, there is $k_i \in \mathbb{Z}^+$ such that $c_i^{k_i} \in F(c_1, c_2, \ldots, c_{i-1})$ (if $i = 1$, then we require $c_1^{k_1} \in F$).

We say that $a(x) \in F[x]$ is **solvable by radicals** if there is a radical extension of F containing all the roots of F (or equivalently, containing the splitting field for F).

Example 16.14:

1. $\mathbb{Q}(\sqrt{2})$ is a simple radical extension of $\mathbb{Q}$ because $(\sqrt{2})^2 = 2 \in \mathbb{Q}$. Since $\mathbb{Q}(\sqrt{2})$ is the splitting field for $x^2 - 2 \in \mathbb{Q}[x]$, we see that the polynomial $x^2 - 2$ is solvable by radicals.

2. More generally, $\mathbb{Q}(\sqrt[n]{2})$ is a simple radical extension of $\mathbb{Q}$ because $(\sqrt[n]{2})^n = 2 \in \mathbb{Q}$. Since $\mathbb{Q}(\sqrt[n]{2})$ is the splitting field for $x^n - 2 \in \mathbb{Q}[x]$, we see that the polynomial $x^n - 2$ is solvable by radicals.

3. Any quadratic polynomial is solvable by radicals. To see this, first note that the roots of the quadratic polynomial $ax^2 + bx + c$ are $\frac{-b+\sqrt{b^2-4ac}}{2a}$ and $\frac{-b-\sqrt{b^2-4ac}}{2a}$. It follows that the splitting field for $ax^2 + bx + c$ is $\mathbb{Q}(\sqrt{b^2 - 4ac})$. Since $(\sqrt{b^2 - 4ac})^2 = b^2 - 4ac \in \mathbb{Q}$, it follows that $\mathbb{Q}(\sqrt{b^2 - 4ac})$ is a simple radical extension of $\mathbb{Q}$.

4. $\mathbb{Q}(\sqrt{2}, \sqrt{3})$ is a radical extension of $\mathbb{Q}$ that is **not** a simple radical extension of $\mathbb{Q}$.

    To see that $\mathbb{Q}(\sqrt{2}, \sqrt{3})$ is a radical extension of $\mathbb{Q}$, simply note that we have $(\sqrt{2})^2 = 2 \in \mathbb{Q}$ and $(\sqrt{3})^2 = 3 \in \mathbb{Q}(\sqrt{2})$.

    To see that $\mathbb{Q}(\sqrt{2}, \sqrt{3})$ is **not** a simple radical extension, it suffices to show that for any $a, b \in \mathbb{Q} \setminus \{0\}$ and any $k \in \mathbb{Z}^+$ with $k > 1$, $(a\sqrt{2} + b\sqrt{3})^k \notin \mathbb{Q}$. For $k = 2$, we have $(a\sqrt{2} + b\sqrt{3})^2 = 2a^2 + 2ab\sqrt{6} + 3b^2 \notin \mathbb{Q}$ because $2ab \neq 0$ and $\sqrt{6} \notin \mathbb{Q}$. I now leave it to the dedicated reader to show that if we assume that $k \in \mathbb{Z}^+$ with $k > 1$ and $(a\sqrt{2} + b\sqrt{3})^k \notin \mathbb{Q}$, then $(a\sqrt{2} + b\sqrt{3})^{k+1} \notin \mathbb{Q}$. So, by the Principle of Mathematical Induction, for all $n \in \mathbb{Z}^+$ with $n > 1$, $(a\sqrt{2} + b\sqrt{3})^n \notin \mathbb{Q}$.

It is natural to ask if every polynomial is solvable by radicals. This is equivalent to asking if every polynomial can be solved by a formula involving sums, differences, products, quotients, and taking roots. In part 3 of Example 16.14 above, we showed that every quadratic polynomial is solvable by radicals. Although much more tedious to prove, it turns out that every cubic (third degree) and quartic (fourth degree) polynomial is solvable by radicals as well. Up until the early 1800s mathematicians were fairly certain that every polynomial is solvable by radicals. However, in 1824, a mathematician named Niels Henrik Abel provided a proof that there are quintic (fifth degree) polynomials that are not solvable by radicals. This means that there is no hope of ever finding a single formula similar to the quadratic formula that can be used to solve any quintic polynomial equation.

Galois Theory (named for the mathematician Everiste Galois) allows us to provide a more elegant proof that there are quintic polynomials not solvable by radicals. We will close out this book by using Galois Theory to prove this result.

The next theorem allows us to use the Galois Correspondence to convert questions about solvability of polynomials by radicals to questions about solvable groups.

Theorem 16.15: Let $a(x)$ be a polynomial over a field F that is solvable by radicals and let E be the splitting field for $a(x)$ over F. Then $G(E/F)$ is a solvable group.

The proof of Theorem 16.15 is quite tedious, and so, I leave it as an optional (very) challenging exercise for the reader. See Problem 14 below.

Theorem 16.16: There is a fifth degree irreducible polynomial $p(x) \in \mathbb{Q}[x]$ such that $G(E/\mathbb{Q}) \cong S_5$, where E is the splitting field for $p(x)$ over $\mathbb{Q}$.

Proof: Let $p(x)$ be a fifth degree irreducible polynomial over $\mathbb{Q}$ with exactly 3 distinct real roots (see Note 1 below for an example of such a polynomial). By Problem 12 in Problem Set 14, $p(x)$ has exactly one pair of complex conjugate roots.

Let $E = \mathbb{Q}(c_1, c_2, c_3, c_4, c_5)$ be the splitting field for $p(x)$ over $\mathbb{Q}$, where c_1, c_2, c_3, c_4, and c_5 are the roots of $p(x)$. Since every automorphism of the Galois group of E over $\mathbb{Q}$ permutes the five roots of $p(x)$, $G(E/\mathbb{Q})$ is isomorphic to a subgroup of S_5.

We have $[\mathbb{Q}(c_1):\mathbb{Q}] = 5$ because $p(x)$ is the minimal polynomial of c_1 over $\mathbb{Q}$ and $p(x)$ has degree 5. Also, $[E:\mathbb{Q}] = [E:\mathbb{Q}(c_1)][\mathbb{Q}(c_1):\mathbb{Q}] = 5[E:\mathbb{Q}(c_1)]$. So, $[E:\mathbb{Q}]$ is a multiple of 5. By Problem 6 below, $|G(E/\mathbb{Q})| = [E:\mathbb{Q}]$, and so, $|G(E/\mathbb{Q})|$ is a multiple of 5. By Cauchy's Theorem (Problem 24 in Problem Set 10), $G(E/\mathbb{Q})$ has a subgroup of order 5. Since this subgroup must be cyclic, $G(E/\mathbb{Q})$ has an element of order 5. By Problem 11 in Problem Set 12, the only elements of S_5 that have order 5 are the 5-cycles. So, thinking of $G(E/\mathbb{Q})$ as a subgroup of S_5, we see that $G(E/\mathbb{Q})$ contains a 5-cycle.

Let $f \in G(E/\mathbb{Q})$ be the automorphism that swaps the complex conjugate roots and leaves the other roots fixed. Thinking of $G(E/\mathbb{Q})$ as a subgroup of S_5, f is a transposition.

Since $G(E/\mathbb{Q})$ contains both a 5-cycle and a transposition, by Problem 17 in Problem Set 12, we see that $G(E/\mathbb{Q}) \cong S_5$. □

Notes: (1) $p(x) = x^5 - 9x + 3$ is a specific example of a fifth degree irreducible polynomial over $\mathbb{Q}$ with exactly 3 distinct real roots (this is just one of infinitely many possibilities).

(2) To see that $p(x)$ is irreducible, we can use Eisenstein's Irreducibility Theorem (see Problem 21 from Problem Set 14). To apply the theorem, simply observe that 3 divides 3 and -9, 3 does **not** divide 1, and $3^2 = 9$ does **not** divide 3.

(3) To see that $p(x)$ has 3 real roots requires a bit of basic Calculus. To see that there are at least 3 real roots, observe that $p(-2) = -11$, $p(0) = 3$, $p(1) = -5$, and $p(2) = 17$. Since $p(x)$ changes sign 3 times, its graph must pass through the x-axis at least 3 times (by the Intermediate Value Theorem).

To see that there cannot be more than 3 real roots, note that the derivative of $p(x)$ has just two real roots. Indeed, $p'(x) = 5x^4 - 9$, and so, $p'(x) = 0$ has just 2 real solutions: $x = \pm\sqrt[4]{\frac{9}{5}}$. It follows that the graph of $p(x)$ can have at most 2 turning points, and therefore, the graph can pass through the x-axis at most 3 times.

Corollary 16.17: There is a fifth degree irreducible polynomial in $\mathbb{Q}[x]$ that is not solvable by radicals.

Proof: By Theorem 16.16, there is a fifth degree irreducible polynomial $p(x) \in \mathbb{Q}[x]$ such that $G(E/\mathbb{Q}) \cong S_5$, where E is the splitting field for E over $\mathbb{Q}$. By part 4 of Example 16.13, $G(E/\mathbb{Q})$ is **not** a solvable group. By Theorem 16.15, $p(x)$ is not solvable by radicals. □

Problem Set 16

Full solutions to these problems are available for free download here:
www.SATPrepGet800.com/AAFBTDW

LEVEL 1

1. Let F and E be fields with $F \leq E$. Prove that $G(E/F)$ is a subgroup of Aut(E).

2. Prove that if $f: \mathbb{Q} \to \mathbb{Q}$ is a field homomorphism, then f is the identity function.

LEVEL 2

3. Let F be a field, let E be the splitting field for a polynomial $a(x)$ over F, and let H be a subgroup of $G(E/F)$. Prove that $H_* = \{x \in E \mid f(x) = x \text{ for all } f \in H\}$ is an intermediate field of $a(x)$ between F and E.

4. Let F, E, and L be fields such that E and L are extensions of F and let $f: E \to L$ be an isomorphism that fixes F. Prove that $[E:F] = [L:F]$.

LEVEL 3

5. Let F, E, and L be fields such that E and L are extensions of F and E is the splitting field for a polynomial $a(x)$ over F. Let $f: E \to L$ be an isomorphism that fixes F. Prove that $f \in G(E/F)$.

6. Let E be the splitting field for a polynomial $a(x)$ over a field F of characteristic 0. Prove that $|G(E/F)| = [E:F]$. (This is Theorem 16.9.)

7. Let G be a group, let N be a normal subgroup of G, and let H be the commutator subgroup of G ($H = \langle xyx^{-1}y^{-1} \mid x, y \in G \rangle$). Prove that G/N is commutative if and only if $H \subseteq N$.

LEVEL 4

8. Let F, K, L, E be fields of characteristic 0 such that K and L are extensions of F and E is a common extension of K and L with E the splitting field for some polynomial over F. Prove that any isomorphism $h: K \to L$ that fixes F can be extended to an element of $G(E/F)$. (This is Theorem 16.5.)

9. Let p be prime and let G be a p-group. Prove that G is solvable.

10. Let G and H be groups with G solvable and let $f: G \to H$ be a homomorphism. Prove that $f[G]$ is solvable.

Level 5

11. Let F_1 and F_2 be fields, let $f: F_1 \to F_2$ be an isomorphism, and let $p(x) \in F_1[x]$ be irreducible. Furthermore, let c and d be roots of $p(x)$ and $g(p(x))$, respectively (where $g: F_1[x] \to F_2[x]$ is defined by $g(a_0 + a_1 x + \cdots + a_n x^n) = f(a_0) + f(a_1)x + \cdots + f(a_n)x^n$, c lies in an extension field of F_1, and d lies in an extension field of F_2). Prove that f can be extended to an isomorphism $h: F_1(c) \to F_2(d)$ such that $h(c) = d$. (This is Theorem 16.3.)

12. Let F be a field of characteristic 0, let E be the splitting field for some polynomial $p(x)$ over F, let H be a subgroup of $G(E/F)$, and let $K = H_*$. Prove that $|H| = [E:K]$.

Challenge Problems

13. Prove that a finite group G is solvable if there exist subgroups $G_0, G_1, \ldots, G_n$ such that

$$\{e\} = G_0 \triangleleft G_1 \triangleleft \cdots \triangleleft G_n = G,$$

where for each $i = 1, 2, \ldots, n$, the quotient G_i/G_{i-1} is cyclic.

14. Let $a(x)$ be a polynomial over a field F that is solvable by radicals and let E be the splitting field for $a(x)$ over F. Prove that $G(E/F)$ is a solvable group.

15. Prove that every cubic polynomial in $\mathbb{Q}[x]$ is solvable by radicals.

INDEX

Abelian, 41
Abelian group, 44
Absolute value, 30, 118, 121, 126
Absorb, 175
Addition modulo n, 45, 81
Additive function, 161
Algebra, 164
Algebraic, 147, 313
Algebraic closure, 324
Algebraic extension, 313
Algebraically closed, 325
Almost ring, 53
Almost semiring, 68
Almost subring, 173
Alternating group, 227, 268, 269
Angle, 122
Antireflexive relation, 70
Antisymmetric relation, 70
Arbitrary but specific element, 15
Argument, 16
Associate, 304
Associativity, 28, 38, 43
Automorphism, 159
Automorphism group, 178
Axiom, 44
Axiom of Choice, 154, 225
Axiom of Extensionality, 21
Baby Chinese Remainder Theorem, 217
Baby Division Algorithm, 185
Base case, 108
Basis, 149, 151
Biconditional, 20
Bijection, 88
Bijective function, 88
Binary expansion, 101
Binary operation, 36, 37
Binary relation, 68, 134

Cancellation laws, 48, 55
Canonical form, 192
Canonical representation, 192
Cantor's Theorem, 97
Cantor-Schroeder-Bernstein Theorem, 99
Cardinality, 13, 95
Cartesian plane, 23
Cartesian product, 23, 24
Cauchy sequence, 118
Cauchy's Theorem, 238
Cayley's Theorem, 160
Center (of a circle), 122
Center (of a group), 138
Center (of a ring), 139
Centralizer, 152, 278
Chain, 154
Characteristic, 82
Chinese Remainder Theorem, 217
Circle, 122
Circumference, 122
Class equation, 279
Clock addition, 45
Clock arithmetic, 45
Clock multiplication, 45
Closed downwards, 40
Closed interval, 75
Closing statement, 16
Closure, 21, 38, 44
Codomain, 85
Coefficient (of a polynomial), 292
Common divisor, 189
Common divisor (of polynomials), 306
Common factor, 189
Common factor (of polynomials), 306
Common multiple, 189
Commutative group, 44

Commutative semigroup, 41
Commutative ring, 53
Commutativity, 28, 41
Commutator subgroup, 237
Comparability condition, 73
Comparable, 73
Compatibility of scalar and vector multiplication, 164, 298
Complete prime factorization, 194
Complex number, 12, 120
Complex Plane, 13
Complex valued function, 88
Composite function, 91
Composite number, 54, 182
Composition, 91
Conditional, 16
Congruence modulo 2, 77
Congruence modulo n, 77, 81
Conjugacy class, 278
Conjugate, 174
Conjugate subgroup, 283
Conjunction, 20
Constant, 134
Constant function, 85
Constant polynomial, 292
Constant sequence, 119
Contradiction, 47
Correspondence Theorem, 250
Coset, 220
Cosine, 124
Countable, 96
Counterexample, 14, 38
Cubic polynomial, 292
$\mathbb{C}$-valued function, 88
Cycle, 94, 263
Cycle diagram, 94
Cycle notation, 94
Cyclic group, 49, 145
Cyclic ring, 147
Cyclic subgroup, 145

339

Degree function, 303
Degree (of a field extension), 319
Degree (of a polynomial), 292
De Moivre's Theorem, 126
De Morgan's Law, 28
Decimal expansion, 101
Dependence relation, 150
Denumerable, 96
Derivative, 324
Determinant, 271, 272
Diagonal matrix, 275
Diamond Isomorphism Theorem, 242, 245
Dictionary order, 72
Digit, 12
Dihedral group, 145, 146
Direct product of groups, 44
Disjoint sets, 26, 30
Disjoint permutations, 264
Disjunction, 20
Distance, 118
Distributivity, 21, 28, 52
Divides, 181
Divides (polynomial), 299
Divisible, 77, 130, 181
Divisible (polynomials), 299
Division, 55
Division Algorithm, 186
Division Algorithm for Polynomials, 301
Divisor, 181
Divisor (polynomial), 299
Domain (of a function), 85
Domain (of a relation), 70
Domain (of a structure), 134
Domain (structure), 54
Domain of polynomials, 298
Eisenstein's Irreducibility Theorem, 311
Element, 9
Ellipses, 9
Empty set, 13

Endomorphism, 159
Epimorphism, 159
Equal sets, 19
Equinumerosity, 95
Equivalence class, 77
Equivalence relation, 76
Euclidean Algorithm, 195
Euclidean Domain, 303
Euclidean function, 303
Euclid's Principle, 191
Euclid's Principle for Polynomials, 307
Euler phi function, 209
Euler's formula, 126
Euler's totient function, 209
Evaluation function, 313
Even, 110, 180
Even permutation, 268
Existential quantifier, 11, 70, 98
Exponent, 112
Exponential form (of a complex number), 126
Extension, 312
Extension field, 312
External direct product, 252
Factor, 181
Factor (polynomial), 299
Factor group, 228
Factor ring, 232
Factor Theorem, 308
Factorial, 184
Factorization, 182, 305
Faithful action, 290
Fence-post formula, 14
Fermat's Little Theorem, 208
Field (of a relation), 70
Field (structure), 56
Field axioms, 57
Field extension, 312
Field homomorphism, 157
Field of polynomials, 298
Field of quotients, 179
Finitary operation, 134

Finitary relation, 134
Finite-dimensional vector space, 171
Finite extension, 319
Finite field extension, 319
Finite sequence, 87
Finitely generated group, 145
First Isomorphism Theorem, 240
Fixed field, 331
Fixer, 331
Fourth Isomorphism Theorem, 250
Free left R-module, 151
Free module, 151
Freshman Theorem, 246, 248
Function, 37, 85
Fundamental Homomorphism Theorem, 240
Fundamental Theorem of Arithmetic, 182, 192
Fundamental Theorem of Field Extensions, 318
Fundamental Theorem of Finite Commutative Groups, 251, 257
Fundamental Theorem of Galois Theory, 332
Galois Correspondence, 332
Galois group, 326
Gaussian integers, 147
GCD, 189
General linear group, 275
Generalized associativity, 114, 115
Generator, 49, 145, 146, 147, 315
Generators and relations, 145
Greatest common divisor, 189, 190
Greatest common divisor (of polynomials), 306
Greatest common factor, 189, 190
Greatest common factor (of polynomials), 306

Group, 43
Group action, 285
Group axioms, 44
Group homomorphism, 155
G-set, 285
Half-open interval, 75
Homogenous, 161
Homomorphism, 155
Ideal, 175
Ideal generated by a set, 176
Idempotent laws, 28
Identity, 42, 43
Identity, function, 93
Image, 93, 171
Imaginary axis, 13
Imaginary part, 120
Implication, 16
Inclusion map, 100
Index, 224
Indirect proof, 27
Inductive hypothesis, 109
Inductive step, 109
Infinite closed interval, 75
Infinite extension, 319
Infinite field extension, 319
Infinite interval, 75
Infinite open interval, 75
Infinite sequence, 87
Injection, 88
Injective function, 88
Intermediate field, 331
Interval, 74
Infinite order, 49
Infinite set, 9
Initial ray, 122
Inner automorphism group, 178
Integer, 9, 116
Integer-valued function, 88
Integral domain, 54
Internal direct product, 253
Intersection, 25, 29
Inverse, 43

Inverse function, 88
Inverse image, 93
Invertible, 43
Irrational number, 13
Irreducible, 300, 304
Irreducible factorization, 305
Isomorphic, 48, 143, 159
Isomorphism, 159, 165
Isomorphism Theorems, 239
Iterated extension, 321
Iterated field extension, 321
Kernel, 171
Klein four group, 160
k-tuple, 22, 63
Lagrange's Theorem, 225
LCM, 189
Leading coefficient (of a polynomial), 292
Leading term (of a polynomial), 292
Least common multiple, 189, 190
Least element, 107
Left cancellation law, 48
Left coset, 220
Left distributivity, 52, 53
Left group action, 285
Left inverse, 46
Left module, 65
Left R-algebra, 297
Left submodule, 140, 142
Length (of a permutation), 263
Length (of a sequence), 87
Linear algebra, 164
Linear combination, 147, 148, 190
Linear combination (of polynomials), 307
Linear congruence, 210
Linear transformation, 157, 161
Linearly dependent, 149, 150
Linearly independent, 149, 150
Linearly ordered set, 73

Linear polynomial, 292
Logical Connective, 16
Logically equivalent, 16, 20
Lower triangular matrix, 275
Matrix, 63, 166
Matrix group, 273, 275
Matrix of a Linear Transformation, 170
Matrix multiplication, 166
Maximal element, 154
Maximal ideal, 234
Member, 9
Membership relation, 9, 69
Minimal polynomial, 315
Modular arithmetic, 206
Module, 65
Modulus, 121, 126
Monic polynomial, 301
Monoid, 42
Monoid homomorphism, 155
Monomorphism, 159
Monotonic function, 157
Multiple, 181
Multiple (polynomial), 299
Multiple of an element, 49
Multiplication modulo n, 45, 81
Multiplication table, 36
Mutually exclusive, 26
Mutually relatively prime, 190
n-ary operation, 134
n-ary relation, 71, 134
Natural number, 9, 105
n-dimensional vector space, 171
Negation, 20
Negative identity, 125
Noetherian ring, 311
Noncommutative linear algebra, 169
Normal subgroup, 175
Normalizer, 287
nth root, 128
nth root of unity, 129

n-tuple, 22, 63
Odd, 185
Odd permutation, 268
One-to-one, 88
Onto, 88
Open Interval, 75
Opening statement, 16
Orbit, 265, 288
Orbit-Stabilizer Theorem, 288
Order diagram, 253
Order homomorphism, 157
Ordered k-tuple, 22, 63
Order of an element, 145
Order of a group, 49
Ordered pair, 22
Ordered triple, 22
Origin, 1, 23
Pairwise disjoint, 30
Pairwise relatively prime, 190
Parity, 77
Partial binary operation, 38
Partial ordering, 70, 71
Partially ordered set, 71
Partition, 78
Partition (of a positive integer), 254
Permutation, 93, 260
Permutation matrix, 269, 270
p-group, 280
PID, 303
PIR, 303
p-subgroup, 281
Polar form (of a complex number), 125
Polynomial, 64, 292
Polynomial addition, 293
Polynomial function, 308
Polynomial multiplication, 294
POMI, 107
Poset, 71
Positive square root, 127
Power of an element, 49
Power set, 18

Predecessor, 105
Prime factorization, 182
Prime ideal, 233
Prime number, 54, 97, 181
Prime triple, 198
Principal argument, 126
Principal ideal, 176, 303
Principal Ideal Domain, 303
Principal Ideal Ring, 303
Principal square root, 127
Principle of Mathematical Induction, 107
Product, 106, 180
Proof by contradiction, 27, 98, 108
Proper subgroup, 226
Proper subset, 15
Propositional logic, 15
Propositional variable, 15
Pure imaginary number, 13, 121
Pythagorean Identity, 125
Quadrantal angle, 124
Quadratic polynomial, 292
Quartic polynomial, 292
Quintic polynomial, 292
Quotient, 55, 186
Quotient group, 228
Quotient ring, 232
Quotient space, 236
$\mathbb{Q}$-valued function, 88
Radian measure, 122
Radical extension, 334
Radius, 122
Range (or a function), 85
Range (or a relation), 70
Rational number, 12, 117
Rational valued function, 88
Ray, 122
Real axis, 13
Real line, 12
Real number, 12, 118, 121
Real part, 120, 121

Real valued function, 89
Recursive definition, 106
Reducible, 300, 304
Reflexive relation, 33, 70
Relatively prime, 190
Relatively prime (polynomials), 307
Remainder, 186
Representative (of an equivalence class), 80
Right distributivity, 52, 53
Right group action, 285
Right module, 65
Right submodule, 140
Ring, 52
Ring axioms, 53
Ring homomorphism, 156
Ring of Gaussian integers, 147
Ring of polynomials, 297
Ring with identity, 53
Right cancellation law, 48
Right coset, 220
Right inverse, 46
Rng, 53
Root field, 322
Root (of a polynomial), 308
Root of unity, 129
$\mathbb{R}$-valued function, 88
SACT, 21
Scalar, 59
Scalar multiplication, 59, 60, 61
Second Isomorphism Theorem, 242, 245
Selector, 225
Selfish set, 133
Semigroup, 38
Semigroup homomorphism, 155
Semiring, 53
Sequence, 87
Set, 9
Set-builder notation, 10
Set difference, 25

Sigma notation, 167
Simple extension, 316
Simple field extension, 316
Simple radical extension, 334
Sine, 124
Solution (of a polynomial), 308
Solvable by radicals, 334
Solvable group, 259, 333
Span, 147, 148, 150
Splitting field, 322
Square matrix, 166
Square root, 127
Stabilizer, 286
Standard Advanced Calculus Trick, 21
Standard form (of a complex number), 125
Statement, 15
Strict linearly ordered set, 73
Strict partially ordered set, 71
Strict partial ordering, 71
Strict poset, 71
String induction, 133
Structure, 134
Subfield, 140, 312
Subgroup, 137
Submodule, 140, 142
Submonoid, 137
Subring, 138
Subrng, 173
Subsemigroup, 137
Subset, 14
Subspace, 140, 141
Substructure, 136
Subtraction, 55
Successor, 105

Sum, 180
Sum identity, 125
Surjection, 88
Surjective function, 88
Sylow p-subgroup, 281
Sylow Theorems, 281, 282, 283
Symmetric group, 93
Symmetric relation, 33, 70
Symmetric difference, 25
Tail of 9's, 12
Tangent, 124
Tautology, 17
Third Isomorphism Theorem, 246, 248
Term (of a polynomial), 292
Terminal ray, 122
Ternary expansion, 101
Ternary relation, 71, 134
Totally ordered set, 73
Transcendental, 147, 313
Transcendental extension, 313
Transitive action, 288
Transitive relation, 19, 70
Transitive set, 34
Transitivity, 19
Transposition, 263
Trichotomy, 73
Triangle Inequality, 119
Tree diagram, 18
Trivial equivalence relation, 76
Trivial group, 50
Trivial matrix group, 275
Trivial subgroup, 138
Truth assignment, 15
Type, 136, 155
UFD, 305

Upper bound, 154
Upper triangular matrix, 275
Unary relation, 71, 134
Uncountable, 96
Union, 24, 29
Unique Factorization Domain, 305
Unit circle, 122
Unital ring, 53
Universal quantifier, 15
Universal set, 15
Universal statement, 40
Unordered pair, 22
Vacuous truth, 17
Vector, 60
Vector addition, 60, 61
Vector space, 60, 61
Vector space homomorphism, 157, 161
Venn diagram, 15
Weight, 148, 190
Well-defined, 81
Well Ordering Principle, 107
Without loss of generality, 90
WOP, 107
Wrapping function, 124
Zero divisor, 54
Zero (of a polynomial), 308
Zero property, 54
Zero ring, 54
Zero polynomial, 292
Zorn's Lemma, 154
$\mathbb{Z}$-valued function, 88

About the Author

Dr. Steve Warner, a New York native, earned his Ph.D. at Rutgers University in Pure Mathematics in May 2001. While a graduate student, Dr. Warner won the TA Teaching Excellence Award.

After Rutgers, Dr. Warner joined the Penn State Mathematics Department as an Assistant Professor and in September 2002, he returned to New York to accept an Assistant Professor position at Hofstra University. By September 2007, Dr. Warner had received tenure and was promoted to Associate Professor. He has taught undergraduate and graduate courses in Precalculus, Calculus, Linear Algebra, Differential Equations, Mathematical Logic, Set Theory, and Abstract Algebra.

From 2003 – 2008, Dr. Warner participated in a five-year NSF grant, "The MSTP Project," to study and improve mathematics and science curriculum in poorly performing junior high schools. He also published several articles in scholarly journals, specifically on Mathematical Logic.

Dr. Warner has nearly two decades of experience in general math tutoring and tutoring for standardized tests such as the SAT, ACT, GRE, GMAT, and AP Calculus exams. He has tutored students both individually and in group settings.

In February 2010 Dr. Warner released his first SAT prep book "The 32 Most Effective SAT Math Strategies," and in 2012 founded Get 800 Test Prep. Since then Dr. Warner has written books for the SAT, ACT, SAT Math Subject Tests, AP Calculus exams, and GRE. In 2018 Dr. Warner released his first pure math book called "Pure Mathematics for Beginners." Since then he has released several more books, each one addressing a specific subject in pure mathematics.

Dr. Steve Warner can be reached at

steve@SATPrepGet800.com

BOOKS BY DR. STEVE WARNER

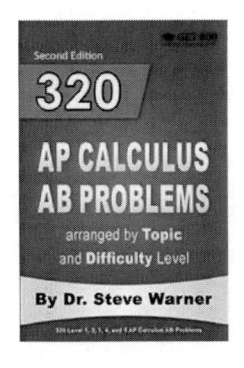

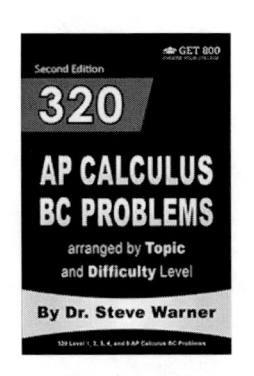

Made in United States
North Haven, CT
27 May 2024